Edited by
Javier Garcia-Martinez

**Nanotechnology for the
Energy Challenge**

Related Titles

Ozawa, K. (ed.)

Lithium Ion Rechargeable Batteries

Materials, Technology, and New Applications

2009

ISBN: 978-3-527-31983-1

Züttel, A., Borgschulte, A., Schlapbach, L. (eds.)

Hydrogen as a Future Energy Carrier

2008

ISBN: 978-3-527-30817-0

Centi, Gabriele / Trifiró, Ferruccio / Perathoner, Siglinda / Cavani, Fabrizio (eds.)

Sustainable Industrial Processes

2009

ISBN: 978-3-527-31552-9

Hirscher, Michael (ed.)

Handbook of Hydrogen Storage

New Materials for Future Energy Storage

2010

ISBN: 978-3-527-32273-2

Pagliaro, Mario

How Nanotechnology Changes our Future

2010

ISBN: 978-3-527-32676-1

Pagliaro, Mario / Palmisano, Giovanni / Ciriminna, Rosaria

Flexible Solar Cells

2008

ISBN: 978-3-527-32375-3

Mitsos, Alexander / Barton, Paul I. (eds.)

Microfabricated Power Generation Devices

Design and Technology

2009

ISBN: 978-3-527-32081-3

Edited by Javier Garcia-Martinez

Nanotechnology for the Energy Challenge

WILEY-VCH Verlag GmbH & Co. KGaA

The Editor

Prof. Javier Garcia-Martinez
Inorganic Chem. Department
University of Alicante
Carretera San Vicente s/n.
03690 Alicante
Spain

All books published by Wiley-VCH are carefully produced. Nevertheless, authors, editors, and publisher do not warrant the information contained in these books, including this book, to be free of errors. Readers are advised to keep in mind that statements, data, illustrations, procedural details or other items may inadvertently be inaccurate.

Library of Congress Card No.: applied for

British Library Cataloguing-in-Publication Data
A catalogue record for this book is available from the British Library.

Bibliographic information published by the Deutsche Nationalbibliothek
The Deutsche Nationalbibliothek lists this publication in the Deutsche Nationalbibliografie; detailed bibliographic data are available on the Internet at http://dnb.d-nb.de.

© 2010 WILEY-VCH Verlag GmbH & Co. KGaA, Weinheim

All rights reserved (including those of translation into other languages). No part of this book may be reproduced in any form – by photoprinting, microfilm, or any other means – nor transmitted or translated into a machine language without written permission from the publishers. Registered names, trademarks, etc. used in this book, even when not specifically marked as such, are not to be considered unprotected by law.

Cover Formgeber, Eppelheim
Typesetting Toppan Best-set Premedia Limited
Printing and Binding T.J. International Ltd, Padstow, Cornwall

Printed in Great Britain
Printed on acid-free paper

ISBN: 978-3-527-32401-9

Foreword

Technology has been mankind's enabler for providing such essentials as food production, clean water, and energy at the scale needed to support over six billion people. During the next several decades, we can expect another three billion or so people, increasing expectations for economic growth and associated services, and concomitantly severe strains on the planet's environment. An unprecedented pace of beneficial technology innovation is needed.

Much of this innovation will be driven by the discovery and use of novel materials. In particular, next generation, clean, efficient energy technologies will emerge from the new materials made possible by emerging nanoscience and nanotechnology – materials that perform at extreme conditions, materials that self-heal in harsh environments, materials much lighter than steel but also stronger, materials that convert sunlight or produce light more efficiently, materials that separate different gasses and fluids more effectively, materials that permit storage at much higher energy and power densities, and many more. And, of course, these materials must meet the economic tests that are essential for large-scale deployment in both industrialized and emerging economies. Success will require the multidisciplinary efforts of physicists, chemists, biologists, engineers, and others, all contributing crucial capabilities for studying, manipulating, and fabricating new materials with atomic level precision and with properties optimized to energy technology requirements.

This book, "Nanotechnology for the Energy Challenge", organizes an important collection of papers describing nanotechnology opportunities across the spectrum of energy production, storage, and use. It provides an excellent reference for researchers, both established and early career, who wish to gain an overview or to explore more deeply in particular application areas, but can also serve as an introduction to the subject for a more general reader. "Nanotechnology for the Energy Challenge" brings together a central development in materials science and engineering with the preeminent challenge of meeting clean energy

Nanotechnology for the Energy Challenge. Edited by Javier Garcia-Martinez
© 2009 WILEY-VCH Verlag GmbH & Co. KGaA, Weinheim
ISBN: 978-3-527-32401-9

imperatives. As such, it can contribute importantly to the acceleration of energy technology innovation that is necessary for a prosperous and sustainable future.

Ernest J. Moniz
Cecil and Ida Green Professor of Physics and Engineering Systems
Director, MIT Energy Initiative
Massachusetts Institute of Technology

Contents

Introduction XV
List of Contributors XVII

Part One Sustainable Energy Production 1

1 Nanotechnology for Energy Production 3
  Elena Serrano, Kunhao Li, Guillermo Rus, and Javier García-Martínez
1.1 Energy Challenge in the 21st Century and Nanotechnology 3
1.2 Nanotechnology in Energy Production 6
1.2.1 Photovoltaics 6
1.2.2 Hydrogen Production 14
1.2.3 Fuel Cells 20
1.2.4 Thermoelectricity 26
1.3 Summary 28
  Acknowledgment 28
  References 29

2 Nanotechnology in Dye-Sensitized Photoelectrochemical Devices 33
  Agustin McEvoy and Michaël Grätzel
2.1 Introduction 33
2.2 Semiconductors and Optical Absorption 34
2.3 Dye Molecular Engineering 38
2.4 The Stable Self-Assembling Dye Monomolecular Layer 40
2.5 The Nanostructured Semiconductor 41
2.6 Conclusions 43
  References 44

3 Thermal-Electrical Energy Conversion from the Nanotechnology Perspective 47
  Jian He and Terry M. Tritt
3.1 Introduction 47
3.2 Established Bulk Thermoelectric Materials 48

Nanotechnology for the Energy Challenge. Edited by Javier Garcia-Martinez
© 2009 WILEY-VCH Verlag GmbH & Co. KGaA, Weinheim
ISBN: 978-3-527-32401-9

3.3	Selection Criteria for Bulk Thermoelectric Materials	51
3.4	Survey of Size Effects	53
3.4.1	Classic Size Effects	54
3.4.2	Quantum Size Effects	55
3.4.3	Thermoelectricity of Nanostructured Materials	56
3.5	Thermoelectric Properties on the Nanoscale: Modeling and Metrology	58
3.6	Experimental Results and Discussions	60
3.6.1	Bi Nanowire/Nanorod	60
3.6.2	Si Nanowire	62
3.6.3	Engineered "Exotic" Nanostructures	64
3.6.4	Thermionics	66
3.6.5	Thermoelectric Nanocomposites: a New Paradigm	68
3.7	Summary and Perspectives	73
	Acknowledgments	74
	References	74

4 Nanomaterials for Fuel Cell Technologies 79
Antonino Salvatore Aricò, Vincenzo Baglio, and Vincenzo Antonucci

4.1	Introduction	79
4.2	Low-Temperature Fuel Cells	80
4.2.1	Cathode Reaction	80
4.2.2	Anodic Reaction	83
4.2.3	Practical Fuel Cell Catalysts	85
4.2.4	Non-Precious Catalysts	90
4.2.5	Electrolytes	90
4.2.6	High-Temperature Polymer Electrolyte Membranes	91
4.2.7	Membrane-Electrode Assembly (MEA)	96
4.3	High-Temperature Fuel Cells	98
4.3.1	High-Temperature Ceramic Electrocatalysts	101
4.3.2	Direct Utilization of Dry Hydrocarbons in SOFCs	103
4.4	Conclusions	106
	References	106

5 The Contribution of Nanotechnology to Hydrogen Production 111
Sambandam Anandan, Jagannathan Madhavan, and Muthupandian Ashokkumar

5.1	Introduction	111
5.2	Hydrogen Production by Semiconductor Nanomaterials	113
5.2.1	General Approach	113
5.2.2	Need for Nanomaterials	114
5.2.3	Nanomaterials-Based Photoelectrochemical Cells for H_2 Production	115
5.2.4	Semiconductors with Specific Morphology: Nanotubes and Nanodisks	117

5.2.5	Sensitization	*123*
5.3	Summary	*131*
	Acknowledgments	*132*
	References	*132*

Part Two Efficient Energy Storage *137*

6 Nanostructured Materials for Hydrogen Storage *139*
Saghar Sepehri and Guozhong Cao

6.1	Introduction	*139*
6.2	Hydrogen Storage by Physisorption	*140*
6.2.1	Nanostructured Carbon	*141*
6.2.2	Zeolites	*142*
6.2.3	Metal–Organic Frameworks	*143*
6.2.4	Clathrates	*143*
6.2.5	Polymers with Intrinsic Microporosity	*144*
6.3	Hydrogen Storage by Chemisorption	*144*
6.3.1	Metal and Complex Hydrides	*144*
6.3.2	Chemical Hydrides	*147*
6.3.3	Nanocomposites	*148*
6.4	Summary	*151*
	References	*151*

7 Electrochemical Energy Storage: the Benefits of Nanomaterials *155*
Patrice Simon and Jean-Marie Tarascon

7.1	Introduction	*155*
7.2	Nanomaterials for Energy Storage	*158*
7.2.1	From Rejected Insertion Materials to Attractive Electrode Materials	*158*
7.2.2	The Use of Once Rejected Si-Based Electrodes	*160*
7.2.3	Conversion Reactions	*161*
7.3	Nanostructured Electrodes and Interfaces for the Electrochemical Storage of Energy	*163*
7.3.1	Nanostructuring of Current Collectors/Active Film Interface	*163*
7.3.1.1	Self-Supported Electrodes	*163*
7.3.1.2	Nano-Architectured Current Collectors	*163*
7.3.2	NanoStructuring of Active Material/Electrolyte Interfaces	*168*
7.3.2.1	Application to Li-Ion Batteries: Mesoporous Chromium Oxides	*168*
7.3.2.2	Application to Electrochemical Double-Layer Capacitors	*169*
7.4	Conclusion	*174*
	Acknowledgments	*175*
	References	*175*

8		**Carbon-Based Nanomaterials for Electrochemical Energy Storage** *177*
		Elzbieta Frackowiak and François Béguin
8.1		Introduction *177*
8.2		Nanotexture and Surface Functionality of sp^2 Carbons *177*
8.3		Supercapacitors *180*
8.3.1		Principle of a Supercapacitor *180*
8.3.2		Carbons for Electric Double Layer Capacitors *182*
8.3.3		Carbon-Based Materials for Pseudo-Capacitors *185*
8.3.3.1		Pseudocapacitance Effects Related with Hydrogen Electrosorbed in Carbon *185*
8.3.3.2		Pseudocapacitive Oxides and Conducting Polymers *188*
8.3.3.3		Pseudo-Capacitive Effects Originated from Heteroatoms in the Carbon Network *190*
8.4		Lithium-Ion Batteries *194*
8.4.1		Anodes Based on Nanostructured Carbons *195*
8.4.2		Anodes Based on Si/C Composites *196*
8.4.3		Origins of Irreversible Capacity of Carbon Anodes *199*
8.5		Conclusions *201*
		References *202*
9		**Nanomaterials for Superconductors from the Energy Perspective** *205*
		Claudia Cantoni and Amit Goyal
9.1		Overcoming Limitations to Superconductors' Performance *205*
9.2		Flux Pinning by Nanoscale Defects *207*
9.3		The Grain Boundary Problem *208*
9.4		Anisotropic Current Properties *210*
9.5		Enhancing Naturally Occurring Nanoscale Defects *212*
9.6		Artificial Introduction of Flux Pinning Nanostructures *215*
9.7		Self-Assembled Nanostructures *216*
9.8		Control of Epitaxy-Enabling Atomic Sulfur Superstructure *221*
		Acknowledgments *223*
		References *224*
		Part Three Energy Sustainability *229*
10		**Green Nanofabrication: Unconventional Approaches for the Conservative Use of Energy** *231*
		Darren J. Lipomi, Emily A. Weiss, and George M. Whitesides
10.1		Introduction *231*
10.1.1		Motivation *232*
10.1.2		Energetic Costs of Nanofabrication *233*
10.1.3		Use of Tools *234*
10.1.4		Nontraditional Materials *236*

10.1.5	Scope *236*	
10.2	Green Approaches to Nanofabrication *238*	
10.2.1	Molding and Embossing *238*	
10.2.1.1	Hard Pattern Transfer Elements *238*	
10.2.1.2	Soft Pattern Transfer Elements *240*	
10.2.1.3	Outlook *243*	
10.2.2	Printing *244*	
10.2.2.1	Microcontact Printing *244*	
10.2.2.2	Dip-Pen Nanolithography *245*	
10.2.2.3	Outlook *246*	
10.2.3	Edge Lithography by Nanoskiving *246*	
10.2.3.1	The Ultramicrotome *248*	
10.2.3.2	Nanowires with Controlled Dimensions *248*	
10.2.3.3	Open- and Closed-Loop Structures *248*	
10.2.3.4	Linear Arrays of Single-Crystalline Nanowires *249*	
10.2.3.5	Conjugated Polymer Nanowires *252*	
10.2.3.6	Nanostructured Polymer Heterojunctions *253*	
10.2.3.7	Outlook *258*	
10.2.4	Shadow Evaporation *259*	
10.2.4.1	Hollow Inorganic Tubes *259*	
10.2.4.2	Outlook *261*	
10.2.5	Electrospinning *263*	
10.2.5.1	Scanned Electrospinning *264*	
10.2.5.2	Uniaxial Electrospinning *265*	
10.2.5.3	Core/Shell and Hollow Nanofibers *265*	
10.2.5.4	Outlook *267*	
10.2.6	Self-Assembly *267*	
10.2.6.1	Hierarchical Assembly of Nanocrystals *268*	
10.2.6.2	Block Copolymers *269*	
10.2.6.3	Outlook *271*	
10.3	Future Directions: Toward "Zero-Cost" Fabrication *271*	
10.3.1	Scotch-Tape Method for the Preparation of Graphene Films *271*	
10.3.2	Patterned Paper as a Low-Cost Substrate *272*	
10.3.3	Shrinky-Dinks for Soft Lithography *272*	
10.4	Conclusions *274*	
	Acknowledgments *275*	
	References *275*	
11	**Nanocatalysis for Fuel Production** *281*	
	Burtron H. Davis	
11.1	Introduction *281*	
11.2	Petroleum Refining *282*	
11.3	Naphtha Reforming *282*	
11.4	Hydrotreating *289*	
11.5	Cracking *293*	

11.6	Hydrocracking	295
11.7	Conversion of Syngas	296
11.8	Water-Gas Shift	296
11.9	Methanol Synthesis	298
11.10	Fischer–Tropsch Synthesis (FTS)	302
11.11	Methanation	307
11.12	Nanocatalysis for Bioenergy	308
11.13	The Future	312
	References	314

12 Surface-Functionalized Nanoporous Catalysts Towards Biofuel Applications 319

Hung-Ting Chen, Brian G. Trewyn, and Victor S.-Y. Lin

12.1 Introduction 319
12.1.1 "Single-Site" Heterogeneous Catalysis 320
12.1.2 Techniques for the Characterization of Heterogeneous Catalysts 321
12.2 Immobilization Strategies of Single-Site Heterogeneous Catalysts 322
12.2.1 Supported Materials 322
12.2.2 Conventional Methods of Functionalization on Silica Surface 324
12.2.2.1 Non-Covalent Binding of Homogeneous Catalysts 324
12.2.2.2 Immobilization of Catalysts on the Surface through Covalent Bonds 327
12.2.2.3 Post-Grafting Silylation Method 328
12.2.2.4 Co-Condensation Method 330
12.2.3 Alternative Synthesis of Immobilized Complex Catalysts on the Solid Support 333
12.3 Design of more Efficient Heterogeneous Catalysts with Enhanced Reactivity and Selectivity 335
12.3.1 Surface Interaction of Silica and Immobilized Homogeneous Catalysts 335
12.3.2 Reactivity Enhancement of Heterogeneous Catalytic System Induced by Site Isolation 337
12.3.3 Introduction of Functionalities and Control of Silica Support Morphology 338
12.3.4 Selective Surface Functionalization of Solid Support for Utilization of Nanospace Inside the Porous Structure 342
12.3.5 Cooperative Catalysis by Multi-Functionalized Heterogeneous Catalyst System 346
12.3.6 Tuning the Selectivity of Multi-Functionalized Hetergeneous Catalysts by Gatekeeping Effect 348
12.3.7 Synergistic Catalysis by General Acid and Base Bifunctionalized MSN Catalysts 351
12.4 Other Heterogeneous Catalyst System on Non-Silica Support 354
12.5 Conclusion 354
References 356

13 Nanotechnology for Carbon Dioxide Capture 359
Richard R. Willis, Annabelle Benin, Randall Q. Snurr, and Özgür Yazaydın

13.1 Introduction 359
13.2 CO_2 Capture Processes 364
13.3 Nanotechnology for CO_2 Capture 366
13.4 Porous Coordination Polymers for CO_2 Capture 371
      References 395

14 Nanostructured Organic Light-Emitting Devices 403
Juo-Hao Li, Jinsong Huang, and Yang Yang

14.1 Introduction 403
14.2 Quantum Confinement and Charge Balance for OLEDs and PLEDs 405
14.2.1 Multilayer Structured OLEDs and PLEDs 405
14.2.2 Charge Balance in a Polymer Blended System 406
14.2.3 Interfacial Layer and Charge Injection 411
14.2.3.1 I–V Characteristics 412
14.2.3.2 Built-in Potential From Photovoltaic Measurement 413
14.2.3.3 XPS/UPS Study of the Interface 415
14.2.3.4 Comparison with Cs/Al Cathode 420
14.3 Phosphorescent Materials for OLEDs and PLEDs 421
14.3.1 Fluorescence and Phosphorescent Materials 421
14.3.2 Solution-Processed Phosphorescent Materials 422
14.4 Multi-Photon Emission and Tandem Structure for OLEDs and PLEDs 428
14.5 The Enhancement of Light Out-Coupling 429
14.6 Outlook for the Future of Nanostructured OLEDs and PLEDs 431
14.7 Conclusion 432
      References 432

15 Electrochromic Materials and Devices for Energy Efficient Buildings 435
Claes-Göran Granqvist

15.1 Introduction 435
15.2 Electrochromic Materials 437
15.2.1 Functional Principles and Basic Materials 437
15.2.2 The Role of Nanostructure 439
15.2.3 The Cause of Optical Absorption 443
15.3 Electrochromic Devices 445
15.3.1 Data on Foil-Based Devices with W Oxide and Ni Oxide 445
15.3.2 Au-Based Transparent Conductors 449
15.3.3 Thermochromic VO_2-Based Films for Use with Electrochromic Devices 451
15.4 Conclusions and Remarks 452
      References 455

Index 459

Introduction

Mankind faces daunting energy challenges in the 21st century, i.e., its over-reliance on the quickly diminishing fossil fuel-based energy sources and the consequent negative impacts to the global environment and climate. Although evolutionary improvements in existing technologies will continue to play important roles in addressing some of the challenges, revolutionary new technology will be the key to a clean, secure and sustainable energy future. Nanotechnology, by manipulating matter at the nanoscale with unprecedented accuracy, holds the promise of providing new materials with distinctly different properties. In recent years, breakthroughs in nanotechnology, especially in their applications in the energy sector, open up the possibility of moving beyond our current alternatives by introducing technologies that are more efficient, environmentally sound and cost effective.

This book, *Nanotechnology for the Energy Challenge*, is a collection of 15 chapters written by some of the world's leading experts in nanotechnology and its applications in the energy fields, each covering a specific subject that falls within three general aspects: production, storage and use of energy, correspondingly the three parts of the book.

Part I Sustainable Energy Production covers the main developments of nanotechnology in clean energy production and conversion. Following a general overview on the contributions of nanomaterials in selected specific areas of energy production, such as photovoltaics, photochemical hydrogen production, fuel cells and thermoelectricity, the remaining individual chapters within this part take these topics, i.e. dye-sensitized photoelectrochemical devices, nanostructured thermoelectric materials, nano-sized electrodes and electrolytes for fuel cells, and nanomaterials-based photoelectrochemical water splitting, into in-depth discussion.

Part II Efficient Energy Storage is concerned with the potential use of nanomaterials in more efficient energy storage systems. Batteries, superconductors, hydrogen storage for fuel cell applications are the main foci, which exemplify the three main families of energy storage systems in which "going-nano" is found to be especially beneficial. Firstly, hydrogen storage by physical and chemical adsorption is reviewed with an emphasis on how the use of nanomaterials helps improve its performance. Then the subsequent two chapters, with different focal points, discuss the impacts of nanostructuring on the performance of batteries and supercapacitors. The final chapter in this part describes the use of extrinsic nano-sized defects to produce advanced superconducting materials with minimal dissipation.

The last part of the book, *Part III Energy Sustainability* wraps around how nanotechnology helps to use energy more efficiently, and mitigate its impact to the environment. While energy-efficient, or "green", nanofabrication of the nanomaterials themselves constitutes an important component of energy sustainability, nanocatalysis in petroleum refining and biofuel production also contribute significantly to the conservation of energy. Carbon dioxide capture by nanoporous materials is another area that where nanotechnology may offer breakthrough opportunity. Nanostructured light emitting diodes (LED) and organic LEDs (OLEDs) provide higher efficiency in energy conversion from electricity to light. Lastly, electrochromic materials with nanofeatures and their use in energy-efficient buildings are discussed. This book is intended to provide a balanced treatment of the various topics of nanotechnologies in the energy and related areas with both general overviews and in-detail discussions to suit for a broader audience. I sincerely hope it will attract your attention to nanotechnology and its applications in energy related areas.

Javier Garcia Martinez
Alicante, Spain
December 2009

List of Contributors

Sambandam Anandan
National Institute of Technology
Nanomaterials & Solar Energy
Conversion Laboratory
Department of Chemistry
Trichy, 620 015
India

Vincenzo Antonucci
CNR-ITAE Institute
Via Salita S. Lucia sopra
Contesse
5-98126, Messina
Italy

Antonino Salvatore Aricò
CNR-ITAE Institute
Via Salita S. Lucia sopra
Contesse
5-98126, Messina
Italy

Muthupandian Ashokkumar
University of Melbourne
School of Chemistry
Melbourne
Vic. 3010
Australia

Vincenzo Baglio
CNR-ITAE Institute
Via Salita S. Lucia sopra Contesse
5-98126, Messina
Italy

François Béguin
CRMD
CNRS/Orléans University
1b rue de la Férollerie
45071, Orléans
France

Annabelle Benin
UOP Research Center
50 East Algonquin Road
Des Plaines, IL 60017
USA

Claudia Cantoni
Oak Ridge National Laboratory
P.O. Box 2008
Oak Ridge, TN 37831
USA

Guozhong Cao
University of Washington
Department of Materials Science and
Engineering
Seattle, WA 98195
USA

Nanotechnology for the Energy Challenge. Edited by Javier Garcia-Martinez
© 2009 WILEY-VCH Verlag GmbH & Co. KGaA, Weinheim
ISBN: 978-3-527-32401-9

Hung-Ting Chen
Iowa State University
Department of Chemistry
U.S. Department of Energy
Ames Laboratory
Ames, IA 50011-3111
USA

Burtron H. Davis
University of Kentucky
Center for Applied Energy Research
2540 Research Park Drive
Lexington, KY 40511
USA

Elzbieta Frackowiak
ICTE
Poznan University of Technology
ul. Piotrowo 3
60-965, Poznan
Poland

Javier García-Martínez
Universidad de Alicante
Molecular Nanotechnology Laboratory
Ctra. Alicante–San Vicente s/n
03690, Alicante
Spain
Rive Technology Inc.
1 Deer Park Drive
Monmouth Junction, NJ 08852
USA

Amit Goyal
Oak Ridge National Laboratory
P.O. Box 2008
Oak Ridge, TN 37831
USA

Claes-Göran Granqvist
The Ångström Laboratory
Department of Engineering Sciences
P.O. Box 534
75121, Uppsala
Sweden

Michaël Grätzel
Ecole Polytechique Fédérale de Lausanne
Laboratory for Photonics & Interfaces
1015, Lausanne
Switzerland

Jian He
Clemson University
Department of Physics and Astronomy
Clemson, SC 29634-0978
USA

Jinsong Huang
University of California
Los Angeles
Department of Materials Science and Engineering
405 Hilgard Avenue
Los Angeles, CA 90095
USA

Juo-Hao Li
University of California
Los Angeles
Department of Materials Science and Engineering
405 Hilgard Avenue
Los Angeles, CA 90095
USA

Kunhao Li
Rive Technology Inc.
1 Deer Park Drive
Monmouth Junction, NJ 08852
USA

Victor S.-Y. Lin
Iowa State University
Department of Chemistry
U.S. Department of Energy
Ames Laboratory
Ames, IA 50011-3111
USA

Darren J. Lipomi
Harvard University
Department of Chemistry and
Chemical Biology
12 Oxford Street
Cambridge, MA 02138
USA

Agustin McEvoy
Dyesol Ltd
3 Dominion Place
Queanbayan
NSW
Australia

Jagannathan Madhavan
University of Melbourne
School of Chemistry
Melbourne
Vic. 3010
Australia

Guillermo Rus
University of Granada
Department of Structural
Mechanics
Politécnico de Fuentenueva
18071, Granada
Spain

Saghar Sepehri
University of Washington
Department of Materials Science
and Engineering
Seattle, WA 98195
USA

Elena Serrano
Universidad de Alicante
Molecular Nanotechnology Laboratory
Ctra. Alicante–San Vicente s/n
03690, Alicante
Spain

Patrice Simon
Université Paul Sabatier
CIRIMAT–UMR CNRS 5085
118 route de Narbonne
31062, Toulouse Cedex
France

Randall Q. Snurr
Northwestern University
Department of Chemical and Biological
Engineering
Evanston, IL 60208
USA

Jean-Marie Tarascon
Université de Picardie Jules Verne
LRCS–UMR CNRS 6007
33 rue Saint Leu
80 039, Amiens
France

Brian G. Trewyn
Iowa State University
Department of Chemistry
U.S. Department of Energy Ames
Laboratory
Ames, IA 50011-3111
USA

Terry M. Tritt
Clemson University
Department of Physics and Astronomy
Clemson, SC 29634-0978
USA

Emily A. Weiss
Harvard University
Department of Chemistry and
Chemical Biology
12 Oxford Street
Cambridge, MA 02138
USA

George M. Whitesides
Harvard University
Department of Chemistry and
Chemical Biology
12 Oxford Street
Cambridge, MA 02138
USA

Richard R. Willis
UOP Research Center
50 East Algonquin Road
Des Plaines, IL 60017
USA

Yang Yang
University of California
Los Angeles
Department of Materials Science and
Engineering
405 Hilgard Avenue
Los Angeles, CA 90095
USA

Ozgur Yazaydin
Northwestern University
Department of Chemical and Biological
Engineering
Evanston, IL 60208
USA

Part One
Sustainable Energy Production

1
Nanotechnology for Energy Production
Elena Serrano, Kunhao Li, Guillermo Rus, and Javier García-Martínez

1.1
Energy Challenge in the 21st Century and Nanotechnology

One of the greatest challenges to mankind in the 21st century is its over-reliance on the diminishing fossil fuels (coal, oil, natural gas) as primary energy sources. Fossil fuels are non-renewable resources that take millions of years to form. Consequently their reserves are depleted much faster than new ones are formed and/or discovered. While concerns over fossil fuel supplies are often the direct or indirect causes to regional and global conflicts, the production, transmission and use of fossil fuels also lead to environmental degradation. Combustion of carbon-based fossil fuels generates not only air pollutants, for example, sulfur oxides and heavy metals, but also CO_2, the infamous greenhouse gas widely believed to be the culprit of global climate change. One of the solutions to this energy challenge, on the one hand, relies upon increasing the efficiency in production, transmission and utilization of the remaining fossil fuels while reducing their negative impacts to the environment. On the other hand, technologies and infrastructures have to be developed or improved in preparation for the smooth transition to the alternative and renewable energy sources, that is, nuclear power, solar energy, wind power, geothermal energy, biomass and biofuels, hydropower, etc.

Technological advancement, as shown by history, will play a pivotal role in this path to a more sustainable energy future. As defined by the International Energy Agency (IEA) [1], there are three generations of renewable energy technologies. The first generation technologies, including hydropower, biomass combustion and geothermal power and heat, emerged from the industrial revolution at the end of the 19th century. The second-generation technologies include solar heating and cooling, wind power, bioenergy and solar photovoltaics. These technologies are now entering markets as a result of continuous investments in their research and development since the 1980s, which were largely driven by energy security concerns linked to the oil crisis in the 1970s. The third-generation technologies currently under development include advanced biomass gasification, biorefinery technologies, concentrating solar thermal power, geothermal power, ocean tide and wave energy, etc. (Figure 1.1).

Nanotechnology for the Energy Challenge. Edited by Javier Garcia-Martinez
© 2009 WILEY-VCH Verlag GmbH & Co. KGaA, Weinheim
ISBN: 978-3-527-32401-9

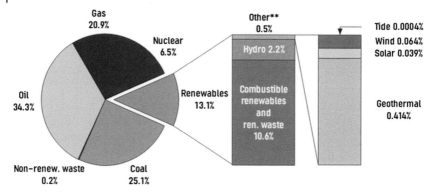

Figure 1.1 Fuel shares of world total primary energy supply. Reprinted with permission from ref. [2].

The rapid evolution of nanoscience and nanotechnology (the study and control of materials and phenomena at length scales between 1 nm to 100 nm) during the past two decades has demonstrated that nanotechnology holds the key to many of the technological advancements in the energy sector which rely, at least in part, on having novel materials with superior properties. According to the "Roadmap Report Concerning the Use of Nanomaterials in the *Energy* Sector" from the Sixth Framework Program [3], the most promising application fields of nanotechnology or nanomaterials for the energy production domain will be in photovoltaics, hydrogen conversion (hydrogen production, fuel cells) and thermoelectricity (see Table 1.1 for the time frame of possible industrial applications of a few exemplary nanomaterials).

This book intends to present a comprehensive overview of up-to-date progress in various important fields of nanotechnology research and development that are closely related to sustainable energy production, storage and usage. This chapter, in particular, mainly serves as an "overview" of the book, with an emphasis on sustainable energy production.

Nanotechnology intervenes at a number of stages of the energy flow that starts from the primary energy sources and finishes at the end user. The growing and diffuse limits of what can be considered as energy and the complex flows of energy in society and ecosystem make it impossible to draw an unequivocal definition of energy production. Herein, the term "energy production" encompasses all processes that convert energy from primary energy sources to secondary energy sources or forms. It does not cover energy production processes that are associated with non-renewable energy sources, that is, fossil fuels (and nuclear power) and a few renewable sources, that is, ocean wave energy, wind energy and hydropower. However, we have included a chapter on the contributions of nanocatalysis in fuel production, because many recent advances in more efficient production and transformation of various fuels owe thanks to catalysis with novel nanomaterials.

Table 1.1 Timeframe of possible industrial applications for nanomaterials. Adapted with permission from ref. [3].

Nanomaterials	2006–2008	2009–2011	2012–2016
Core shell nanoparticle			Solar cells
Fullerenes		Photovoltaic energy supply	
Indium phosphide (InP) thin films	Photovaltaic solar cells		
Metal ceramic nanocomposite (coating)		Hyodrogen production Solar	
Nickel (carbon-coated) [Ni–C] powders	Conductors		
Poly(9,9'-dioctylfluorene-co-bithiophene) (F8T2) nanolayers		Solar energy conversion	
Poly(octadecylsiloxane) (PODS) nanolayers		Batteries Photovoltaics	
Polyacrylonitrile (PAN) nanostructures	Phovoltaic devices		
Polyaniline-SnO_2 (PANI-SnO_2) and polyaniline-TiO_2 (PANI-TiO_2) nanostructured films		TiO_2 solid state solar cells	
Polyfluorene/polyaniline (PF/PANI) nanostructures		Photoelectrical devices	
Silicon (Si) nanopowders and nanowires	Solar cells		
Ceria (CeO_2) nanoparticles, coatings		Solid oxide fuel cells	
Platinum (Pt) nanoparticles	Catalysts		
Polymer with carbon nanoparticles/fillers (bulk)		Fuel cells	
Boron (B) nanowires		Thermoelectric energy conversion	
Carbides coatings and related materials	Anticorrosive coatings High absorbing black coatings		

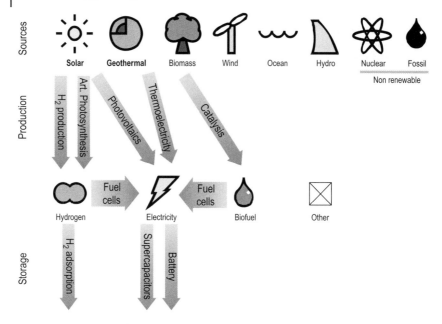

Figure 1.2 Flow chart of the energy production and storage processes where nanotechnology may contribute.

In Figure 1.2, the energy production processes where nanotechnology has potential key roles are grouped by the arrows in the upper half, while the arrows in the lower half depict where nanotechnology may play significant parts in the energy storage or transportation processes. Parts One and Two of this book deal with these two topics respectively, while Part Three revolves around how nanotechnology may positively impact on the efficient usage and sustainability of energy, as well as the environmental protection aspects.

1.2
Nanotechnology in Energy Production

1.2.1
Photovoltaics

Photovoltaic cells, more commonly known as solar cells, are devices that absorb energy radiated from the sun that reaches the earth in the form of light and convert it directly into electricity. The phenomenon and principle of the photoelectric effect, that is, direct generation of electricity from light, was first discovered by the French physicist A. Becquerel in 1839 [4]. In a solar cell, when photons are absorbed by semiconducting materials, mobile electron-hole pairs (excitons) are generated. Once the excitons are separated, via drift or diffusion, the electrons can be directed to power an external load.

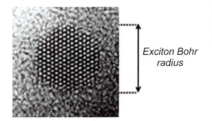

Figure 1.3 A TEM micrograph of a 7-nm crystalline silicon quantum dot. Reprinted with permission from ref. [7].

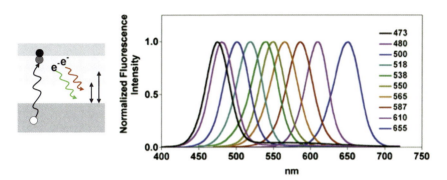

Figure 1.4 Emission spectra and color as a function of particle size of a quantum dot. Reprinted with permission from ref. [8].

There have been three generations of photovoltaic technologies [5]. First-generation solar cells build on high-quality single crystalline silicon wafers and consist of large-area, single p-n junction diodes which can achieve very high efficiency (close to the theoretical efficiency of 33%) but their production costs are prohibitively high. Second-generation solar cells, represented by thin-film devices based on cadmium telluride (CdTe), copper indium gallium selenide (CIGS), amorphous silicon and micromorphous silicon, require lower energy and production costs. Unfortunately, they suffer from much reduced energy conversion efficiencies compared to the first generation because of the defects inherent in the lower-quality processing methods. Third-generation solar cell technologies aim to increase the efficiency of second generation solar cells while maintaining low production costs. There are generally three approaches adopted: (i) the use of multi-junction photovoltaic cells, (ii) light concentration and (iii) the use of excess thermal generation to enhance voltages or carrier collection (e.g., photovoltaic thermal hybrid cells). Nano-structured semiconducting materials hold the promise of achieving high efficiency at low cost.

Nanocrystal quantum dots (NQDs) [6] are nanometer-scale single crystalline particles of semiconductors, for example, Si (Figure 1.3). Due to the quantum confinement effect, their light absorption and emission wavelengths can be controlled by tailoring the size of NQDs (Figure 1.4). In addition, researchers at the National Renewable Energy Laboratory (NREL), in Golden (Colorado, USA)

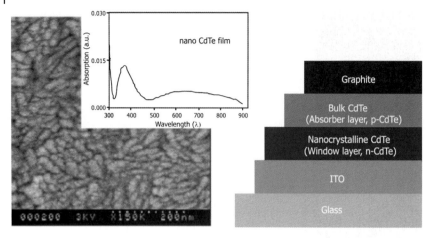

Figure 1.5 Example of nanomaterials for photovoltaic cells fabrication. Left: FE-SEM image of a nanocrystalline CdTe film on ITO-coated glass substrate. The inset shows the absorption spectrum of a nanocrystalline CdTe film on ITO-coated glass substrate. Right: Device configuration of a glass/ITO/n-nano-CdTe/p-bulk CdTe/graphite solar cell. Adapted with permission from ref. [9].

recently showed that silicon nanocrystals can produce two or three excitons per high-energy photon of sunlight absorbed, in comparison to bulk Si which can only produce one exciton per high-energy photon. Using nanocrystals as the light-absorbing materials, typically on a supporting matrix of conductive polymers or high surface area mesoporous metal oxides (e.g., TiO_2), the nanocrystal solar cells have a theoretical energy conversion efficiency of 40% in contrast to that of the typical solar panels based on bulk Si of around 20% [7].

Nanocrystal solar cells can be tuned to absorb light of different wavelengths by adjusting the size of nanocrystals [9]. In addition, the absorber layer thickness in the nanostructured cells can be reduced down to 150 nm, ten times lower than that of the classic thin film solar cells. For thin-film multilayered cells, using nanocrystalline materials also help achieve a regular crystalline structure, which further enhances the energy conversion efficiency.

An example of nanostructured layers in thin-film solar cells has been recently reported by Singh et al. [9]. Nanocrystalline CdTe and CdS films on indium tin oxide (ITO)-coated glass substrates have been synthesized as potential n-type window layers in p-n homo(hetero)junction thin-film CdTe solar cells. CdTe nanocrystals of around 12 nm in diameter exhibit an effective bandgap of 2.8 eV, an obvious blue shift from the 1.5 eV of bulk CdTe (Figure 1.5). As shown in Figure 1.6, films of CdS nanofibers (formed inside mesoporous TiO_2 films) are deposited over a plastic or an ITO-coated glass substrate.

Dye-sensitized photoelectrochemical solar cells (PES or Grätzel cells) represent a new class of low-cost thin-film solar cells (Figure 1.7) [10–12]. Nano-structured TiO_2, CeO_2, CdS and CsTe are of great interest as the windowing and light-

Figure 1.6 FE-SEM image of porous CdS film on plastic substrate (left) and CdS fibers on ITO-coated glass substrate (right). Adapted with permission from ref. [9].

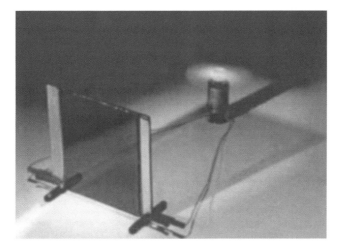

Figure 1.7 Grätzel cell. Reprinted with permission from ref. [10].

absorbing layers [13–15]. Different aspects of the materials have been studied, with a focus on the size and shape control of the nanocrystals, for their use in dye-sensitized solar cells [16]. By far, among all the PES studied, TiO_2-based cells provide the best results, albeit the absolute efficiency is still quite low [17]. Two different approaches to increased energy harvesting in TiO_2-based PES cells have been demonstrated by Corma's group, which are based on photonic crystals (PC) and photonic sponge (PhS) new architectures, respectively [18, 19]. Much better photoelectrical performances are achieved when the titania (TiO_2) electrode is structured to localize photons. In fact, the total efficiency of dye-sensitized PES cells is enhanced by a factor of five thanks to the PhS topology (see Figure 1.8), which also expands the cell photocurrent response to the 300–800 nm range.

By replacing the TiO_2 nanoparticles with arrays of TiO_2 nanotubes, Grims et al. [20] showed that the resulting PES cells (Figure 1.9) exhibit promising photoconversion efficiencies, probably up to the theoretical limit. It is believed that the nanotubes provide the benefit of reducing the resistance to electron and charge transfer. However, the cell size needs to be scaled up.

Following a similar approach, the efficiency of solar cells can be considerably increased by facilitating charge transfer through the incorporation of carbon

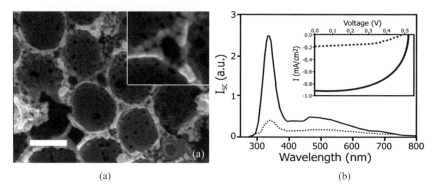

Figure 1.8 (a) SEM images of titania nanoparticles arranged in a PhS architecture, smaller cavities are arranged in the regions between the larger ones (scale bar: 1 μm). (b) Photocurrent spectra for standard (dashed line) and PhS (continuous line) titania electrodes. The inset shows the corresponding I–V spectra of PES cells under AM 1.5 conditions. Adapted with permission from ref. [19].

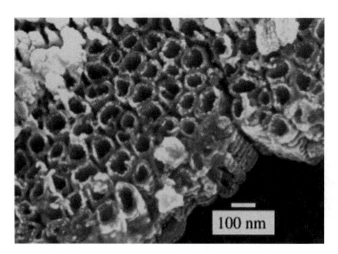

Figure 1.9 Cross-sectional FE-SEM image of TiO_2 nanotube array (scale bar: 100 nm). Reprinted with permission from ref. [20].

nanotubes and fullerenes in matrices of other semiconductor materials. In 2005, Guldi et al. [21] attached molecules of porphyrin to the walls of single-wall carbon nanotubes (SWNT) and showed, for the first time, a photoinduced electron transfer process in the hybrid nanotube complex (Figure 1.10). The authors proved that electrons from the porphyrin molecules were transferred to nanotube walls when the latter were exposed to light in the visible range. As noted by the authors, "this separation of charge is sufficiently long-lived for us to divert and use the electrons" [22]. The results, Guldi remarks, meet "the first criteria for the development of solar cells based on modified carbon nanotubes" [22].

More recently, scientists at Georgia Tech Research Institute (GTRI) constructed a new 3D solar cell composed of carbon nanotube arrays. The researchers first grow micrometer-scale towers of multi-walled carbon nanotubes on top of silicon wafers patterned with a thin layer of iron by photolithography (Figure 1.11) [23]. Afterwards, the microtowers are coated with photoactive materials, CdTe and CdS,

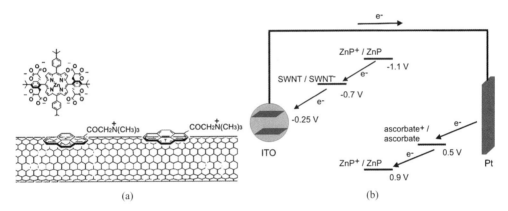

Figure 1.10 (a) Structure of SWNT/pyrene$^+$/ZnP^{8-}. (b) Photocurrent generation in ITO electrodes covered with a single SWNT/pyrene$^+$/ZnP^{8-} stack. Adapted with permission from ref. [21].

Figure 1.11 Left: Cross-sectional image shows cadmium telluride coating around a carbon nanotube microtower (scale bar: 10 μm). Right: 3D image of nanotube-based solar cell (scale bar: 20 μm). Adapted with permission from ref. [23].

which serve as p-type and n-type photovoltaic layers, respectively. Finally, the cell is coated with transparent ITO as top electrode. The carbon nanotube microtowers serve both as a conductor connecting the photovoltaic materials to the silicon (bottom junction) and as a support of the 3D arrays that more effectively absorb sunlight from different angles.

Organic or polymer solar cells are another relatively new class of low-cost solar cells. Typical efficiencies in 1.5–6.5% range have been achieved by using evaporated bilayer [24, 25], bulk heterogeneous conjugated polymer–fullerene [26–34], co-evaporated molecular[35–38] and organic/inorganic hybrid devices [39–52]. Compare to silicon-based devices, the efficiency of organic solar cells is fairly low. However, they are much less expensive to fabricate. In addition, they offer advantages in applications where lightweight, flexibility and disposability are desirable.

The photophysics of organic photovoltaic devices is based on the photo-induced charge transfer from donor-type conjugated polymers to acceptor-type polymers or molecules rather than semiconductor p-n junctions. Improvement in the nanoscale morphology control, together with development of novel low-bandgap materials with better light absorption and faster charge carrier mobilities, are believed to lead to organic solar cells that can achieve power conversion efficiencies close to and beyond 10%. Nanoscale morphology control strongly depends on various parameters, such as the choice of materials and solvents, the donor–acceptor weight ratio, the method of deposition, the drying and the annealing. Two approaches have been developed to form well-controlled nanomorphology. The first is utilizing self-assembly of semiconducting block copolymers [42–49], which contain an acceptor block structure (with fullerene pendants) attached to n-type conjugated polymers (Figure 1.12). The second relies upon a "mini-emulsion"

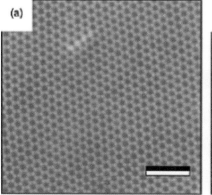
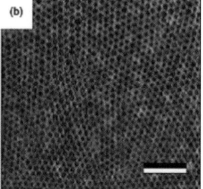

Figure 1.12 Luminescence micrographs showing the honeycomb structure of thin films of di-block copolymers (scale bars: 20 µm). In (a), the local photoluminescence (PL) of a structure without the fullerenes is compared with (b) the one including pendant fullerenes, where a strong PL-quenching is observed. Reprinted with permission from ref. [49].

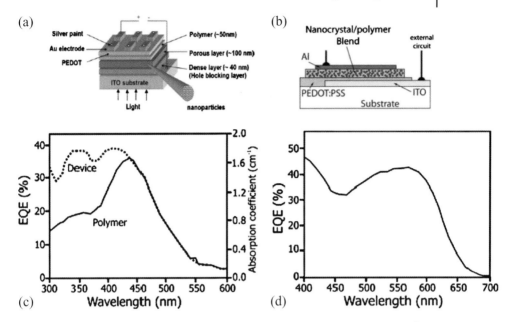

Figure 1.13 Schematic of the device structures of: (a) a multilayer polymer–porous metal oxide device and (b) hybrid nanocrystal–polymer blend photovoltaic devices. (c, d) Their corresponding EQE spectra (the corresponding absorption spectrum of the TPD (4M)–MEH–M3EH–PPV polymer has been also included in (c; solid line). Adapted with permission from ref. [53].

process to control the size of polymer nanospheres, Dispersions of nanospheres of two individual polymers or of a blend of two polymers can be spin-coated to form thin films with the dimension of phase separation well-controlled by the particle size [50].

An alternative to achieve better control of nanoscale phase separation is to involve inorganic nanostructures. Mesoporous metal oxides with regular nano-sized pores, which are determined by the organic templates used in their syntheses, are utilized as scaffolds for the formation of ordered heterojunctions by infiltrating polymers into the ordered pores through melt infiltration, dip-coating or polymerization. TiO_2 is, by far, the most commonly employed material since it is abundant, non-toxic and transparent to visible light [51, 52]. This hybrid solar cell approach gives high control over the pore size and the scale of phase separation, whereas a recent comparison study by Nelson and colleagues found that a metal oxide nanocrystal/polymer blend film achieved the best photovoltaic performance so far (Figure 1.13), suggesting that "the problems of interparticle transport can be overcome more easily than those of polymer infiltration". [53]

In summary, nanomaterials are expected to play important roles in solar cell technology in the coming decade, as shown in Figure 1.14 and in Tables 1.1 and 1.2 [3].

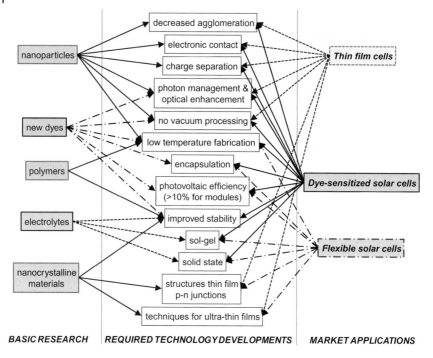

Figure 1.14 Basic research underway with the technology developments required to achieve the desired applications. Reprinted with permission from ref. [54].

1.2.2
Hydrogen Production

Hydrogen, the most abundant element in the universe, is proposed to be used as the energy carrier for the distribution of energy in the hydrogen economy (Figure 1.15). Unlike fossil fuels, which are both primary energy sources and energy carriers, hydrogen serves only for the latter role. In other words, hydrogen has to be produced by using other energy sources, from fossils fuels, from nuclear power or from the renewable sources. Hydrogen economy [56] is very attractive because the use of hydrogen as a fuel, either in combustion mode or in fuel cell mode, can achieve much higher energy conversion efficiencies than the current fossil fuel internal combustion engines. Furthermore, the use of hydrogen does not generate CO_2 "green-house" gas or other pollutants. However, hydrogen production currently relies upon fossil fuels (natural gas, oil, coal) and only ~5% of the commercial hydrogen is produced from renewable energy sources [57]. Only when hydrogen is fully produced from renewable energy sources can its lifecycle be considered clean and renewable.

Hydrogen is produced by steam reforming and water electrolysis. Only the latter can use electricity from renewable sources to produce hydrogen. However, the electricity needed constitutes almost the same value of the hydrogen produced (>US$20/GJ assuming a cost of electricity at about US$0.05 kWh) [55]. Hydrogen

Table 1.2 Expected time frame for the development of nanomaterials in the solar photovoltaics market segment. TI, Technology Inversion; LP, Laboratory Prototype; ID, Industrial Demonstrator; I, Industrialization; ME, Market Entry. Reprinted with permission from ref. [3].

	2006	2007	2008	2009	2010	2011	2012	2013	2014	2015	2016
Core shell nanoparticles	I				ME						
Fullerenes	ME										
Indium phosphide (InP) thin films	I		ME								
Metal ceramin nanocomposite (coating)	ID	ME									
Nickel (carbon-coated) powders	ME										
Poly (9,9′-dioctylfluorene-co-bithiophene) (F8T2) nanolayers	LP	ID	I	ME							
POSS nanolayers	ID	I	ME								
PAN nanostructures	I		ME								
PANI–SnO$_2$ and PANI–TiO$_2$	I		ME								
Polyfluorene/PANI nanostructures	TI		LP		ID	I	ME				
Silicon nanopowders and nanowires	I		ME								

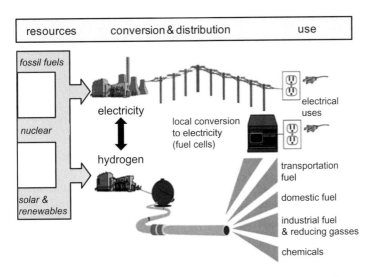

Figure 1.15 Hydrogen energy system. Adapted with permission from ref. [55].

can also be produced by other ways, such as solar thermal, thermochemical or photoelectrochemical water splitting, as well as biological hydrogen production. Clean, low-cost and environmentally friendly production of hydrogen based on the use of solar energy through photocatalytic water splitting is currently under intensive research [57, 58]. This section describes the devices and mechanisms for photocatalytic water splitting, along with examples where nanomaterials help to increase the efficiency of hydrogen production.

Photoelectrochemical water splitting [59], also named artificial photosynthesis, directly uses electricity produced by photovoltaic systems to generate hydrogen and oxygen within one monolithic device, the photoelectrochemical cell (PEC). Therefore, in general, the contributions from nanotechnology to the increased efficiency of solar energy-to-electricity conversion described in Section 1.2.1 also apply here. In addition, further improvements can be gained by enhancing the activity and stability of the photocatalysts, for example, TiO_2.

Mainly, there are three approaches to solar water splitting: a solid-state photovoltaic cells (PV) coupled with a water electrolyzer, a semiconducter liquid junction and a combined PV-semiconductor liquid junction [60]. Strictly speaking, the first one is not a PEC and it suffers from high cost. Alternatively, metal/semiconductor Schottky junction cells with single or multiple p-n junctions immersed in the aqueous system as one electrode can help reduce the cost. In a semiconductor liquid junction cell, the photopotential needed to drive the water-splitting reaction is generated directly at the semiconductor/liquid interface. In a combined PV-liquid junction system, the solar cell can be connected to either the reduction or oxidation photocatalyst to provide the additional electrical bias needed for the generation of hydrogen or oxygen.

Semiconductor-based artificial photosynthesis shares some similar issues with semiconductor-based photovoltaics. For example, TiO_2 is the favorite photoelectrode material because it is highly stable over a wide range of pH and it is abundant and non-toxic. But it absorbs less than 12% of the incident sunlight due to its large bandgap. Although other materials have been studied, such as WO_3, SnO_2, CdS, $CuInSe_2$ or SiC [57, 61–63], they also suffer from various problems such as solar energy matching, long-term stability and loss of activity in aqueous solution. Nanosized semiconductor particles, despite their larger bandgaps compared to the corresponding bulk materials, have very high surface areas that, in principle, allow faster charge capture by the solution species and lower probability of bulk charge recombination.

Dye sensitization is a convenient way to improve the utilization of solar spectrum (Figure 1.16). Some of the frequently used dyes are thionine, methylene blue, rhodamin, etc.; and usually, some redox systems or sacrificial agents (i.e., the I^{3-}/I^- pair and EDTA) are added to the solution to regenerate the dyes during the reaction cycle [57]. While dye sensitizers can expand the light absorption to a broader range, the very small thickness required for efficient electron injection limits the overall light harvesting efficiency. The use of electrodes with a large surface area as the substrate for photosensitizers, for example, highly porous TiO_2, helps maintain high quantum efficiency.

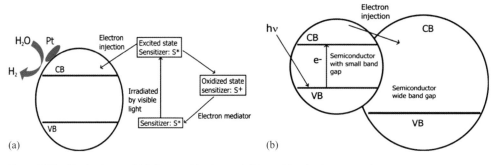

Figure 1.16 Mechanisms of (a) dye sensitization and (b) sensitization by composite semiconductors. Adapted with permission from ref. [57].

Sensitization by another semiconductor is also being developed (Figure 1.16), for example, composite systems of $CdS-SnO_2$, $CdS-TiO_2$ or $CdS-ZnS$. The efficiency on hydrogen production by using $CdS-ZnS$ composite semiconductors was proved first by De et al. in 1996 [64] and later by Koca et al. [65]. Alternatively, Tennakone et al. [66] combined dye-sensitized SnO_2 nanocrystallites (10–15 nm) to platinized ZnO (600 nm) particles and obtained 25 times larger hydrogen production with a $Pt/ZnO/SnO_2/dye$ than with a $Pt/ZnO/dye$ system. More recently, So et al. conducted photocatalytic hydrogen production using $CdS-TiO_2$ composite semiconductors. [67] Optical absorption spectra analysis showed that $CdS-TiO_2$ could absorb photons with wavelength up to 520 nm and exposure to visible light illumination revealed that $CdS-TiO_2$ composite semiconductors produce hydrogen at a higher rate than neat CdS or TiO_2.

Cation or anion doping is another effective way of shifting the bandgaps of the photoelectrodes towards the visible light region. Metal ions should be doped near the surface of TiO_2 for better electron transfer from hole to surface. Deep dopings, however, are prone to become electron-hole recombination sites. When TiO_2 is bombarded with high-energy transition metal ions (accelerated by high voltage), these high-energy ions are injected into the lattice and interact with TiO_2. This process modifies the TiO_2 electronic structure and shifts its photo-response to the visible region (up to 600 nm). Currently, TiO_2 implanted with metal ions, such as V, Cr, Mn, Fe or Ni, are believed to be among the most effective photocatalysts for solar energy utilization and are in general referred as "second-generation photocatalysts" [57]. Anion doping is less investigated but it is believed to have the advantage of not increasing the recombination probability as cation doping usually does. Asahi et al. [68] showed that mixing half the states of N rather than O in anatase TiO_2 narrows the bandgap of TiO_2.

Doping has also been applied to nanostructured photocatalysts. Wang et al. [69] used a metal plasma ion implantation (MPII) technique to incorporate trace amounts of transition metal ions of Cu, Ni, V, and Fe into the as-synthesized nanoscale anatase TiO_2 thin films. The authors observed that the bandgap energy of TiO_2 was reduced by 0.9–1.0 eV after the metal loading (Fe and Cu ions showing

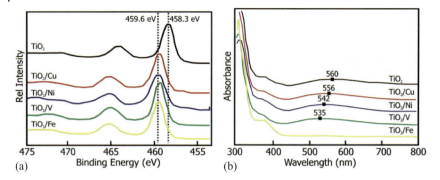

Figure 1.17 (a) The Ti (2P) narrow scan from the metal-doped TiO_2 revealing a common shift of 1.2–0.8 eV toward higher binding energy. (b) UV–Vis absorption spectra of various metal-doped TiO_2 nanoscale thin films. Adapted with permission from ref. [68].

a slight advantage over the others). The red-shift results in an increased photosensitivity of TiO_2 in the visible light regime, which further widens the applicability of the photocatalytic TiO_2 in low-UV environments (Figure 1.17).

In another case, Corma et al. [13] doped mesostructured CeO_2 nanoparticles with La^{3+} or Zr^{4+}. CeO_2 nanoparticles give a red-shift of the absorption spectrum of about 80 nm as compared to TiO_2, which leads to a considerably better response in the visible region of the solar spectrum. Moreover, the solar cells constructed upon 50 μm films of La^{3+}- or Zr^{4+}-doped CeO_2 nanoparticles (Figure 1.18) achieved higher efficiencies than the ones based on non-doped ceria and titania nanoparticles.

Similarly, loading the photocatalysts with noble metals, such as Pt, Au, Pd, Rh, Ni, Cu and Au, that have lower Fermi levels than TiO_2, reduces the E_g bandgap and the recombination probability. In 1988, Maruthamuthu and Ashokkumar showed that the hydrogen production efficiency of WO_3 was enhanced by loading the samples with metal by thermal deposition [70]. This was later verified by several research groups [71–73]. Anpo et al. observed by ESR that TiO_2 transfers electrons to Pt particulates [73]. The deposition method chosen strongly affects the efficiency, which suggests that nanostructured materials could provide crucial improvements.

In addition to the modification of photocatalysts, the addition of chemical additives to photoelectrochemical cells also helps to increase the solar-to-hydrogen efficiency. Electron donors such as methanol, ethanol, lactic acid or EDTA are proved to be the most efficient electron donors for TiO_2 photocatalytic water-splitting hydrogen production [29, 57, 71, 74–78]. Inorganic ions (S^{2-}/SO_3^{2-}, Ce^{4+}/Ce^{3+}, IO_3^-/I^-) can also be used (Figure 1.19) [57, 79–81]. The addition of carbonate salts inhibits the backward reaction by scavenging the photogenerated holes to form carbonate radicals [82, 83]. Sayama et al. observed the highest rate of photocatalytic production of hydrogen and oxygen (H_2 rates: 378 and 607 μmol/h, respectively) when $NaHCO_3$ and Na_2CO_3 are used as additives to the ZrO_2 suspen-

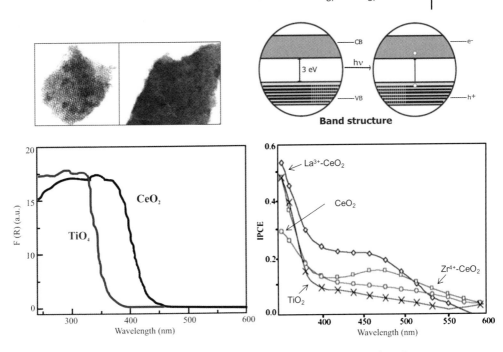

Figure 1.18 Photocatalytic activity of hierarchically mesostructured doped CeO_2. Adapted with permission from ref. [13].

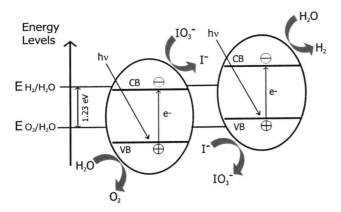

Figure 1.19 Photocatalytic hydrogen production under the mediation of I^-/IO_3^-. Adapted with permission from ref. [57].

sion versus 72–242 µmol/h when chloride, phosphate, borane or sulfate are added.

As a summary, photoelectrochemical water splitting for hydrogen production as described above has the potential to integrate the most abundant primary energy source, solar energy, and the cleanest fuel, hydrogen, into a fully renewable cycle.

Table 1.3 Expected costs of nanomaterials and the corresponding expected market size within the hydrogen conversion category (2006–2015). Reprinted with permission from ref. [3].

Nanomaterials	Estimated market size (t/year)	Estimated material costs (€/kg)
Ceria (CeO_2) nanoparticles, coatings	10 000 (2007–2010)	3000
Metal ceramic nanocomposite (coating)	–	50
Nickel (carbon-coated) powders	7500 (2008–2010)	1500 (2007–2010)
	15 000 (2011–2014)	1200 (2011–2014)
Platinum nanoparticles	–	–

However, the technologies are still in early stages of development. Significant research is still required before they become competitive with conventional processes, both technologically and economically [58].

"The transition to a hydrogen economy may be the biggest infrastructure project of the 21st century", points out the Royal Society of Chemistry (RSC) in a recent review [84]. The expected cost of nanomaterials for hydrogen production and the corresponding market size in the period 2006–2015 [3] are summarized in Table 1.3.

1.2.3
Fuel Cells

A fuel cell is an electrochemical energy converter in which a fuel (hydrogen, methanol, ethanol, methane, etc.) reacts with an oxidant (oxygen, air, etc.) on electrodes (anode and cathode, respectively) that are separated by an electrolyte (e.g., a proton-conducting polymer membrane). The electrodes act as catalysts, promoting the oxidation half-reaction of the fuel on the anode and reduction half-reaction of oxygen on the cathode. The electrons released at the anode are directed through an outer circuit to do useful work. The electrolyte acts as a charge (H^+, OH^-, O^{2-}, etc.) transporter that links the two half-redox reactions. Figure 1.20 depicts a generic fuel cell and Table 1.4 lists the half-reactions of several representative types of hydrogen fuel cells. In contrast to batteries, which store electrical energy chemically, fuel cells extract chemical energy from external reactants (fuels, oxidants) and convert it to electrical energy as long as the reactants are available.

The fuel cells listed in Table 1.5 can be classified, based on their operating temperatures, as low-temperature and high-temperature fuel cells. The first three: proton exchange membrane fuel cell (PEMFC, also known as polymer electrolyte membrane fuel cell under the same acronym), alkaline fuel cell (AFC) and phosphoric acid fuel cell (PAFC) are typical low-temperature fuel cells (operating temperature < 200 °C). The last two: molten carbonate fuel cell (MCFC) and solid oxide fuel cell (SOFC) belong to the high-temperature fuel cell class, with operating

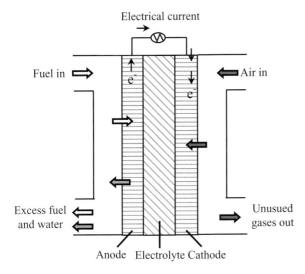

Figure 1.20 Scheme of a generic fuel cell.

Table 1.4 Electrochemical reactions in different types of fuel cells.

Fuel cell type	Anode reaction	Cathode reaction
Proton exchange	$H_2 \rightarrow 2H^+ + 2e^-$	$\frac{1}{2}O_2 + 2H^+ + 2e^- \rightarrow H_2O$
Alkaline	$H_2 + 2OH^- \rightarrow 2H_2O + 2e^-$	$\frac{1}{2}O_2 + H_2O + 2e^- \rightarrow 2OH^-$
Phosphoric acid	$H_2 \rightarrow 2H^+ + 2e^-$	$\frac{1}{2}O_2 + 2H^+ + 2e^- \rightarrow H_2O$
Molten carbonate	$H_2 + CO_3^{2-} \rightarrow H_2O + CO_2 + 2e^-$	$\frac{1}{2}O_2 + CO_2 + 2e^- \rightarrow CO_3^{2-}$
Solid oxide	$H_2 + O^{2-} \rightarrow H_2O + 2e^-$	$\frac{1}{2}O_2 + 2e^- \rightarrow O^{2-}$

temperatures above 600 °C. A brief comparison of the five types of fuel cells is given in Table 1.5 and Figure 1.21.

In general, low-temperature fuel cells (PEMFC, PAFC, AFC) feature a quicker start-up, which makes them more suitable for portable applications (e.g., vehicles, small portable devices). However, they require as fuel relatively pure hydrogen, and consequently an external fuel processor, which increases the complexity and cost and decreases the overall efficiency. They also require a higher loading of the precious metal catalysts. In contrast to low-temperature fuel cells, MCFC and SOFC are more flexible regarding fuel because they can reform various fuels (methanol, ethanol, natural gas, gasoline, etc.) inside the cells to produce hydrogen. They are also less prone to catalyst "poisoning" by carbon monoxide and carbon dioxide [86]. However, their slower start-up limits them to more stationary applications [87].

Table 1.5 A comparison of fuel cell technologies. Adapted with permission from ref. [85].

Fuel cell type	Common electrolyte	Combined heat and power (CHP) efficiency	Applications	Advantages
Polymer electrolyte membrane (PEM) Operating temperature: 50–100 °C	Solid organic polymer poly-perfluorosulfonic acid	70–90%	☐ Backup power ☐ Portable power ☐ Small distributed generation ☐ Transportation ☐ Specialty vehicles	☐ Solid electrolyte reduces corrosion and electrolyte management problems ☐ Low temperature ☐ Quick start-up
Alkaline (AFC) Operating temperature: 90–100 °C	Aqueous solution of potassium hydroxide soaked in a matrix	>80%	☐ Military ☐ Space	☐ Cathode reaction faster in alkaline electrolyte, leads to higher performance ☐ Can use a variety of catalysts
Phosphoric acid (PAFC) Operating temperature: 150–200 °C	Liquid phosphoric acid soaked in a matrix	>85%	☐ Distributed generation	☐ Higher overall efficiency with CHP ☐ Increased tolerance to impurities in hydrogen
Molten carbonate (MCFC) Operating temperature: 600–700 °C	Liquid solution of lithium, sodium and/or potassium carbonates, soaked in a matrix	>80%	☐ Electric utility ☐ Large distributed generation	☐ High efficiency ☐ Fuel flexibility ☐ Can use a variety of catalysts ☐ Suitable for CHP
Solid oxide (SOFC) Operating temperature: 600–1000 °C	Yttria-stabilized zirconia	<90%	☐ Auxiliary power ☐ Electric utility ☐ Large distributed generation	☐ High efficiency ☐ Fuel flexibility ☐ Can use a variety of catalysts ☐ Solid electrolyte reduces electrolyte management problems ☐ Suitable for CHP ☐ Hybrid/GT cycle

In parallel to the development of classic fuel cells, a new and promising type of fuel cell, the microbial fuel cell, is currently under intensive research [88]. Examples are shown in Figure 1.22.

Despite their overall higher efficiency than traditional electricity generation systems, there are still many critical barriers to the widespread utilization of fuel cells for portable and stationary electricity generation. Steel and Heinzel, in their

1.2 Nanotechnology in Energy Production | 23

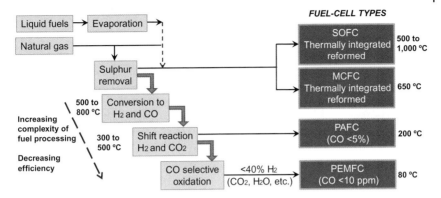

Figure 1.21 Fuel processing: comparison between the different fuel cell types. Adapted with permission from ref. [86].

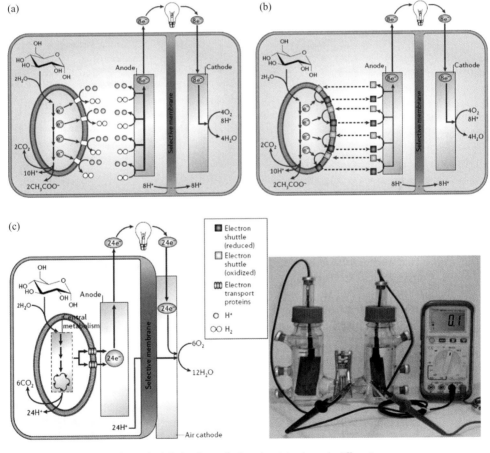

Figure 1.22 Examples of microbial fuel cells producing electricity through different mechanisms of electron transfer to the anode. Reprinted with permission from ref. [88].

Table 1.6 Probable applications and critical materials in the near future in fuel cells. PEN, Positive/electrolyte/negative; UPS, uninterruptible power systems; LPG, liquid petroleum gas; CO, carbon monoxide. Adapted with permission from ref. [86].

Application	Size (kW)	Fuel cell	Fuel	Critical material issues
Power systems for portable electronic devices	0.001–0.05	PEMFC DMFC SOFC	H_2 CH_3OH CH_3OH	Membranes exhibiting less permeability to CH_3OH, H_2O Novel PEN structures
Micro-CHP	1–10	PEMFC SOFC	CH_4 LPG CH_4 LPG	CO-tolerant anodes, novel membranes, bipolar plates More robust thick-film PENs operating at 500–700 °C
APU, UPS, remote locations, scooters	1–10	SOFC	LPG Petrol	More robust thick-film PENs operating at 500–700 °C; rapid start-up
Distributed CHP	50–250	PEMFC MCFC SOFC	CH_4 CH_4 CH_4	CO-tolerant anodes, novel membranes, bipolar plates Better thermal cycling characteristics Cheaper fabrication processes; redox-resistant anodes
City buses	200	PEMFC	H_2	Cheaper components
Large power units	10^3–10^4	SOFC/GT	CH_4	Cheaper fabrication processes for tubular SOFC system

review "*Materials for fuel-cell technologies*" [86], summarized the critical materials issues associated with each type of fuel cell (Table 1.6).

The materials issues listed in Table 1.6 generally fall into two categories: performance of electrodes/catalysts and performance of electrolytes/membranes. In addition, issues also exist with H_2 storage, which are covered in Chapter 7. Accordingly, the application of nanotechnology in the field of fuel cells also falls mainly within these three categories. In this chapter, only the first two areas are discussed, that is, the development of nanocatalysts for both anode and cathode reactions and nanostructured electrolyte membranes. Some examples are given in the following paragraphs.

Nanostructuring fuel cell electrodes increases the surface area (per unit weight) of catalysts and enhances the contact between fuels and catalysts, which leads to improved cell efficiency. The preparation of nanoscale electrocatalysts for fuel cells typically starts from either supported or unsupported colloidal nanomaterial precursors, for example, colloidal platinum (Pt) sols. Chen's group [89] prepared a Pt-based nanoelectrocatalyst, PaniNFs, for direct methanol fuel cells (DMFCs), using polyaniline nanofibres as the support. In comparison to typical Pt catalysts supported on high surface area carbon, PaniNFS was found to have three

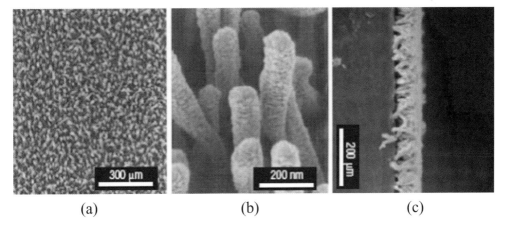

Figure 1.23 Platinum-coated nanostructured whisker supports (0.25 mg/cm^2): (a) Plane view, (b) 45° view (higher magnification) and (c) nanostructured film of the MEA. Reprinted with permission from ref. [87].

advantages: a higher electrochemical active surface area, a smaller Pt particle diameter with a narrower particle size distribution and a higher methanol oxidation reaction catalytic activity.

Although Pt-based catalysts are the most effective materials for low-temperature fuel cells, the amount of Pt required is very high, which increases the cost of the cells. Consequently, the trend is to reduce the Pt content. One approach is to improve the membrane-electrode assemblies (MEA). In PEMFCs, similar to the composite cathode approach in lithium-ion batteries, a composite layer is formed by deeply mixing the Pt/C catalysts with the electrolyte ionomers, forming a three-phase reaction zone, which leads to the reduction of Pt loading without a loss of cell performance (Figure 1.23) [87]. The 3M Corporation developed new durable multi-level MEAs, in which part of the MEA is a nanostructured thin film catalyst based on platinum-coated nanowhiskers [90]. As can be seen in Figure 1.23c, the Pt-coated nanowhiskers are sandwiched between the PEM and the gas-diffusion layer.

The development of nanoelectrodes also applies to intermediate/high-temperature fuel cells. The most common anode material used for SOFCs is a composite material known as nickel (Ni)-cermet, composed of nickel (Ni, typically 40–60 vol%) and yttria stabilized zirconia ceramic material [91]. However, Ni-cermet anodes have a few limitations that impede their commercialization. Ni-cermets are sensitive to poisoning by sulfur present in fuels [92]. They cannot tolerate being oxidized during start-up and shut-down cycles [93]. And they can be severely damaged by carbon formation if exposed to dry hydrocarbons [94]. For these reasons, several alternatives are being studied. Gross *et al.* [95] suggests two strategies for the development of anodes for SOFCs with improved performance: (i) the use of bimetallic compositions with layered microstructures, in which one metal is used for thermal stability while the other provides the required carbon tolerance, and (ii) the separation of the anode into two layers, a thin functional layer for electrocatalysis and a thicker conduction layer for current collection. As an example, the authors obtained an anode impedance of $0.26\,\Omega.cm^2$ was obtained at 500 °C in humidified H_2 using a Ag-paste conduction layer and a 12 μm thick functional layer made from 1 wt% Pd and 40 wt% ceria in YSZ. Interestingly, replacing the Ag paste with a 100 μm layer of porous $La_{0.3}Sr_{0.7}TiO_3$ (LST) had a minimal effect on cell performance [96].

Many efforts have also been dedicated to making solid proton-conducting electrolytes with uniform porosity in order to address corrosion problems and temperature constraints with liquid electrolytes [97]. Crystalline structures such as zeolites could be uniformly arranged to provide alternative solid electrolytes [98]. Slater et al. recently discovered apatite-type silicate/germanate oxide-ion conductors for SOFCs [99]. In these apatite systems, the conductivity involves interstitial oxide-ions, in contrast to the traditional perovskite- and fluorite-based oxide-ion conductors, in which conduction proceeds via oxygen vacancies.

A key performance limitation in polymer electrolyte fuel cells (PEMFCs) is the so-called "mass transport loss". Typically, perfluorosulfonic polymer electrolyte membranes (for example, Nafion™) are used in PEMFCs because of their excellent electrochemical stability and conductivity. Unfortunately, methanol cross-over and membrane dehydration processes hinder the efficiency of the cells [100, 101]. Nanostructured membranes can help address some of these issues [87, 102]. For example, Kerres et al. observed an improvement in both thermal and mechanical stability in their nanoseparated sulfonated poly(ether ketone) polymer blends [103]. Other examples are membranes based on poly(arylene ether sulfone) and block co-polymer ion-channel-forming materials or acid-doped polyacrylamide and polybenzoimidazole, exhibiting, in some cases, conductivities at least comparable to Nafion™ [104, 105].

1.2.4
Thermoelectricity

Thermoelectricity (TE), also known as the Peltier–Seebeck effect, refers to the direct conversion of temperature differences to electric potential or vice versa. One well-known application of thermoelectric effect is the thermocouple for temperature measurement. Currently, there are two primary areas of research: (i) electricity generation from waste heat and (ii) thermoelectric refrigeration, where thermoelectric effect can be utilized to increase energy efficiency and reduce pollutants.

The Figure of merit for thermoelectric materials is defined as $Z = \sigma S^2/k$, where σ is the electrical conductivity, k is the thermal conductivity, and S is the Seebeck coefficient. More commonly, the dimensionless Figure of merit ZT (product of the Figure of merit Z and the absolute temperature T) is used to describe the thermodynamic efficiency because Z is normally temperature-dependent. Efforts to improve the ZT of thermoelectric materials, especially through a nanostructuring approach, mainly focus on increasing the Seebeck coefficient (S) and/or decreasing the thermal conductivity (k).

Typically, semiconductors have higher Seebeck coefficients than metals. Among numerous thermoelectric materials, bismuth telluride (Bi_2Te_3) and its alloys are the most important TE materials used in state-of-the-art devices near room temperature, while lead telluride (PbTe)-based alloys and SiGe alloys are extensively used in power supplies for space exploration and generators for medium to high temperatures [106]. Nanoscaled multi-layered bulk materials manufactured by repeated pressing and rolling of alternately stacked thin metallic foils in the Cu–Fe system were synthesized by Shinghu et al. in 2001 [107], who observed a significant change in thermoelectricity depending on the layer thickness. In 2002, researchers at the Research Triangle (North Carolina, USA) reported the development of nanostructured layers of bismuth telluride and antimony telluride (SbTe), supplying twice the efficiency of market TE materials (ZT = ~1). The ultrathin layers (<5 nm) of these "superlattices" impeded atomic vibrations and thus heat flow, while not affecting electron flow, thus increasing ZT.

Recently, cobalt antimonide ($CoSb_3$) has been extensively studied because of its promising properties for thermoelectric applications. The handicap of $CoSb_3$ is that it possesses too high a thermal conductivity to be an efficient thermoelectric. Several approaches have been developed to lower the thermal conductivity of cobalt antimonide, that is, through doping [108, 109], through nanostructuring [110] and through inclusion/dispersion [111–113]. He et al. [114] prepared $CoSb_3/ZrO_2$ nanocomposites in different ratios and found an improvement of 11% on ZT for the composite compared to non-dispersed $CoSb_3$. Interestingly, Tropak et al. [115] observed a diminution of thermal conductivity of the nanostructured cobalt antimonide of one order of magnitude compared to annealed polycrystalline or single crystalline cobalt antimonide, while having a resistance typical of semiconductors and comparable to those of Si–Ge alloys. Skutterudites are a promising group of materials for thermoelectrics since these materials have the electrical properties of a crystal but thermal properties like those of an amorphous material. Tritt et al. [113] synthesized nanothermoelectric skutterudite materials from $CoSb_3$ by means of inclusion of $CoSb_3$ nanoparticles in a $CoSb_3$ polycrystalline matrix, which results in a favorable overall reduction in thermal conductivity. Thus, even if the nanoparticles seem to aggregate throughout the sample, they are promising candidates for thermoelectric materials.

In addition, ultrathin bismuth (Bi) nanorods were recently synthesized by Wang et al. [116], who demonstrated that they can be very useful for thermoelectric devices and applications (Figure 1.24). When the diameter of the nanorods is <10 nm, a semimetal-to-semiconductor transition is observed. Similarly, heterostructured nanorods of Bi_2S_3 [117], nanowires of Bi_2Se_3 and nanorods of PbTe [118] and CdSe [119] were also prepared, proving that nanostructured materials are promising building blocks for high-performance thermoelectrical devices.

28 | 1 Nanotechnology for Energy Production

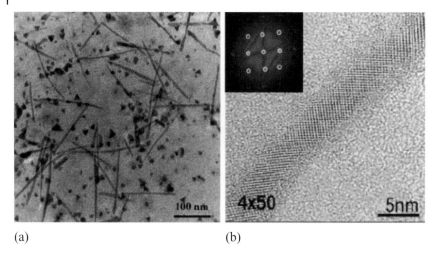

Figure 1.24 (a) TEM image of as-prepared bismuth nanowires. (b) HRTEM images and the corresponding FFT patterns of individual bismuth nanowires. Reprinted with permission from ref. [117].

1.3
Summary

While nanotechnology, just like any other new technology, still faces many challenges such as reliability, safety, lifetime and costs, its applications in the energy production sector have shown great promise for increasing energy conversion efficiency and reducing emissions of "green house" gas and pollutants. The examples presented in this chapter are intended to be introductory instead of exhaustive or even representative in some cases. It is one of the objectives of the authors to demonstrate how the manipulation and arrangement of nanoscale components can lead to new and/or superior materials properties, which may provide wide-ranging solutions to the energy challenge mankind is facing in the 21st century. Readers are strongly encouraged to refer to the following chapters and the references therein for a more comprehensive understanding of the roles of nanotechnology in the various sectors of the energy domain.

Acknowledgment

This research has been funded by the Spanish Ministerio de Educación y Ciencia (CTQ2005-09385-C03-02). J.G.M. is grateful for the financial support under the Ramón y Cajal Program. E.S. acknowledges financial support from Spanish MICINN through the Juan de la Cierva Program (JCI-2008-2165) and G.R. financial support from Spanish Generalitat Valenciana through grant INV05-10. The authors are also grateful to Dr. David H. Olson at Rive Technology Inc. for insightful discussions. The Department of Applied Physics of the University of Alicante, and especially Dr. Carlos Untiedt, are kindly thanked for their continuous support.

References

1 Renewable Policy Network for the 21st Century (2006) Renewables Global Status Report 2006 Update, http://www.ren21.net/pdf/RE_GSR_2006_Update.pdf (accessed 14 July 2008).
2 Sutter, U., and Loeffler, J. (2006) Use of Nanomaterials in the Energy Sector, https://eed.llnl.gov/flow/images/LLNL_Energy_Chart300.jpg (accessed 14 July 2008).
3 Sutter, U., and Loeffler, J. (2006) Roadmap Report Concerning the Use of Nanomaterials in the Energy Sector, http://www.nanoroad.net/download/roadmap_e.pdf (accessed 13 July 2008).
4 Becquerel, A.E. (1839) *C. R. Acad Sci. Paris*, **9**, 561–567.
5 Hirshman, W.P., Hering, G., and Schmela, M. (2008) *Photon Int.*, 152.
6 Stockman, M. (2004) *Nat. Mater.*, **3**, 423.
7 Bullis, K. (2007) *Technol. Rev. Energy*, 19256, 1–2.
8 http://www.microbiology.emory.edu/altman/images/qdotSpectra.gif (accessed 19 February 2009).
9 Singh, R.S., Rangari, V.K., Sanagapalli, S., Jayaraman, V., Mahendra, S., and Singh, V.P. (2004) *Solar Energy Solar Cells*, **82**, 315.
10 Grätzel, M. (2003) *Proceedings of The 1st International Symposium on Sustainable Energy System, Kyoto, Japan 2003*.
11 O'Regan, B., and Grätzel, M. (1991) *Nature*, **353**, 737.
12 Nazeerudin, M.K., Kay, A., Rodicio, I., Humphry Backer, R., Mueller, E., Liska, P., Vlachopoulos, N., and Grätzel, M. (1993) *J. Am. Chem. Soc.*, **115**, 6382.
13 Corma, A., Atienzar, P., Garcia, H., and Chane-Ching, J.-Y. (2004) *Nat. Mater.*, **3**, 394.
14 Singh, V.P., Singh, R.S., Thompson, G.W., Jayaraman, V., Sanagapalli, S., and Rangari, V.K. (2004) *Solar Energy Mater. Solar Cells*, **81**, 293.
15 Mathew, X., Enriquez, J.P., Sebastian, P.J., McClure, J.C., and Singh, V.P. (2000) *Solar Energy Mater. Solar Cells*, **63**, 355.
16 Neale, N.R., and Frank, A.J. (2007) *J. Mater. Chem.*, **17**, 3216.
17 Rodriguez, I., Ramiro-Manzano, F., Atienzar, P., Martinez, J.M., Meseguer, F., Garcia, H., and Corma, A. (2007) *J. Mater. Chem.*, **17**, 3205.
18 Rodriguez, I., Atienzar, P., Ramiro-Manzano, F., Meseguer, F., Corma, A., and Garcia, H. (2005) *Photonics Nanostruct. Fundam. Appl.*, **3**, 148.
19 Ramiro-Manzano, F., Atienzar, P., Rodriguez, I., Meseguer, F., Garcia, H., and Corma, A. (2007) *Chem. Commun.*, 3, 242.
20 Mor, G.K., Shankar, K., Paulose, M., Varghese, O.K., and Grimes, C.A. (2005) *Nano Lett.*, **5**, 191.
21 Guldi, D.M., Rahman, G.M.A., Prato, M., Jux, N., Qin, S., and Ford, W. (2005) *Angew. Chem. Int. Ed.*, **44**, 2015.
22 Graham, S. (2003) Carbon Nanotubes for Solar Cells, http://www.sciam.com/article.cfm?id=carbon-nanotubes-for-sola (accessed 13 July 2008).
23 Toon, J. (2007) Nano-Manhattan: 3D Solar Cells Boost Efficiency while Reducing Size, Weight and Complexity of Photovoltaic Arrays, http://gtresearchnews.gatech.edu/newsrelease/3d-solar.htm (accessed 13 July 2008).
24 Peumanns, P., and Forrest, S.R. (2001) *Appl. Phys. Lett.*, **79**, 126.
25 Xue, J., Uchida, S., Rand, B.P., and Forrest, S.R. (2004) *Appl. Phys. Lett.*, **84**, 3013.
26 Shaheen, S.E., Brabec, C.J., Sariciftci, N.S., Padinger, F., Fromherz, T., and Hummelen, J.C. (2001) *Appl. Phys. Lett.*, **78**, 841.
27 Kroon, J.M., Wienk, M.M., Verhees, W.J.H., and Hummelen, J.C. (2002) *Thin Solid Films*, **403–404**, 223.

28 Munters, T., Martens, T., Goris, L., Vrindts, V., Manca, J., Lutsen, L., Ceunick, W.D., Vanderzande, D., Schepper, L.D., Gelan, J., Sariciftci, N.S., and Brabec, C.J. (2002) *Thin Solid Films*, **403–404**, 247.
29 Aernouts, T., Geens, W., Portmans, J., Heremans, P., Borghs, S., and Mertens, R. (2002) *Thin Solid Films*, **403–404**, 297.
30 Schilinsky, P., Waldauf, C., and Brabec, C.J. (2002) *Appl. Phys. Lett.*, **81**, 3885.
31 Padinger, F., Rittberger, R.S., and Sariciftci, N.S. (2003) *Adv. Funct. Mater.*, **13**, 85.
32 Svensson, M., Zhang, F., Veenstra, S.C., Verhees, W.J.H., Hummelen, J.C., Kroon, J.M., Inganäs, O., and Andersson, M.R. (2003) *Adv. Mater.*, **15**, 988.
33 Wienk, M.M., Kroon, J.M., Verhees, W.J.H., Knol, J., Hummelen, J.C., van Hall, P.A., and Janssen, R.A.J. (2003) *Angew. Chem. Int. Ed.*, **42**, 3371.
34 Brabec, C.J., Sariciftci, N.S., and Hummelen, J.C. (2001) *Adv. Funct. Mater.*, **11**, 15.
35 Geens, W., Aernouts, T., Poortmans, J., and Hadziioannou, G. (2002) *Thin Solid Films*, **403–404**, 438.
36 Peumans, P., Uchida, S., and Forrest, S.R. (2003) *Nature*, **425**, 158.
37 Maennig, B., Drechsel, J., Gebeyehu, D., Simon, P., Kozlowski, F., Werner, A., Li, F., Grundmann, S., Sonntag, S., Koch, M., Leo, K., Pfeiffer, M., Hoppe, H., Meissner, D., Sariciftci, S., Riedel, I., Dyakonov, V., and Parisi, J. (2004) *Thin Solid Films.*, **79**, 1.
38 Gebeyehu, D., Pfeiffer, M., Maennig, B., Drechsel, J., Werner, A., and Leo, K. (2004) *Thin Solid Films*, **451–452**, 29.
39 Krüger, J., Plass, R., Cevey, L., Piccirelli, M., Grätzel, M., and Bach, U. (2001) *Appl. Phys. Lett.*, **79**, 2085.
40 Krüger, J., Plass, R., Grätzel, M., and Matthieu, H.-J. (2002) *Appl. Phys. Lett.*, **81**, 367.
41 Huynh, W.U., Dittmer, J.J., and Alivisatos, A.P. (2002) *Science*, **295**, 2425.
42 Stalmach, U., de Boer, B., Videlot, C., van Hutten, P.F., and Hadziioannou, G. (2000) *J. Am. Chem. Soc.*, **122**, 5464.
43 Hadziioannou, G. (2002) *MRS Bull.*, **27**, 456.
44 Sun, S.S., Fan, Z., Wang, Y., Taft, C., Haliburton, J., and Maaref, S. (2003) *Proceedings of SPIE-Organic Photovoltaics III* (eds Z.H. Kafafi and D. Fichou), SPIE, Bellingham, WA, USA, p. 114.
45 Possamai, G., Camaioni, N., Ridolfi, G., Franco, L., Ruzzi, M., Menna, E., Casalbore-Miceli, G., Fichera, A.M., Scorrano, G., Corvaja, C., and Maggini, M. (2003) *Synth. Methods*, **139**, 585.
46 Loi, M.A., Denk, P., Hoppe, H., Neugebauer, H., Winder, C., Meissner, D., Brabec, C., Sariciftci, N.S., Gouloumis, A., Vazquez, P., and Torres, T. (2003) *J. Mater. Chem.*, **13**, 700.
47 Zhang, F., Svensson, M., Andersson, M.R., Maggini, M., Bucella, S., Menna, E., and Inganäs, O. (2001) *Adv. Mater.*, **13**, 1871.
48 Cravino, A., Zerza, G., Maggini, M., Bucella, S., Svensson, M., Andersson, M.R., Neugebauer, H., Brabec, C.J., and Sariciftci, N.S. (2003) *Monat. Chem.*, **134**, 519.
49 Hoppe, H., and Sariciftci, N.S. (2004) *J. Mater. Res.*, **19**, 1924.
50 Kietzke, T., Neher, D., Landfester, K., Montenegro, R., Güntner, R., and Scherf, U. (2003) *Nat. Mater.*, **2**, 408.
51 Coakley, K.M., Liu, Y., McGehee, M.D., Frindell, K., and Stucky, G.D. (2003) *Adv. Funct. Mater.*, **13**, 301.
52 Coakley, K.M., and McGehee, M.D. (2003) *Appl. Phys. Lett.*, **83**, 3380.
53 Bouclé, J., Ravirajanac, P., and Nelson, J. (2007) *J. Mater. Chem.*, **17**, 3141.
54 AIRA/Nanotec IT (2006) Roadmaps at 2015 on Nanotechnology Application in the Sectors of Materials, Health & Medical Systems, Energy, http://www.nanoroadmap.it/roadmaps/NRM_SYNTHESIS.pdf (accessed 20 June 2008).
55 Sherif, S.A., Barbir, F., and Veziroglu, T.N. (2005) *Solar Energy*, **78**, 647.
56 Committee on Alternatives and strategies for Future Hydrogen Production and Use, National Research Council, National Academy of Engineering (2004) The Hydrogen

Economy: Opportunities, Costs, barriers, and R&D Needs, http://www.nap.edu/openbook.php?isbn=0309091632 (accessed 9 May 2008).

57 Ni, M., Leung, M.K.H., Leung, D.Y.C., and Sumathy, K. (2007) *Renew. Sustain. Energy Rev.*, **11**, 401.

58 Service, R.F. (2004) *Science*, **305**, 958.

59 Currao, A. (2007) Photoelectrical water splitting. *Chimia*, **61**, 815.

60 Grimes, C.A., Varghese, O.K., and Ranjan, S. (2008) *Light, Water, Hydrogen. The Solar Generation of Hydrogen by Water Photoelectrolysis* (eds Z.A. Grimes, O.K. Varghese, and S. Ranjan), Springer Science, New York, USA, p. 115.

61 Jang, J.S., Kim, H.G., Joshi, U.A., Jang, J.W., and Lee, J.S. (2008) *Int. J. Hydrogen Energy*, **33**, 5975.

62 Sebastian, P.J., Castaneda, R., Ixtlilco, L., Mejia, R., Pantoja, J., and Olea, A. (2008) *Proc. SPIE*, **7044**, 704405.

63 Silva, L.A., Ryu, S.Y., Choi, J., Choi, W., and Hoffmann, M.R. (2008) *J. Phys. Chem. C*, **112**, 12069.

64 De, G.C., Roy, A.M., and Bhattacharya, S.S. (1996) *Int. J. Hydrogen Energy*, **21**, 19.

65 Koca, A., and Sahin, M. (2002) *Int. J. Hydrogen Energy*, **27**, 363.

66 Tennakone, K., and Bandara, J. (2001) *Appl. Catal. A: Gen.*, **208**, 335.

67 So, W.W., Kim, K.J., and Moon, S.J. (2004) *Int. J. Hydrogen Energy*, **29**, 229.

68 Asahi, R., Morikawa, T., Ohwaki, T., Aoki, K., and Taga, Y. (2001) *Science*, **293**, 269.

69 Wang, D.-Y., Lin, H.-C., and Yen, C.-C. (2006) *Thin Solid Films*, **515**, 1047.

70 Maruthamuthu, P., and Ashokkumar, M. (1988) *Solar Energy Mater.*, **17**, 433.

71 Bamwenda, G.R., Tsubota, S., Nakamura, T., and Haruta, M. (1995) *J. Photochem. Photobiol. A: Chem.*, **89**, 177.

72 Subramanian, V., Wolf, E.E., and Kamat, P. (2004) *J. Am. Chem. Soc.*, **126**, 4943.

73 Anpo, M., and Takeuchi, M. (2003) *J. Catal.*, **216**, 505.

74 Gurunathan, K., Maruthamuthu, P., and Sastri, V.C. (1997) *Int. J. Hydrogen Energy*, **22**, 57.

75 Lee, S.G., Lee, S.W., and Lee, H.I. (2001) *Appl. Catal. A: Gen.*, **207**, 173.

76 Li, Y.X., Lu, G.X. and Li, S.B. (2003) *Chemosphere*, **52**, 843.

77 Kida, T., Guan, G.Q., Yamada, N., Ma, T., Kimura, K., and Yoshida, A. (2004) *Int. J. Hydrogen Energy*, **29**, 269.

78 Wu, N.L., and Lee, M.S. (2004) *Int. J. Hydrogen Energy*, **29**, 1601.

79 Abe, R., Sayama, K., Domen, K., and Arakawa, H. (2001) *Chem. Phys. Lett.*, **344**, 339.

80 Sayama, K., Mukasa, K., Abe, R., Abe, Y., Arakawa, H., and Photochem, J. (2002) *Photobiol. A: Chem.*, **148**, 71.

81 Lee, K., Nam, W.S., and Han, G.Y. (2004) *Int. J. Hydrogen Energy*, **29**, 1343.

82 Sayama, K., and Arakawa, H. (1992) *J. Chem. Soc., Chem. Commun.*, **2**, 150.

83 Sayama, K., and Arakawa, H. (1994) *J. Photochem. Photobiol. A: Chem.*, **77**, 243.

84 Royal Society of Chemistry (2005) Chemical Science Priorities for Sustainable Energy Solutions, http://www.rsc.org/images/ChemicalSciencePrioritiesSustainableEnergySolutions_tcm18-12642.pdf (accessed 13 July 2008).

85 U.S. Department of Energy Hydrogen Program (2008) Comparison of Fuel Cell Technologies, http://www1.eere.energy.gov/hydrogenandfuelcells/fuelcells/pdfs/fc_comparison_chart.pdf (accessed 1 April 2009).

86 Steele, B.C.H., and Heinzel, A. (2001) *Nature*, **414**, 345.

87 Aricó, A.S., Bruce, P., Scrosati, B., Tarascon, J.-M., and Schalkwijk, W.V. (2005) *Nat. Mater.*, **4**, 366.

88 Lovley, D.R. (2006) *Nat. Rev.*, **4**, 497.

89 Chen, Z., Xu, L., Li, W., Waje, M., and Yan, Y. (2006) *Nanotechnology*, **17**, 5254.

90 Atanasoski, R. (2004) *Proceedings of the 4th International Conference Applications of Conducting Polymers, ICCP-4, Como, Italy*, Abstract.

91 Atkinson, A., Barnett, S., Gorte, R.J., Irvine, J.T.S., McEvoy, A.J., Mogensen,

M.B., Singhal, S., and Vohs, J. (2004) *Nat. Mater.*, **3**, 17.

92 Matsuzaki, Y., and Yasuda, I. (2000) *Solid State Ionics*, **132**, 261.

93 Rietveld, G., Nammensma, P., and Ouweltjes, J.P. (2001) *Proceedings of the Seventh International Symposium on Solid Oxide Fuel Cells (SOFCVII), The Electrochemical Society Proceedings Series* (eds H. Yokokawa and S.C. Singhal), PV, Pennington, NJ, p. 125.

94 McIntosh, S., and Gorte, R.J. (2004) *Chem. Rev.*, **104**, 4845.

95 Gross, M.D., Vohs, J.M., and Gorte, R.J. (2007) *J. Mater. Chem.*, **17**, 3071.

96 Gross, M.D., Vohs, J.M., and Gorte, R.J. (2007) *Electrochem. Solid State Lett.*, **10**, B65.

97 Dai, Y., Wang, Y., Bajue, S., Greenbaum, S.G., Golodnitsky, D., Ardel, G., Strauss, E., and Peled, E. (1998) *Electrochim. Acta*, **10–11**, 1557.

98 Kelemen, G., Lortz, W., and Schoen, G. (1989) *J. Mater. Sci.*, **24**, 333.

99 Kendrick, E., Islam, M.S., and Slater, P.R. (2007) *J. Mater. Chem.*, **17**, 3104.

100 Srinivasan, S., Mosdale, R., Stevens, P., and Yang, C. (1999) *Annu. Rev. Energy Environ.*, **24**, 231.

101 Aricò, A.S., Srinivasan, S., and Antonucci, V. (2001) *Fuel Cells*, **1**, 133.

102 Kreuer, K.D. (2001) *J. Membr. Sci.*, **185**, 29.

103 Kerres, J., Ullrich, A., Meier, F., and Haring, T. (1999) *Solid State Ionics*, **125**, 243.

104 Alberti, G., and Casciola, M. (2003) *Annu. Rev. Mater. Res.*, **33**, 129.

105 Li, Q., He, R., Jensen, J.O., and Bjerrum, N.J. (2003) *Chem. Mater.*, **15**, 4896.

106 Tritt, T.M., and Subramanian, M.A. (2006) *MRS Bull.*, **31**, 188.

107 Shingu, P.H., Ishihara, K.N., Otsuki, A., and Daigo, I. (2001) *Mater. Sci. Eng.*, **A304–306**, 399.

108 Nolas, G.S., Takizawa, H., Endo, T., Sellinschegg, H., and Johnson, D.C. (2000) *Appl. Phys. Lett.*, **77**, 52.

109 Stiewe, C., Bertini, L., Toprak, M., Christensen, M., Platzek, D., Williams, S., Gatti, C., Meuller, E., Iversen, B.B., Muhammed, M., and Rowe, M. (2005) *J. Appl. Phys.*, **97**, 044317/1.

110 Toprak, M., Stiewe, C., Platzek, D., Williams, S., Bertini, L., Müller, E., Gatti, C., Zhang, Y., Rowe, M., and Muhammed, M. (2004) *Adv. Funct. Mater.*, **14**, 1189.

111 Katsuyama, S., Watanabe, M., Kuroki, M., Maehata, T., and Ito, M. (2003) *J. Appl. Phys.*, **93**, 2758.

112 Shi, X., Chen, L., Yang, J., and Meisner, G.P. (2004) *Appl. Phys. Lett.*, **84**, 2301.

113 Alboni, P.N., Ji, X., He, J., Gothard, N., Hubbard, J., and Tritt, T.M. (2007) *J. Electron. Mater.*, **36**, 711.

114 He, Z., Stiewe1, C., Platzek, D., Karpinski, G., Müller, E., Li, S., Toprak, M., and Muhammed, M. (2007) *Nanotechnology*, **18**, 235602.

115 Scoville, N., Bajgar, C., Rolfe, J., Fleurial, J.P., and Vandersande, J. (1995) *Nanostruct. Mater.*, **5**, 207.

116 Wang, Y., Kim, J.-S., Lee, J.Y., Kim, G.H., and Kim, K.S. (2007) *Chem. Mater.*, **19**, 3912.

117 Liufu, S.-C., Chen, L.-D., Yao, Q., and Wang, C.-F. (2007) *Appl. Phys. Lett.*, **90**, 112106./1.

118 Qiu, X., and Burda, C. (2005) *Proceedings of The Physical Chemistry of Interfaces and Nanomaterials IV, SPIE-The International Society for Optical Engineering* (eds C. Burda and R.J. Ellingson), SPIE, San Diego, CA, USA, p. 592918/1.

119 Kale, R.B., and Lokhande, C.D. (2005) *Semicond. Sci. Technol.*, **20**, 1.

2
Nanotechnology in Dye-Sensitized Photoelectrochemical Devices
Agustin McEvoy and Michaël Grätzel

2.1
Introduction

The basic requirement of any solar-to-electric conversion system is obviously the absorption of a significant part of the incident solar radiation with a consequent utilization of the energy of the photons in an electrical process. The first reported observation of such an effect was by Edmond Becquerel in 1839, reported in his *Memoire sur les effets électriques produits sous l'influence des rayons solaires* [1] in which he used a contact to what we now recognize as a semiconductor, a silver halide. His work was inspired by that of his contemporary, Daguerre, who was initiating the science and art of photography at that time. Notably, he used as photosensitive material the same silver halide used for the photographic effect. To use modern terminology, his photocell was a photoelectrochemical device, in which the contact to the silver halide solid, a semiconductor, was a liquid electrolyte. The first solid-state photocell only followed in 1877 and used selenium [2]. When the modern era of photovoltaics began, essentially with Rappaport and others at the RCA Laboratories in Princeton (New Jersey, USA) [3], silicon was the preferred semiconductor. In consequence the silicon cell remains the dominant device for photovoltaic solar energy applications. However since solar energy is essentially a dilute and intermittent resource, with a maximum input of $1\,kW/m^2$ at midday under perfect atmospheric conditions, being subject to the diurnal day and night variation, and with a limited conversion efficiency, applications under typical European conditions yield only about 900 kW-hr of electricity per year per installed rated kilowatt of photovoltaic generation capacity, or some $5–10\,m^2$ of module surface area. The present rapid expansion of photovoltaic capacity in Europe, particularly in Germany, is directly related to public promotion policies. Economic competition with other electrical power systems is dependent on cost minimization. The imperative is a large-area device, with low cost per unit area, associating efficiency in conversion of sunlight into electricity with ruggedness and reliability, thereby providing a service lifetime sufficient to amortize the capital investment and provide power in direct competition with other technologies. Several technical options as well as the conventional silicon photovoltaic cell are now under develop-

Nanotechnology for the Energy Challenge. Edited by Javier Garcia-Martinez
© 2009 WILEY-VCH Verlag GmbH & Co. KGaA, Weinheim
ISBN: 978-3-527-32401-9

ment to achieve this goal, including thin film devices using compound semiconductors. One such option revives the photoelectrochemical route established by Becquerel in 1839, but of course with the application of the latest materials and procedures, including newly developed electroactive dyes. This device, the dye-sensitized solar cell (DSC) is attracting intense research interest and increasing industrial investment, and involves the appropriate application of nanoscale techniques.

2.2
Semiconductors and Optical Absorption

Until the introduction of dye sensitization, all photovoltaic devices relied on the absorption of photons of sufficient energy by semiconductors. The essential characteristic of a semiconductor in this context is that at normal temperature it has only a very few mobile charges able to conduct electricity, but on absorption of a photon of appropriate energy an electron can be promoted from immobility in the valence band to mobility in the conduction band, leaving a corresponding hole, equivalent to a mobile positive charge, in the valence band. Since a mobile electron occupies a more energetic state, the material is transparent to lower energy photons but fits the requirement stated above for higher energy photons in that, once absorbed, their energy provides mobile charges in the semiconductor. The minimum energy for optical absorption by silicon, for example is 1.1 eV, corresponding to the bandgap between its valence and conduction states. It can therefore absorb the entire visible spectrum and that part of the near infrared with wavelength shorter than 1.13 μm. Narrower bandgap materials absorb a wider section of the solar spectrum, and as a consequence of absorbing more photons, can provide a higher current density. However, there is a corresponding disadvantage in that the voltage difference between the separated charge carriers, being a consequence of the bandgap, is lower for the narrower-gap semiconductors. In a necessary compromise, for a single-junction cell there is an optimal bandgap. This in practice indicates a bandgap of 1.4 eV as optimal, corresponding for example to the III–V compound gallium arsenide, GaAs. However the loss of efficiency in using silicon is relatively minor, and the question of cost, engineering experience and other materials parameters then dominate the choice.

As well as silicon, a periodic table group IV element, modern materials science has provided a wide range of other semiconductors, so that electronic device designers have available for selection a range of bandgaps, charge carrier mobilities and other optoelectronic properties giving specific advantages in photovoltaics, and equally and more recently in light-emitting diodes which are the inverse of the photovoltaic process. Typical of these are the III–V materials such as gallium arsenide, the II–VI semiconductors like cadmium sulfide and particularly the I–III–VI$_2$ series including copper–indium selenide, increasingly used in thin-film cells.

2.2 Semiconductors and Optical Absorption

After optical absorption of a photon and production of a charge-carrier pair, the photovoltaic effect requires the separation of the electron and hole so that an electric current can be generated. This occurs at a junction where two materials of different conduction mechanism contact. In a solid state device, the semiconductor may contact a metal to form a Schottky junction, a chemically different semiconductor to form a heterojunction, or the same semiconductor material but with a majority of carriers of the opposite polarity for a homojunction. The silicon homojunction for example has contacting zones of n- and p-type silicon, the former having electrons as majority carriers due to the insertion into the silicon lattice structure of elements of higher valency which donate electrons, whereas the p-type contains lower valency electron-accepting elements, providing a majority of holes. In the photoelectrochemical cell the contacting phase of different conductivity mechanism is an electrolyte, and the device is therefore fully analogous to the metal–semiconductor contact in a Schottky junction.

In principle, therefore, a semiconductor can form a large-area photosensitive cell simply by contact with a liquid electrolyte, in principle a low-cost solution for junction formation.

In practice, there is an unavoidable contradiction in materials selection for a simple photoelectrochemical cell. The bandgap is a measure of the strength of chemical binding of the elements in a semiconductor and is related to the heat of formation of the solid. Strongly bonded semiconductors tend therefore to have wider bandgaps, and therefore restricted photoresponse to the incident solar spectrum. In contrast, an electrolyte is a severe environment for the operation of a semiconductor device, so the less stable narrow gap materials with potentially a wide spectral response are subject to a rapid photocorrosion. Taking the example of n-type cadmium sulfide with bandgap of 2.2 eV, the photoexcited electrons are delivered to an external circuit while the holes accumulate at the semiconductor–electrolyte interface. Rather than exchange charge with the electrolyte, the most immediately available electron donor is the lattice sulfur ion, S^{2-}. This is rapidly oxidized to free sulfur, while the corresponding cadmium ion, Cd^{2+}, is released into solution in the electrolyte. In practice therefore the only available intrinsically stable photoelectrodes are the wide bandgap oxides of reactive metals, such as titanium dioxide, TiO_2. However this material has a bandgap of 3.1 eV and is therefore totally insensitive to the visible spectrum and photoresponds only to ultraviolet radiation.

Photography had been confronted with this dilemma a century earlier. Silver bromide, for example, has a bandgap of 2.5 eV, and is therefore sensitive only to photons in the green, blue and ultraviolet spectral regions. The original photographs, being "orthochromic", therefore did not provide a realistic representation of scenes viewed in the visible spectrum. In 1873 Vogel, Professor of "Photochemistry, Spectroscopy and Photography" in Berlin, associated a dye with the halide photographic emulsion to provide "panchromatic" response extending through the whole visible spectrum [4]. Just as Becquerel followed the photographic achievements of Daguerre, the principle of sensitization by dye was applied to photoelectrochemistry in 1887 by Moser [5]. In this system, the optical absorption

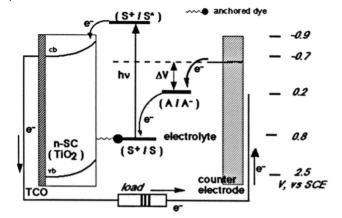

Figure 2.1 Schematic of operation of the dye-sensitized electrochemical photovoltaic cell. The photoanode, made of a mesoporous dye-sensitized semiconductor, receives electrons from the photoexcited dye S*, which is thereby oxidized to a cation and which in turn transfers its positive charge to a redox species dissolved in the electrolyte. The redox system is regenerated by reduction at the counter-electrode by the electrons circulated through the external circuit.

is by the dye, so the spectral response is not determined by the semiconductor bandgap. This sub-bandgap sensitization of the semiconductor with an electroactive dye divides the light absorption and charge separation functions. The wide-bandgap, intrinsically stable semiconductor, such as titanium dioxide with its bandgap of 3.1 eV, can then photorespond to visible light of wavelength 400–750 nm, or 1.6–3.0 eV photons. The work of Gerischer and Tributsch [6, 7] on ZnO definitively established the mechanism of the dye sensitization process and indicated its significance for photoelectrochemistry, finally rendering compatible effective wideband visible spectral absorption with the stability of a semiconductor substrate, otherwise sensitive only to ultraviolet light. It is now evident that the process involves the excitation of the dye from its charge-neutral ground state to an excited state by the absorption of the energy of a photon, followed by relaxation through electron loss to the semiconductor substrate. The dye is left as a surface-adsorbed, positively charged cation. In a redox electrolyte two dissolved ion species are interconvertible by exchange of electrons. Here the dye cation is easily neutralized in such a redox reaction. The redox system in turn recovers an electron from a counter-electrode. A closed regenerative cycle for the conversion of incident light into an electric current is therefore established. This process is presented schematically in Figure 2.1. The standard redox system is the iodide/tri-iodide couple, $I^-/I_3^-: I_3^- + 2e^- \Leftrightarrow 3I^-$

This regenerative photoelectrochemical device is evidently functionally equivalent to a conventional solid-state photovoltaic cell. However, as only a monomolecular adsorbed film of the sensitizing dye can transfer charge to the substrate, the original sensitized devices on single-crystal substrates had low

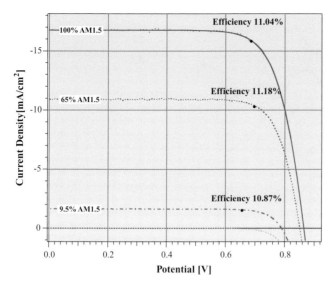

Figure 2.2 Current–voltage characteristics of an actual dye-sensitized solar cell, under test with different levels of illumination, and indicating the maximum power point and efficiency in each case (EPFL, Lausanne).

optical absorption, and therefore low photovoltaic efficiency. It remained therefore to associate the dye with a nanostructured semiconductor, so that the active interface area for light absorption and charge transfer greatly exceeded the projected geometrical area of the surface and gave the required opacity and light absorption, to provide a device capable of challenging the efficiency of its solid-state counterpart [8].

Nanometric structuring is applied on three increasing scales in an efficient dye-sensitized photovoltaic cell: (1) in the molecular engineering of the dye, (2) in the self-assembly of the chemisorbed monomolecular dye layer in the semiconductor surface and (3) in the mesoporous structure of the sensitized semiconductor. Despite the origin of the principle of sensitization in the 19th century, the first convincing application to a photovoltaic device was presented only in the past 20 years [9]. Further development work has enabled these dye-sensitized cells to reach a white-light conversion efficiency exceeding 11%, as shown by the cell characteristics in Figure 2.2. That such a performance is attainable with a nanostructured semiconductor is due to the ultra-fast kinetics of electron injection from the excited dye into the solid. That separation of charge carriers, associated with the majority carrier nature of the device where the electrons enter an n-type material, strongly inhibits charge carrier recombination losses. Electrons can be lost only through recapture by the dye cation or by the redox electrolyte after crossing the phase boundary between the semiconductor and the electrolyte, an inherently slower process than when electrons and holes are co-sited within a semiconductor solid lattice.

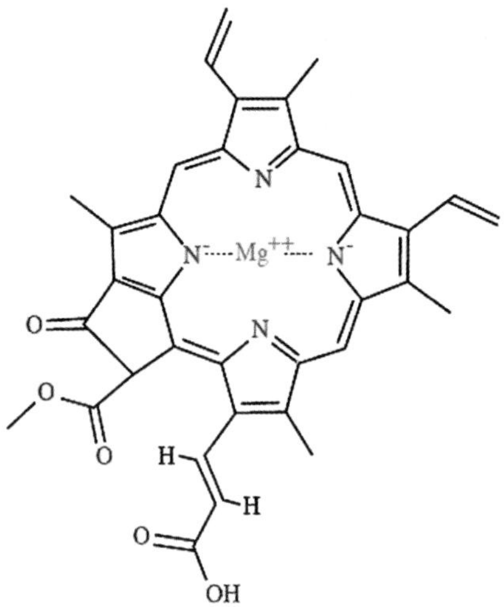

Figure 2.3 Nature's solar energy conversion dye, chlorophyll c, the version of the molecule found in green algae.

2.3
Dye Molecular Engineering

It is evident that to achieve this level of photovoltaic performance and maintain it over an adequate in-service life requires optimization of each component in the system and attention to their synergies. Of particular importance is the choice of a suitable electroactive dye. The development was inspired by a natural prototype, the chlorophyll molecule as in Figure 2.3, with its metal ion, magnesium, within a nitrogen "cage" constituting a chromophore [10]. As the basis for photosynthesis in plants, chlorophyll requires a photoexcited state attainable by the absorption of photons widely across the visible spectrum, exactly as does a photosensitizing dye. The energy level of the excited state of the molecule, or LUMO – lowest unoccupied molecular orbital – must lie higher than the conduction band edge of the semiconductor, so that an electron can be injected during the relaxation process. For optimum absorption of white light, therefore, the HOMO – highest occupied molecular level – should be about 1.5 eV lower. In terms of photoexcitation, the molecular HOMO–LUMO gap is completely analogous to the bandgap of a semiconductor. An appropriate molecular structure for the sensitization of a titanium dioxide semiconductor had been identified already by Clark and Suttin [11], using a ruthenium tris-bipyridyl complex. It is essentially the bond between the metal center and the bipyridyl ligand which constitutes the chromophore, whereby the molecular orbitals are established, and consequently the optical characteristics.

However with this trisbipyridyl molecular sensitizer, the HOMO–LUMO gap is about 2.0 eV, and as a result the photovoltaic response is limited to wavelengths

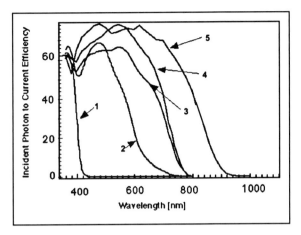

Figure 2.4 The photocurrent action spectra of cells containing various sensitizers, where the incident photon to current conversion efficiency is plotted as a function of wavelength. 1, Terpyridyl panchromatic "black dye"; 2, standard N3 dye; 3, cyanide-substituted trimer dye; 4, original tris-bipyridyl ruthenium [11] dye; 5, unsensitised titanium dioxide.

below 600 nm as shown in Figure 2.3 (curve RuL_3; L = dicarboxylbipyridyl, or dcbpy). However, in comparison with the photoresponse of the unsensitized titanium dioxide, limited to the ultraviolet at wavelengths shorter than 400 nm, there is already a significant extension of light absorption into the visible part of the spectrum. With detailed molecular engineering, the electronic structure of the molecule can then be modified to raise the HOMO level, narrowing the HOMO–LUMO gap and thereby extending the response towards the red end of the spectrum. Figure 2.4 presents this progress with removal of one of the pyridyl ligands and modification of the molecular orbitals, first by insertion of cyanide (CN) in a trimer structure, and later with thiocyanide (SCN) substituents in place of one of the bipyridyl rings. The carboxylic groups on the pyridyl rings permit a stable chemical attachment of the molecule to a titanium dioxide surface, producing a reliable photoelectrochemical anode. This is of course part of the self-assembly process of the monomolecular dye layer on the semiconductor substrate. For many years this molecule was accepted as the standard dye for photoelectrochemical sensitization of titanium dioxide photoanodes. Formally this dye, coded N3 at EPFL where it was developed, is EPFL N3 dye, or dithiocyanato bis(4,4'dicarboxylic acid-2,2'bipyridine) ruthenium (II), and its molecular structure is given in Figure 2.5 [12]. Ultimately a "black dye", absorbing across the whole visible spectrum and providing a panchromatic response [13], was evolved by associating a single terpyridyl complex and three spectrum-modifying thiocyanide groups with the central charged ruthenium atom.

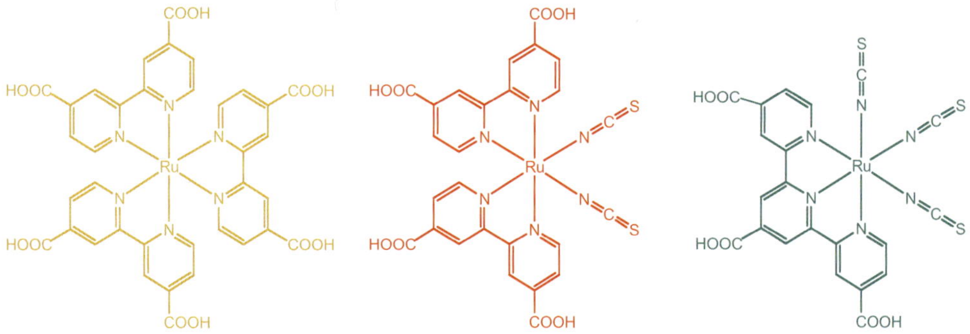

Figure 2.5 Some stages in the development of the sensitizer dye are illustrated by the corresponding molecular structures. From left to right: the original trisbypyridyl ruthenium dye of Clark and Suttin [11], the industry standard N3, ruthenium dithiocyanato bis(4,4'dicarboxylic acid-2,2'bipyridine) and the panchromatic "black dye", ruthenium trithiocyanato-terpyridyl.

2.4
The Stable Self-Assembling Dye Monomolecular Layer

Only an optically excited dye molecule in intimate contact with the titanium dioxide semiconductor can transfer an electron and relax to a lower-energy charged cation state. It follows that only a molecular monolayer of dye can be photoactive. Dye aggregates or multilayers therefore significantly reduce the efficiency of a cell; while they are absorbers of light, relaxation from the photoexcited state must take place by recombination processes as electron transfer is impossible. The incorporation of the carboxylate component, or other acidic groups such as phosphonate, into the dye structure promotes chemisorption to the oxide surface and induces self-assembly of the molecular monolayer.

The stability of the molecular monolayer is crucial to the service life of the dye-sensitized photovoltaic system. To emphasize this rigorous requirement, while the sensitized system in photography is called upon to function only once, on exposure of the film, a reasonably competitive service life of 20 years involves some 2×10^8 cycles of photon absorption, dye excitation and relaxation by electron transfer. This has implications, not only for the stability of the dye molecule itself, but also for the reliability of its chemisorptive bond to the underlying titanium dioxide semiconductor. State of the art stable photodevices use an organic ionic liquid as electrolyte [14] because of its low vapor pressure, wide temperature range, chemical and electrochemical stability. Aqueous electrolytes in contrast, as used in the earliest DSC demonstrations, have high vapor pressure, limited effective temperature range and are liable to dissociation by electrolysis under relatively low potential differences. There is also the possibility of desorption of the dye through hydrolysis of the carboxylate bond to the metal oxide. However, even traces of water in an ionic liquid electrolyte can induce this failure mechanism, limiting

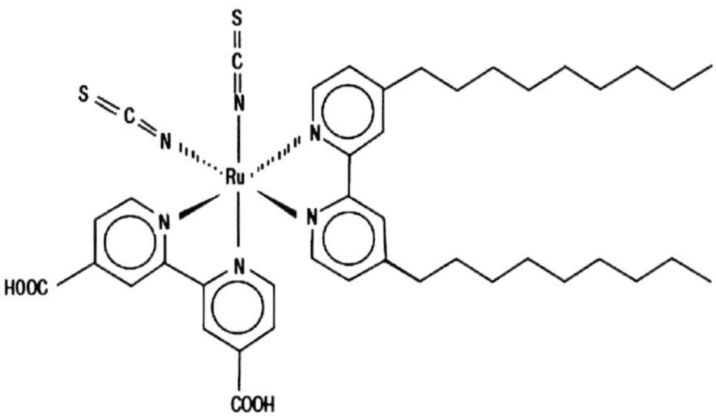

Figure 2.6 Structure of the Z-907 amphiphilic dye [cis-Ru(H₂dcbpy)(dnbpy)(NCS)₂; where the ligand H₂dcbpy is 4,4·dicarboxylic acid-2,2·bipyridine and dnbpy is 4,4·dinonyl-2,2·bipyridine]. The addition of a hydrocarbon side chain ensures hydrophobicity of the photoelectrode surface after sensitization and enhances stability against aqueous contaminants in the electrolyte.

service life to months rather than decades. Further dye molecular engineering can eliminate sensitivity to contamination of the ionic liquid electrolyte.

As an example of this type of dye molecular engineering, the molecule (laboratory code Z-907) shown in Figure 2.6 has of course the two bipyridyl ligands bonded to ruthenium to establish photosensitivity, the two thiocyanide ligands to modify the spectral response and the two carboxyl groups to chemically adsorb to the semiconductor oxide surface, but in addition it has two hydrocarbon chains to thereby present a hydrophobic surface to the electrolyte [15]. The consequence of this structure is the self-assembly of a monomolecular dye layer of specific orientation to provide the required physical properties, uniformly chemisorbed on the nanostructured semiconductor, but essentially presenting a surface to the electrolyte actively rejecting contaminant water by a physical process. Recent experimental data project towards a 20-year lifetime for stability-optimized DSC photovoltaic systems under typical European exposure conditions [16].

A recent further improvement to the dye in Figure 2.6 is the insertion of a thiophene, or aromatic sulfur, component into the hydrophobic chain; without loss of any other advantageous properties of the dye, the optical absorption coefficient is significantly increased, allowing a thinner photoanode with enhanced electrode kinetics and hence better performance.

2.5
The Nanostructured Semiconductor

A similar development process and attention to detail was required for the semiconductor substrate. If molecular design and engineering has underpinned the evolution of efficient stable sensitizer dyes, for the semiconductor substrate it is a matter of the materials science of nanoporous ceramic films. The nanoporous structure permits the specific surface concentration of the sensitizing dye to be sufficiently high for total absorption of the incident light necessary for efficient solar energy conversion, since the area of the monomolecular dye layer is two or three orders of magnitude larger than the geometric area of the substrate. As already noted, this high roughness in a DSC photoanode still does not promote charge carrier loss by recombination, since the electron and the positive charge find themselves within picoseconds on opposite sides of the electrolyte–solid interface, much faster than any possible electron escape and redox or cation capture process. The original semiconductor structure used for early photosensitization experiments was a fractal derived by hydrolysis of an organo-titanium compound [8]. Later, suspensions of commercially available anatase titania powders were found to be similarly effective. Hydrothermal techniques are particularly appropriate for the synthesis of an optimized monodisperse nanoparticulate anatase TiO_2 powder which is then used in suspension in a liquid medium [17].

A specific advantage of the hydrothermal technique is the ease of control of the particle size, and hence of the nanostructure and porosity of the resultant semiconductor layer. The microstructure of the semiconductor is of course a compromise, to achieve an optimal optical absorption and photovoltaic performance. Nanosize grains give the greatest surface area, but pores must remain sufficiently large so that the mobility of the redox ions, as the charge carriers in the electrolyte, is not unduly inhibited. Also some degree of optical scattering by larger particles in the semiconductor film is desirable, particularly for devices which function under indirect illumination, such as vertical building facades. Processing parameters such as precursor chemistry, hydrothermal growth temperature for the titania powder and sintering conditions are varied in the optimization procedure. A flowsheet for hydrothermal processing is presented in Figure 2.7 as an example of the required development work. Figure 2.8 shows the control of porosity of the final film which results from it, as determined by a nitrogen adsorption method. The procedure involves the hydrolysis of the titanium alkoxide precursor in an aqueous medium to produce a sol, which is then subjected to the hydrothermal Ostwald ripening in an autoclave. The temperature of the hydrothermal treatment is decisive for the ultimate particle size. A standard sol, treated for 12 h at 230 °C has a mean particle diameter in the order of 10 nm. The colloidal suspension is applied to a transparent conducting oxide film on a glass or plastic, or to a metal foil, by one of several standard coating processes, of which tape casting, spraying and screen printing are examples. The film porosity is controlled by the addition of an organic filler, such as carbowax, to the suspension prior to deposition. A firing protocol to dry the film, then to pyrolyze binders and organics ensures a coherent low-resistance ohmic contact to the conducting substrate. Figure 2.9 shows a scanning electron micrograph of a typical mesoporous TiO_2 film. Typical final film thicknesses are 5–20 μm, with TiO_2 mass 1–4 mg/cm^2, film porosity 50–65%, average pore size 15 nm and particle diameter 15–20 nm. Other oxide semiconductors which have been studied in the context of dye-sensitized photovoltaics include ZnO, SnO_2, Nb_2O_5 and $SrTiO_3$ [18–22].

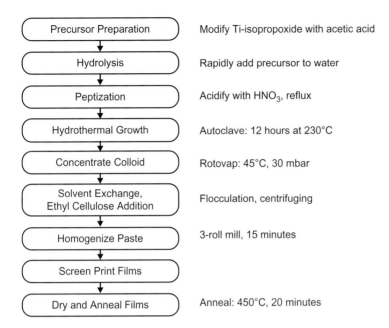

Figure 2.7 Process for the hydrothermal preparation of titanium dioxide for the production of photoanodes on a conducting glass substrate.

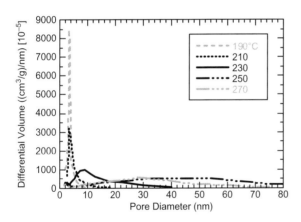

Figure 2.8 Control of semiconductor substrate porosity by the temperature of hydrothermal processing.

2.6
Conclusions

While this review is limited to the nanotechnology aspects of sensitized photoanode development for dye solar cells, it does indicate clearly the synergy of the different parameters which must be optimized together to achieve a credible

Figure 2.9 Scanning electron micrograph of the surface of a mesoporous anatase film prepared from a hydrothermally processed TiO_2 colloid.

device. Photoconversion efficiency of solar light into electricity may be the final goal, but reliability and service life, as well as economic factors, also enter in. Other technical issues which contribute to the function of the device are of course the electrolyte and the passive cathode. For the electrolyte some relevant considerations are mentioned, including a wide temperature range of stability, a low vapor pressure, electrochemical stability under the potential differences likely to be encountered in photovoltaic service, as well as the mobility of the ions for the redox cycling process between the electrodes and finally viscosity. For the cathode, to minimize losses a catalytic surface is required; here again nanotechnology is relevant. Finally, for a photovoltaic technology to succeed it is inadequate to validate a concept in single laboratory cells. There are questions of sealing and alignment, series connection and system management, as well as production engineering. All these are at present being addressed industrially, and production on a significant scale is beginning. As a result a real contribution to the energy economy can be expected from dye-sensitized photovoltaic technology, in association with, and also in competition with, other renewable and conventional methods for the provision of electricity.

References

1 Becquerel (Edmond), A.E. (1839) *C. R. Acad Sci. Paris*, **9**, 561.
2 Adams, W.G., and Day, R.E. (1877) *Proc. R. Soc. Lond.*, **A25**, 113.
3 Rappaport, P. (1959) *RCA Rev.*, **20**, 373.
4 West, W. (1974) *Photogr. Sci. Eng.*, **18**, 35.
5 Moser, J. (1887) *Monat. Chem. (Vienna)*, **8**, 373.
6 Gerischer, H., and Tributsch, H. (1968) *Ber. Bunsenges. Phys. Chem.*, **72**, 437.

7 Tributsch, H. (1968). Techn. Ph.D. Thesis, Hochschule München, Germany.
8 Zakeeruddin, S.M, Klein, C., Wang, P., Graetzel, M. (1998) Patents CH 674596, EP 0333641, US 492772.
9 O'Regan, B., and Grätzel, M. (1991) *Nature*, **353**, 737.
10 Grätzel, M. (1999) *Cattech*, **3**, 3.
11 Clark, W.D.K., and Suttin, N. (1977) *J. Am. Chem. Soc.*, **99**, 4676.
12 Zakeeruddin, S.M, Klein, C., Wang, P., Graetzel, M. (1993) Patents EP 0613466, US 5463057.
13 Nazeeruddin, M.K., Pechy, P., Graetzel, M. (2001) Patents EP 0983282, US 6245988.
14 Papageorgiou, N., Athanassov, Y., Armand, M., Bonhôte, P., Pettersson, H., Azam, A., and Grätzel, M. (1996) *J. Electrochem. Soc.*, **143**, 3099.
15 Wang, P., Zakeeruddin, S.M., Moser, J.E., Nazeeruddin, M.K., Sekiguchi, T., and Grätzel, M. (2003) *Nat. Mater.*, **2**, 402.
16 Desilvestro, H. (2008) Abs. 409, *Proc. 17th Intern. Conf. Photochem. Conversion Storage Solar Energy, Sydney, Australia.*
17 Brooks, K.G., Burnside, S.D., Shklover, V., Comte, P., Arendse, F., McEvoy, A.J., and Grätzel, M. (1999) *Proc. Am. Ceram. Soc.*, Indianapolis, U.S.A. April 1999 / *Electrochem. Mater. Devices*, 115.
18 Bedja, I., Hotchandani, S., and Kamat, P.V. (1994) *J. Phys. Chem.*, **98**, 4133.
19 Sayama, K., Sugihara, H., and Arakawa, H. (1998) *Chem. Mater.*, **10**, 3825.
20 O'Schwartz, B., and Regan, D.T. (1996) *J. Appl. Phys.*, **80**, 4749.
21 Rensmo, H., Keis, K., Lindstrom, H., Sodergren, S., Solbrand, A., Hagfeldt, A., Lindquist, S.E., Wang, L.N., and Muhammed, M. (1997) *J. Phys. Chem. B*, **101**, 2598.
22 Dabestani, R., Bard, A.J., Campion, A., Fox, M.A., Mallouk, T.E., Webber, S.E., and White, J.M. (1988) *J. Phys. Chem.*, **92**, 1872.

3
Thermal-Electrical Energy Conversion from the Nanotechnology Perspective
Jian He and Terry M. Tritt

3.1
Introduction

The demand for alternative energy technology to reduce our reliance on fossil fuels along with their environmental impact leads to important regimes of research, including that of direct thermal-electrical energy conversion via thermoelectricity [1]. Heat from different sources such as solar heat, geothermal heat, or the exhaust gases of automobiles can be directly converted into electricity by a thermoelectric device. Alternatively, a thermoelectric device can run in reverse as a heat pump. Being all solid-state, lightweight and compact, responsive, without moving parts or hazardous working fluids and feasible for miniaturization, a thermoelectric device can easily work in tandem with other alternative energy and energy conversion technologies.

However, all these salient features come at a price: the conversion efficiency of thermoelectricity is much less than that of equivalent mechanical systems. For this reason thermoelectricity has been long confined to various *niche* applications, where the issue of efficiency is less of a concern than the issues of energy availability and reliability, compactness of the device and quiet operation. Historically some of these *niche* applications include thermoelectrically powered radios, which were first reported in Russia around the 1920s and in World War II. And, a proof of principle thermoelectric climate-control system was installed in a 1954 Chrysler automobile. The first long-term test of thermoelectric devices was in NASA's deep-space missions, where fuel cells, solar cells and nuclear power are not feasible [2]. During the *Apollo* mission, thermoelectric devices served as the power supply. Currently, radioisotope thermoelectric generators (RTGs) are the primary power source (~350 W for each RTG) used in deep-space missions beyond Mars: the *Voyager* and *Cassini* spaceships are equipped with RTGs using ^{238}Pu as the thermal energy source and SiGe as the thermoelectric conversion material.

Over the past decade there has been a heightened interest in the field of thermal-electrical energy conversion driven by the need for more efficient and environmentally benign solid-state refrigeration and power generation. Refrigeration aspects include important applications such as cooling microelectronics (e.g., CPU

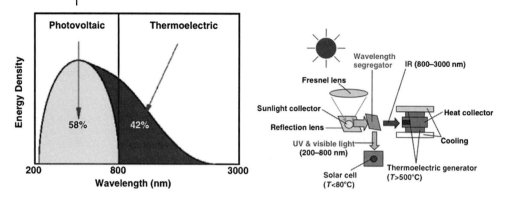

Figure 3.1 Left: The wavelength distribution of the energy density of the solar radiation, which is equivalent to a 6000 K black body. Right: Schematic drawing of a solar-thermoelectric hybrid power generator [6].

chips) and optoelectronics (e.g., infrared detectors, laser diodes) [3]. Most microelectronics and optoelectronics devices require responsive small-scale or localized spot cooling that does not impose a large heat load, which is best satisfied by the thermoelectric refrigeration. The gain from temperature stabilization and also device performance can be significant. Another emerging application is in relation to the refrigeration of biological specimens. Currently, millions of thermoelectric climate-controlled seats are being installed in luxury cars, while millions of thermoelectric coolers are used to provide cold beverages.

Concerning thermoelectric power generation, the wristwatches marketed by Seiko and Citizen and biothermoelectric pacemakers are powered by the small temperature differences between a body and its surroundings or within the body. One emerging market is to harvest the large amount of waste heat (about two-thirds of the generated power) from an automobile's exhaust or engine and convert it into "on-board" electrical energy using thermoelectric devices [4]. The other important application is to work in tandem with solar energy conversion technologies. Solar energy provides the most sustainable and worldwide available energy source [5]. A solar collector/concentrator-thermoelectric hybrid device can be fabricated (Figure 3.1), with the photovoltaic cell and thermoelectric device each covering the ultraviolet spectrum (200–800 nm wavelength) and the infrared spectrum (800–3000 nm wavelength) [6]. Forthcoming applications include remote "self-powered" systems for wireless data communications in the microwatt power range and waste heat recovery from industrial furnace and power plant in the kilowatt power range.

3.2
Established Bulk Thermoelectric Materials

Thermoelectricity is based on two basic effects: (i) the Seebeck effect, generating electricity from a temperature gradient ("power generation mode") [7] and (ii) the

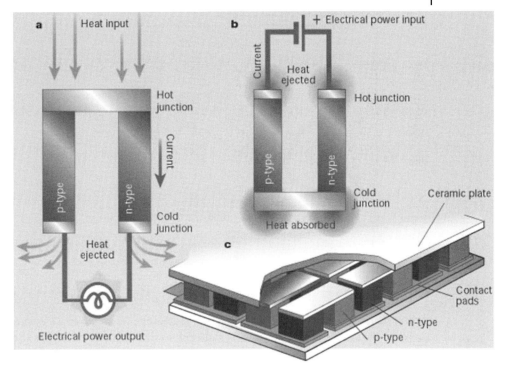

Figure 3.2 The basic thermoelectric module, made of n-type and p-type materials, can work either in the (a) power generation mode or (b) refrigeration mode. A thermoelectric device is made of many basic modules, as shown in (c) [9].

Peltier effect, generating a temperature gradient when electrical current is applied ("refrigeration mode") [8]. As shown in Figure 3.2, a basic thermoelectric couple consists of two legs made of n-type and p-type material; and many of these couples are connected electrically *in series* and thermally *in parallel* in order to make a thermoelectric module or device. The performance of a thermoelectric material is gauged by the dimensionless figure of merit, ZT, defined as:

$$ZT = \frac{\sigma \alpha^2 T}{\kappa} = \frac{\sigma \alpha^2 T}{(\kappa_{ph} + \kappa_e)} \tag{3.1}$$

where σ is the electrical conductivity, α is the Seebeck coefficient, κ is the thermal conductivity (including the lattice thermal conductivity κ_{ph} and the carrier thermal conductivity κ_e) and T is the temperature in Kelvin. The Wiedemann–Franz relationship, $\kappa_e = L\sigma T$, is often used to estimate κ_e, where L is the Lorentz number. Furthermore, the efficiency of thermal-electrical energy conversion η is given by:

$$\eta = \eta_c \eta_{TE} = \left[\frac{T_{hot} - T_{cold}}{T_{hot}}\right]\left[\frac{\sqrt{1+ZT_m} - 1}{\sqrt{1+ZT_m} + \left(\frac{T_{cold}}{T_{hot}}\right)}\right] \tag{3.2}$$

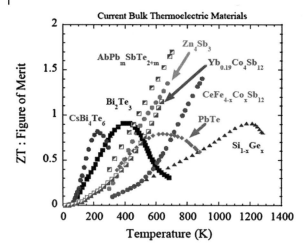

Figure 3.3 ZT as a function of temperature for current bulk thermoelectric materials.

where the Carnot efficiency η_c is given as the ratio of the temperature difference between the hot-end temperature T_{hot} and the cold-end temperature T_{cold}, T_m is the average temperature and a thermoelectric component η_{TE} is given as a function of ZT_m.

Equations 3.1 and 3.2 shift the fundamental challenge or issue of thermal electrical energy conversion primarily to the optimization of the physical properties of the thermoelectric material. Per Equation 3.1, central to improving ZT is to manage the electrical and thermal transport such that the material is a "phonon-glass electron-crystal" (PGEC) [10], that is, the electrical transport should behave like that of a crystal and the thermal transport behave like that of a glass. In addition, the material must simultaneously possess a high Seebeck coefficient, $\alpha > 100\,\mu V\,K^{-1}$. It is difficult to satisfy these criteria in a simple crystalline bulk material since all three quantities (σ, α, κ) that govern ZT are inter-related, and a modification to any of these quantities often adversely affects the others. For instance, σ and α are in large part determined by the electron band structure; an increase of σ usually results in a decrease of α. Meanwhile, the thermal conductivity κ is determined by both electrical conductivity σ via the Wiedemann–Franz relationship and the lattice thermal conductivity κ_{ph}. The mechanisms that decrease κ_{ph} (e.g., alloying, grain boundary scattering) decrease the carrier mobility μ, thus deteriorating σ.

As a result, the current bulk thermoelectric materials have ZT values of 1–2 in their respective temperature range of usage (Figure 3.3) [11]. These materials ideally have conversion efficiencies of 7–15%, depending on the specific materials and the temperature differences involved. Figure 3.4 presents the thermoelectric efficiencies for power generation and refrigeration as a function of ZT_m and the (T_{hot}/T_{cold}) ratio, in comparison with other energy conversion technologies

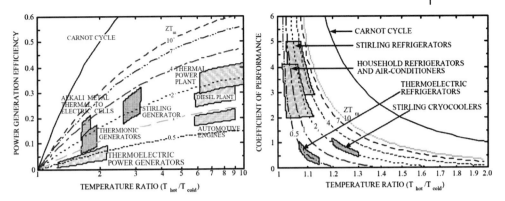

Figure 3.4 The competitiveness of thermoelectricity in power generation (left) and refrigeration (right) [12].

[10]. It is clear that in order to compete in efficiency, the next generation thermoelectric material should have a ZT value on the order of 2–3 (η = 15–20%) for both n-type and p-type materials with low parasitic losses (e.g., contact resistance, radiation effects, inter-diffusion of the metals) and low manufacturing costs.

Over the past decade, the efforts of developing the next generation thermoelectric materials have culminated into a two-pronged strategy, depending on the type of material studied. One approach is to continue working on novel bulk materials in light of the PGEC strategy, while the other is to search for high ZT in systems that possess reduced dimensionality or very small characteristic length scales. These two approaches can be pursued independently; however, they recently came together via the concept of a nanocomposite thermoelectric material. The present chapter is organized in accord with this research direction, with the emphasis on low-dimensional and nanostructured thermoelectric materials.

3.3
Selection Criteria for Bulk Thermoelectric Materials

Modern thermoelectrics research dates back to the 1950s, wherein the newly developed theory of semiconductors and semiconductor transport helped establish the basic principles of thermoelectricity and imposed several guidelines for bulk materials suitable for thermoelectrics. It is instructive to briefly review these existing guidelines so that the reader can better comprehend how, and to what extent, the advent of nanotechnology impacts the field of thermoelectricity.

The first two guidelines are concerned with two basic parameters of semiconductor: the carrier concentration n and the band gap Δ, respectively. The first guideline states that degenerate semiconductors with $n \sim 10^{18}$–$10^{20}\,\mathrm{cm}^{-3}$ make good thermoelectrics because such n values maximize the term $\sigma\alpha^2$ [13]. The second guideline states that the semiconductors with band gap $\Delta \sim 10k_B T$ make good thermoelectrics, where k_B is Boltzmann constant and T is the operating temperature. The reason is because such a band gap size ensures a reasonable value for the carrier mobility μ and also minimizes the detrimental contribution from the minority carrier to the Seebeck coefficient (i.e., *bipolar* effect). According to Mahan, this "$10k_B T$" rule holds for either a direct or an indirect gap and for both phonon and impurity scattering mechanisms [14]. Many years ago Goldsmid showed that Z is proportional to $\mu(m^*)^{1.5}$ [15], where m^* is effective mass. Enhancing m^* without deteriorating μ thus leads to the third guideline: it is thermoelectrically favorable for material to adopt high-symmetry crystal structure (to generate degenerate bands) and to have small electro-negativity difference among the constituent elements (to minimize the scattering of charge carriers by optical phonons). These three guidelines are thus concerned with several inter-related band structure parameters: n, Δ and μ. In view of the concept of PGEC, good electrical properties are only half of the story: the material should simultaneously possess a low lattice thermal conductivity. Low lattice thermal conductivities are often found, in conjunction with low Debye temperature and large anharmonic vibration, in materials made up of heavy elements or with many atoms per unit cell.

These guidelines cover most of the established bulk thermoelectric materials, which are heavily doped semiconductor bulk materials containing bismuth, antimony, tellurium and lead. There are several notable exceptions, such as the cobalt-containing layer-structured oxides made up of light atoms with a large electronegativity difference [16]. The large electronegativity difference and the resulting strong scattering of carriers by optical phonons is one reason for low carrier mobility in oxides. Even for the metallic oxides, the Seebeck coefficient is often found to be small due to their complex band structure. The surprising discovery of *"good thermoelectrics"* in cobalt-containing oxides led to important new concepts such as the "hybrid crystal". The basic idea is to decouple and control the electrical and thermal transport in complex crystal systems that are composed of building blocks or crystal modules with different compositions, structural symmetries and thermoelectric functions. The misfit-layered cobalt oxides [16, 17], partially filled skutterudites and clathrates [18–20] and novel Zintl-phase compounds [21, 22] provide examples of such control.

Despite the progress in identifying novel bulk thermoelectric materials, the inherent inter-relation of σ, α and κ still imposes restrictions on further improvement of ZT in bulk material. The future expansion of thermoelectricity is more tied to novel concepts and ideas that implement decoupling these inter-related quantities. In this regard the recent advent of nanomaterials and nanotechnology hold much promise. As shown in Figure 3.5, many of recent advances in enhancing ZT are linked to nanoscale phenomena in either bulk material containing nanoscale constituents (e.g., $AgPb_{18}SbTe_{20}$, nanocrystalline Bi_2Te_3 alloy) or the "custom-engineered" low-dimensional system (e.g., PbSeTe/PbTe quantum dot [QD], Bi_2Te_3/Sb_2Te_3 superlattice systems).

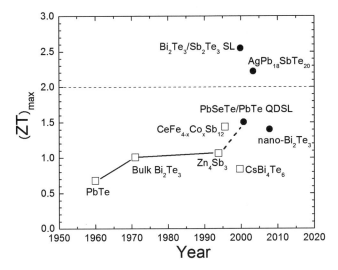

Figure 3.5 The timeline of the maximum ZT value, $(ZT)_{max}$, attained from the 1950s to the 2000s. The advances linked to nanoscale phenomena are denoted by black dots. Note that the ZT is considerably enhanced in nanocrystalline Bi_2Te_3 alloy as compared to the bulk value, despite the same composition and crystal structure.

3.4
Survey of Size Effects

In the previous discussion, all thermoelectric properties are presumed to be those of the bulk materials, invariant of the dimensionality D or the characteristic length scale d of the system. This is *not* the case in a low-dimensional system and nanostructure. In the nanostructured systems,[1] one has a new material parameter: the characteristic length scale d, which can be the quantum dot size, the superlattice period, the nanowire diameter, etc.

The introduction of this new parameter d eases the limitations that arise from the inter-related quantities in a bulk material because of the classic and quantum size effects. The classic size effects are concerned with the limitation of the mean free path (electron, phonon) due to scattering mechanisms related to their transport, whereas the quantum size effects arise from the confinement-induced change in the electronic band structure and vibration properties of the system. When the value of d is reduced to the order of nanometers, it is difficult to distinguish the classic and quantum size effects, from the thermoelectric perspective. The classic and quantum size effects make the physical properties of a nanostruc-

---

1) The reduced dimensionality is intimately related to the constraint of d along one or more directions and the underlying mechanisms remain the same, so we use the generic term "nanostructured systems" to represent both low-dimensional systems and nanostructured systems.

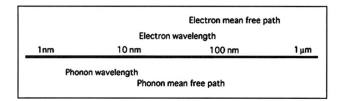

Figure 3.6 Characteristic length scales of the mean free path and wavelength of electrons and phonons in solids. The variation in the mean free path is usually larger than that in the wavelength.

tured system so different from that of its bulk counterpart that the nanostructured system is eligible to be treated as a *new* material.

To help elucidate the mechanisms through which the classic and quantum size effects work, we present, in Figure 3.6, the characteristic values of the mean free path and wavelength of electrons and phonons in solids. These physical quantities have wide variation in their values, and they are even more different in their energy- and momentum-dependence, thus offering the opportunity to decouple the thermal and electrical transport via controlling the value of d.

3.4.1
Classic Size Effects

The energy carriers involved in thermoelectricity are basically electrons and phonons, although the contribution from the spin degree of freedom is important in a few cases [23]. These energy carriers are subject to various scattering mechanisms in their transport. Let us first discuss phonons. The lattice thermal conductivity is less infinite because of the phonon scattering with defects, impurities, electrons, boundaries and other phonons, which limits the phonon mean free path. The dependence of the lattice thermal conductivity on sample size was first pointed out by Casimir in the 1930s [24], while the interface thermal resistance on the boundary between liquid helium and solids was explored by Kapitza in the 1940s [25]. In nanostructures the phonon scattering by the boundary/interface can dominate other scattering mechanisms. For instance, the thermal conductivity of silicon nanowires can be reduced by several orders of magnitude from that of the bulk counterpart due to boundary scattering [26]. In a more complex case, the mixed heat transport both parallel and perpendicular to interfaces in the Si/Ge superlattice nanowire has been modeled (Figure 3.7). The thermal conductivity could be reduced below the alloy limit [28] by the effects of diameter, superlattice period and interface transmissivity [27]. For the strategy of phonon mean free path reduction to be effective, it is important to specify the relative contribution of phonons to the lattice thermal conductivity according to their mean free path and wavelength [29].

The classic size effects on electrons are in large part due to grain boundary/interface scattering. As Sharp *et al.* pointed out, the grain boundary could scatter

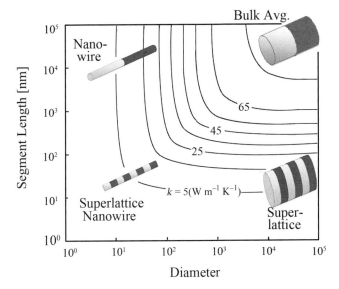

Figure 3.7 Contours of constant thermal conductivity for Si/Ge superlattice nanowire as a function of diameter and segment length, assuming a diffusive boundary scattering [27].

phonon more than electron by choosing the appropriate grain size [30]. The approach is in principle applicable to nanostructured systems. If the grain size is chosen to be smaller than the mean free path of phonons but larger than that of electrons, one can reduce the thermal conductivity by boundary scattering without seriously degrading the electrical transport. Another mechanism is that the energetic barriers at boundary/interface can filter out low-energy electrons and enhance the power factor, $\alpha^2 \sigma T$ [31, 32]. The boundary scattering also affects the Seebeck coefficient due to the phonon-drag effect. The phonon drag contribution is mainly associated with long-wavelength phonons that are sensitive to boundary scattering.

3.4.2
Quantum Size Effects

A second type of size effect, quantum size effects or quantum confinement, becomes important when the characteristic length scale d is comparable with the wavelength of the electron and the phonon. We first address electrons in this case. As shown in Figure 3.8, at a given carrier concentration, the Seebeck coefficient α is enhanced with reducing d due to the size-quantization effect [33]. Hicks and Dresselhaus first proposed that the reduced dimensionality could be used to enhance the electronic density of states near the conduction band edge to improve the power factor [34, 35]. Recently the thermoelectric potential of graphene-based interference device has been suggested, owing to a giant Seebeck coefficient on the order of ~ 300 mV K^{-1} [36].

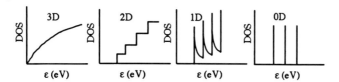

Figure 3.8 Schematic energy dependence of the electron density of states (DOS) in 3-dimensional (3D) bulk, 2D quantum well, 1D quantum wire and 0D quantum dot systems. The sharp DOS maxima near Fermi level may lead to a large Seebeck coefficient.

The quantum size effects also alter the vibration properties and thermal transport in nanostructure. In a phenomenological model, $\kappa_{ph} = C_v v l_{ph}$, where C_v is the specific heat, v is the sound velocity and l_{ph} is the phonon mean free path. In bulk solids and nanostructures, the phonon mean free path *cannot* be shorter than the average inter-atomic spacing [10], so new reduction mechanisms in C_v and v must be implemented in order to beat the *alloy limit* of thermal conductivity. At temperatures far below the Debye temperature, the phonon density of states is dominated by acoustic phonons that usually obey a dispersion relation $\varepsilon \propto k^\alpha$. For acoustic modes in D-dimension this contribution to the specific heat is expected to be $C_{ph} \propto T^{D/\alpha}$, that is, the low temperature lattice specific heat should scale with the increasing temperature to the dimensionality of the system [37]. Indeed, it has been experimentally confirmed in the single wall *nanotube bundle* sample [38]. Another exciting finding is the quantum of thermal conductance in a low temperature regime dominated by ballistic, massless, phonon modes. Schwab *et al.* demonstrated that the phonon thermal conductance in a 1-dimensional (1D) nanostructure was quantized in units of the quantum of thermal conductance G_{th}, where $G_{th} = (\pi^2 k_B T)/3h = 9.456 \times 10^{-13}\,\text{W K}^{-2} \times T$ and h is Plank's constant. G_{th} corresponds to the maximum value of energy that can be transported by a single phonon mode at T, the 1D nanostructure thus behaves like a phonon guide similar to the optical one for light [39]. In superlattices, Ren and Dow introduced the concept of mini-Umklapp scattering to describe the change of phonon dispersion relation [40]. The modulated Umklapp scattering arising from the new mini-reciprocal lattice vectors has several effects: (i) the group velocities of phonons are reduced, especially the heat-carrying high-energy acoustic phonons and (ii) the scattering rates of the "Umklapp" process are increased.

3.4.3
Thermoelectricity of Nanostructured Materials

In this section we specifically discuss how ZT is improved in nanostructured system. Recall Equation 3.1, the power factor $\alpha^2 \sigma T$ in ZT depends on the exact form of the electron band structure (in particular, position of Fermi level, gap size, density of states, mobility) and the scattering parameter. The other factor in ZT is the thermal conductivity κ. The thermal conductivity depends on the detailed phonon dispersion relations and the scattering parameter as well. Accordingly the nanostructured material approach serves a two-fold strategy: (i) using the quantum size effects to optimize the power factor via somewhat independently tuning α and σ; (ii) using numerous interfaces associated with the nanostructures to scatter phonon more effectively than electron. The validity of this twofold strategy was first tested in model periodic 2D quantum well systems [41] and in 1D quantum wire systems [42].

We start with the first strategy. The Seebeck coefficient in the nanostructured materials can be enhanced through three avenues: quantum size effects, electron energy filtering and carrier-pocket engineering. In a degenerate semiconductor the Seebeck coefficient can be described by the Mott relation as that of a metallic-like material [43]

$$\alpha \sim \frac{\pi^2 k_B^2 T}{3e} \left(\frac{\partial \ln \sigma(\varepsilon)}{\partial \varepsilon} \right)_{\varepsilon = E_F} \quad (3.3)$$

where $\sigma(\varepsilon)$ is the electrical conductivity when Fermi energy $E_F = \varepsilon$, e the elementary charge and k_B the Boltzmann's constant. In the Drude model, $\sigma(\varepsilon)$ can be expressed in terms of carrier concentration n, mobility μ, relaxation time τ and effective mass m^*:

$$\sigma(\varepsilon) = n(\varepsilon) e \mu(\varepsilon) = n(\varepsilon) e^2 \frac{\tau(\varepsilon)}{m^*} \quad (3.4)$$

Note that $n(\varepsilon)$ is a function of the density of states (DOS), per Equations 3.3 and 3.4, α can be enhanced by increasing the energy dependence of the differential conductivity $\partial(\sigma(\varepsilon))/\partial \varepsilon$ at Fermi level. As such, the quantum size effects can enhance α via generating sharp DOS maxima at Fermi level (Figure 3.8).

Also, one can increase the energy-dependence of $\mu(\varepsilon)$ or $\tau(\varepsilon)$ in order to enhance α via preferentially scattering electrons depending on their energy. With low energy electrons being filtered out, it ensures that the reduction of σ is more than compensated by the increased α, thus increasing the power factor. Following this approach, Humphery and Linke proposed utilizing potential energy barriers in 2D structures to generate minibands in the band structure to implement energy filtering [44]. They also proposed an interesting concept, namely, "energy-specific" equilibrium. Placing electrons in such an "energy-specific" equilibrium, and given a fined-tuned DOS that might possibly be realized in a double-barrier-embedded nanowire, the efficiency of such a thermoelectric device can, in principle, approach the fundamental Carnot efficiency at low temperatures. Shakouri and Bowers proposed a different energy-filtering scenario based on the hot electron injections from a potential energy barrier [45]. The carrier-pocket engineering can be used additionally to optimize band structure. It was originally designed for the superlattices, in which one type of carrier is quantum confined in the quantum well region and another type of carrier is quantum confined in the barrier region [46]. It has been applied to Si/SiGe superlattice system [47], and latterly extended to self-assembled nanostructured composites.

We now turn to the second strategy: the reduction of thermal conductivity primarily, the lattice thermal conductivity. The thermal conductivities measured

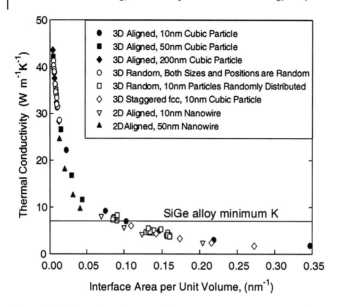

Figure 3.9 Thermal conductivity of Si–SiGe nanocomposites as a function of interfacial area per volume [51].

on Si/Ge [48] and InAs/AlSb superlattices [49] were found to be significantly lower than that of the compositionally equivalent alloy and the predictions based on Fourier's law. A complementary theoretical study [50] suggested that the thermal conductivity reduction in a superlattice comes from the sequential interface scattering of phonons rather than the coherent superposition of phonon waves. The transmission, partially specular and partially diffusive scattering of phonons at the boundary/interface play major roles in the thermal resistance. In a theoretical simulation of Si nanowires and nanoparticles in a SiGe matrix, the calculated thermal conductivity of the composite scale well with the interfacial area per unit volume, irrespective of the shape, size and orientation of the nanoinclusions (Figure 3.9) [51]. The above discussion on the thermal conductivity reduction mechanism suggests that the periodicity of superlattices is *not* a necessary condition for such thermal conductivity reduction.

3.5
Thermoelectric Properties on the Nanoscale: Modeling and Metrology

Understanding the thermoelectric properties on the nanoscale necessitates a combined effort of modeling and measurement. The major challenge is to derive the intrinsic thermal transport properties on the nanoscale. Depending on the characteristic length scale of the system from μm to nm, a variety of methods, from the Fourier conduction model, the Boltzmann transport equation, the molecular dynamics simulation to the *ab initio* atomistic models, are available for computational analysis of thermodynamics and thermal transport in solids [52, 53]. However, there is so far no model that adequately treats the wave nature of phonon in a length scale comparable with the phonon mean free path and wavelength. In addition, the definition of *temperature* remains a fundamental issue on the nanoscale and in the non-equilibrium state [54]. Another methodological challenge is how to incorporate the insights of atomic-level approaches into the macroscopic continuum model, a multi-scale modeling seems to be the option.

There have been efforts to experimentally verify the theoretical predictions made on the nanoscale thermoelectric properties. From the metrology point of view, characterizing nanoscale phenomena requires nanometer spatial resolution, which can be provided by nanofabricated test apparatus, atomic force microscope and scanning tunneling microscope. Progress in the miniaturization of microelectronic mechanical system (MEMS) makes it possible to measure the thermoelectric properties on an individual nanoscale object in some cases. Yet, advances in the metrology deserve mentioning, for example, the 3-ω method, coherent phonon source, coherent optical method, time-domain thermoreflectance [55], scanning thermoelectric probe (SThEM) [56] and transmission electron microscope-based hot-wire probe [57]. In particular, the SThEM technique allows measurements of the spatial profiles of the thermoelectric voltage (Figure 3.10), dopant concentration and calorimetry on the nanoscale. However, these nanoscale metrology methods are still subject to problems such as the (electrical and thermal) contact

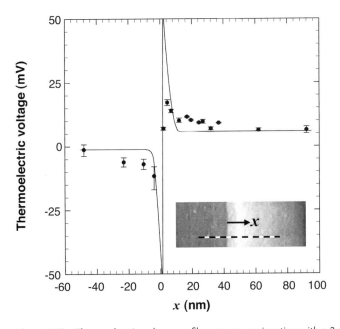

Figure 3.10 Thermoelectric voltage profile across a *p-n* junction with a 2 nm resolution [56].

resistance, interaction of measurement probe and substrate (if any) with the nanostructure, oxide layer removal and precise temperature measurement, etc. The reader is referred to the references [54, 58, 59] and the references listed therein for more details of nanoscale thermoelectric metrology.

3.6
Experimental Results and Discussions

3.6.1
Bi Nanowire/Nanorod

Bulk Bi is an interesting material from the thermoelectric perspective due to its highly anisotropic band structure, high carrier mobility and very light effective mass [60]. In many aspects Bi nanowire is also an excellent scientific platform in which the concepts of nanostructure thermoelectricity, such as quantum confinement, enhanced DOS by 1D nature, boundary scattering of electrons and phonons, can be tested. Lin et al. first considered the Bi nanowire along the crystallographic direction [1–12], and found that decreasing diameter of Bi nanowire increased the band gap and the system underwent a semimetal to semiconductor transition at ~ 49 nm (Figure 3.11) [61]. In the same work the ZT was calculated as a function of the position of the Fermi level and the electron density. The ZT values for wire diameter smaller than 10 nm are significantly higher than that of bulk Bi along the trigonal direction (Figure 3.12).

Pressure-injection of molten Bi into nano-sized pores of an alumina template has been used for preparing oriented arrays of Bi nanowires [62, 63]. A semimetal to semiconductor transition has been predicted for Bi nanowires with an average wire diameter <200 nm [64] and experimentally confirmed through temperature and magnetic field dependencies of the electrical resistance [42, 62] and room

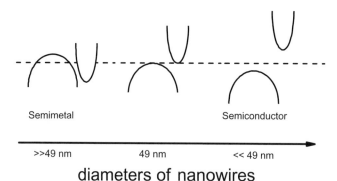

Figure 3.11 Schematic diagram of the confinement-induced semimetal-semiconductor transition in Bi nanowire. The transition occurs at the nanowire diameter ~ 49 nm.

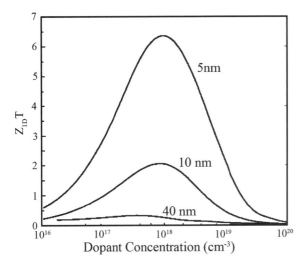

Figure 3.12 Calculated 1D ZT for n-type doped bismuth nanowires, as a function of the electron density. The calculations are made at 77 K, for wires oriented along the trigonal direction, for the three values of the nanowire diameter [61].

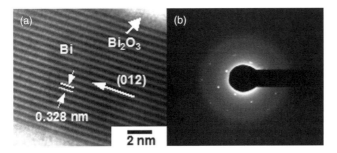

Figure 3.13 (a) High resolution transmission electron microscopy lattice image of a 10 nm diameter Bi nanorod showing the excellent ordering of the lattice planes and a thin bismuth oxide sheath; (b) The corresponding selected area diffraction pattern for the Bi nanorod in (a) [69].

temperature infrared absorption spectroscopy study [64–66]. Wang et al. [67] confirmed the semimetal to semiconductor transition in 5–500 nm diameter Bi nanoparticles. Huber et al. [68] reported the lack of semiconductor nature in their 30 nm diameter Bi nanowires and have interpreted their experimental results in terms of surface-induced charge carriers in a spherical Fermi surface pocket. Recently, Reppert et al. [69] utilized the pulsed laser vaporization (PLV) method for synthesizing ~ 10 nm diameter Bi nanorods in bulk quantities (Figure 3.13). The results of room temperature infrared and UV-visible absorption studies supported the

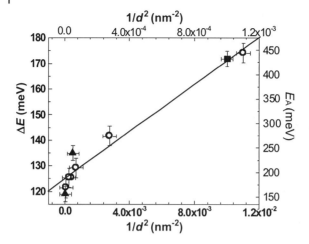

Figure 3.14 Dependence of the band gap energy (ΔE) and absorption threshold energy (E_A) on the diameter of Bi nanorods and nanowires. The open circles represent E_A values (right y-axis) reported by Cornelius et al. [66] for their nanowire diameters (top x-axis). The triangles and square data points represent the diameter-dependence (bottom x-axis) for ΔE (left y-axis) reported by Black et al. [65] and Reppert et al. [69], respectively. The dependence of the band gap on nanowire diameter in the semiconductor regime is quite consistent with the $1/d^2$ dependence (solid line).

semimetal to semiconductor transition (Figure 3.14); and the optical limiting measurements detected unusual nonlinear absorption [70].

Cronin et al. reported a four-probe resistivity measurement on a single Bi nanowire (left panel of Figure 3.15) [71]. An oxide layer formed on the Bi nanowires, resilient to acid etching, and was removed by reduction in high temperature hydrogen or ammonia gas or by the focused ion beam method. The temperature dependence of the resistivity of the Bi nanowire was found to decrease monotonically with increasing temperature, in contrast to that of bulk behavior. Also, the magnitude of resistivity systematically increased with decreasing nanowire diameter, suggesting the effects of boundary scattering of electrons (right panel of Figure 3.15). The Seebeck coefficient has been measured in the freely suspended 240–620 nm diameter Bi nanowires between 4 K and 300 K and for stress as high as 1 GPa [72]. The peaks of up to $80\,\mu V\,K^{-1}$ in the Seebeck coefficient observed around 40 K were interpreted in terms of diffusive Seebeck coefficient under strong electron and hole-boundary scattering.

3.6.2
Si Nanowire

An exciting proof of the concept of nanostructure thermoelectricity came from Si, a poor thermoelectric material in its bulk form. Li et al. [26] reported the thermal conductivity of single Si nanowire with diameters from 22 to 115 nm. The measurement set up is shown in Figure 3.16. The measured thermal conductivity values

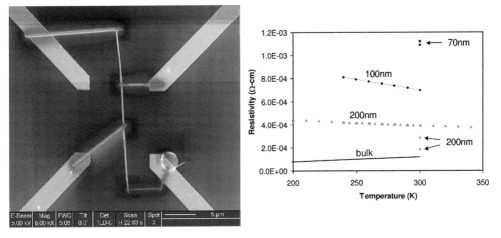

Figure 3.15 Left: Scanning electron microscopy image of a 100 nm Bi nanowire with four Pt electrodes prepared using a focused ion beam. Right: Temperature dependences of the resistivity of Bi nanowire with different diameters as compared with that of bulk Bi [71].

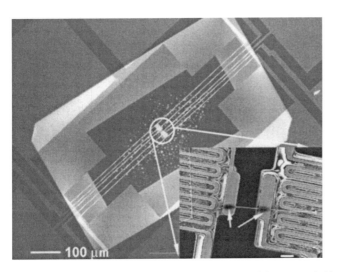

Figure 3.16 The scanning electron microscopy image of the suspended heater. The inset shows a 100 nm Si nanowire bridging the heater pads, the arrows point to the wire-pad junctions wrapped with amorphous carbon deposit, the scale bar is 2 μm [26].

were more than two orders of magnitude lower than the bulk value. The strong diameter dependence of thermal conductivity has been ascribed to the boundary scattering and possible modification of phonon spectrum.

In addition, the surface roughness of the Si nanowires was found to strongly affect the thermal conductivity. As shown in Figure 3.17, in contrast to the smooth surfaces of vapor–liquid–solid (VLS)-grown Au-catalyzed Si nanowires [74], the

Figure 3.17 Smooth surfaces of VLS-Si nanowire (left) [73] are in contrast to those rough surfaces of EE-Si nanowire (right) [74]. The insets show the sharp selected area electron diffraction pattern, indicative of single crystalline nature of the wire.

electroless etching (EE) growth technique yielded Si nanowires with a rough surface. The mean roughness height and roughness period are on the order of few nanometers. While the VLS and EE nanowires showed similar diameter dependence of thermal conductivity, the magnitude of thermal conductivity of the EE nanowire was a few times smaller than that of VLS wire of comparable diameter (Figure 3.18). The thermal conductivity peak of the EE wire also shifted to higher temperature, suggesting the phonon mean free path was limited by boundary scattering as opposed to the intrinsic Umklapp process. It is clear that the surface roughness provides an extra scattering channel, but the observed extent of the thermal conductivity reduction is beyond the estimate of any existing theory. Nonetheless, the thermal conductivity reduction by 100-fold from the bulk value yielded a $ZT \sim 0.6$ at room temperature. A slightly different work on an array of 10×20 and $20 \times 20\,nm^2$ cross-section Si nanowires showed a 100-fold improvement in ZT, including $ZT \sim 1.0$ at 200 K, by varying the size and impurity doping level. The authors attributed it to the phonon effects on the Seebeck coefficient and thermal conductivity [75]. A recent theory paper by Vo et al. [76] addressed the thermoelectric performance of both p- and n-type Si nanowires in relation to the crystallographic directions and surface reconstruction.

3.6.3
Engineered "Exotic" Nanostructures

The best performing thermoelectric materials demonstrated to date are based on the "nanoengineered" quantum dot and superlattice systems. Harman et al. reported molecular beam epitaxial growth of an array of quantum dots of composi-

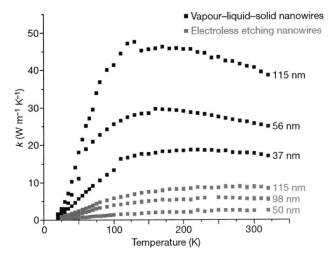

Figure 3.18 The temperature-dependence of thermal conductivity measured on the vapor–liquid–solid nanowires [26] and electroless etching nanowires [74].

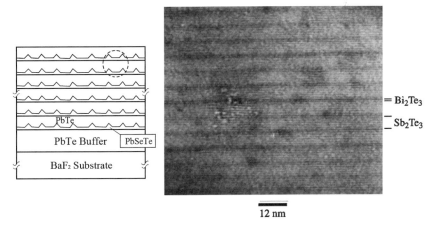

Figure 3.19 Left: Schematic drawing of a PbTe/PbSe$_{0.98}$Te$_{0.02}$ quantum dot superlattice system [77]. Right: Scanning electron microscopy image of a Bi$_2$Te$_3$/Sn$_2$Te$_3$ superlattice system [78].

tion PbSe$_{0.98}$Te$_{0.02}$ sandwiched between PbTe layers, the growth is for thousands of nanometer periods on a BaF$_2$ substrate with a PbTe buffer layer (left panel of Figure 3.19) [77]. Along the direction parallel to the superlattice plane, a $ZT \sim 3.5$ was reported at 570 K in the Bi-doped n-type sample, and encouraging results were also obtained in the Na-doped p-type sample [26, 79, 80] Along the direction perpendicular to the superlattice plane, Venkatasubramanian *et al.* reported [81] $ZT \sim 2.4$ and 1.4 at 300 K in the p-type and n-type Bi$_2$Te$_3$/Sn$_2$Te$_3$ superlattice systems (right panel of Figure 3.19), respectively.

As mentioned earlier, bulk oxides typically make poor thermoelectric materials. As one of the best oxide thermoelectric materials, bulk $SrTi_{0.8}Nb_{0.2}O_3$ has a $ZT \sim 0.37$ at 1000 K [82]. Can the quantum confinement effect in the narrowest possible 2D space, that is, one unit cell layer, yield some improvement? Ohta *et al.* fabricated a superlattice system composed of high density ($n \sim 10^{21}\,cm^{-3}$) 2D electron gas (2DEG) $SrTi_{0.8}Nb_{0.2}O_3$ confined within a unit cell layer in the insulating $SrTiO_3$. A significant enhancement of the Seebeck coefficient was observed when the thickness of the $SrTi_{0.8}Nb_{0.2}O_3$ layer was reduced below 1.56 nm. In the best case, a ~ fivefold enhancement of the Seebeck coefficient along with a fairly high 2D electrical conductivity have been attained, the authors estimated a $ZT \sim 2.4$ for the 2DEG and $ZT \sim 0.24$ for the superlattice system at 300 K [83]. Furthermore, Mune *et al.* showed that the critical barrier thickness for the quantum electron confinement was 6.25 nm (16 lattice constants of $SrTiO_3$) [84].

Advances have been also made in making more complicated nanoscale architecture. For instance, a Si/SiGe superlattice nanowire (Figure 3.20a) was fabricated using a hybrid pulsed laser ablation/chemical vapor deposition process [86] and studied by thermal conductivity measurement (Figure 3.20b) [26]. The alloy scattering of phonons in the SiGe segments was found to be the dominant phonon scattering mechanism while the boundary scattering also contributed to the thermal conductivity reduction [85]. Further, the chemical bean epitaxy growth technique enabled the fabrication of a prototype InP/InAs superlattice nanowire (Figure 3.21) [87], which is the candidate to substantialize the aforementioned "energy-specific" equilibrium state and the "reversible thermoelectricity" [44]. The thermoelectric potential of superlattice nanowire has been theoretically discussed by Lin *et al.* [88].

3.6.4
Thermionics

In a superlattice the energy barrier at the junctions of different materials has been used as the energy filter to boost its thermoelectricity. The electron motion across the barrier is described by the thermionic emission theory, which underlies the other important avenue of direct thermal-electrical energy conversion, *thermionics*. Despite the different nomenclature, the thermodynamics and the material parameters underlying both thermoelectrics and thermionics are remarkably similar. The basic difference between a thermoelectric device and a thermionic device lies in the manner that the current flow moves: in a thermionic device the electrons ballistically transport across the energy barrier, while in thermoelectric devices the motion of electrons is quasi-equilibrium and diffusive [89].

The application of a traditional vacuum-based thermionic energy converter is restricted by the high work function of known materials and the space charge effect. Shakouri and Bowers suggested that these could be circumvented using double heterojunction structures to fabricate an all solid state device [45]. Thermionic refrigeration was demonstrated by Shakouri *et al.* using a single In–Ga–AsP barrier heterostructure between the cathode and anode that were both made

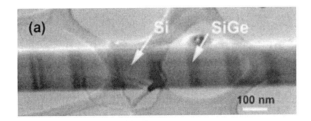
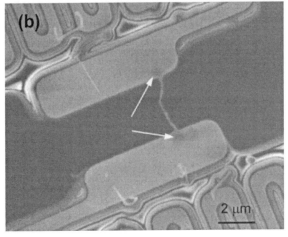
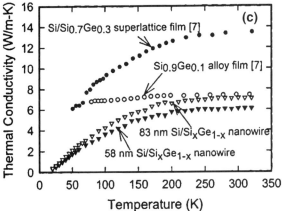

Figure 3.20 (a) Transmission electron microscopy image of a Si/SiGe superlattice nanowire, (b) a scanning electron microscopy of an 83 nm Si/SiGe sample bridging the two suspended heater pads for thermal conductivity measurements (the arrows point to the carbon deposits) and (c) thermal conductivities of Si/SiGe samples, along with the results of SiGe thin film and thin film superlattice for comparison [85].

3 Thermal-Electrical Energy Conversion from the Nanotechnology Perspective

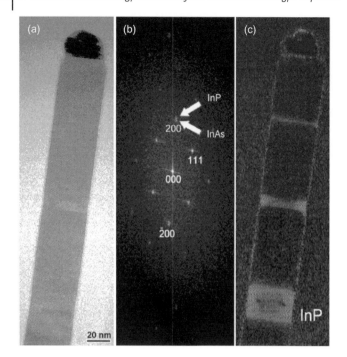

Figure 3.21 A prototype 40 nm diameter InP/InAs superlattice nanowire is presented in form of (a) high resolution transmission electron microscopy image, (b) power spectrum, (c) inverse Fourier transform using the vicinity of the InP part of split reflection along the crystallographic direction [200] [87].

from n^+-In–GaAs [90]. In the same work, it was mentioned that the cooling of the cathode, ~0.5 K at a temperature of 293 K, was due to thermionic emission rather than the Peltier effect. In a later work in the SiGeC/Si superlattice system [91], cooling by as much as 2.8 and 6.9 K was reached at 298 and 373 K, respectively. The advent of nanotechnology also has impact on thermionics study. For instance, the thermionic emission from nanocrystalline diamond-coated tips [92], carbon nanotubes [93] and nanocrystalline diamonds [94] have been measured. In these cases, the field-enhancing structures appear to result in significantly enhanced emissions when compared to similar flat surfaces without field enhancements.

3.6.5
Thermoelectric Nanocomposites: a New Paradigm

The promising results obtained in the quantum dot and superlattice systems carry two important messages: (i) the power factor enhancement and thermal conductivity reduction can be simultaneously achieved in nanostructures and probably only in nanostructures [51] and (ii) different from the PGEC examples mentioned in

Table 3.1 The room temperature thermoelectric properties of several best performance superlattice systems and those of their corresponding bulk materials [95].

Thermoelectric properties at 300 K	PbTe–PbSeTe quantum dot SLs	PbTe–PbSe bulk alloy	Bi_2Te_3–Sb_2Te_3 SLs	Bi_2Te_3–Sb_2Te_3 bulk alloy
S^2s ($\mu W\,cm^{-1}\,K^{-2}$)	32	28	40	50
k ($W\,m^{-1}\,K^{-2}$)	0.6	2.5	0.5	1.45
ZT	1.6	0.34	2.4	1.0

Section 3.3, the quantum dot and superlattice systems are heterostructured PGEC systems: the thermal insulation arises from less-disperse phonon modes and interface scattering of phonons while the electron transport is facilitated by optimal choice of band-offsets in these semiconductor heterostructures.

Although the nanostructure approach was initially inspired by the idea of utilizing quantum size effects to improve the electrical performance, until now the main gains in enhancing ZT are from the thermal conductivity reduction. As shown in Table 3.1, the superior ZT values of "nanoengineered" quantum dot and superlattice systems are mainly due to the thermal conductivity reduction. As discussed earlier, the periodicity of the superlattice is suggested *not* as a necessary condition for such thermal conductivity reduction (Section 3.4.2). It naturally leads to the idea of using "self-assembled" nanocomposites as a potential cost-effective and scalable alternative to the "custom-engineered" superlattice systems for high ZT materials [50]. In one of the earliest proof of principle nanocomposite approaches, Zhao et al. reported a 25% increase in ZT due primarily to thermal conductivity reduction in the nanocomposite sample made by hotpressing a mixture of Bi_2Te_3 nanowire/nanotube and Bi_2Te_3 microsize powders [96].

By definition, a nanocomposite material is a composite material containing nanoscale constituents; by design, it is a PGEC system with nanoscale heterogeneity. It should be noted that the macroscopic composite approach is restricted from the theoretical point of view. Within the framework of the effective medium approximation model [97], which is commonly used as a theoretical tool to analyze the transport property of composite, the ZT of a composite system can not exceed that of the best performing constituent of the composite, if there is no contribution from the interface/grain boundary [97, 98]. In the nanocomposite approach, the role of the interface becomes increasingly important with the diminishing characteristic length scale of one or more constituents. The interface effect can be a dominant factor given a typical value of 10^{19} interfaces cm^{-3}.

Recently, we conducted a proof of principle grain boundary/interface engineering study in a pulverized p-Bi_2Te_3 system [99, 100]. An alkali-metal-containing nanolayer was coated onto the surface of the bulk matrix grain, which then became part of the grain boundary upon hotpressing. The layer introduced extra carriers into the system, thus compensating for the mobility loss. As compared to the commercial ingot, the same ZT and a better compatibility factor [101] were

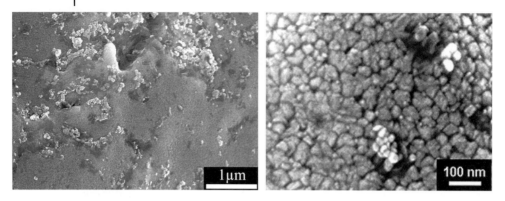

Figure 3.22 The scanning electron microscopy images at two different scales show the surface roughness over a wide range [102].

attained. In a different study, the surface of PbSnTe coarse grain was roughened, instead of coating, by consecutive hydrothermal and hotpressing treatments. As shown in Figure 3.22, the surface roughness height and roughness periodicity were on the order of a few nanometers to 100 nm. The roughness resulted in strong phonon scattering at the grain boundary and the thermal conductivity was reduced by a factor of two without significantly affecting the power factor [102]. These results are reminiscent of the observations in the Si nanowire with a rough surface [74]. Such a surface-roughness approach is important for those systems where the grain size can not be used as a control parameter.

There is an important difference between the electrical transport and the heat transport in thermoelectricity: the charge transport is nearly mono-energetic (energy levels within a few $k_B T$ around the Fermi level), whereas the thermal transport by phonons is broad in energy and momentum (at room temperature all phonon states in the Brillouin zone are involved in the transport). In addition, the phonon spectrum is more temperature dependent. Accordingly one needs to cover a wide range of phonon scattering wavelength to effectively reduce the thermal conductivity. The aforementioned examples of the Si nanowires and PbSbTe materials with a coarse grain and rough boundary suggest that the size of phonon scatters should be as low as 10 nm. Therefore nanostructuring is the only option. Following this line, nanoparticles have been embedded in bulk matrix as phonon-scattering centers in conjunction with other phonon-scattering mechanisms. We developed a nano-plating technique [103] in order to grow a layer of $CoSb_3$ nanoparticles directly on the surface of the La-filled $CoSb_3$-based skutterudite grain. The resulting nanocomposite material with 5% nanoinclusion showed a 15% improvement of ZT due to the combined phonon scattering by nanoparticles and "rattlers" [104]. One advantage of this nano-plating technique is the homogeneous distribution of nanoparticles in the bulk matrix. Using nanoparticles as phonon scattering centers was also reported by Kim *et al.* in an $ErAs/In_{0.53}Ga_{0.47}As$ system as the cause for the thermal conductivity below the alloy limit [105] and in a $C_{60}/CoSb_3$ composite for the enhanced ZT [106].

At present stage modeling of nanocomposites is very difficult because of the complexity and heterogeneous nature of the system. It is a materials design problem involving the intricate tuning of structure–property relationships down to the nanoscale in heterogeneous solids. In this regard the concept of micromorphology is important. The micro-morphology has three major aspects: (i) the composition, structure and characteristic dimension of local nanoscale constituents, (ii) the physical inter-connectiveness of these nanoscale constituents and (iii) interface properties. In a self-assembled nanocomposite the periodicity is absent but the collective behavior still exists. For instance, when the phonon coherence lengths are larger than the characteristic dimensions of local nanoscale constituent, the effect of phonon interference alters the phonon dispersion relation. Similar considerations apply to the electron transport in nanostructure. Therefore, aspects (ii) and (iii) often matter more than (i) in determining the thermoelectric properties of the nanocomposite; in other words, it is less important whether the individual component has a good thermoelectric performance.

Hence, controlling one or more of these micro-morphological aspects offers an additional new *"tuning knob"* in nanocomposites. There are some preliminary efforts along this line. Iketa *et al.* reported utilizing the decomposition of $Pb_2Sb_6Te_{11}$ near the eutectic point to form a nanoscale superlattice of PbTe and Sb_2Te_3 in which the interlamellar spacing could be controlled by the temperature and time of the decomposition process [107]. In contrast, the $Mg_2(Si,Sn)$-based solid solution is a well known "green" thermoelectric material, the previous optimization study was mainly focused on the doping and alloying approach [108]. Recently, we utilized the miscibility gap and peritectic point in the pseudo-binary phase diagram of Mg_2Si and Mg_2Sn to control the micro-morphology of the as-grown composite samples. Depending on the starting composition, the resulting micro-morphology varied from a coarse grain coated by a thin layer [109] to a coarse grain with embedded nanodomains [110].

Investigation of nanocomposite thermoelectric materials is becoming one of the most active directions in current thermoelectric investigations and study. For a more comprehensive coverage of the early efforts, the reader is referred to the review papers by Dresselhaus *et al.* [51, 111]. Here we only address a selected number of examples. The first example is the $AgPb_mSbTe_{2+m}$ (namely, LAST-m) system made by Kanatzidis' group. As shown in Figure 3.23, the *in situ* formed Ag_2Te nanodomains in the matrix of PbTe were identified by microscopy study and proposed as the cause for the state-of-the-art $ZT \sim 1.7$ in the intermediate temperature range 600–700 K [112–114]. Another PbTe-based nanocomposite material was prepared with bulk PbTe enriched with 6% Pb subject to proper heating treatment, and a large amount of the 30–40 nm Pb precipitates were formed in the matrix of PbTe [115]. The enhanced Seebeck coefficient was interpreted in terms of the carrier-energy filtering mechanism. In the $(GeTe)_x(AgSbTe_2)_{100-x}$ (namely, TAGS-x) compounds that have chemical similarity to the LAST compounds, Yang *et al.* reported a correlation between the presence of the nanodomains and the corresponding ZT values [116]. These nanodomains were of same composition but had different crystallographic orientation from the

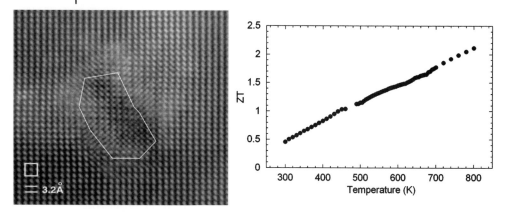

Figure 3.23 Left: Ag$_2$Te nanodomains formed in the matrix of PbTe; Right: ZT versus T of the LAST-18 compound [112].

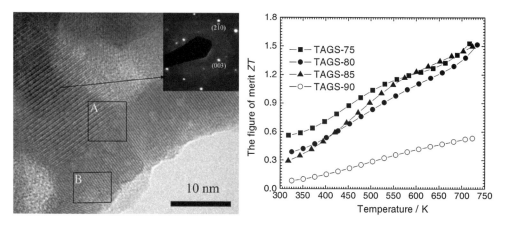

Figure 3.24 The nanodomains have been identified in TAGS-75, -85 and -90 compounds with high ZT but absent in the low ZT TAGS-90 compounds [116].

matrix, and the domain boundary appeared to be clean (left panel of Figure 3.24). In another promising thermoelectric material β-Zn$_4$Sb$_3$ with a simple formula and nominal stoichiometry, atomic pair distribution function analysis of X-ray and neutron diffraction data revealed nanoscaled local structures [117]. Even for p- and n-type Bi$_2$Te$_3$ commercial thermoelectric materials, the nanoscale structural modulation has been reported, and this is believed to contribute to the low lattice thermal conductivity in these alloys [118]. There is accumulating evidence supporting the view that nanoscale inhomogeneity (or in a more generic way, complexity on multiple length scales) is necessary for high ZT. In fact, the partially filled skutterudite and clathrate bulk material systems also fits into this description.

The nanocomposite materials can be directly made from pure nanomaterials. Poudel et al. [119] reported an important advance in the study of nanocomposites.

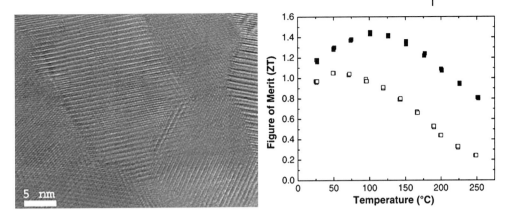

Figure 3.25 Left: Nanocrystals of a few nanometers size with coherent grain boundary. Right: The ZT of bulk nanocrystalline material (solid square) is significantly enhanced as compared to the commercial ingot (open square) [119].

The authors ball-milled commercial p-Bi_2Te_3 bulk thermoelectric material in an inert atmosphere and hotpressed the "as-obtained nanopowder" into a bulk sample. As shown in the left panel of Figure 3.25, the sample consisted of large number of crystalline nanodomains with clean grain boundaries and an average size of 20 nm; an enhanced peak ZT of 1.4 at 100 °C was attained (more than a 50% improvement). The ZT enhancement came mainly from a large reduction in the phonon thermal conductivity, but also benefited from a reduction of detrimental *bipolar* contributions at high temperatures due to defects on the grain boundary. The inert atmosphere helped avoid the oxidation that readily occurs in intermetallic nanopowder, and there appeared to be minimal grain growth during the hotpressing process. In a different work Martin *et al.* reported the thermoelectric performance of a PbTe nanocomposite made up of 100–150 nm PbTe nanoparticles, and the thermoelectric properties showed strong dependence on the stoichiometry, surface oxidation and porosity [120].

3.7
Summary and Perspectives

The decreasing fossil fuel supplies and increasing energy demand worldwide imposes a pressing need for improved direct thermal-electrical energy conversion. Thermoelectricity is the simplest technology that can be applicable to such energy conversion. It is important to recognize that no single technology can meet the world's energy needs in the 21st century; one needs a combination of many technologies. Given the ubiquitous heat sources and the modular aspects of thermoelectric devices, thermoelectricity is certainly guaranteed a position as an alternative energy technology of the 21st century.

Over the past decade, the development of novel thermoelectric materials containing nanostructure or being composed of nanostructures with almost double the figure of merit ZT of the best existing thermoelectric materials opens several new classes of applications for direct thermal-electrical energy conversion. Another 50% increase in ZT (to ZT ~ 3) will position thermoelectricity to be an important contributor to our energy needs, especially in the areas of waste heat recovery and solar energy conversion. Given the progress in understanding the nanoscale phenomena as well as in the materials design and fabrication, the likelihood of achieving these goals appears to be within reach in the relatively near future.

Acknowledgments

The authors acknowledge support from a DOE/EPSCoR Implementation Grant (No. DE-FG02-04ER-46139) and, in addition, support from the SC EPSCoR Office/ Clemson University cost sharing.

References

1 Nolas, G.S., Sharp, J., and Goldsmid, H.J. (2001) *Thermoelectrics Basic Principles and New Materials Developments*, Springer, Berlin.
2 Angrisit, S.W. (1977) *Direct Energy Conversion*, 3rd edn, Allyn and Bacon, Boston.
3 Allen, A.W. (1997) Detector handbook. *Laser Focus World*, **33** (March Issue), S15.
4 Yang, J., and Caillat, T. (2006) Thermoelectric materials for space and automotive power applications. *MRS Bull.* (Special Edition), **31**, 224.
5 Lewis, N.S. (2007) Power the planet. *MRS Bull.*, **32**, 808.
6 Tritt, T.M., Bottner, H., and Chen, L. (2008) *MRS Bull.*, **33**, 366 and the references therein.
7 Seebeck, T.J. (1823) *Abh. K. Akad. Wiss. Berlin*, 265.
8 Peltier, J.C. (1834) *Ann. Chem.*, **LVI**, 371.
9 Vining, C.B. (2001) *Nature*, **413**, 577.
10 Slack, G.A. (1995) *CRC Handbook of Thermoelectrics*, Taylor and Francis, p. 407.
11 Subramanian, M.A., and Tritt, T.M. (2006) *MRS Bull.*, **31**, 188.
12 Chen, G., Yang, B., Liu, W.L., and Zeng, T. (2001) *Proceedings of International Conference on Energy Conversion and Applications*, vol. 1, p. 287.
13 Ioffe, A.F. (1957) *Semiconductor Thermoelements, and Thermoelectric Cooling*, Infosearch Ltd.
14 Mahan, G.D. (1989) *J. Appl. Phys.*, **65**, 1578.
15 Goldsmid, H.J. (1986) *Electronic Refrigeration*, Pion Limited, London.
16 Terasaki, I., Sasago, Y., and Uchinokura, K. (1997) *Phys. Rev. B*, **56**, R12658.
17 Shikano, M., and Fanahashi, R. (2003) *Appl. Phys. Lett.*, **82**, 1851.
18 Sales, B.C., Mandrus, D., and Williams, R.K. (1996) *Science*, **272**, 1325.
19 Nolas, G.S., Cohn, J.L., Slack, G.A., and Schujman, S.B. (1998) *Appl. Phys. Lett.*, **73**, 178.
20 Nolas, G.S., Kaeser, M., Littleton, R.T., IV, and Tritt, T.M. (2000) *Appl. Phys. Lett.*, **77**, 1855.
21 Kauzlarich, S.M., Brown, S.R., and Jeffery Snyder, G. (2007) *Dalton Trans.*, **21**, 2099.
22 Cui, Y., Zhu, T.J., Zhang, S.N., Zhao, X.B., He, J., Su, Z., Ji, X.H., and Tritt, T.M., (2008) *J. Appl. Phys.*, **104**, 013705.

23 Wang, Y., Rogado, N.S., Cava, R.J., and Ong, N.P. (2003) *Nature*, **423**, 425.
24 Casimir, H.B.G. (1938) *Physica*, **5**, 495.
25 Kapitza, P.L. (1941) *J. Phys. Moskow*, **4**, 181.
26 Li, D., Wu, Y., Kim, P., Shi, L., Yang, P., and Marjumdar, A. (2003) *Appl. Phys. Lett.*, **83**, 2934.
27 Dames, C., and Chen, G. (2004) *J. Appl. Phys.*, **95**, 682.
28 Slack, G.A. (1979) *Solid State Physics*, vol. **34** (eds D. Turnbull and H. Ehrenreich), Academic Press, New York, p. 1.
29 Dames, C., and Chen, G. (2006) *CRC handbook of Thermoelectrics*, Taylor and Francis, pp. 42–41.
30 Sharp, J., Poon, J., and Goldsmid, H.J. (2001) *Phys. Stat. Sol.*, **187**, 507.
31 Moyzhes, B.Y., and Nemchinsky, V.A. (1992) *Proceedings of Eleven International Conference on Thermoelectrics, Arlington, TX*, p. 232.
32 Rowe, D.M., and Gao, M. (1994) *Proceedings of Thirteen International Conference on Thermoelectrics, Kansas City*, p. 339.
33 Heremans, J.P. (2005) *Acta Phys. Pol.*, **4**, 609.
34 Hicks, L.D., and Dresselhaus, M.S. (1993) *Phys. Rev. B*, **47**, 16631.
35 Hicks, L.D., and Dresselhaus, M.S. (1993) *Phys. Rev. B*, **47**, 12727.
36 Dragoman, D., and Dragoman, M. (2007) *Appl. Phys. Lett.*, **91**, 203116.
37 Dresselhaus, M.S., Dresselhaus, G., and Avouris, P. (eds) (2000) *Carbon Nanotubes Synthesis, Structure, Properties and Applications*, Topics in Appl. Phys., vol. **80**, Springer-Verlag, Heidelberg.
38 Hone, J., Batlogg, B., Benes, Z., Johnson, A.T., and Fisher, J.E. (2000) *Science*, **289**, 1730.
39 Schwab, K., Hendrikson, E.A., Worlock, J.M., and Roukes, M.L. (2000) *Nature*, **404**, 974.
40 Ren, S.R., and Dow, J.D. (1981) *Phys. Rev. B*, **25**, 3750.
41 Hicks, L.D., Harman, T.C., Sun, X., and Dresselhaus, M.S. (1996) *Phys. Rev. B*, **53**, R10493.
42 Lin, Y.-M., Cronin, S.B., Ying, J.Y., Dresselhaus, M.S., and Heremans, J.P. (2000) *Appl. Phys. Lett.*, **76**, 3944.

43 Culter, M., and Mott, N.F. (1969) *Phys. Rev.*, **181**, 1336.
44 Humphery, T.E., and Linke, T. (2005) *Phys. Rev. Lett.*, **94**, 096601.
45 Bowers, J.D., and Shakouri, A. (1997) *Appl. Phys. Lett.*, **71**, 1234.
46 Koga, T., Sun, X., Cronin, S.B., and Dresselhaus, M.S. (1998) *Appl. Phys. Lett.*, **73**, 2950.
47 Koga, T., Sun, X., Cronin, S.B., and Dresselhaus, M.S. (1999) *Appl. Phys. Lett.*, **75**, 2348.
48 Borca-Tasciuc, T., Liu, W.L., Zeng, T., Song, D.W., Moore, C.D., Chen, G., Wang, K.L., Goorsky, M.S., Radetic, T., Gronskt, R. Koga, T., and Dresselhaus, M.S. (2000) *Superlett. Microstruc.*, **28**, 119.
49 Borca-Tasciuc, T., Achimov, D., Liu, W.L., Chen, G., Ren, H.-W., Lin, C.-H., and Pei, S.S. (2001) *Microsc. Thermophys. Eng.*, **5**, 225.
50 Yang, R., and Chen, G. (2004) *Phys. Rev. B*, **69**, 195316.
51 Dresselhaus, M.S., Chen, G., Tang, M.Y., Yang, R., Lee, H., Wang, D., Ren, Z., Fleurial, J.-P., and Gogna, P. (2007) *Adv. Mater.*, **19**, 1.
52 Li, G. (2008) *Comp. Mech.*, **42**, 593.
53 Chen, G. (2006) *IEEE Trans. Comp. Pack. Tech.*, **29** (2), 238.
54 Yu, C., Shi, L., Yao, Z., Li, D., and Marjumdar, A. (2005) *Nano Lett.*, **5** (9), 1842.
55 Zhang, Y., Christofferson, J., Shakouri, A., Li, D., Marjumdar, A., Wu, Y., Fan, R., and Yang, P. (2006) *IEEE Trans. Nanotechnol.*, **5**, 67.
56 Lyeo, H., Khajetoorians, A.A., Shi, L., Pipe, K.P., Ram, R.J., Shakouri, A., and Shih, C.K. (2004) *Science*, **303**, 5659.
57 Dames, C., Chen, S., Harris, C.T., Huang, J.Y., Ren, Z.F., Dresselhaus, M.S., and Chen, G. (2007) *Rev. Sci. Instrum.*, **78**, 104903.
58 Yang, C.C., Armellin, J., and Li, S. (2008) *J. Phys.Chem. B*, **112**, 1482.
59 Cahill, D., Ford, W.K., Goodson, K.E., Mahan, G.D., Marjumdar, A., Maris, H.J., and Phillpot, S.R. (2003) *J. Appl. Phys.*, **93** (2), 793 and the references therein.

60 Goldsmid, H.J. (2006) *Proceedings of Twenty Fifth International Conference on Thermoelectrics*, p. 5.
61 Lin, Y.-M., Sun, X., and Dresselhaus, M.S. (2000) *Phys. Rev. B*, **62**, 4610.
62 Heremans, J.P., Thrush, C.M., Lin, Y.-M., Cronin, S.B., Zhang, Z., Dresselhaus, M.S., and Mansfield, J.F. (2000) *Phys. Rev. B*, **61**, 2921.
63 Zhang, Z., Ying, J.Y., and Dresselhaus, M.S. (1998) *J. Mater. Res.*, **13**, 1745.
64 Black, M.R., Lin, Y.-M., Cronin, S.B., Rabin, O., and Dresselhaus, M.S. (2002) *Phys. Rev. B*, **65**, 195417.
65 Black, M.R., Hagelstein, P.L., Cronin, S.B., Lin, Y.-M., and Dresselhaus, M.S. (2003) *Phys. Rev. B.*, **68**, 235417.
66 Cornelius, T.W., Toimil-Molares, M.E., Neumann, R., Fahsold, G., Lovrincic, R., Pucci, A., and Karim, S. (2006) *Appl. Phys. Lett.*, **88**, 103114.
67 Wang, Y.M., Kim, J.S., Kim, G.H., and Kim, K.S. (2006) *Appl. Phys. Lett.*, **88**, 143106.
68 Huber, T.E., Nikolaeva, A., Gitsu, D., Konopko, L., Foss, C.A., Jr., and Graf, M.J. (2004) *Appl. Phys. Lett.*, **84**, 1326.
69 Reppert, J., Rao, R., Skove, M., He, J., Tritt, T.M., and Rao, A.M. (2007) *Chem. Phys. Lett.*, **442**, 334.
70 Sivaramakrishnan, S., Muthukumar, V.S., Sai, S.S., Venkataramaniah, K., Reppert, J., Rao, A.M., Anija, M., Philip, R., and Kuthirummal, N. (2007) *Appl. Phys. Lett.*, **91**, 093104.
71 Cronin, S.B., Lin, Y.-M., Gai, P.L., Rabin, O., Black, M.R., Dresselhaus, G., and Dresselhaus, M.S. (2000) *MRS Symp. Proc.*, **635**, C.5.7.1. Cronin, S.B., Lin, Y.-M., Gai, P.L., Rabin, O., Black, M.R., Dresselhaus, G., and Dresselhaus, M.S. (2000) *MRS Symp. Proc.*, **691**, G.10.4.1.
72 Gitsu, D., Komopko, L., Nikolaeva, A., and Huber, T.E. (2005) *Appl. Phys. Lett.*, **86**, 102105.
73 Wu, Y., Yan, H., Huang, M., Messer, B., Song, J.H., and Yang, P. (2002) *Chem. Eur. J.*, **8**, 1260.
74 Hochbaum, A.I., Chen, R., Delgado, R.D., Liang, W., Garnett, E.C., Najarian, M., and Majumdar, A. (2008) *Nature*, **451**, 163.
75 Boukai, A.I., Tahir-Kheli, Y.B.J., Yu, J.K., Goddard, W.A., 3rd, and Heath, J.R.(2008) *Nature*, **451** (10), 168.
76 Vo, T.T.M., Williamson, A.J., Lordi V., and Galli, G. (2008) *Nano Lett.*, **8** (4), 1111.
77 Harman, T.C., Taylor, P.J., Spears, D.L., and Walsh, M.P. (2000) *J. Electron. Mater.*, **29**, L1.
78 Venkatasubramanian, R., Colpitts, T., O'Quinn, B., Liu, S., El-Masry, N., and Lamvik, M. (1999) *Appl. Phys. Lett.*, **75**, 1104.
79 Harman, T.C., Walsh, M.P., LaForge, B.E., and Turner, G.W. (2005) *J. Electron. Mater.*, **34**, L19.
80 Harman, T.C., Taylor, P.J., Walsh, M.P., and LaForge, B.E. (2002) *Science*, **297**, 2229.
81 Venkatasubramanian, R., Silvola, E., Colpitts, T., and O'Quinn, B. (2001) *Nature*, **413**, 597.
82 Ohta, S., Nomura, T., Ohta, H., Hirano, M., Hosono, H., and Koumoto, K. (2005) *Appl. Phys. Lett.*, **87**, 092108.
83 Ohta, H., Kim, S., Mune, Y., Mizoguhi, T., Nomura, K., Ohta, S., Momura, T., Naanishi, Y., Ikuhara, Y., Hirano, M., Hosono, H., and Koumoto, K. (2007) *Nat. Mater.*, **6**, 129.
84 Mune, Y., Ohta, H., Mizoguchi, T., Ikuhara, Y., and Koumoto, K. (2007) *Mater. Res. Soc. Sym. Proc.*, **1044**, 369.
85 Li, D., Wu, Y., Fan, R., Yang, P., and Marjumdar, A. (2003) *Appl. Phys. Lett.*, **83**, 3186.
86 Wu, Y., Fan, R., and Yang, P. (2002) *Nano Lett.*, **2**, 83.
87 BjÖrk, M.T., Ohlsson, B.J., Sass, T., Persson, A.I., Thelander, C., Magnusson, M.H., Deppert, K., Wallenberg, L.R., and Samuelson, L. (2002) *Nano Lett.*, **2**, 87.
88 Lin, Y.-M., and Dresselhaus, M.S. (2003) *Phys. Rev. B*, **68**, 075304.
89 Mahan, G.D., Sofo, J.O., and Bartkowiak, M. (1998) *J. Appl. Phys.*, **83**, 4683.
90 Shakouri, A., Labounty, C., Peprik, J., Abraham, P., and Bowers, J.E. (1999) *Appl. Phys. Lett.*, **74**, 88.
91 Fan, X.F., Zeng, G.H., LaBounty, C., Bowers, J.E., Croke, E., Ahn, C.C.,

Huxtable, S., Marjumdar, A., and Shakouri, A. (2001) *Appl. Phys. Lett.*, **78**, 1580.

92 Garguilo, J.M., Koeck, F.A.M., Nemanich, R.J., Xiao, X.C., Carlisle, J.A., and Auciello, O. (2005) *Phys. Rev. B*, **72**, 165404.

93 Wang, Y.Y., Gupta, S., Garguilo, J.M., and Nemanich, R.J. (2005) *Appl. Phys. Lett.*, **86**, 063109.

94 Koeck, F.A.M., Garguilo, J.M., and Nemanich, R.J. (2005) *Diamond Relat. Mater.*, **14**, 704.

95 Yang, R., and Chen, G. (2005) *Mater. Integr.*, **18**, 31.

96 Zhao, X.B., Ji, X.H., Zhang, Y.H., Zhu, T.J., Tu, J.P., and Zhang, X.B. (2005) *Appl. Phys. Lett.*, **86 (6)**, 2111.p

97 Sahimi, M. (2003) *Heterogeneous Materials I: Linear Transport and Optical Properties*, Springer.

98 Bergman, D.J., and Levy, O. (1991) *J. Appl. Phys.*, **70**, 6821.

99 Ji, X., He, J., Su, Z., Gothard, N., and Tritt, T.M. (2008) *J. Appl. Phys.*, **103**, 054314.

100 He, J., Ji, X., Su, Z., Gothard, N., Edwards, J., and Tritt, T.M. (2008) *MRS Symp. Proc.*, **1044**, 21.

101 Snyder, G.J., and Ursell, T.S. (2003) *Phys. Rev. Lett.*, **91**, 148301.

102 Ji, X., Zhang, B., Su, Z., Holgate, T., He, J., and Tritt, T.M. (2009) *Phys. Stat. Sol. A*, **206**, 221.

103 Ji, X., He, J., Su, Z., Alboni, P., Gothard, N., Zhang, B., and Tritt, T.M. (2007) *Phys. Sol. Stat.*, **1**, 229.

104 Alboni, P.N., Ji, X., He, J., Gothard, N., and Tritt, T.M., (2008) *J. Appl. Phys.*, **103**, 113207.

105 Kim, W., Zide, J., Gossard, A., Klenov, D., Stemmer, S., Shakouri, A., and Marjumdar, A. (2006) *Phys. Rev. Lett.*, **96**, 045901.

106 Shi, X., Chen, L., Yang, J., and Meisner, G.P. (2004) *Appl. Phys. Lett.*, **84**, 2301.

107 Iketa, T., Collins, L.A., Ravi, V.A., Gascoin, F.S., Hile, S.M., and Snyder, G.J. (2007) *Chem. Mater.*, **19**, 763.

108 Zaitsev, V.K., Fedorov, M.I., Gurieva, E.A., Eremdin, I.S., Konstantinov, P.P., Samunin, A.Y., and Vedernikov, M.V. (2006) *Phys. Rev. B*, **74**, 045207.

109 Zhang, Q., He, J., Zhao, X.B., Zhang, S.N., Zhu, T.J., Yin, H., and Tritt, T.M. (2008) *J. Phys. D*, **41**, 185103.

110 Zhang, Q., He, J., Zhu, T.J., Zhang, S.N., Zhao, X.B., and Tritt, T.M. (2008) *Appl. Phys. Lett.*, **93**, 102109.

111 Dresselhaus, M.S., Chen, G., Ren, Z., Fleurial, J.-P., Cogna, P., Tang, M.Y., et al. (2008) *MRS Symp. Proc.*, **1044**, 29.

112 Hsu, K.F., Loo, S., Guo, F., Chen, W., Dyck, J.S., Uher, C., Hogan, T., Polychroniadis, E.K., and Kanatzidis, M.G. (2004) *303, Science*, 818.

113 Androulakis, J., Hsu, K.-F., Pcionek, R., Kong, H., Uher, C., D'Angelo, J.J., Downey, A., Hogan, T., and Kanatzidis, M.G. (2006) *Adv. Mater.*, **18**, 1170.

114 Bilc, D., Mahanti, S.D., Quarez, E., Hsu, K.-F., Pciomek, R., and Kanatzidis, M.G. (2004) *Phys. Rev. Lett.*, **93**, 146403.

115 Heremans, J.P., Thrush, C.M., and Morelli, D.J. (2005) *J. Appl. Phys.*, **98**, 063703.

116 Yang, S.H., Zhu, T.J., Sun, T., He, J., Zhang, S.N., and Zhao, X.B. (2008) *Nanotechnology*, **19 (24)**, 5707.

117 Kim, H.J., Bozin, E.S., Haile, S.M., Snyder, G.J., and Billinge, S.J.L. (2007) *Phys. Rev. B*, **75**, 134103.

118 Peranio, N., and Eibl, O. (2008) *J. Appl. Phys.*, **103**, 024314.

119 Poudel, B., Hao, Q., Ma, Y., Lan, Y., Minnich, A., Yu, B., Yan, X., Wang, D., Muto, A., Vashaee, D., Chen, X., Liu, J., Dresselhaus, M.S., Chen, G. and Ren, Z. (2008) *Science*, **320**, 634.

120 Martin, J., Nolas, G.S., Zhang, W., and Chen, L. (2007) *Appl. Phys. Lett.*, **90**, 222112.

4
Nanomaterials for Fuel Cell Technologies
Antonino Salvatore Aricò, Vincenzo Baglio, and Vincenzo Antonucci

4.1
Introduction

Fuel cells have reached a mature level of technology. These systems now appear ready for electro-traction, portable power sources, distributed power generation and stationary applications [1]. High thermodynamic efficiency and near-zero emission levels make them an attractive alternative to internal combustion engines, batteries and thermal combustion power plants. Like storage batteries, fuel cells deliver energy by consuming electroactive chemicals, but they differ significantly in that these chemicals are delivered on-demand to the cell. As a result, a fuel cell can generate energy continuously and for as long as the electroactive chemicals are provided. Typically, these chemicals consist of hydrogen-rich fuel supplied to the anode and air supplied to the cathode.

One of the main drawbacks limiting the commercial exploitation of these devices is the sluggishness of electrochemical reactions when they occur close to their reversible potentials [1].

The electrochemical processes involved in low temperature fuel cells need noble metal-based electro-catalysts (Pt or Pt-alloys) if they are to occur at significant rates; this is one of the reasons fuel cells are less competitive than other power sources currently on the market [1, 2]. There are different ways to address this drawback: (i) increase operation temperature from the present 80–120 °C while maintaining a solid polymer electrolyte configuration, (ii) search for low-cost non-noble metal based electro-catalysts, (iii) decrease the noble metal loading.

Research on high-temperature fuel cells, such as solid oxide fuel cells, addresses several issues. A decrease in the operating temperature from 800–950 °C to 500–750 °C can allow the use of cheap ferritic stainless steels as interconnectors and increase life-time; direct utilization of hydrocarbon fuels can reduce fuel processing, and new cell architectures capable of sustaining thermal and redox cycles are highly amenable [2].

Polymer electrolyte membrane fuel cells (PEMFCs) recently gained momentum in applications for transportation and as portable power sources, whereas phosphoric acid fuel cells (PAFCS), solid oxide fuel cells (SOFCs) and molten carbonate

Nanotechnology for the Energy Challenge. Edited by Javier Garcia-Martinez
© 2009 WILEY-VCH Verlag GmbH & Co. KGaA, Weinheim
ISBN: 978-3-527-32401-9

fuel cells (MCFCs) still offer advantages for stationary applications, especially in terms of co-generation.

Nanostructured materials have always attracted a great deal of interest in the field due to: (i) the possibility of tailoring electrochemical, thermal and mechanical properties when the dimensions of fuel cell systems are confined and (ii) the strong influence of surface properties on overall behavior [2]. There is a significant effect of spatial confinement and the surface on the physico-chemical characteristics due to small particle size. Nanomaterials, in particular, offer unique properties or combinations of properties in terms of electrodes and electrolytes in fuel cell devices, and new nanostructured materials may allow fundamental advances in fuel cell performance [2]. A new generation of highly efficient and non-polluting energy conversion systems is vital to meeting the challenges of global warming and the finite reality of fossil fuels. This chapter addresses, particularly, the properties of fuel cell components using innovative nanostructured materials and the practical application of such technology.

4.2
Low-Temperature Fuel Cells

Presently, nanostructured materials and processing methods have a strong impact on the development of low-temperature fuel cells (T < 200 °C), particularly in regard to the dispersion of precious metal catalysts, the development and dispersion of non-precious catalysts, fuel reformation and hydrogen storage and the fabrication of membrane-electrode assembly (MEA) [2]. Platinum-based catalysts are the most active materials for low-temperature fuel cells fed with hydrogen, reformate, or methanol [1, 2]. To reduce costs, the platinum loading must be decreased (while maintaining or enhancing MEA performance) and continuous processes for fabricating MEAs in high volume must be developed. A few methods to improve the electrocatalytic activity of Pt-based catalysts are being actively investigated. Pt utilization can be enhanced by increasing either its dispersion on the support or the interfacial region with the electrolyte [1]. Catalyst activity improves by either tailoring the particle size or alloying Pt with transition metals [2].

4.2.1
Cathode Reaction

The oxygen reduction reaction (ORR) limits the performance of low-temperature fuel cells significantly, whereas hydrogen oxidation is much faster and occurs close to the reversible potential [1]. One approach to increase catalyst dispersion involves the deposition of Pt nanoparticles onto a carbon black support. Studies carried out on carbon supported Pt-based electrocatalysts for oxygen reduction in phosphoric acid and PEM fuel cells have shown that electrocatalytic activity (mass activity, $mA\,g^{-1}$ Pt; specific activity, $mA\,cm^{-2}$ Pt electrochemical surface area) depends on the mean particle size [3, 4]. Mass activity reaches a maximum for Pt/C catalysts with a mean particle size of approximately 2.5–3.0 nm. In contrast, specific activity increases gradually with increasing Pt particle size.

These aspects were investigated by Kinoshita [3]. He observed that the mass activity and specific activity for oxygen electro-reduction in acid electrolytes varies with Pt particle size, according to the relative fraction of Pt surface atoms on the (111) and (100) faces [3] (Figure 4.1). The mass-averaged distribution of surface

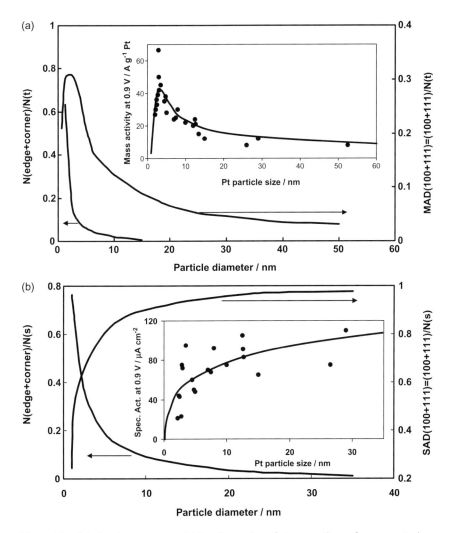

Figure 4.1 Calculated mass-averaged (a) and surface-averaged distributions (b) as a function of particle size in Pt particles with cubo-octahedral geometry. N(t) and N(s) indicate the total number of atoms and the number of atoms on the surface respectively (ref. [3]). The variation of mass activity (a) and specific activity (b) for oxygen reduction in acid electrolyte versus particle size is shown in the inset (ref. [2]).

atoms on the (111) and (100) planes reaches a maximum (~3 nm), whereas the total fraction of surface atoms at the edge and corner sites decreases rapidly with an increase in particle size. In contrast, the surface-averaged distribution for the (111) and (100) planes shows a rapid increase with particle size, which accounts for the increase in experimentally determined specific activity with particle size (Figure 4.1). A dual-site reaction is assumed as the rate-determining step (rds):

$$O_2 + Pt \rightarrow Pt-O_2$$

$$Pt-O_2 + H^+ + e^- \rightarrow Pt-HO_2$$

$$Pt-HO_2 + Pt \rightarrow Pt-OH + Pt-O \text{ (r.d.s.)}$$

$$Pt-OH + Pt-O + 3H^+ + 3e^- \rightarrow 2Pt + 2H_2O$$

This mechanism accounts for the role of dual sites of proper orientation [3].

On the basis of this evidence, surface atoms on the (111) and (100) planes are, thus, more electro-catalytically active than Pt atoms located on high Miller index planes [3]. Similarly, platinum atoms at the edge and corner sites are considered less active than Pt atoms on the crystal faces. Accordingly, in electrocatalysts containing Pt particles smaller than 1.5–2.0 nm in size, both mass and specific activity decrease significantly as the relative fraction of atoms at the edge and corner sites increases. Thus, the maximum in mass activity at about a 2.5–3.0 nm mean particle size is due to an excellent compromise between intrinsic catalytic activity and dispersion [3].

The electrocatalysis of O_2 reduction is also enhanced by alloying Pt with transition metals. The specific activity of Pt–Co, Pt–Fe, Pt–Cr and Pt–Cr–Co alloy electrocatalysts for oxygen reduction is higher than that of platinum in low-temperature fuel cells [1, 4, 5]. This enhancement in electrocatalytic activity has been interpreted differently, and several studies have carried out in-depth analyses of the surface properties of proposed alloy combinations [1, 4, 5]. Although a comprehensive understanding of much reported evidence has not yet been attained, the electrocatalytic effects observed have been ascribed to several factors, such as interatomic spacing, preferred orientation and electronic interactions.

For example, many investigations into PEMFCs have proven that enhanced electrocatalytic activity for the ORR of several binary Pt-based alloy catalysts, such as Pt–M, (where M = Co, Fe, etc.), compared to that of pure Pt [4, 5], can be interpreted in terms of increased Pt d-band vacancy (electronic factor) and a favorable Pt–Pt interatomic distance (geometric effect). Accordingly, a lattice contraction due to alloying would result in a more favorable Pt–Pt distance for the dissociative adsorption of O_2. Since it has already been observed that the rate determining step involves a rupture of the O–O bond through a dual site mechanism, a decrease in Pt–Pt distance favors dual site O_2 adsorption. In addition, the interplay between electronic and geometric factors (Pt d-band vacancy, Pt coordination numbers) and its relative effect on OH chemisorption from the electrolyte occurs [6].

In the case of Pt–Co–Cr, upon thermal treatment the formation of an ordered tetragonal structure (fct) leads to a more active electrocatalyst than a Pt face-centered cubic (fcc) structure [4, 5]. However, due to the thermal treatments required to form alloys (700–800 °C), the state-of-art Pt–Co–Cr electrocatalysts have a particle size of 6 nm [2, 5]. Thus, an increase in intrinsic electrocatalytic activity is often counteracted by a decrease in the surface area of the electrocatalyst. This effect is less significant in the Pt_3Co_1 catalyst, for which small particle size (<3 nm) and a suitable degree of alloying (75 : 25 atm solid solution, fcc structure) can be obtained.

4.2.2
Anodic Reaction

For some applications, PEMFCs are fed with hydrogen produced by a reforming process (e.g., on board methanol reforming to H_2 for transportation). The fuel stream, thus, contains traces of both CO_2 and CO compounds. CO_2 adsorbs on the Pt surface as CO-like species under reducing potential conditions. Linearly adsorbed CO also forms due to residual carbon monoxide in the reformate stream and the direct electro-oxidation of methanol in direct methanol fuel cells (DMFCs). These CO-like species are adsorbed irreversibly on the surface of the electrocatalyst and severely poison Pt, which reduces the electrical efficiency and power density of the fuel cell significantly [6, 7]. It is a known fact that the electrocatalytic activity of Pt is promoted by the presence of a second metal, such as Ru, Sn, or Mo [6, 7]. The mechanism by which such a synergistic promotion of H_2/CO and methanol oxidation reactions occur has been the subject of numerous studies and is still open to debate. The first hypothesis suggested that metal promoters either alter the electronic properties of Pt or act as redox intermediates [8, 9]. A second hypothesis based on bifunctional theory proposed a mechanism by which the oxidation reaction of the poisoning intermediate is enhanced by the adsorption of oxygen or hydroxyl radicals on promoters or adatoms adjacent to the reacting species [10]. Combining the electronic (ligand effect) and bifunctional theories, one can reasonably deduce that the role of the second element is to increase OH adsorption on the catalyst surface at lower overpotentials and to decrease the adsorption strength of the poisoning CO species.

In the case of oxygen reduction, particle size is important for structure-sensitive reactions such as CH_3OH and CO electro-oxidation. The effect of particle size on the anodic reaction should be similar to that of ORR since catalytic activity for Pt–Ru surfaces is also maximized by the presence of (111) crystallographic planes [11]. Data reported by Ren *et al.* [12] shows an increase in mass activity for methanol oxidation with a decrease in particle size and may also be fitted with a "volcano-like" relationship with a maximum of approximately 3 nm and a shape similar to that generally observed in the oxygen reduction reaction. Unfortunately, in high surface area electrocatalysts, scattering due to variable compositions of the Pt–Ru surface does not allow clarity on behavior patterns. Generally, most Pt–Ru fuel cell electrocatalysts have a particle size of above 2 nm and are crystalline with a fcc structure.

For particles 1.0–1.5 nm in size, approximately 50% of the atoms are on the surface and there is no ordered structure. In a recent study, Pt–Ru catalysts characterized by a particle size smaller than 2 nm were investigated for methanol oxidation [13]. The catalytic performance of catalysts whose primary particles had a mean size of approximately 1.0–1.5 nm was quite poor compared to conventional catalysts; also, in that particle size range, the structure was mainly amorphous [13].

According to bifunctional theory, the role of Ru in these processes is to promote water discharge and the removal of strongly adsorbed CO species at low potentials through the following reaction mechanism [6, 10]:

$$Ru + H_2O \rightarrow Ru-OH + H^+ + 1e^-$$

$$Ru-OH + Pt-CO \rightarrow Ru + Pt + CO_2 + H^+ + 1e^-$$

Optimal Ru content in carbon supported Pt–Ru catalysts for methanol oxidation reaction at high temperatures (90–130 °C) was 50 atm% (Figure 4.2). Optimum Ru surface composition is related to relevant synergism accomplished by a Pt–Ru surface with 50% atomic Ru maximizing the product of θ_{OH} (OH coverage) and k

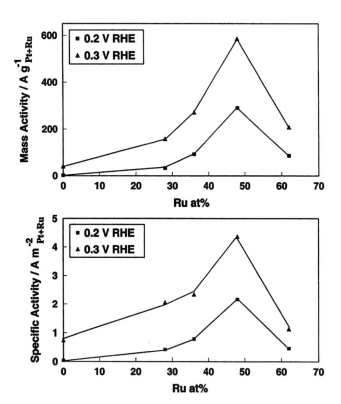

Figure 4.2 Mass and specific activities for methanol (2M, 2 atm back pressure) electro-oxidation on unsupported Pt–Ru catalysts at 130 °C in the presence of a Nafion 112 electrolyte versus XRD-determined bulk composition. The activities were calculated on the basis of the Pt+Ru total loading and XRD-determined metal surface area (ref. [6]).

(intrinsic rate constant) according to an assumed surface reaction between CO_{ads} and OH_{ads} as rds [6].

4.2.3
Practical Fuel Cell Catalysts

In general, preparation procedures for practical fuel cell catalysts, such as impregnation, colloidal deposition and surface reaction, involve the adsorption step of active compounds on a carbon black surface [6]. Synthesis of a highly dispersed electrocatalyst phase in conjunction with a high metal loading on carbon support is one of the goals in the preparation of fuel cell catalysts. X-ray diffraction patterns and transmission electron micrographs for highly dispersed–high metal concentration fuel cell catalysts are shown in Figures 4.3–4.5, respectively.

A large broadening of the X-ray diffraction peaks of the fcc structure clearly indicates small crystallite size for the supported metal phase. Crystallite size varies with metal or alloy compositions, the concentration of metal phase on the support and the type of carbon black support (Figure 4.3). The most commonly used carbon blacks are acetylene black (BET area: $50\,m^2\,g^{-1}$), Vulcan XC-72 (BET area: $250\,m^2\,g^{-1}$) and Ketjen black (BET area: $\sim900\,m^2\,g^{-1}$). All of these materials have optimal electronic conductivity but, as noted above, they differ in BET surface area, thus, most probably in morphology as well. A low surface area carbon black (such as acetylene black or Vulcan) does not allow high dispersion of the metal phase,

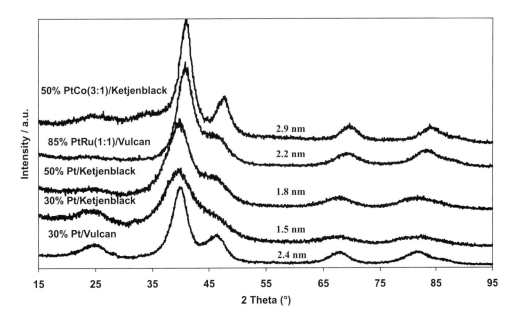

Figure 4.3 XRD patterns of various fuel cell catalysts (refs. [6, 14]).

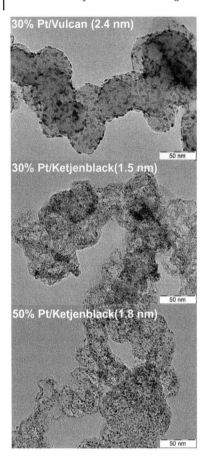

Figure 4.4 TEM micrographs of different carbon-supported Pt catalysts (refs. [6, 14]).

especially with a high metal loading, that is, low carbon content (Figure 4.4); however, a low surface area carbon does not generally have micropores in its structure that could hinder mass transport through the electrocatalytic layer. A high surface area carbon black can accommodate a high amount of metal phase (Figure 4.4) with a high degree of dispersion easily, but at the same time a significant amount of micropores on the carbon support does not allow homogeneous distribution of the electrocatalytic phase through the support, which could lead to mass transport limitations for the reactant as well as limited access to inner electrocatalytic sites. In the same preparation method, synthesis of Pt-M alloys often requires high-temperature treatments for suitable solid solutions to form. Such treatments have a strong influence on catalyst morphology (Figure 4.5).

Recently, carbon nanotubes and nanostructured mesoporous carbon supports have been considered a suitable alternative to conventional carbon blacks. Carbon nanotubes (CNTs) have been used as catalyst supports in electrocatalysis as

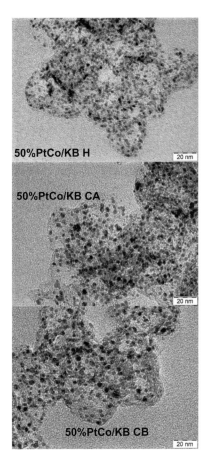

Figure 4.5 TEM micrographs of 50% Pt–Co catalysts supported on Ketjenblack (Pt$_3$Co; refs. [6, 14]).

well as in fuel cell research due to their unique properties, such as high external surface area, good mechanical strength and resistance to corrosion phenomena. Some CNT-supported catalysts, for example, Ru, Pt, Rh, Pd and Pt-based alloys, have shown good activity and/or selectivity in a variety of chemical reactions [15]. Generally, this has been attributed to improved metal support interaction, mass transfer, or catalyst chemical states induced by the CNTs [15]. One example of a Pt–Fe alloy catalyst supported on multiwalled carbon nanotubes (MWCNTs) for the oxygen reduction reaction (ORR) is shown in Figure 4.6. The main advantage of using carbon nanotubes is improved mass transfer characteristics for electrochemical reactions and enhanced stability with respect to carbon blacks containing significant amounts of amorphous carbon. However, there is a significant increase in costs associated wtih the use of CNTs and the performance achieved with them is not significantly better than with conventional carbon supports.

Recent developments in the field of CO tolerant catalysts include the synthesis of new catalyst nanostructures by the spontaneous deposition of Pt submonolayers

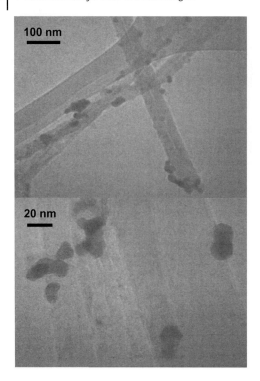

Figure 4.6 XRD patterns of Pt-based electro-catalysts supported on MWCNs (ref. [15]).

on carbon supported and unsupported Ru nanoparticles [16]. This also appears to be an efficient approach to reducing the Pt loading [17] (Figure 4.7).

Further advancements concern a better understanding of the surface chemistry of electrocatalyst nanoparticles and the effects of strong metal–support interaction on the electronic nature of platinum sites. Surface functional groups on carbon may also act as anchoring centers for metal particles, limiting their growth and enhancing dispersion. In the last few decades, Pt loading in low-temperature fuel cells has been reduced by almost one order of magnitude. This was made possible by the development of high surface area catalysts and by increasing the catalyst–electrolyte interface [1, 2]. As discussed above, the preparation of high surface area catalysts implies dispersion of catalytically active nanoparticles on electrically conductive supports such as high surface area carbon blacks or carbon nanotubes which, at the same time, assure good electrical conductivity through their π-graphitic basal planes. Because of the peculiar nature of Pt or Pt-alloy nanoparticles and the synthesis procedures used for catalyst preparation, the chemical interaction between the active phase and support modifies electro-catalytic properties with respect to the bulk metal. The metal–support interaction involves a chemical bond between the metal atoms and carbon functional groups. This charge transfer mechanism can be described in terms of electron accepting/donat-

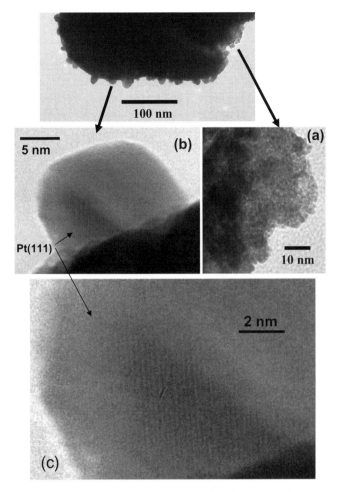

Figure 4.7 Transmission electron micrographs of Pt-decorated Ru catalyst with specific enlargements of the supporting Ru phase (a) and decorating Pt particles (b, c; ref. [17]).

ing characteristics, and the effects of this interaction can be referenced in the Bronsted–Lewis acid–base theory.

An alternative approach to carbon blacks and carbon nanotubes is the fabrication of porous silicon catalyst support structures with 5 μm pore diameters and a thickness of about 500 μm. These structures have high surface areas and are of interest to miniature PEM fuel cells [18]. Finely dispersed, uniform distribution of nanometer-scale catalyst particles deposited on the walls of the silicon pores creates an efficient, three-dimensional electrode capable of high power generation in direct methanol fuel cells (DMFC) [18]. Established silicon wafer fabrication methods can be used to create catalyzed, porous silicon electrodes.

4.2.4
Non-Precious Catalysts

Only a few formulations without noble metals have been tested as anodes. These have been based mainly on transition metal alloys like NiZr, transition metal oxides and tungsten-based compounds [6]. All of these materials showed lower reaction rates than Pt-based electrocatalysts; such unsatisfactory preliminary results did not stimulate much more work in this direction.

For cathode reaction, as alternatives to platinum, organic transition metal complexes, for example, iron or cobalt organic macrocycles from the phenylporphyrin family [19] and nanocrystalline transition metal chalcogenides [20], are being actively investigated for oxygen reduction reaction, the latter compounds in relation, especially, to their high selectivity for ORR and tolerance to methanol cross-over in DMFC. Mainly, these materials consist of the Chevrel-phase type ($Mo_4Ru_2Se_8$), transition metal sulfides ($Mo_xRu_yS_z$, $Mo_xRh_yS_z$), or other transition metal chalcogenides [$(Ru_{1-x}Mo_x)SeO_z$] [6, 20]. Some of the materials possess semiconducting properties, thus they need to be dispersed as nanoparticles on electron-conducting supports such as carbon blacks, acting as current collectors. The metal–organic macrocyclic is supported on high surface area carbon and treated at high temperatures (500–800 °C); the nanostructured residue, which possesses a specific geometrical arrangement of atoms, exhibits suitable electrocatalytic activity without any degradation in performance even though the level of activity is lower than Pt [19].

4.2.5
Electrolytes

Perfluorosulfonic polymer electrolyte membranes (e.g., Nafion®) are presently employed in H_2/air, methanol/air fuel cells due to their excellent conductivity and electrochemical stability [1, 6]. The main problems connected with these electrolytes are membrane dehydration at high temperatures and large methanol/ethanol cross-over in liquid fuel-fed systems. The first of these drawbacks severely hinders fuel cell operation above 100 °C, a pre-requisite temperature for the suitable oxidation of small organic molecules. In fact, this oxidation process involves the formation of strongly adsorbed reaction intermediates such as CO-like species [6]. High temperature operation is also a solution to water and thermal management constraints in automotive applications. In response to this, alternative membranes, including sulfonated poly (ether ether ketone) and poly (ether sulfone) electrolytes and acid-doped polyacrylamid and polybenzoimidazole, have been suggested [6]. However, at present, there is no suitable polymer electrolyte membrane that fulfils the automotive requirements of high performance and stability in a wide temperature range (from −20 °C to 150 °C).

Various relationships between membrane nanostructure and transport characteristics, such as conductivity, diffusion, permeation and electro-osmotic drag, have been observed [2, 21]. Interestingly, compared to Nafion, the presence of less connected hydrophilic channels and a wider separation of sulfonic groups in sulphonated poly(ether ketone) reduces water permeation and electro-osmotic drag while maintaining high protonic conductivity [21]. Furthermore, an improvement in thermal and mechanical stability has been shown in nano-separated acid–base polymer blends obtained by combining polymeric N-bases and polymeric sulfonic acids [22]. In the past decade, considerable efforts have addressed the development of composite membranes [23–27]. These include ionomeric membranes modified by dispersing insoluble acids, oxides, zirconium phosphate and so on inside their polymeric matrix; other examples are ionomers or inorganic solid acids with high proton conductivity embedded in porous non-proton-conducting polymers [23–27]. Alberti and Casciola [27] prepared nanocomposite electrolytes by *in situ* formation of insoluble layered Zr phosphonates in ionomeric membranes. Such compounds, for example $Zr(O_3P-OH)(O_3P-C_6H_4-SO_3H)$, show levels of conductivity that are much higher than Zr phosphates and comparable to Nafion. In an attempt to reduce the drawbacks of perfluorosulphonic membranes, nanoceramic fillers have been included in polymer electrolyte networks. Stonehart, Watanabe and co-workers [23] successfully reduced H_2 cross-over and humidification constraints in PEMFCs by the inclusion of small amounts of SiO_2 and Pt/TiO_2 (~7 nm) nanoparticles that retain electrochemically produced water inside the membrane and favor a recombination of permeated gases. Similarly modified membranes containing nanocrystalline ceramic oxide filler have been demonstrated in operating conditions up to approximately 150 °C [24–26]. Enhancement of mechanical properties and a decrease in methanol cross-over have also been envisaged [24–26]. The inorganic filler induces structural changes in the polymer matrix, improving mechanical properties, whereas the water retention mechanism appears more likely in the presence of acidic functional groups on the surface of nanoparticle fillers [26]. These aspects are discussed in further detail in the following section.

4.2.6
High-Temperature Polymer Electrolyte Membranes

A low-cost, high-temperature membrane with suitable ionic conductivity and stability in a wide temperature range (from −20 °C to 150 °C) would be a potential solution to some of the drawbacks presently affecting hydrogen or reformate-fueled polymer electrolytes (PEMFCs) and direct methanol fuel cells (DMFCs) [2, 6]. Fuel cell operation at elevated temperatures can limit the effects of electrode poisoning by adsorbed CO molecules, increase both methanol oxidation and oxygen reduction kinetics and simplify water and thermal management. Furthermore, high-temperature operation can reduce the complexity of the reforming reactor employed for PEMFCs [1]. A temperature range between 130 and 150 °C is the ideal operating condition for the application of these systems in electric vehicles and for distributed power generation. However, for automotive application, an efficient cold start-up as well as a rapid one is also appropriate.

Various proton-conducting polymer electrolyte materials have been investigated for high-temperature operation. Two categories of membranes can be proposed depending on whether water is required for proton conduction or not. Polymer electrolytes involving water molecules in the proton mobility mechanism (e.g., perfluorosulfonic membranes) require humidification to maintain suitable conductivity characteristics. The amount of humidification may vary depending on the operating temperature and membrane properties; it influences the size and complexity of the device. Some other electrolytes do not necessarily involve water molecules in the proton conduction mechanism (e.g., PBI/H_3PO_4 [28], blends of PBI and polysulfone [29], hybrids of polymers, proton conducting inorganic compounds such as $Zr(HPO_4)_2$ [27], etc.); these systems do not necessarily need humidification. However, there are some drawbacks such as: (i) short-term stability of the system, (ii) phosphoric acid leakage from the membrane during operation, (iii) poor extension of the three-phase reaction zone inside the electrodes due to the absence of a proper ionomer, (iv) reduced conductivity levels for inorganic proton conductors. These problems decrease the utilization of water-free protonic electrolytes in low temperature fuel cells. Alternatively, as noted in the previous section, composite perfluorosulfonic membranes containing different types of inorganic nano-fillers, such as hygroscopic oxides [24, 26, 30, 31], surface modified oxides [32], zeolites [33], inorganic proton conductors, [34] etc., show increased conductivity with respect to bare perfluorosulfonic membranes at high temperatures. The mechanism that enhances proton conduction at such temperatures is presently a subject of debate.

There is evidence that such an effect is due mainly to the water retention capability of nano-fillers [31], which adsorb and retain water physically on the surface at temperatures close to those ideal for PEMFC operation in automotive applications. In the adsorption process, the first layer involves a chemical interaction between the surface sites of the nanofiller and water. Generally, this causes water displacement on the surface with the formation of a chemical bond between water residues and filler functional groups [26]. Additional layers of adsorbed water may form subsequently by physical interaction involving Van der Waals bonds. In this case, no displacement of water should occur. Such bonds become weaker as the distance between physically adsorbed water and the surface increases. Chemically adsorbed water can involve up to a monolayer, whereas physical adsorption and water condensation in the pores may build up a shell of water molecules surrounding the primary particles and agglomerates (Figure 4.8) of the inorganic filler [31]. Most of these inorganic materials have intrinsically low proton conduction up to 150 °C. They can be loaded with proper dispersion in amounts of up to 3–5% inside the membrane without affecting conductivity at or below 90 °C significantly. However, an increase in operating temperature is made possible by the presence of the filler [31]. Proper distribution of the nanoparticle filler in the membrane water channels can maximize the effect of water retention in the conduction path at high temperatures.

Besides extending the operation of perfluorosulfonic membranes (e.g., Nafion®) in high temperature ranges, the presence of hygroscopic inorganic oxides inside

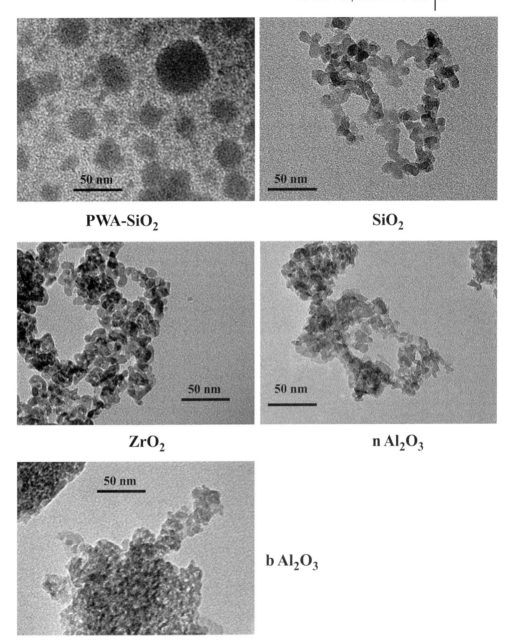

Figure 4.8 Transmission electron micrographs of various inorganic fillers employed in the preparation of composite Nafion recast membranes (refs. [26, 31]).

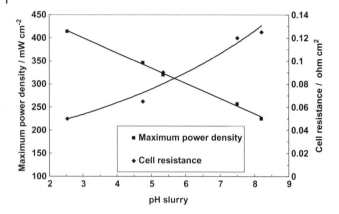

Figure 4.9 Variation of the maximum power density and cell resistance of composite membrane-based DMFC at 145 °C as a function of the pH of slurry of the inorganic filler. Methanol feed 2 M, 2.5 atm; oxygen feed 2.5 atm. Pt loading 2 ± 0.1 mg cm^{-2}. (refs. [26, 31]).

the composite membrane reduces cross-over effects by increasing the "tortuosity factor" in the permeation path [23–25]. Such effects are particularly important in DMFC systems.

A series of composite membranes based on recast Nafion ionomer containing different inorganic nanoparticle fillers (SiO$_2$, phosphotungstic acid-impregnated SiO$_2$, ZrO$_2$, Al$_2$O$_3$) and varying mainly in terms of their acid–base characteristics have been prepared and investigated in direct methanol fuel cells (Figure 4.8). It was observed that the acid–base properties of inorganic fillers play a key role in the water uptake of composite Nafion based-membranes at temperatures close to 150 °C by influencing the proton conductivity of the electrolyte. The presence of acidic OH groups on the filler surface facilitates water coordination, which acts as a vehicle molecule for proton migration. The DMFC performance of various M&E assemblies based on composite membranes containing fillers with different acid-base characteristics increases as the pH of the slurry of the inorganic filler decreases (Figures 4.9 and 4.10). This evidence indicates that the ionic conductivity of the composite membranes and their range of operation may be increased by the appropriate tailoring of the surface characteristics of ceramic oxides inside the membrane.

For materials characterized by the same type of surface functional groups, the effect of the filler surface area determines the water retention properties of composite membranes at high temperatures. This effect appears to be associated with the larger number of water-adsorbing acidic sites on the filler surface [31]. As expected, surface properties play a more important role than the crystalline structure of the filler since the water molecules, acting to promote proton migration, are effectively coordinated by the surface groups.

The conductivity and performance of composite perfluorosulfonic membranes in DMFCs is strongly related to surface acidity which, in turn, influences the

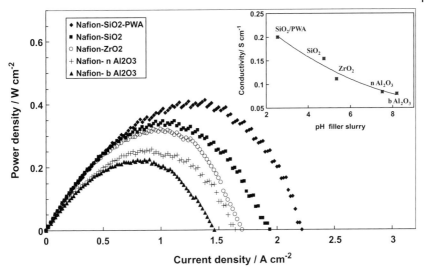

Figure 4.10 DMFC power density curves at 145 °C for MEAs containing different inorganic fillers. Methanol feed 2 M, 2.5 atm; oxygen feed 2.5 atm. Pt loading 2 ± 0.1 mg cm^{-2}. The inset shows the variation of membrane conductivity at 145 °C as a function of the pH of slurry of the filler (refs. [26, 31]).

characteristics of the water physically adsorbed on the inorganic filler surface. It has been observed that the more acidic the nanofiller surface is, the higher is the capability of a strong interaction with water through the formation of hydrogen bonds is. This produces a decrease in O–H stretching and bending frequencies in the physically adsorbed water (Figure 4.11). Furthermore, both an increase in water uptake in the composite membrane and enhanced proton conductivity are observed in the presence of acidic fillers. Proton migration inside the membrane appears to be assisted by water molecules on the surface of the nanofiller particles and could be also encouraged by the formation and breaking of hydrogen bonds.

Conventional ion-exchange perfluoropolymer membranes, such as the well-known Nafion® membrane, are based on long-side-chain polymers (LSC). In the past decade, Solvay Solexis developed a new short-side-chain (SSC) proton conducting perfluoropolymer membrane, that is, Hyflon® ion, characterized by excellent chemical stability and equivalent weight (850 g/eq.) lower than conventional Nafion 117 (1100 g/eq.) [35]. Besides improved conductivity due to the higher degree of sulfonation, the short-side-chain Hyflon® Ion ionomer is characterized, in protonic form, by a primary transition at approximately 160 °C, whereas the conventional Nafion shows this transition at about 110 °C. This characteristic of the Hyflon® ion membrane ensures proper operation at high temperatures (100–150 °C) provided that a sufficient amount of water is supplied to the membrane or retained inside the polymer under these conditions [35]. In principle, the water uptake properties of sulfonic acid-based membranes may be modulated by select-

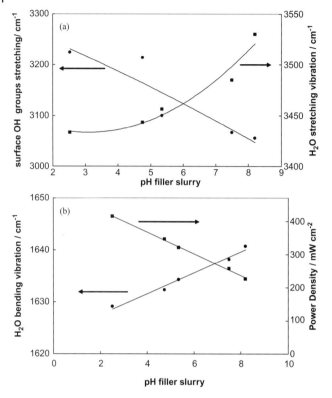

Figure 4.11 Variation of O–H stretching vibration frequencies for surface OH functionalities and physically adsorbed water versus the pH of slurry of the fillers (a); variation of O–H bending vibration frequencies of physically adsorbed water and DMFC maximum power density versus the pH of slurry of the fillers (b) (ref. [26]).

ing the proper concentration and distribution of sulfonic groups inside the polymer. In this regard, the Hyflon® ion membrane, due to its equivalent weight lower than conventional polymers, is favored for high temperature operation. Hyflon® ion membranes were investigated in PEFCs for automotive applications in the framework of the Autobrane FP6 EU project. PEFC assemblies based on these membranes showed low cell resistance at high temperature (110 °C) and more promising performance than conventional membranes (Figure 4.12).

4.2.7
Membrane-Electrode Assembly (MEA)

In PEMFCs, the Pt/C catalyst is intimately mixed with the electrolyte ionomer to form a composite catalyst layer extending the three-phase reaction zone. Networks for both electron and ion conduction are present in the layer, and the interfacial region between catalyst particles and polymer electrolyte micelles (e.g., Nafion

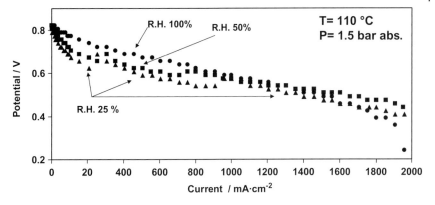

Figure 4.12 Steady-state polarization curves at different relative humidities for a high temperature polymer electrolyte fuel cell based on Hyflon (Solvay-Solexis) membrane and CNR-ITAE catalysts. The effect of internal humidification is observed at high current densities and low RH (Autobrane FP6 EU Project).

Figure 4.13 SEM micrograph of a composite (catalyst, ionomer) catalytic layer for PEFCs.

ionomer) is enhanced [36]. As shown above, the carbon black support is generally composed of spherical particles with an average size of 20 nm (Vulcan XC), allocating Pt nanoparticles with a mean size ranging from 2 to 4 nm [14], whereas the size of Nafion micelles ranges between 50 and 200 nm [6] (Figure 4.13). Using such an approach, a reduction in total Pt content to less than $0.5\,mg\,cm^{-2}$ without degrading cell performance or life-time has been demonstrated [36]. However, one of the main disadvantages of this method is that the ionomer does not soak deeply into the smaller pores of the active layer as it would in the case of a liquid electrolyte. Thus, the reaction area is limited to an interface between Pt particles distributed on the outer surface of carbon agglomerates and the ionomer [6]. As an

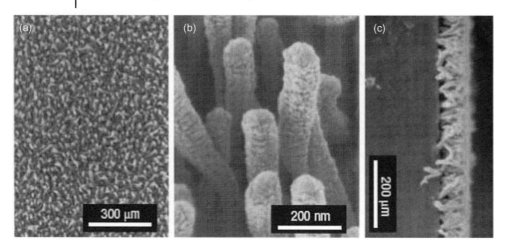

Figure 4.14 3M Corporation Pt-coated nanostructured whisker supports (0.25 mg cm^{-2}), at 10 000× plane view (a) and 150 000×, 45° view (b). The nanostructured film of the MEA (c) shows the Pt-coated nanowhiskers sandwiched between the PEM and the gas diffusion layer (refs. [2, 37]).

alternative approach, durable multi-level MEAs (including gaskets) have been developed (3M Corporation) using high-speed precision coating technologies and an automated assembling process [37]. Part of the MEA is a nanostructured thin-film catalyst (NSTF) using highly oriented high aspect ratio single crystalline whiskers of an organic pigment material with a density of 3×10^9 to 5×10^9 whiskers cm^{-2}. This support allows for high specific activity of the applied catalysts and facilitates processing and manufacturing. Platinum-coated nano-whiskers and a cross-section of the MEA are shown in Figure 4.14. However, electrocatalytic activities thus obtained are comparable to catalyst–ionomer inks.

4.3
High-Temperature Fuel Cells

The trend towards nanomaterials is not limited to low-temperature fuel cells. Nanostructured electro-ceramic materials are increasingly employed in intermediate temperature solid oxide fuel cells (IT-SOFCs). Although primary nanosized particles in SOFC materials are generally not stable due to the thermal treatments required for cell fabrication (>1000 °C), they often form nanostructured microstructure fuel cell components with electrocatalytic and ion conduction properties that are different from those typical of polycrystalline materials [38]. Nanosized YSZ (8% Y_2O_3–ZrO_2) and ceria-based (CGO, SDC, YDC) powders allow significant reduction in firing temperatures during the membrane forming step in SOFC cell fabrication procedures due to their sintering facility with respect to polycrystalline powders [39].

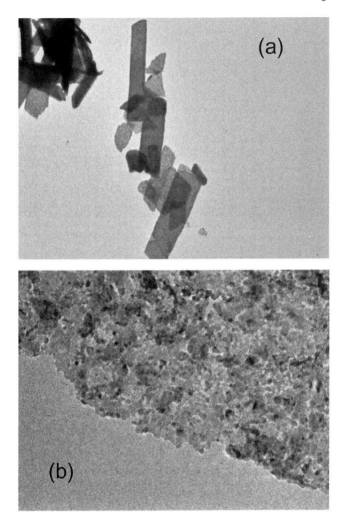

Figure 4.15 TEM micrographs of gadolinia doped ceria nanopowders obtained by co-precipitation (a) and sol-gel method (b).

For the manufacturing of bulk ceramics, coatings, films and composites, the use of fine ceramic powders as precursors is of fundamental interest. Various approaches to synthesis using wet chemistry, often called the chemical route, lead to good homogeneity of the raw materials due to reagent mixing at the molecular level. The resulting oxide powders have a high specific surface area and, consequently, high beneficial reactivity. However, for large scale production of nano-sized ceramic powders, chemical combustion synthesis is widely used [40]. Figure 4.15 shows raw ceramic nanopowders used to manufacture dense thin film electrolyte membranes (Figure 4.16).

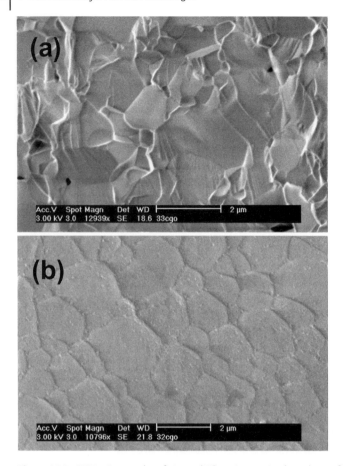

Figure 4.16 SEM micrographs of sintered (dense) ceramic electrolytes of gadolinia doped ceria obtained from a nanopowder firing process. (a) and (b) refer to raw nanopowders obtained by co-precipitation and sol-gel, respectively.

The nanocrystalline ceria, which is characterized by mixed electronic–ionic conducting properties, promotes charge transfer reactions at the electrode–electrolyte interface [39]. By decreasing the particle size in the electroceramics, the quantum confinement effect causes an increase in the band gap and, thus, favors the occurrence of a purely ionic domain. However, such effects no longer occur with grain sizes below approximately 10 nm for YSZ and 50 nm for CeO_2 [38]. In addition, the presence of non-stoichiometric effects (e.g., reduction of Ce^{4+} to Ce^{3+} in the presence of fuel at the anodic compartment) produces a significant increase in the electronic conduction in ceria, resulting in mixed ionic–electronic conduction. These properties provide benefits in terms of promoting electrochemical activity in the electrode, especially for the direct electrochemical oxidation of hydrocarbons [41, 42], yet the decrease in ion transport number for the electrolyte reduces fuel cell efficiency due to the presence of a parasitic electron drain through the ceramic membrane. The occurrence of Ce^{3+} ions under reducing conditions is favoured by an increase in temperature [39]. Accordingly, the electro-catalytic domain of ceria-based electrolytes spans a smaller range of partial pressures and temperatures with respect to YSZ, thus restricting the practical operating range of ceria electrolyte-based SOFCs to less than 650–700 °C.

The ionic charge carriers in electro-ceramic materials are due, essentially, to the presence of point defects. Thus, an increase in the density of mobile defects in the space charge region as a consequence of a significantly larger interface area and grain boundaries in nanostructured systems leads to completely different electrochemical behavior from that of bulk polycrystalline materials [43, 44]. According to Maier [43], these "trivial" size effects, which are due simply to increased proportions of the interface, are distinguished from the "true" size effects that occur when the particle size is four times smaller than the Debye length. In the latter case, local properties in terms of ionic and electronic charge carrier transport change also.

4.3.1
High-Temperature Ceramic Electrocatalysts

Similar to low-temperature fuel cells, electrochemical reaction in high-temperature solid oxide fuel cells occurs at the triple-phase boundary [39, 45]. This is the interface formed by the electronic phase, ionic conducting ceramics and the gas phase. A suitable extension of the three-phase reaction zone is obtained by the fabrication of a porous Ni metal–ceramic electrolyte (YSZ) composite (cermet) and perovskite ($La_{0.6}Sr_{0.4}MnO_3$)–ceramic electrolyte (YSZ) layer for the anode and the cathode, respectively. These composite electrodes are manufactured in order to allow both ionic and electronic percolation [39]. The use of mixed electronic–ionic conductivity perovskites, for example, $La_{0.6}Sr_{0.4}Fe_{0.8}Co_{0.2}O_3$ (LSFCO), favors the percolation of charge carriers within the electrode layer. In SOFCs, electrochemical reactions occur, essentially, at the electrode functional layers (a few tens of microns thick) that are in contact with the ceramic electrolyte membrane. The functional layers contain fine ceramic particles of both electronic and ionic conducting materials. These functional layers can be distinguished from the thick support or current collector layers that contain large-size ceramic particles. An example of the structure of an electrode-supported SOFC cell is shown in Figure 4.17.

A high-surface-area electrocatalyst in the functional layer is necessary for enhancement of the reaction rate at intermediate temperatures (500–750 °C) and promotion of the direct electrochemical oxidation of hydrocarbons [45]. High-surface-area electrocatalysts for SOFCs can also be manufactured by infiltration or impregnation methods [42]. For example, the skeleton of an SOFC cell can be manufactured in the conventional mode by high-temperature firing (1200–1400 °C) to obtain a full-density ceramic electrolyte capable of avoiding gas permeation. A porous structure for the electrodes is obtained by using pore formers and, for the

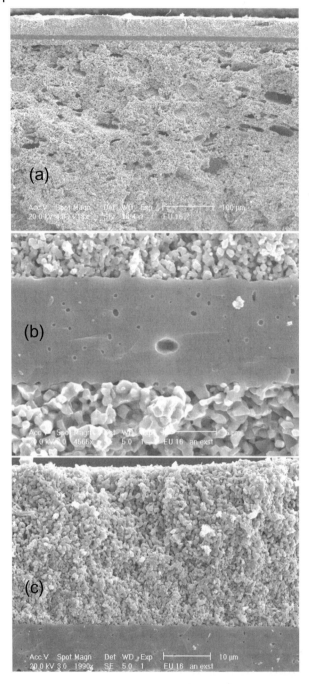

Figure 4.17 SEM micrographs of an electrode supported solid oxide fuel cell. (a) Complete cell, (b) electrode-electrolyte interface, (c) functional (electrolyte interface) and current collector (external) layers characterized by different particle sizes.

anode only, by subsequent *in situ* reduction as well. The porous electrodes can be infiltrated by nanosized ceramic electrocatalysts that are subsequently fired at temperatures significantly lower than those required to achieve a full-density ceramic electrolyte or a good electrode–electrolyte interface [42]. Using another approach, highly dispersed electrocatalyst nanoparticles are impregnated on electronically conducting ceramic materials and stabilized by interaction with the support [46].

Due to the fact that SOFCs operate at high temperatures, the particle size of ceramic electrocatalysts is more than one order of magnitude larger than Pt catalysts used in low-temperature fuel cells. Nevertheless, since the electrochemical reaction rate is greatly enhanced by high-temperature operation, the effect of the high temperatures compensates for the lower number of catalytic sites. Thus, the electrochemical performance of state of the art PEFCs fed with hydrogen, in comparison, does not differ much. However, in the case of SOFCs, no noble metals catalysts are used (essentially Ni at the anode, LSM or LSFCO at the cathode) and a large variety of fuels, including natural gas, can be fed to the anode using internal reforming or direct oxidation processes. Unlike PEFCs, SOFCs are quite limited in terms of cold and rapid start-up as well as thermal and redox cycles. These limitations have reduced perspectives on their application in transportation and portable power generation, but they are quite appealing for the stationary and distributed generation of electrical energy.

4.3.2
Direct Utilization of Dry Hydrocarbons in SOFCs

The direct oxidation process for dry hydrocarbons in SOFCs shows higher intrinsic efficiency than internal reforming [41, 42, 45, 46]. Steam reforming requires an excess of water injected into the stream to avoid the formation of carbon deposits in the presence of Ni-cermet anodes [46]. However, at intermediate temperatures, a high steam:carbon (S/C) ratio reduces the process efficiency and increases water-management constraints [46]. The cermets used in conventional SOFC anodes contain up to 65vol% of metallic Ni as the electron-conducting phase and yttria-stabilized zirconia as the ion-conducting phase, to form a three-phase reaction zone with reactant gas [39]. It is well known that Ni catalyzes the cracking of hydrocarbons, causing the formation of carbon fibers that poison the anode surface [41, 42]. Thus, for the direct oxidation of hydrocarbons in intermediate temperatures (500–800 °C) solid oxide fuel cells (IT-SOFCs), alternative SOFC electrocatalysts are required [45, 46]. These include low-Ni-load electrodes and various catalyst formulations, for example, Cu/CeO_2, Ni alloys such as Ni–Cu and various perovskites [41, 42, 45, 46]. Most efforts have focused on the direct oxidation of methane even though some attempts have addressed the direct oxidation of larger molecular weight hydrocarbons [46]. However, because reaction rates in the direct oxidation process are much slower than those in internal reforming, it is necessary to use nanosized electrocatalysts characterized by high surface area. An example of nanosized electrocatalysts used in direct oxidation processes is shown in Figures

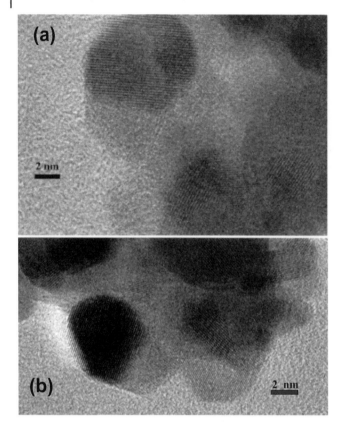

Figure 4.18 TEM micrographs of NiCu precursor (a) and catalyst (b) used in IT-SOFC anodes for the direct oxidation of methane (ref. [45]).

4.18 and 4.19. To avoid carbon deposition at the Ni sites, Ni is alloyed with Cu [45] to mitigate the formation of poisoning carbon nanofibers (Cu is quite inert in this process), or Ni is used in the form of an oxide-phase like La_2NiO_4, [46] highly dispersed on a conductive support. The main requirement is stablility in the reducing environment at the anode compartment.

The carbon formation mechanism on Ni-based catalysts involves the deposition of carbon species onto the metal surface, the dissolution of carbon in the bulk and, subsequently, the growth of carbon fibers at the metal particles. The retardation of carbon deposition in the presence of alkaline earth oxide additives was recently investigated for SOFCs. This effect is ascribed mainly to the greater availability of oxygen species at the metal surface due to spillover from the support [47]. In this regard, tailoring of the ceramic electrocatalyst surface could mitigate the carbon deposition process while maintaining high reaction rates as a result of a high surface area.

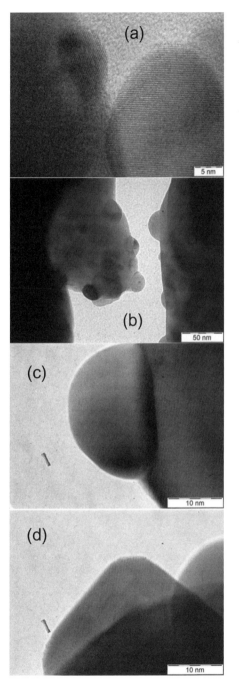

Figure 4.19 TEM micrographs of Ni-modified perovskite anode catalyst for the direct oxidation of propane in IT-SOFC (ref. [46]). (a) Perovskite electronic conducting support, (b) Electrocatalyst, (c) and (d) Ni-phase nanoparticles of different shapes deposited on the perovskite support.

4.4
Conclusions

The development of efficient, non-polluting, inexpensive fuel cell devices is fundamental to future sustainable energy conversion processes. Although performance levels as well as system efficiency require further improvement, recent advances in the development of nanostructured material-based fuel cell components may provide the solution to combining this technology with efficient hydrogen storage systems or fuel distribution infrastructures. This technology may find application in next generation electro-traction vehicles or portable, distributed and stationary electrical energy generation.

Numerous studies indicate that the preparation of appropriate nanocomposite and nanostructured materials allows a fine tuning of electrocatalytic and ion transport properties for both low and high temperature fuel cell components. As discussed above, clearly, there is an optimum particle size for each application. Materials with extremely small particle size do not possess the specific geometric arrangements of atoms required for active sites in electrocatalysts, and they are often unstable during fuel cell operation and, therefore, subject to quick growth or dissolution. In ceramic fuel cell devices, an excessive number of grain boundaries, which often contain impurities, produces blocking effects on ion transport or causes undesired electronic conductivity in materials in which a purely ionic domain is required. Notwithstanding the paramount importance of nanostructured materials, one should also consider that the performance of practical fuel cell stack devices is still significantly affected by issues such as scale-up problems, stack housing design and materials, and gas manifold and sealing. Thus, the performance of practical systems still deviates considerably from that of small ideal single cells. Nevertheless, passing from bulk materials to the nanoscale, it is possible to change electrode and electrolyte properties significantly and, consequently, improve performance in devices for energy conversion and storage.

References

1 Srinivasan, S., Mosdale, R., Stevens, P., and Yang, C. (1999) Fuel cells: reaching the era of clean and efficient power generation in the twenty-first century. *Annu. Rev. Energy Environ.*, **24**, 281–238.

2 Aricò, A.S., Bruce, P., Scrosati, B., Tarascon, J.-M., and Van Schalkwijk, W. (2005) Nanostructured materials for advanced energy conversion and storage devices. *Nat. Mater.*, **4** (5), 366–377.

3 Kinoshita, K. (1990) Particle size effects for oxygen reduction on highly dispersed platinum in acid electrolytes. *J. Electrochem. Soc.*, **137**, 845–848.

4 Freund, A., Lang, J., Lehman, T., and Starz, K.A. (1996) Improved Pt alloy catalysts for fuel cells. *Cat. Today*, **27**, 279–283.

5 Mukerjee, S., Srinivasan, S., Soriaga, M.P., and McBreen, J. (1995) Role of structural and electronic properties of Pt and Pt alloys on electrocatalysis of oxygen reduction. An in-situ XANES and EXAFS investigation. *J. Electrochem. Soc.*, **142**, 1409–1422.

6 Aricò, A.S., Srinivasan, S., and Antonucci, V. (2001) DMFCs: from fundamental aspects to technology development. *Fuel Cells*, **1**, 133–161.

7 Schmidt, T.J., Noeske, M., Gasteiger, H.A., Behm, R.J., Britz, P., Brijoux, W., and Bonnemann, H. (1997) Electrocatalytic activity of Pt–Ru alloy colloids for CO and CO/H_2 electrooxidation- stripping voltammetry and rotating-disk measurements. *Langmuir*, **14**, 2591–2595.

8 Mc Breen, J., and Mukerjee, S. (1995) In-situ X-ray absorption studies of a Pt–Ru electrocatalys. *J. Electrochem. Soc.*, **142**, 3399–3404.

9 Lin, W.F., Iwashita, T., and Viestich, W. (1999) Catalysis of CO electrooxidation at Pt, Ru and PtRu alloy- an in-situ FTIR study. *J. Phys. Chem. B*, **103**, 3250–3257.

10 Watanabe, M., and Motoo, S. (1975) Electrocatalysis by ad-atoms: Part III. Enhancement of the oxidation of carbon monoxide on platinum by ruthenium ad-atoms. *J. Electroanal. Chem.*, **60**, 275–283.

11 Chrzanowski, W., and Wieckowski, A. (1998) Enahncement in methanol oxidation by spontaneously deposited ruthenium on low-index platinum-electrodes. *Catal. Lett.*, **50**, 69–75.

12 Ren, X., Zelenay, P., Thomas, S., Davey, J., and Gottesfeld, S. (2000) Recent advances in direct methanol fuel cells at Los Alamos national laboratory. *J. Power Sources*, **86**, 111–116.

13 Garcia, G., Baglio, V., Stassi, A., Pastor, E., Antonucci, V., and Aricò, A.S. (2007) Investigation of Pt–Ru nanoparticle catalysts for low temperature methanol electro-oxidation. *J. Solid State Electrochem.*, **11**, 1229–1238.

14 Aricò, A.S., Stassi, A., Modica, E., Ornelas, R., Gatto, I., Passalacqua, E., and Antonucci, V. (2008) Performance and degradation of high temperature polymer electrolyte fuel cell catalysts. *J. Power Sources*, **178** (2), 525–536.

15 Baglio, V., Di Blasi, A., D'Urso, C., Antonucci, V., Aricò, A.S., Ornelas, R., Morales-Acosta, D., Ledesma-Garcia, J., Godinez, L.A., Arriaga, L.G., and Alvarez-Contreras, L. (2008) Development of Pt and Pt–Fe catalysts supported on multiwalled carbon nanotubes for oxygen reduction in direct methanol fuel cells. *J. Electrochem. Soc.*, **155** (8), B829–B833.

16 Brankovic, S.R., Wang, J.X., and Adzic, R.R. (2001) Pt submonolayers on Ru nanoparticles– a novel low Pt loading, high Co tolerance fuel-cell electrocatalyst. *Electrochem. Solid-State Lett.*, **4**, A217–A220.

17 Aricò, A.S., Baglio, V., Di Blasi, A., Modica, E., Manforte, G., and Antonucci, V. (2005) Electrochemical analysis of high temperature methanol electrooxidation at Pt-decorated Ru catalysts. *J. Electroanal. Chem.*, **576**, 161–169.

18 Hockaday, R.G., DeJohn, M., Navas, C., Turner, P.S., Vaz, H.L., and Vazul, L.L. (2000) A better power supply for portable electronics: microfuel cells. *Proceedings Fuel Cell Seminar, Portland, Oregon, USA, 30 Oct.–2 Nov. 2000*, pp. 791–794.

19 Sun, G.R., Wang, J.T., and Savinell, R.F. (1998) Iron(III) tetramethoxyphenylporphyrin (fetmpp) as methanol tolerant electrocatalyst for oxygen reduction in direct methanol fuel cells. *J. Appl. Electrochem.*, **28**, 1087–1093.

20 Reeve, R.W., Christensen, P.A., Hamnett, A., Haydock, S.A., and Roy, S.C. (1998) Methanol tolerant oxygen reduction catalysts based on transition metal sulfides. *J. Electrochem. Soc.*, **145**, 3463–3471.

21 Kreuer, K.D. (2001) On the development of proton conducting polymer membranes for hydrogen and methanol fuel cells. *J. Membr. Sci.*, **185**, 29–39.

22 Kerres, J., Ullrich, A., Meier, F., and Haring, T. (1999) Synthesis and characterization of novel acid-base polymer blends for application in membrane fuel cells. *Solid State Ionics*, **125**, 243–249.

23 Watanabe, M., Uchida, H., Seki, Y., Emori, M., and Stonehart, P. (1996) Self-humidifying polymer electrolyte membranes for fuel cells. *J. Electrochem. Soc.*, **143**, 3847–3852.

24 Aricò, A.S., Creti, P., Antonucci, P.L., and Antonucci, V. (1998) Comparison of ethanol and methanol oxidation in a liquid feed solid polymer electrolyte fuel cells at high temperature. *Electrochem. Solid-State Lett.*, **1**, 66–68.

25 Antonucci, V., and Aricò, A.S. (1999) Polymeric membrane electrochemical cell

operating at temperatures above 100 °C Europ. Patent, EP 0 926 754 A1, 1–5.

26 Aricò, A.S., Baglio, V., Di Blasi, A., and Antonucci, V. (2003) FTIR spectroscopic investigation of inorganic fillers for composite DMFC membranes. *Electrochem. Commun.*, **5**, 862–866.

27 Alberti, G., and Casciola, M. (2003) Composite membranes for medium-temperature PEM fuel cells. *Annu. Rev. Mater. Res.*, **33**, 129–154.

28 Wang, J., Wasmus, S., and Savinell, R.F. (1995) Evaluation of ethanol, 1-propanol, and 2-propanol in a direct oxidation polymer-electrolyte fuel cell. *J. Electrochem. Soc.*, **142**, 4218–4224.

29 Hasiotis, C., Deimede, V., and Kontoyannis, C. (2001) New polymer electrolytes based on blends of sulfonated polysulfones with polybenzimidazole. *Electrochim. Acta*, **46**, 2401–2406.

30 Dimitrova, P., Friedrich, K.A., Vogt, B., and Stimming, U. (2002) Transport properties of ionomer composite membranes for direct methanol fuel cells. *J. Electroanal. Chem.*, **532**, 75–83.

31 Aricò, A.S., Baglio, V., Di Blasi, A., Modica, E., Antonucci, P.L., and Antonucci, V. (2004) Surface properties of inorganic fillers for application in composite membranes-direct methanol fuel cells. *J. Power Sources*, **128**, 113–118.

32 Staiti, P., Aricò, A.S., Baglio, V., Lufrano, F., Passalacqua, E., and Antonucci, V. (2001) Hybrid Nafion-silica membranes doped with heteropolyacids for application in direct methanol fuel cells. *Solid-State Ionics*, **145**, 101–107.

33 Jung, D.H., Cho, S.Y., Peck, D.H., Shin, D.R., and Kim, J.S. (2003) Preparation and performance of a Nafion®/montmorillonite nanocomposite membrane for direct methanol fuel cell. *J. Power Sources*, **118**, 205–211.

34 Yang, C., Srinivasan, S., Aricò, A.S., Cretì, P., Baglio, V., and Antonucci, V. (2001) Composite Nafion/Zirconium Phosphate membranes for direct methanol fuel cell operation at high temperature. *Electrochem. Solid-State Lett.*, **4**, A31–A34.

35 Aricò, A.S., Baglio, V., Di Blasi, A., Antonucci, V., Cirillo, L., Ghielmi, A., and Arcella, V. (2006) Proton exchange membranes based on the short-side-chain perfluorinated ionomer for high temperature direct methanol fuel cells. *Desalination*, **199**, 271–273.

36 Debe, M.K. (2003) Novel catalysts, catalyst supports and catalyst coated membrane methods, in *Handbook of Fuel Cells – Fundamentals, Technology and Applications*, vol. **3** (eds W. Vielstich, H.A. Gasteiger, and A. Lamm), John Wiley and Sons, Ltd., Ch 45, pp. 576–589.

37 Atanasoski, R. (2004) *Recent Advances on the 3M MEA Technology for PEMFs*, 4th International Conference on Applications of Conducting Polymers, ICCP-4, Como, Italy, February, pp. 18–20.

38 Schoonman, J. (2003) Nanoionics. *Solid-State Ionics*, **157**, 319–326.

39 Steele, B.C.H. (2000) Current status of intermediate temperature fuel cells (IT-SOFCs). *Eur. Fuel Cell News*, **7**, 16–19.

40 Montinaro, D., Sglavo, V.M., Bertoldi, M., Zandonella, T. Aricò, A.S., Lo Faro, M., and Antonucci, V. (2006) Tape casting fabrication and co-sintering of solid oxide "half cells" with a cathode-electrolyte porous interface. *Solid-State Ionics*, **177** (19–25 SPEC. ISS.), 2093–2097.

41 Perry Murray, E., Tsai, T., and Barnett, S.A. (1999) A direct-methane fuel cell with a ceria-based anode. *Nature*, **400**, 649–651.

42 Park, S., Vohs, J.M., and Gorte, R.J. (2000) Direct oxidation of hydrocarbons in a solid-oxide –fuel cell. *Nature*, **404**, 265–267.

43 Maier, J. (2003) Defect chemistry and ion transport in nanostructured materials. Part II Aspects of nanoionics. *Solid-State Ionics*, **157**, 327–334.

44 Knauth, P., and Tuller, H.L. (2002) Solid-state ionics – roots, status, and future-prospects. *J. Am. Ceram. Soc.*, **85**, 1654–1680.

45 Sin, A., Kopnin, E., Dubitsky, Y., Zaopo, A., Aricò, A.S., La Rosa, D., Gullo, L.R., and Antonucci, V. (2007) Performance and life-time behaviour of NiCu-CGO anodes for the direct electro-oxidation of

methane in IT-SOFCs. *J. Power Sources*, **164** (1), 300–305.

46 Lo Faro, M., La Rosa, D., Nicotera, I., Antonucci, V., and Arico, A.S. (2009) Electrochemical investigation of a propane-fed solid oxide fuel cell based on a composite ni-perovskite anode catalyst. *Appl. Catal. B: Environ.* doi: 10.1016/j.apcatb.2008.11.019

47 La Rosa, D., Sin, A., Lo Faro, M., Monforte, G., Antonucci, V., and Aricò, A.S. (2009) Mitigation of carbon deposits formation in intermediate temperature solid oxide fuel cells fed with dry methane by anode doping with barium. *J. Power Sources*. doi: 10.1016/j.jpowsour.2009.01.096

5
The Contribution of Nanotechnology to Hydrogen Production

Sambandam Anandan, Jagannathan Madhavan, and Muthupandian Ashokkumar

5.1
Introduction

The provision of an abundant, clean and secure renewable energy source is one of the key technological challenges facing the mankind. Our current energy infrastructure is dominated by fossil fuels, one of the major sources for greenhouse gas emission. The resurgence in the chemistry and biochemistry of hydrogen, the world's simplest molecule, has been spurred by the recent scientific and technological interest in hydrogen as an energy carrier and potential transportation fuel. One of the challenges facing the widespread adoption of hydrogen as an energy vector is the lack of an efficient, economical and sustainable method of hydrogen production. A number of key challenges must be overcome for hydrogen to be used broadly in a sustainable future energy infrastructure to solve the global energy problems [1].

Biomass and biomass wastes are promising sources for the sustainable production of hydrogen in an age of diminishing fossil fuel reserves. However, conversion of biomass into hydrogen remains a challenge, since processes such as enzymatic decomposition of sugars, steam reforming of bio-oils and gasification suffer from low hydrogen production rates and/or complex processing requirements [2, 3]. Figure 5.1 schematically shows the "ideal" methodologies that could be used for the production and uses of hydrogen.

However, further scientific advances are necessary in order to develop more energy efficient and cost-effective methods for purification and delivery, to design higher energy density hydrogen storage systems, especially for vehicle on-board storage, and to enable more durable fuel cells for converting hydrogen into electrical energy. Plants capture the energy from sunlight and thus grow. During this process, they produce oxygen by redox reactions involving water and carbon dioxide. In other words, the oxidation of water and the reduction of CO_2 are achieved with solar energy. Research work continues on photolysis, electrolysis, and thermal processes for H_2 production. Single-step thermal dissociation of water [4, 5], although conceptually simple, has been impeded by the need of a high-temperature heat source for achieving a reasonable degree of dissociation,

Nanotechnology for the Energy Challenge. Edited by Javier Garcia-Martinez
© 2009 WILEY-VCH Verlag GmbH & Co. KGaA, Weinheim
ISBN: 978-3-527-32401-9

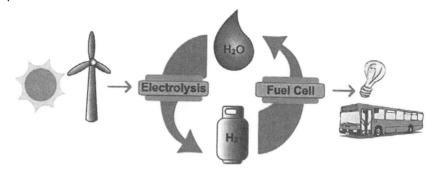

Figure 5.1 Ideal and sustainable production of hydrogen (generation and application) using renewable energy sources.

and by the need for an effective technique for separating H_2 and O_2 to avoid their recombination. The electrolysis of water is a technique that uses a direct current to split water into hydrogen and oxygen [6, 7]. The minimum potential difference between the cathode and anode must be near 1.5 V for the efficient electrolysis and hence water electrolysis is also not yet economically feasible.

The method of the hydrogen production by water photolysis using semiconductor photocatalysts has been realised as a clean technology after the pioneering photoelectrochemical work published by Fujishima and Honda [8] in 1972. A number of research groups have studied the photoelectrolysis of water using light energy, which is considered as a possible source of energy for the future [9–17]. It has been shown that both non-oxide and oxide semiconductors are useful for the photodissociation of water into H_2 and O_2 due to the direct absorption of photons by these materials [18–24]. Non-oxide materials harvest light in the visible region, whereas most of the stable oxides are photochemically active only in the ultraviolet (UV) region of the solar spectrum [24]. However, oxide semiconductors show good stability towards photocorrosion.

Semiconductors (SC) are characterized by an energy gap between valence and conduction bands. Upon excitation by a photon, an electron is promoted to the conduction band (CB) from the valence band (VB). The separated electron and hole pair can move to the surface of the semiconductor particle and react with water or other adsorbed substrates (e.g., H_2S, CH_3OH, other sacrificial agents) leading to the production of hydrogen. Theoretically, when the conduction band potential level of a semiconductor is more negative than that of hydrogen evolution and the valence band potential level is more positive than that of oxygen evolution, it is possible to decompose water into H_2 and O_2 [8], as shown in Reactions (5.1)–(5.4).

$$\text{Semiconductor photocatalyst} + h\nu \rightarrow e^- + h^+ \tag{5.1}$$

$$H_2O \rightarrow OH^- + H^+ \tag{5.2}$$

$$2e^- + 2H^+ \rightarrow H_2 \tag{5.3}$$

$$2h^+ + 2OH^- \rightarrow 2H^+ + O_2 \tag{5.4}$$

Obviously, a potential difference of more than 1.23 eV is necessary in photocatalytic reactions to generate hydrogen from water. In general, various semiconductors have been used for the photocatalytic decomposition of water [25–31]. However, semiconductors with a small band gap energy that could absorb the visible region of the solar spectrum are limited. Semiconductor materials with a small band gap are continuously being investigated in order to use the less energetic but more abundant visible light [32–44].

Hydrogen production by the photoelectrolysis of water using nanocrystalline semiconductors in the form of electrodes, colloids, powders and thin films has drawn significant attraction in recent years [45–48]. There has been an intensive research effort in the synthesis of numerous nanostructural inorganic oxides, such as TiO_2, MnO_2, Co_3O_4, MoO_3, V_2O_5, Al_2O_3, ZrO_2, SnO_2 and SiO_2 [49–57]. In particular, nanosized titanium dioxide is of great interest due to its diverse applications in catalysis, photovoltaic cells and other semiconductor devices.

There are several reviews available on the use of semiconductors for hydrogen production in the literature. A review in 1998 by Ashokkumar [24] provided a coherent description of the semiconductor bulk particles for photosplitting of water molecules. Nowotny *et al.* [58] and Fukuzumi [59] summarized the necessity of new photosensitive materials that convert solar energy into chemical energy (e.g., hydrogen). The current review primarily focuses on the recent development in the field of semiconductor nanomaterials for hydrogen production.

5.2
Hydrogen Production by Semiconductor Nanomaterials

5.2.1
General Approach

An intense research activity is seen in recent years in advancing the synthesis and functionalization of semiconductor and metal nanoparticles with various sizes and shapes. The goal of these activities is to improve the performance and utilization of nanoparticles in many applications. The size and shape dependent optical and electronic properties of these nanoparticles make an interesting case for exploiting them in light-induced chemical reactions. The most popular choice of a semiconductor photocatalyst is TiO_2 (E_g = 3.2 eV) and much of the published work on photocatalysis uses this material. This is due to its ability to photoelectrolyze water into hydrogen and oxygen.

The schematic representation shown in Figure 5.2 [10] illustrates the photoelectrochemical principle involved in water-splitting reactions. While the scheme is based on a photoelectrochemical cell, similar principles are valid for n-type

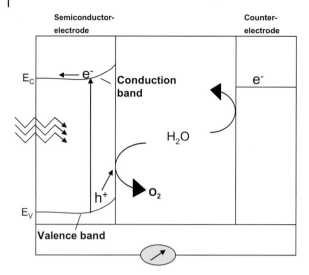

Figure 5.2 Operation of a photoelectrochemical cell based on a n-type semiconductor that generates hydrogen through the photocleavage of water [10].

semiconductors (including TiO_2) used as powders and colloids. In the case of powdered semiconductors, each particle can be considered as a micro/nano-photoelectrochemical cell. The electron-hole redox reactions occur at the particle/solution interface.

Hydrogen evolution from water was achieved by irradiating TiO_2 suspensions with UV radiation [60]. The solar radiation on earth's surface is poor in UV light. The development of photocatalytic systems capable of using the visible zone of the solar spectrum by semiconductor suspensions can be achieved in three ways: (i) SC of large band gap aided by photosensitizers, (ii) SC of small band gap and (iii) mixtures of SC of large and small band gaps. These aspects are discussed in a later section of this review.

5.2.2
Need for Nanomaterials

Semiconductors in the nanometer size range exhibit completely new and improved properties than bulk materials, based on their specific characteristics, such as size, size distribution, morphology, phase, etc. The newly developing field of nanotechnology opens new avenues for synthesizing improved semiconductor materials for hydrogen production with large surface-to-volume ratios, that is, a decrease in the particle size increases the surface area and hence the photocatalytic efficiency significantly. Also, nano-sized particles possess local electron polarization and the ability to spatially separate photogenerated electrons and holes [61]. During the past 15 years, many studies have been conducted with semiconductor particles in the form of colloids or particle suspensions in aqueous solution [10, 23, 43]. Nanocrystals also attract special interest because of the quantum size effects. Semiconductor particles that exhibit size-dependent optical and electronic properties are termed as quantized particles (or Q particles). The particles with a diameter of ≥15 nm behave as bulk semiconductors. However, similarly prepared crystallites of ~ 2–10 diameters, which possess some bulk lattice structures, differ considerably in their electronic properties. When the size of the semiconductor particles are comparable to or smaller than the de Broglie's wavelength, the density of electronic states (i.e., the allowed energy states for the electrons and holes in the conduction and valence band become discrete) decreases leading to an increase in the effective bandgap. The energy spacing between the quantized levels is inversely proportional to the effective mass and the square of the particle diameter. Further, as a consequence of the blue shift in the absorption edge with decreasing particle size, the redox potentials of the photogenerated electrons and holes in quantized semiconductor particles are changed [62]. It has been reported that the Q particles are more photoactive than the macrocrystalline semiconductor particles [63].

5.2.3
Nanomaterials-Based Photoelectrochemical Cells for H_2 Production

The nanoparticle-based semiconductor thin film photoelectrodes start a new era in the development of efficient photoelectrochemical (PEC) solar cells. Nanocrystalline thin films of metal oxides such as, TiO_2, ZnO, WO_3 and SnO_2 with high porous structure have drawn considerable interest in recent years for their use in PEC cells as photosensitive semiconductor electrodes. To make use of the properties of nanometer-sized semiconductor particles for light to electrical/chemical energy conversion, a pathway for electrical conduction between particles must be provided. This can be achieved by either of the two methods:

1) Particles are applied to a conducting substrate from a suspension and then sintered to form electrical contact between the particles and to allow for charge transport to the substrate [64].

2) Nanocrystalline particles are formed directly on the substrate by an electrochemical or chemical deposition process [65].

The nanocrystalline electrodes distinguish themselves from macrocrystalline electrodes by their porosity and high surface to volume ratio. Hence many authors [66–70] attempted to synthesize TiO_2 particles of different sizes and shapes, which can enhance the rate of electron diffusion through the porous structure in order to elicit the maximal conversion efficiencies under solar irradiation. Further discussion on the effect of the shape of nanocrystalline materials is provided later.

TiO_2 is mainly used for water/air purification (photodegradation of pollutants) as well as for photocatalytic hydrogen production [24, 71–73]. Both applications require the photogeneration of hole/electron pairs. However, their use and mech-

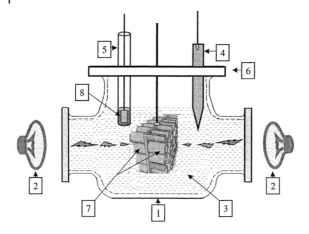

1. Quartz cell, 2. Light source, 3. 1M KOH solution containing 5 vol. % ethylene glycol, 4. Ag/AgCl reference electrode, 5. cathode compartment, 6. Teflon lid, 7. Photoanode containing double sided carbon doped titania nanotubular arrays, and 8. Cathode containing platinum nanoparticles and TiO_2 nanotube composites.

Figure 5.3 A schematic diagram of the photoelectrochemical cell to generate hydrogen by water splitting using double-side-illuminated titania nanotube arrays [75].

anisms are different. In photocatalytic degradation of organics, the holes are key species in the overall activity: the addition of electron-sink (e.g., molecular oxygen, metal, etc.) [74] to reduce the recombination losses is crucial for the photocatalytic degradation of the organic compounds. However, in a photoelectrochemical system the electrons are key to reduce protons to generate molecular hydrogen: hence, the addition of hole traps (e.g., methanol) to reduce the electron-hole recombination losses is required.

Mohapatra et al. [75] designed a special photoelectrochemical cell (Figure 5.3) to generate hydrogen by water splitting at a rate of $38\,ml\,h^{-1}$ using double-sided $TiO_2/Ti/TiO_2$ material as photoanode and platinum nanoparticles dispersed on TiO_2 nanotubes as photocathode ($PtTiO_2/Ti/PtTiO_2$).

In continuation, Mor et al. [76] also designed a special photoelectrochemical system, that is a n-TiO_2/p-Cu–Ti–O coupled system (Figure 5.4) to generate fuel with oxygen evolved from the n-TiO_2 side of the diode and hydrogen from the p-Cu–Ti–O side under global AM 1.5 illumination.

They concluded that this system operates similar to photosynthesis. The p-type Cu–Ti–O nanotube array films are used in combination with n-type TiO_2 nanotube array films to achieve a pn-junction photochemical diode capable of generating hydrogen by water splitting.

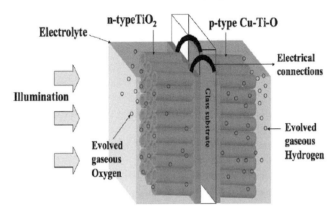

Figure 5.4 Illustration of the photoelectrochemical diode for water splitting comprised of n-type TiO$_2$ and p-type Cu–Ti–O nanotube array films, with their substrates connected through an ohmic contact. The oxygen-evolving TiO$_2$ side of the diode absorbs UV light, passing the visible light to the hydrogen-evolving Cu–Ti–O side [76].

5.2.4
Semiconductors with Specific Morphology: Nanotubes and Nanodisks

Many research groups [66–70, 77, 78] have attempted to prepare different semiconductor nanomaterials. TiO$_2$ was synthesized in various forms (spherical, rod, tube, disk, etc.) and widely used for the photoelectrolysis of water to generate hydrogen and oxygen. TiO$_2$ in nanotubular form has a large surface area. Several recent studies have indicated that TiO$_2$ nanotubes have improved properties compared to any other form of TiO$_2$ for application in photocatalysis [79, 80], sensing [81–84], photoelectrolysis [85–87] and photovoltaics [88–91]. Recently, Shankar et al. published a fascinating review on the anodized TiO$_2$ tube arrays [92] and also reported [93] on the fabrication of a highly ordered TiO$_2$ nanotube array up to 220 mm in length (Figure 5.5). Longer tube lengths result in a larger surface area for reactions. In a porous film consisting of nanometer-sized TiO$_2$ particles, the effective surface area is enhanced by a thousand-fold, thus making the light absorption very efficient. Nature uses a similar means of absorption enhancement in leaves by stacking the chlorophyll-containing thylakoid membranes of the chloroplast to form the grain structures. A characteristic feature in the nanocrystalline TiO$_2$ films is that the charge transport of the photoinjected electrons passing through all the particles and grain boundaries is highly efficient, resulting in very high quantum yields, close to unity [94].

Recently Mor et al. [86] found a remarkable photoconversion efficiency (~6.8%) while using highly ordered transparent TiO$_2$ nanotube arrays prepared by anodization at 5 °C (Figure 5.6). Using this system, they performed photosplitting of water and achieved a hydrogen oxygen ratio of 2:1. With the nanotube array photoanodes held at constant voltage bias, 48 mmol of hydrogen gas was generated. They also suggested that the oxygen bubbles evolving from the nanotube array

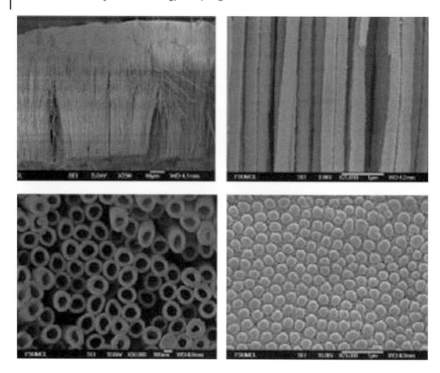

Figure 5.5 Illustrative FESEM cross-sectional, top and bottom images of nanotube array grown at 60 V in a DMSO electrolyte containing 2% HF [93].

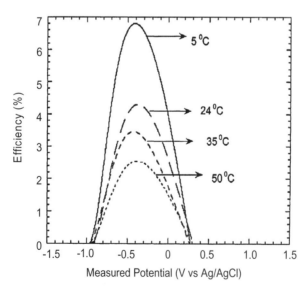

Figure 5.6 Photoconversion efficiency as a function of measured potential for 10 V samples anodized at four different temperatures [86].

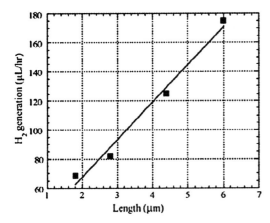

Figure 5.7 The rate of hydrogen generation from crystallized nanotube arrays of different lengths annealed at 530 °C. An electrode area of 1 cm^2 was exposed to 100 mW cm^{-2} AM 1.5 illumination in all cases [98].

photoanode do not remain on the sample, hence the output remains stable with time irrespective of the duration of hydrogen production. This is due to the light scattering within a porous structure – incident photons are more effectively absorbed than that on a flat electrode [95].

The reason for high quantum efficiency in the case of TiO$_2$ nanotube arrays prepared by anodization at 5 °C compared to anodization at 50 °C may be due to the effective utilization of the incident light by the nanotube arrays for charge carrier generation at the lower temperature. Further, the relevant dimensional features of the titania nanotube arrays, that is, half the wall thickness, are all smaller than 10 nm, which is less than the retrieval length of crystalline titania [96], hence the bulk recombination is greatly reduced leading to an enhancement in the quantum yield [96]. A similar substantial enhancement of the quantum yield was also observed by Van de Lagemaat et al. [97] while using nanoporous SiC prepared by anodic etching.

The effect of nanotube array length on the hydrogen evolution rate under AM 1.5 illumination with samples annealed at 530 °C was studied by Paulose et al. (Figure 5.7) [98]. Both the photocurrent magnitude and the photoconversion efficiency were found to increase with increasing length of the nanotube. On exposure of 6 mm nanotube array samples annealed at 600 °C to photons of 337 nm (3.1 mW/cm^2) and 365 nm (89 mW/cm^2), the quantum efficiency was calculated to be 81% and 80%, respectively. The high quantum efficiency clearly indicated that the incident light is effectively utilized by the nanotube arrays for charge carrier generation. After this finding, the preparation of TiO$_2$ nanotubes by the anodization process has caught the attention of the scientific community due to its one-dimensional nature, ease of handling and simple method of preparation. Over the years, several electrolytic combinations have been used for the anodization of titanium. However, the reported titania nanotubes are not well

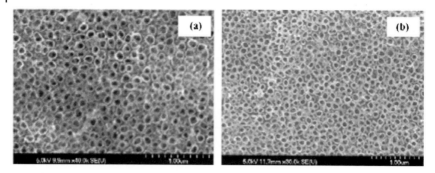

Figure 5.8 FESEM images of nanotubular TiO_2 prepared by sonoelectrochemical method using 0.5 M H_3PO_4 and 0.14 M fluoride salt solution: (a) NH_4F and (b) KF [99].

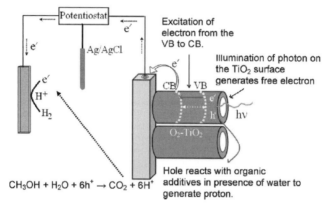

Figure 5.9 Scheme illustrating the process of hydrogen generation using oxygen annealed TiO_2 (O_2–TiO_2) nanotube arrays using methanol as an organic additive in 1 M KOH solution [100].

ordered, and it takes several hours to make micron-length nanotubes in a high-pH electrolyte.

Mohapatra et al. [99] used a sonoelectrochemical technique to anodize titania, which led to the faster synthesis of well-ordered titania nanotubes (Figure 5.8). This anodization approach leads to the formation of self-organized titania nanotubular arrays of controllable tube diameter, good uniformity and conformability over large areas. The self-organized one-dimensional nanotube creates a better opportunity to harvest sunlight more efficiently than the randomly oriented nanoparticles or the nanotubes prepared by sol-gel process. The property of titania nanotubes prepared by the anodization method makes them a versatile catalyst for photoelectrolysis of water.

Mohapatra et al. [100] used oxygen annealed TiO_2 (O_2–TiO_2) nanotube arrays as a photoelectrode in the presence of organic additives (Figure 5.9) for hydrogen generation from water. O_2–TiO_2 nanotubes are n-type semiconductors with a band gap of 3.0–3.1 eV. Exposure of the nanotubular photoanode to simulated sunlight results in the generation of electron and hole pairs (Reaction 5.5).

$$TiO_2 + h\nu \rightarrow (TiO_2 \cdot h_{VB}^+) + (TiO_2 \cdot e_{CB}^-) \tag{5.5}$$

As the band gap is in the higher range, mostly the UV component of the solar spectrum helps to excite the electrons from the VB to the CB. The electron from the CB moves to the external circuit and reaches the cathode (Pt). Meanwhile, the hole comes out from the nanotube surface to the electrolytic solution and oxidizes water to generate the proton. In this process, oxygen (O_2) is evolved in the anodic compartment of the PEC cell (Reaction 5.6). The external potential applied through the potentiostat to the cell helps the proton to reach the cathodic surface, where it combines with the electron to evolve hydrogen (Reaction 5.7). Further, they have suggested that the sunlight consists of 4–5% of UV light in the entire spectrum and hence modification of TiO_2 nanotubes is necessary, so that the full spectrum of the sunlight can be harvested in a more efficient way.

$$2h^+ + 2H_2O \rightarrow 2O_2 + 4H^+ \tag{5.6}$$

$$2e^- + 2H^+ \rightarrow H_2 \tag{5.7}$$

For this reason, they prepared nanotubes functionalized with 2,6-dihydroxyanthraquinone (anthrafavic acid) [101] and doped with foreign elements ($TiO_{2-x}C_x$) [102]. In the earlier one, the functionalization takes place by the chemical condensation of the Ti–OH hydroxyl groups present on the TiO_2 nanotubular surface with the phenolic hydroxyl groups of anthrafavic acid forming an inorganic-organic hybrid material (Figure 5.10). The condensation results in an intramolecular ligand-to-metal charge transfer, which leads to an enhancement in the visible light

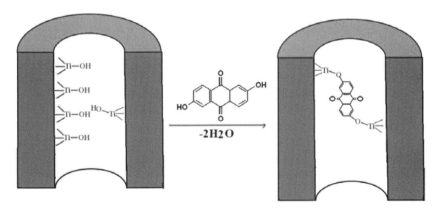

Figure 5.10 Surface sensitization of TiO_2 nanotubes by 2,6-dihydroxyantraquinone [101].

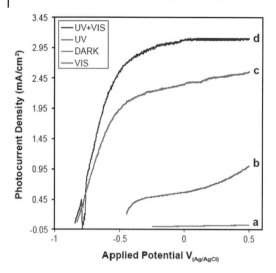

Figure 5.11 Variation of photocurrent density versus applied potential of the $TiO_{2-x}C_x$ nanotubular arrays in (a) dark, (b) illumination of visible (520 ± 46 nm), (c) UV (330 ± 70 nm) and (d) complete solar spectrum [102].

absorption. This hybridization is responsible for an increase in the photoelectrochemical generation of hydrogen from water up to 30%.

For the second case, they synthesized self-organized highly ordered $TiO_{2-x}C_x$ nanotubular arrays with a wide range of nanotube structure by the sonoelectrochemical anodization method using aqueous ethylene glycol and ammonium fluoride solutions. The authors mentioned that the doping of carbon occurs from the reduction of ethylene glycol and the total carbon in the TiO_2 matrix is 63% as shown by XPS analysis. A reduction in the band gap from 3.2 to ~2.2 eV was also observed [102]. Further, they evaluated the water photoelectrolysis activity by using carbon-doped titania nanotube arrays (exposed area = 0.7 cm^2) as photoanode and Pt foil as photocathode under the complete solar spectrum. The prepared nanotubes showed a photocurrent of ~2.8 mA cm^{-2} at 0.4 V Ag/AgCl, where about 20% of the activity is contributed by the visible light. The efficiency of the PEC cell using UV and visible light is found to be 13.3% and 8.5%, respectively (Figure 5.11).

In addition, Park et al. [103] also synthesized $TiO_{2-x}C_x$ nanotube arrays for more efficient water splitting under visible-light illumination (>420 nm) compared to pure TiO_2 nanotube arrays. They found that the total photocurrent was more than 20 times higher than that with P-25 nanoparticulate film under white-light illumination.

There are also a few reports available on the use of nanodisk structures [104]. The SEM images of the ultrathin WO_3 nanodisks are shown in Figure 5.12. Cyclic voltametry studies were conducted with a working electrode of WO_3 nanodisks on a FTO substrate, a Pt wire counter electrode, and a Ag/AgCl reference electrode

Figure 5.12 SEM images of WO_3 nanodisks. (a) A large population of nanodisks that lie atop one another in a thin film on a FTO substrate. (b) Higher-magnification image of less densely populated nanodisks showing their structures to range in size in the hundreds of nanometers along the long and short axes [104].

in 0.5 M $NaClO_4$ electrolyte solution buffered to pH 4.43 with phosphate buffer. The scan rate was fixed at 5 mV s^{-1}.

Under light irradiation, an efficient hydrogen gas evolution was observed at higher reductive potentials. Hence, these WO_3 nanodisks with large flat surface area are found to be useful in PEC cells for hydrogen production via water splitting.

5.2.5
Sensitization

Water splitting is an uphill reaction with a large positive change in the Gibbs free energy. The function of a semiconductor is to convert photo energy into chemical energy to drive this reaction. As mentioned earlier, the basic requirement for semiconductors is that their conduction band should be more negative than the reduction potential of H_2O and their valence band should be more positive than the oxidation potential of H_2O. The photocatalytic efficiencies of semiconductors could be enhanced by using semiconductor couples or by doping the semiconduc-

Figure 5.13 TEM images of NiO/Ta$_2$O$_5$ calcined and NiO/Ta$_2$O$_5$ plasma-treated [113].

tors with metals. For example, Ta$_2$O$_5$ and ZrO$_2$ are oxide semiconductors [105, 106] that exhibit low efficiency since their surface is not active enough to generate H$_2$. However, loading of metals onto the surfaces of these semiconductors can introduce more active H$_2$-generating sites. Park *et al.* constructed Pt/TiO$_2$ PEC arrays for direct hydrogen generation with a solar-to-hydrogen efficiency of 2.2% and with 88% solar-to-electrical conversion efficiency [107].

In other studies, semiconductors with perovskite, layered and tunnel structure were reported as promising photocatalysts [108–112]. Their activity is greatly improved when NiO is loaded as a co-catalyst. Compared with other metals or oxides, NiO was found to be the most effective photocatalyst since a metallic interlayer between the semiconductor and NiO makes the electron transfer easier.

Similarly, a clean metal surface is necessary to reduce recombination reactions. In order to prepare a clean metal surface, plasma treatment was preferred over calcination [113]. The nickel particles treated with calcination process produced particles with a spherical shape (Figure 5.13) with a diameter of 30–40 nm. Whereas, samples treated with plasma method (Figure 5.13) were half-ellipsoidal. The length along the metal support interface orientation was 20–30 nm, two times longer than that along the perpendicular direction (10–15 nm).

This provided a higher metal coverage on the support with a larger metal surface. Hence, the catalysts with the plasma treatment exhibited higher activity than those with the traditional treatment (thermal decomposition). The reaction rate of the plasma treated NiO/Ta$_2$O$_5$ sample was 1.7 times higher than that over the calcined NiO/Ta$_2$O$_5$ sample, which might be due to efficient charge separation and an increased active surface. The smaller size and flatter shape of the metal particles produced a larger metal surface and thus more active sites for H$_2$ generation. The catalysts with the clean metal-support interface showed higher activity, confirming that this interface was more favorable.

Hydrogen can be photocatalytically produced from methane and water by using platinum-loaded semiconductor photocatalysts [114]. Platinum-loaded lanthanum-doped NaTaO$_3$ (Pt/NaTaO$_3$:La) showed higher photocatalytic activity for this reaction than platinum-loaded TiO$_2$. The highest H$_2$ production rate of 4.5 μmol min^{-1} (6.6 ml h^{-1}) was obtained over Pt(0.03)/NaTaO$_3$:La, corresponding to 0.6% methane conversion. These results suggested that high crystallinity of semiconductor is important for the activity in photocatalytic steam reforming of methane.

An overview reported by Luzzi [115] indicates that recent efforts have broadened the field of investigations to include the use of alternative semiconductor materials, exhibiting a lower band gap energy for the photoanode. Also, the spectral sensitization [116] of semiconductors has been actively studied in solar energy conversion systems involving the photolysis of water to generate hydrogen. Sensitization enables the use of semiconductors that are not intrinsic absorbers of visible light and do not suffer destructive photodecomposition [117]. A variety of semiconductor electrodes have been studied, for example, ZnO [118–120], SnO$_2$ [121, 122], SrTiO$_3$ [123–125] and TiO$_2$ [126–130] with sensitization by surface dye layers or metal dopants. Sensitization of TiO$_2$ by various dye compounds, such as Ru(bpy)$_3^{2+}$ [131–136], RuL$_3^{2+}$ (L = 2,2'–bipyridine-4,4'-dicarboxylate) [137–139] and metal-quinolinol [140] have also been employed. Gratzel and co-workers [141–146] and others [147–150] reported the production of hydrogen from aqueous dispersions of sensitized TiO$_2$.

Dhanalakshmi et al. [151] showed that the photoactivity of Pt/TiO$_2$ system in the visible region was improved by the addition of the sensitizer, [Ru(dcbpy)$_2$(dpq)]$^{2+}$ [where dcbpy is 4,4'-dicarboxy 2,2'-bipyridine, dpq is 2,3-bis(2-pyridyl)-quinoxaline] leading to an efficient water reduction. Visible-light-induced water cleavage is based on the concept of light-induced charge injection from the excited sensitizer to the semiconductor catalyst. The photoexcitation of the sensitizer is followed by electron transfer into the conduction band of the photocatalyst and is channeled to the platinum site, where hydrogen evolution occurs. The various steps involved in the mechanism are mentioned as follows (Reactions 5.8–5.12).

1) Absorption of light:

$$\text{Ru}^{2+}\text{-complex}_{\text{surface}} \xrightarrow{\text{visible light}} {}^*\text{Ru}^{2+}\text{-complex}_{\text{surface}} \tag{5.8}$$

2) Sensitization:

$$^*\text{Ru}^{2+}\text{-complex}_{\text{surface}} + \text{TiO}_2 \rightarrow \text{Ru}^{3+}\text{-complex}_{\text{surface}} + \text{TiO}_2(e^-_{\text{CB}}) \tag{5.9}$$

3) Regeneration of complex:

$$2\text{Ru}^{3+}\text{-complex}_{\text{surface}} + \text{H}_2\text{O} \rightarrow 2\text{Ru}^{2+}\text{-complex}_{\text{surface}} + \tfrac{1}{2}\text{O}_2 + 2\text{H}^+ \tag{5.10}$$

4) Hydrogen formation:

$$\text{TiO}_2(e^-_{\text{CB}}) + \text{Pt} \rightarrow \text{TiO}_2 + \text{Pt}(e^-) \tag{5.11}$$

$$\text{Pt}(e^-) + \text{H}_3\text{O}^+ \rightarrow \text{Pt} + \tfrac{1}{2}\text{H}_2 \tag{5.12}$$

Also they observed that Pt doping enhanced the efficiency of TiO_2 in water-cleavage processes. The sensitizer reported by these authors are found to possess greater light absorption at longer wavelengths when anchored onto Pt/TiO_2 and hence increased the conversion efficiency of light energy into chemical energy (hydrogen fuel).

The possibility that solar illuminated inorganic complexes improve the efficiency of the water splitting process has inspired many researchers in this field [12, 152]. Hence, several research groups studied the potential of the ruthenium bipyridyl-methyl viologen system in promoting water photoreduction [19]. It has been suggested that a system consisting of an electron relay [such as methyl viologen (MV^{2+})], a photosensitizer [such as tris(2,2'-bipyridine)ruthenium(II)], a sacrificial donor (such as triethanolamine) and a catalyst (such as colloidal platinum) can be very efficient in water photoreduction [153].

Detellier and Villemure [154] showed that the above system in the absence of platinum and in the presence of a natural clay (Whoming Bentonite) produces hydrogen when irradiated by visible light at natural pH 7.0. Further they extended their study with different homoionic clays but no drastic difference in the hydrogen production was observed when compared to the natural material, montmorillonite [155].

The co-existence of electron acceptor (e.g., aluminum at the crystal edges) and electron donor (e.g., transition metals in the reduced states) sites can enhance the electron transfer efficiency to the protons, generated at the highly acidic clay surface, leading to efficient hydrogen production reactions [156]. This aspect generated interest among many researchers [157–160] in designing molecular systems that mimic photosynthesis, since these may be used to capture sunlight reaching the earth's surface and convert the energy to useful chemical fuels. The above studies [157–160] reported the feasibilty of encapsulating the common photosensitizer, $[Ru(bpy)_3]^{2+}$, within a Y-zeolite supercage. An efficient photoinduced electron transfer to viologen acceptors occupying neighboring supercages followed by H_2 production could be achieved. The photochemical reduction of water via a water oxidation cycle is schematiclly represented in Figure 5.14.

The reason for choosing zeolites is due to the large variety of well defined cavities and channels present within the zeolite matrix. In addition, their lattices can be regarded as an enormous crystalline polyanion, which contains cations for charge compensation [161]. Following this study, Jacobs and Uytterhoeven [162] studied the cleavage of water over zeolite containing silver ions (Reactions 5.13 and 5.14).

$$2Ag^+ + 2ZO^- + H_2O \xrightarrow{h\nu} 2Ag^\circ + 2ZOH + \tfrac{1}{2} O_2 \qquad (5.13)$$

$$Ag^\circ + ZOH \xrightarrow{873K} 2Ag^+ + ZO^- + \tfrac{1}{2} H_2 \qquad (5.14)$$

Also, Kuznicke and Eyring [163] investigated water splitting by titanium(III) exchanged zeolite under visible light illumination. However, this system used thermal energy as well as light energy. To overcome this problem, Anandan *et al.* [164] attempted to prepare heteropolyacid (HPA)-encapsulated TiHY zeolite, a new

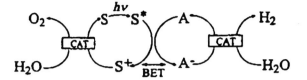

$$S \xrightarrow{h\nu} S^*$$
$$S^* + A \longrightarrow S^+ + A^-$$

Figure 5.14 Scheme for a synthetic water-splitting system using zeolites [159].

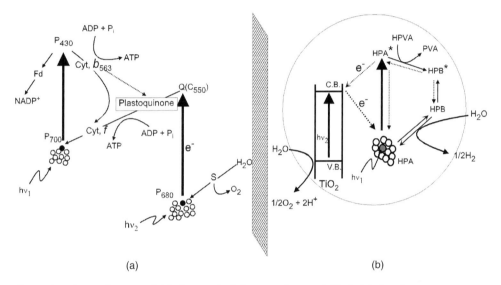

Figure 5.15 Energy diagrams of: (a) photosynthesis in green plants and (b) photoinduced electron transfer at the HPA-encapsulated TiHY zeolite for the generation of H_2 and O_2 [164].

photocatalyst mimicking the plant photosynthetic system. This photocatalyst was observed to generate H_2 and O_2 from aqueous solutions upon illumination, which is quite analogous to the "Z-scheme" mechanism for plant photosynthetic systems (Figure 5.15).

When the samples are illuminated, HPA encapsulated in the zeolite cavities is reduced to the so-called heteropoly blues (HPB) showing new absorption at 750 nm, due to intervalence charge transfer [W(VI) → W(V)] [165]. The formation of HPB is induced by direct electron transfer from the TiO_2 conduction band to the ground state HPA in addition to photoreduction through the excited state of HPA. Thus, all reactions favor the oxidation and reduction of water without thermal energy but by using both the visible and near-UV light.

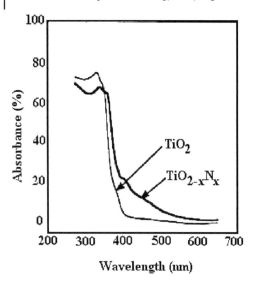

Figure 5.16 Experimental optical absorption spectra of $TiO_{2-x}N_x$ (thick lines) compared with TiO_2 (thin lines) [170].

Several attempts have been made to lower the band gap energy of TiO_2 by transition metal doping [166, 167] and reduction by hydrogen [168, 169], but no noticeable change in band gap energy of TiO_2 could be observed. This is because the doped materials suffered from thermal instability, an increase of carrier-recombination centers, or the requirement of an expensive ion-implantation facility [170, 171]. The reduction of TiO_2 introduces localized oxygen vacancy states located at 0.75 to 1.18 eV below the conduction band minimum of TiO_2. The energy levels of the optically excited electrons are lower than the redox potential of the hydrogen evolution (H_2/H_2O) located just below the conduction band minimum of TiO_2 [168, 169]. Hence the electron mobility in the bulk region is small due to the localization of oxygen vacancy states. Asahi et al. [170] used non-metals (such as C, N, F, P or S) for the doping rather than metals, in order to avoid the d-states deep in the band gap of TiO_2 which result in recombination centers. They found that doping with N was most effective since its p-states contributed to the band gap narrowing and hence a shift of the absorption edge to a lower energy by N doping (Figure 5.16).

This dominant transitions at the absorption edge have been identified with those from N $2p_\pi$ to Ti d_{xy}, instead of O $2p_\pi$ as in TiO_2. Jansen and Letschert [171] also confirmed similar band gap control by N, although they did not discuss the detailed electronic states. Khan et al. [172] investigated the possibility of using chemically modified (CM) n-TiO_2 by flame pyrrolysis of Ti metal sheet for the photogeneration of hydrogen from water. This procedure attempted to lower the band gap energy of TiO_2 so that it could absorb the visible light of the solar spectrum while retaining its stability (Figure 5.17). From the characterization studies,

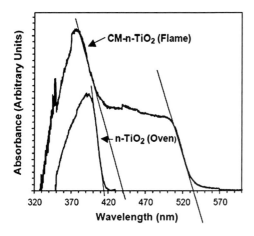

Figure 5.17 The UV-visible spectra of CM-n-TiO$_2$ (flame-made) and reference n-TiO$_2$ (electric tube furnace- or oven-made). The flame-made sample shows threshold wavelength of 414 nm (band gap of 3.0 eV) [172].

they concluded that the as-prepared samples mainly contained rutile structure, whereas the reference n-TiO$_2$ film showed a mixture of rutile and anatase crystalline forms. Using XPS analysis, they reported an average composition of CM-n-TiO$_2$ to be n-TiO$_{2-x}$C$_x$, where x is ~0.15.

Thus this material, in which carbon substitutes for some of the lattice oxygen atoms, absorbs light at wavelength below 535 nm and has a lower band gap energy than rutile (2.32 vs 3.0 eV). This material performs efficient water splitting with a total conversion efficiency of 11% and a maximum photoconversion efficiency of 8.35% when illuminated with visible light whereas under similar conditions n-TiO$_2$ shows a low efficiency (1%; Figure 5.18).

Another way to extend the absorption range of semiconductors is to mix two or more semiconducting materials with different band gaps and band positions so that an effective charge separation could occur between the contacting particles [173]. Various couples semiconductor systems [CdS/CdTe, GaInP$_2$/GaAs, Cu(In, Ga)Se$_2$, etc.] were proved to be effective for photoelectrolysis of water [36–44]. Tambwekar *et al.* [174] investigated (CdS–ZnS)–TiO$_2$ composite photocatalyst particles for the production of hydrogen from H$_2$S. They reported that the photocatalytic activity was related to the metal oxide support. Recently, Jang *et al.* [175] fabricated a composite photocatalyst composed of CdS nanoparticles deposited on TiO$_2$ nanosheets and studied its photoactivity for hydrogen generation from aqueous solution containing hole scavengers such as sulfide and sulfite. The synthesized CdS–TiO$_2$ composite material had nano-sheet morphology with a width of 70–80 nm, a thickness of 10–20 nm and a length of 200 nm. These results clearly showed that the decoration of active CdS nanoparticle onto TiO$_2$ nanosheet yielded a pronounced hydrogen production. Here, the main role of TiO$_2$ is to transfer the electrons generated from CdS, thereby preventing the electron-hole recombination.

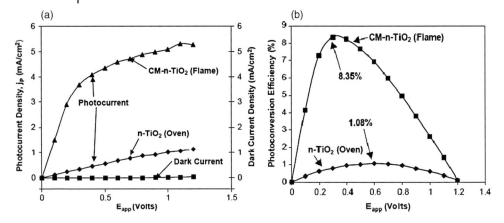

Figure 5.18 (a) Photocurrent density j_p as a function of applied potential E_{app} at CM-n-TiO$_2$ (flame-made) and the reference n-TiO$_2$ (electric tube furnace- or oven-made) photoelectrodes under xenon lamp illumination at an intensity of 40 mW cm^{-2}. Dark current densities at CM-n-TiO$_2$ (flame-made) as a function of applied potential are also shown. (b) Photoconversion efficiency as a function of applied potential potential E_{app} at CM-n-TiO$_2$ (flame-made) and the reference n-TiO$_2$ (electric tube furnace- or oven-made) photoelectrodes under xenon lamp illumination at an intensity of 40 mW cm^{-2} [172].

CdS nanoparticles deposited onto TiO$_2$ nanotube was also compared with trititanate nanotubes for photocatalytic water decomposition by Yao et al. [176], who reported that about 1708 μl g^{-1} of hydrogen was produced in 6 h with an apparent quantum yield of 0.1% at 420 nm. However, the trititanate nanotubes produced a small amount of hydrogen. The increase in the hydrogen production for CdS–TiO$_2$ nanotubes was attributed to a strong visible light absorption at wavelength 541 nm (Figure 5.19). Also, the coupling of semiconductors enhances the charge separation process.

Titanium oxide nanotubes coated with tungsten oxide were prepared to harvest more solar light and employed for water-splitting reactions [177]. The tungsten trioxide coatings significantly enhanced the visible spectrum absorption of the titanium dioxide nanotube array, as well as their photocurrents. Hydrogen gas was generated at an overall conversion efficiency of 0.87% upon white light illumination on a WO$_3$/TiO$_2$ nanotube with a 2.2:1 volume ratio of hydrogen and oxygen.

Susumu et al. [178] reported the hydrogen generation from the nanoparticles of CdS, ZnS, TiO$_2$ or AgI immobilized polyuria (PUA). This surface modification passivated the surface sulfur vacancies and substantially changed the particle surface characteristics. Photocatalytic generation of hydrogen on CdS-PUA or ZnS-PUA composites was studied in presence of propan-2-ol as a sacrificial agent. The results showed that ZnS-PUA was superior to CdS-PUA owing to the greater reducing ability of the conduction band electrons in ZnS than in CdS.

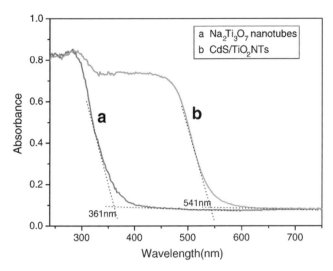

Figure 5.19 UV-Visible absorption spectra of trititanate nanotubes and CdS–TiO$_2$ nanotubes [176].

Yet another modification introduced to improve the efficiency of the TiO$_2$ photoelectrode was by admixing In$_2$O$_3$ and VO$_2$ with the nano-structured TiO$_2$ photoelectrode [179]. On comparing the hydrogen gas evolution kinetics of nano TiO$_2$ with admixtures (In$_2$O$_3$–TiO$_2$, VO$_2$–TiO$_2$), a clear enhancement was reported: from 13.1 l h^{-1} m^{-2} (TiO$_2$) to 14.6 l h^{-1} m^{-2} (for In$_2$O$_3$–TiO$_2$) and 16.0 l h^{-1} m^{-2} (for VO$_2$–TiO$_2$). The enhancement is suggested to be due to the surface modification of TiO$_2$ by thin In$_2$O$_3$ and VO$_2$ islands with high conductivity and a wide absorption range in the visible region.

5.3 Summary

In summary, the analysis of various reports provided in this review may provide an understanding of the uses of nano-sized semiconductor materials for hydrogen production reactions using solar energy. A number of challenges are yet to be met in order to commercialize the photocatalysis technique for fuel generation. However, the continuing efforts of scientists in the area of semiconductor photocatalysis have significantly improved the efficiency of the photocatalytic processes in recent years. The synthesis of efficient nanostructured materials has been shown to improve the photocatalytic efficiency of certain materials. For example, photoelectrochemical cells made of semiconductor nanotubes show higher efficiency towards water-splitting reactions. The doping of semiconductors with metals, using dye sensitizers and the use of mixed semiconductors have been shown to provide significant enhancement to the photocatalytic activities of semiconductors.

Acknowledgments

The authors thank the Department of Science and Technology (DST), New Delhi (INT/AUS/P-1/07 dated 19th September 2007) and the Department of Innovation, Industry, Science and Research (DIISR), Australia for the sanction of an India-Australia strategic research fund for their collaborative research. Also, S.A. thanks Prof. Dr. M. Chidambaram, Director, NIT, Trichy, for sanctioning a TEQIP abroad visit to the University of Melbourne, Australia and also for inviting Prof. M. Ashokkumar to NIT, Trichy under the TEQIP visiting professor program.

References

1. Blanchette, J.S., Jr. (2008) *Energy Policy*, **36**, 522–530.
2. Woodward, J., Orr, M., Cordray, K., and Greenbaum, E. (2000) *Nature*, **405**, 1014–1015.
3. Garcia, L., French, R., Czernik, S., and Chornet, E. (2000) *Appl. Catal. A: Gen.*, **201**, 225–239.
4. Steinfeld, A. (2005) *Sol. Energy*, **78**, 603–615.
5. Lichet, S. (2003) *J. Phys. Chem. B*, **107**, 4253–4260.
6. Lichet, S. (2002) *Semiconductor Electrodes and Photoelectrochemistry*, Wiley-VCH Verlag GmbH, Weinheim.
7. Lichet, S. (2005) *Int. J. Hydrogen Energy*, **30**, 459–470.
8. Fujishima, A., and Honda, K. (1972) *Nature*, **238**, 37–38.
9. Bard, A.J. (1980) *Science*, **207**, 139–144.
10. (a) Gratzel, M. (2001) *Nature*, **414**, 338–344. (b) Gratzel, M. (1983) *Energy Resources Through Photochemistry & Catalysis*, Academic Press, New York.
11. Kawai, T., and Sakata, T. (1980) *Nature*, **286**, 474–476.
12. Balzani, V., Moggi, L., Manfrin, M.F., Bolletta, F., and Gleria, M. (1975) *Science*, **189**, 852–856.
13. Kudo, A., Kato, H., and Nakagawa, S. (2000) *J. Phys. Chem. B*, **104**, 571–575.
14. Amouyal, E., Keller, P., and Moradpour, A. (1980) *J. Chem. Soc. Chem. Commun.*, 1019–1020.
15. Turner, J.A. (1999) *Science*, **285**, 687–689.
16. Hara, M., Waraksa, C.C., Lean, J.T., Lewis, B.A., and Mallouk, T.E. (2000) *J. Phys. Chem. A*, **104**, 5275–5280.
17. Ishikawa, A., Takata, T., Kondo, J.N., Hara, M., Kobayashi, H., and Domen, K. (2002) *J. Am. Chem. Soc.*, **124**, 13547–13553.
18. Amy, L.L., Guangqan, L., John, T., Jr., and Yates, J.T. (1995) *Chem. Rev.*, **95**, 735–758.
19. Harriman A., and West, M.A. (eds) (1982) *Photogeneration of Hydrogen*, Academic Press.
20. Konigstein, C. (1995) *J. Photochem. Photobiol. A: Chem.*, **90**, 141–152.
21. Amouyal, E. (1995) *Sol. Energy Mater. Sol. Cells*, **38**, 249–276.
22. Boltan, J.R. (1996) *Sol. Energy*, **57**, 37–50.
23. Gratzel, M. (2003) *J. Photochem. Photobiol.C: Photochem. Rev.*, **4**, 145–153.
24. Ashokkumar, M. (1998) *Int. J. Hydrogen Energy*, **23**, 427–438.
25. Ritterskamp, P., Kuklya, A., Wustkamp, M.A., Kerpen, K., Weidenthaler, C., and Demuth, M. (2007) *Angew. Chem. Int. Ed.*, **46**, 7770–7774.
26. Neumann, B., Bogdanoff, P., Tributsch, H., Sakthivel, S., and Kisch, H. (2005) *J. Phys. Chem. B*, **109**, 16579–16586.
27. Nakamura, R., and Nakato, Y. (2004) *J. Am. Chem. Soc.*, **126**, 1290–1298.
28. Qu, Z.W., and Kroes, G.J. (2006) *J. Phys. Chem. B*, **110**, 23306–23314.
29. Santato, C., Ulmann, M., and Augustynski, J. (2001) *J. Phys. Chem. B*, **105**, 936–940.

30 Khan, S.U.M., and Akikusa, J. (1999) *J. Phys. Chem. B*, **103**, 7184–7189.

31 Kay, A., Cesar, I., and Gratzel, M. (2006) *J. Am. Chem. Soc.*, **128**, 15714–15721.

32 Zou, Z., Ye, J., and Arakawa, H. (2000) *Chem. Phys. Lett.*, **332**, 271–277.

33 Zou, Z., Ye, J., Sayama, K., and Arakawa, H. (2001) *Nature*, **414**, 625–627.

34 Oshikiri, M., Boero, M., Ye, J., Zou, Z., and Kido, G. (2002) *J. Chem. Phys.*, **117**, 7313–7318.

35 Bharat, K.B., Ook, B.J., Sang Mi, L., Hyunju, C., Sang-Jin, M., and Wee, L.C. (2006) *Adv. Func. Mater.*, **16**, 1349–1354.

36 Hagfeldt, A., and Gratzel, M. (1995) *Chem. Rev.*, **95**, 49–68.

37 Nozik, A.J., and Memming, R. (1996) *J. Phys. Chem.*, **100**, 13061–13078.

38 Ashokkumar, M., and Maruthamuthu, P. (1988) *Int. J. Hydrogen Energy*, **13**, 677–680.

39 Khaselev, O., and Turner, J.A. (1998) *Science*, **280**, 425–427.

40 Lewis, N.S. (1998) *J. Phys. Chem. B*, **102**, 4843–4855.

41 Park, J.H., and Bard, A.J. (2006) *Electrochem. Solid State Lett.*, **9**, E5–E8.

42 de Tacconi, N.R., Chenthamarakshan, C.R., Yogeeswaran, G., Watcharenwong, A., de Zoysa, R.S., Basit, N.A., and Rajeswar, K. (2006) *J. Phys. Chem. B*, **110**, 25347–25355.

43 Ashokkumar, M., and Maruthamuthu, P. (1989) *Int. J. Hydrogen Energy*, **14**, 275–277.

44 Kelly, N., and Gibson, T.L. (2006) *Int. J. Hydrogen Energy*, **31**, 1658–1673.

45 Ekambaram, S., Yanagisawa, M., Uchida, S., Fujishiro, Y., and Sato, T. (2000) *Mol. Crys. Liq. Cryst.*, **341**, 213–218.

46 Kazuyuki, T., Hideyuki, T., and Takatoshi, M. (2007) *Materia*, **46**, 162–165.

47 Anna, K.V., Natalia, E.I., Alexander, S.L., Viktor, B.K., Alexandra, R.E., Valentina, L.I., Stepan, K.Y., Vladimir, I.G., and Piotr, M.A. (2008) *J. Photochem. Photobiol. A: Chem.*, **198**, 126–134.

48 Lu, D., Takata, T., Saito, N., Inoue, Y., and Domen, K. (2006) *Nature*, **440**, 295.

49 Macak, J.M., Tsuchiya, H., Ghicov, A., Yasuda, K., Hahn, R., Bauer, S., and Schmuki, P. (2007) *Curr. Opin. Solid State Mater. Sci.*, **11**, 3–18.

50 Zhuangjun, F., Zhongwei, Q., Tong, W., Jun, Y., and Shanshan, W. (2008) *Mater. Lett.*, **62**, 3345–3348.

51 Li, T., Yang, S., Huang, L., Gu, B., and Du, Y. (2004) *Nanotechnology*, **15**, 1479–1482.

52 Hu, S., and Wang, X. (2008) *J. Am. Chem. Soc.*, **130**, 8126–8127.

53 Wang, Y., Takahashi, K., Shang, H., and Cao, G. (2005) *J. Phys. Chem. B*, **109**, 3085–3088.

54 Huang, J., Liu, S., Wang, Y., and Ye, Z. (2008) *Appl. Surf. Sci.*, **254**, 5917–5920.

55 Tsuchiya, H., Macak, J.M., Ghicov, A., Taveira, L., and Schmuki, P. (2005) *Corros. Sci.*, **47**, 3324–3335.

56 Lai, M., Gonzalez, M., Jose, A., Gratzel, M., and Riley, D.J. (2006) *J. Mater. Chem.*, **16**, 2843–2845.

57 Ruescher, C.H., Bannat, I., Feldhoff, A., Ren, L., and Wark, M. (2007) *Micropor. Mesopor. Mater.*, **99**, 30–36.

58 Nowotny, J., Sorrell, C.C., Sheppard, L.R., and Bak, T. (2005) *Int. J. Hydrogen Energy*, **30**, 521–544.

59 Fukuzumi, S. (2008) *Eur. J. Inorg. Chem.*, **2008**, 1351–1362.

60 Abe, T., Suzuki, E., Nagoshi, K., Miyashita, K., and Kaneko, M. (1999) *J. Phys. Chem.*, **103**, 1119–1123.

61 Kato, H., Asakura, K., and Kudo, A. (2003) *J. Am. Chem. Soc.*, **125**, 3082–3089.

62 Nedeljkovic, J.M., Nenadovic, M.T., Micic, O.I., and Nozik, A.J. (1986) *J. Phys. Chem.*, **90**, 12–13.

63 Hoffmann, A.J., Mills, G., Yec, H., and Hoffmann, M.R. (1992) *J. Phys. Chem.*, **96**, 5546–5552.

64 Reagan, B.O., Moser, J., Anderson, M., and Gratzel, M. (1990) *J. Phys. Chem.*, **94**, 8720–8726.

65 Hodes, G., Howell, I.D.J., and Peter, L.M. (1992) *J. Electrochem. Soc.*, **139**, 3136–3140.

66 Richter, C., Wu, Z., Panaitescu, E., Willey, R.J., and Menon, L. (2007) *Adv. Mater.*, **19**, 946–948.

67 Ghicov, A., Tsuchiya, H., Macak, J.M., and Schmuki, P. (2005) *Electrochem. Commun.*, **7**, 505–509.
68 Macak, J.M., Tsuchiya, H., Ghicov, A., and Schmuki, P. (2005) *Electrochem. Commun.*, **7**, 1133–1137.
69 Vitiello, R.P., Macak, J.M., Ghicov, A., Tsuchiya, H., Dick, L.F.P., and Schmuki, P. (2006) *Electrochem. Commun.*, **8**, 544–548.
70 Bae, S., Shim, E., Yoon, J., and Joo, H. (2008) *Sol. Energy Mater. Sol. Cells*, **92**, 402–409.
71 Fujishima, A., Rao, T.N., and Tryk, D.A. (2000) *J. Photochem. Photobiol. C: Photochem. Rev.*, **1**, 1–21.
72 Bahnemann, D. (2004) *Sol. Energy*, **77**, 445–459.
73 Anandan, S., and Yoon, M. (2003) *J. Photochem. Photobiol. C: Photochem. Rev.*, **4**, 5–18.
74 Kamat, P.V. (2002) *J. Phys. Chem. B*, **106**, 7729–7744.
75 Mohapatra, S.K., Mahajan, V.K., and Misra, M. (2007) *Nanotechnology*, **18**, 445705.
76 Mor, G.K., Varghese, O.K., Wilke, R.H.T., Sharma, S., Shankar, K., Latempa, T.J., Choi, K.S., and Grimes, C.A. (2008) *Nano Lett.*, **8**, 1906–1911.
77 Yin, Y., Jin, Z., and Hou, F. (2007) *Nanotechnology*, **18**, 495608.
78 Yeredla, R.R., and Xu, H. (2008) *Nanotechnology*, **19**, 055706.
79 Adachi, M., Murata, Y., Harada, M., and Yoshikawa, Y. (2000) *Chem. Lett.*, **29**, 942–943.
80 Chu, S.Z., Inoue, S., Wada, K., Li, D., Haneda, H., and Awatsu, S. (2003) *J. Phys. Chem. B*, **107**, 6586–6589.
81 Varghese, O.K., Gong, D., Paulose, M., Ong, K.G., Dickey, E.C., and Grimes, C.A. (2003) *Adv. Mater.*, **15**, 624–627.
82 Mor, G.K., Carvalho, M.A., Varghese, O.K., Pishko, M.V., and Grimes, C.A. (2004) *J. Mater. Res.*, **19**, 628–634.
83 Varghese, O.K., Mor, G.K., Grimes, C.A., Paulose, M., and Mukherjee, N. (2004) *J. Nanosci. Nanotechnol.*, **4**, 733–737.
84 Paulose, M., Varghese, O.K., Mor, G.K., Grimes, C.A., and Ong, K.G. (2006) *Nanotechnology*, **17**, 398–402.
85 Mor, G.K., Shankar, K., Varghese, O.K., and Grimes, C.A. (2004) *J. Mater. Res.*, **19**, 2989–2996.
86 Mor, G.K., Shankar, K., Paulose, M., Varghese, O.K., and Grimes, C.A. (2005) *Nano Lett.*, **5**, 191–195.
87 Varghese, O.K., Paulose, M., Shankar, K., Mor, G.K., and Grimes, C.A. (2005) *J. Nanosci. Nanotechnol.*, **5**, 1158–1165.
88 Uchida, S., Chiba, R., Tomiha, M., Masaki, N., and Shirai, M. (2002) *Electrochemistry*, **70**, 418–420.
89 Adachi, M., Murata, Y., Okada, I., and Yoshikawa, Y. (2003) *J. Electrochem. Soc.*, **150**, G488–G493.
90 Mor, G.K., Shankar, K., Paulose, M., Varghese, O.K., and Grimes, C.A. (2006) *Nano Lett.*, **6**, 215–218.
91 Paulose, M., Shankar, K., Varghese, O.K., Mor, G.K., Hardin, B., and Grimes, C.A. (2006) *Nanotechnology*, **17**, 1446–1448.
92 Mor, G.K., Varghese, O.K., Paulose, M., Shankar, K., and Grimes, C.A. (2006) *Sol. Energy Mat. Sol. Cells*, **90**, 2011–2075.
93 Shankar, K., Mor, G.K., Prakasam, H.E., Yoriya, S., Paulose, M., Varghese, O.K., and Grimes, C.A. (2007) *Nanotechnology*, **18**, 065707.
94 Hagfeldt, A., Bjorksten, U., and Lindquist, S.E. (1992) *Sol. Energy Mat. Sol. Cells*, **27**, 293–304.
95 Marin, F.I., Hamstra, M.A., and Vanmaekelbergh, D. (1996) *J. Electrochem. Soc.*, **143**, 1137–1142.
96 Lubberhuizen, W.H., Vanmaekelbergh, D., and Van Faassen, E. (2000) *J. Porous Mater.*, **7**, 147–152.
97 Van de Lagemaat, J., Plakman, M., Vanmaekelbergh, D., and Kelly, J. (1996) *J. Appl. Phys. Lett.*, **69**, 2246–2248.
98 Paulose, M., Mor, G.K., Varghese, O.K., Shankar, K., and Grimes, C.A. (2006) *J. Photochem. Photobiol. A: Chem.*, **178**, 8–15.
99 Mohapatra, S.K., Misra, M., Mahajan, V.K., and Raja, K.S. (2007) *J. Catal.*, **246**, 362–369.
100 Mohapatra, S.K., Raja, K.S., Mahajan, V.K., and Misra, M. (2008) *J. Phys. Chem. C*, **112**, 11007–11012.

101 Mohapatra, S.K., and Misra, M. (2007) *J. Phys. Chem. C*, **111**, 11506–11510.
102 Mohapatra, S.K., Misra, M., Mahajan, V.K., and Raja, K.S. (2007) *J. Phys. Chem. C*, **111**, 8677–8685.
103 Park, J.H., Kim, S., and Bard, A.J. (2006) *Nano Lett.*, **6**, 24–28.
104 Abraham, W., Kuykendall, T.R., Wei, C., Shaowei, C., and Zhang, Z.J. (2006) *J. Phys. Chem. B*, **110**, 25288–25296.
105 Takahara, Y., Kondo, J.N., Takata, T., Lu, D., and Domen, K. (2001) *Chem. Mater.*, **13**, 1194–1199.
106 Sayama, K., and Arakawa, H. (1994) *J. Photochem. Photobiol. A: Chem.*, **77**, 243–247.
107 Park, J.H., and Bard, A.J. (2005) *Electrochem. Solidstate. Lett.*, **8**, G371–375.
108 Domen, K., Kudo, A., Onishi, T., Kosugi, N., and Kuroda, H. (1986) *J. Phys. Chem.*, **90**, 292–295.
109 Zheng, N., Bu, X., Vu, H., and Feng, P. (2005) *Angew. Chem. Int. Ed.*, **44**, 5299–5303.
110 Ishihara, T., Nishiguchi, H., Fukamachi, K., and Takita, Y. (1999) *J. Phys. Chem. B*, **103**, 1–3.
111 Wu, J., Lin, J., Yin, S., and Sato, T. (2001) *J. Mater. Chem.*, **11**, 3343–3347.
112 Ogura, S., and Inoue, Y. (2000) *Phys. Chem. Chem. Phys.*, **2**, 2449–2454.
113 Ji-Jun, Z., Chang-Jun, L., and Yue-Ping, Z. (2006) *Langmuir*, **22**, 2334–2339.
114 Yoshida, H., Kato, S., Hirao, K., Nishimoto, J., and Hattori, T. (2007) *Chem. Lett.*, **36**, 430–431.
115 Luzzi, A. (ed.)(2004) Photoelectrolytic production of Hydrogen, Final report of Annex 14, International Energy Agency Hydrogen implementing Agreement www.ieahia.org (accessed 26 September 2008).
116 Carroll, B.H. (1977) *Photogr. Sci. Eng.*, **21**, 151–163.
117 Gerischer, H.J. (1977) *J. Electroanal. Chem.*, **82**, 133–143.
118 Broich, B., and Heiland, G. (1980) *Surf. Sci.*, **92**, 247–264.
119 Matsumura, M., Mitsuda, K., and Tsubomura, H. (1983) *J. Phys. Chem.*, **87**, 5248–5251.
120 Shimidzu, T., Iyoda, T., Koide, Y., and Kanada, N. (1983) *Nouv. J. Chim.*, **7**, 21.
121 Ghosh, P.K., and Spiro, T.G. (1980) *J. Am. Chem. Soc.*, **102**, 5543–5549.
122 Memming, R. (1980) *Surf. Sci.*, **101**, 551–563.
123 Mackor, A., and Blasse, G. (1981) *Chem. Phys. Lett.*, **77**, 6–8.
124 Breddels, P.A., and Blasse, G. (1981) *Chem. Phys. Lett.*, **79**, 209–213.
125 Chang, B.T., Campet, G., Claverie, J., and Hagenmuller, P. (1984) *Bull. Chem. Soc. Jpn.*, **57**, 2574–2577.
126 Clark, W.D.K., and Sutin, N. (1977) *J. Am. Chem. Soc.*, **99**, 4676–4682.
127 Spitler, M.T., and Calvin, M.J. (1977) *J. Chem. Phys.*, **66**, 4294–4305.
128 Hamnett, A., Dare-Edwards, M.P., Wright, R.D., Seddon, K.R., and Goodenough, J.B. (1979) *J. Phys. Chem.*, **83**, 3280–3290.
129 Dare-Edwards, M.P., Goodenough, J.B., Hamnett, A., Seddon, K.R., and Wright, R.D. (1980) *Faraday Discuss. Chem. Soc.*, **70**, 285–298.
130 Giraudeau, A., Fan, F.R.F., and Bard, A.J. (1980) *J. Am. Chem. Soc.*, **102**, 5137–5142.
131 Amouyal, E., and Koffi, P. (1985) *J. Photochem.*, **29**, 227–242.
132 Desilvestro, J., Gratzel, M., Kavan, M., and Moser, J. (1985) *J. Am. Chem. Soc.*, **107**, 2988–2990.
133 Loy, L., and Wolf, E.E. (1985) *Sol. Energy*, **34**, 455–461.
134 Minero, C., Lorenzi, E., Pramauro, E., and Pelizzetti, E. (1984) *Inorg. Chim. Acta*, **91**, 301–305.
135 Nakahira, T., and Gratzel, M. (1985) *Macromol. Chem. Rapid Commun.*, **6**, 341–347.
136 Quint, R.M., and Getoff, N. (1988) *Int. J. Hydrogen Energy*, **13**, 269–276.
137 Furlong, D.N., Wells, D., and Sasse, W.H.F. (1986) *J. Phys. Chem.*, **90**, 1107–1115.
138 Duonghong, D., Serpone, N., and Gratzel, M. (1984) *Sci. Pap. Inst. Phys. Chem. Res. Jpn.*, **78**, 232–236.
139 Duonghong, D., Serpone, N., and Gratzel, M. (1984) *Helv. Chim. Acta*, **67**, 1012–1018.
140 Borgarello, E., Pelizzetti, E., Ballardini, R., and Scandola, F. (1984) *Nouv. J. Chim.*, **8**, 567.

141 Borgarello, E., Kiwi, J., Pelizzetti, E., Visca, M., and Gratzel, M. (1981) *Nature*, **289**, 158–160.
142 Duonghong, D., Borgarello, E., and Gratzel, M. (1981) *J. Am. Chem. Soc.*, **103**, 4685–4690.
143 Borgarello, E., Kiwi, J., Pelizzetti, E., Visca, M., and Gratzel, M. (1981) *J. Am. Chem. Soc.*, **103**, 6324–6329.
144 Borgarello, E., Kiwi, J., Gratzel, M., Pelizzetti, E., and Visca, M. (1982) *J. Am. Chem. Soc.*, **104**, 2996–3002.
145 Houlding, V.H., and Gratzel, M. (1983) *J. Am. Chem. Soc.*, **105**, 5695–5696.
146 Moser, J., and Gratzel, M. (1984) *J. Am. Chem. Soc.*, **106**, 6557–6564.
147 Furlong, D.N., Wells, D., and Sasse, W.H.F. (1985) *J. Phys. Chem.*, **89**, 1922–1928.
148 Hashimoto, K., Kawai, T., and Sakata, T. (1984) *J. Phys. Chem.*, **88**, 4083–4088.
149 Maruthamuthu, P., Muthu, S., Gurunathan, K., Ashokkumar, M., and Sastri, M.V.C. (1992) *Int. J. Hydrogen Energy*, **17**, 863–866.
150 Shimidzu, T., Iyoda, T., and Koide, Y. (1985) *J. Am. Chem. Soc.*, **107**, 35–41.
151 Dhanalakshmi, K.B., Latha, S., Anandan, S., and Maruthamuthu, P. (2001) *Int. J. Hydrogen Energy*, **26**, 669–674.
152 Marcus, R.J. (1965) *Science*, **123**, 399–405.
153 Kiwi, J., Kalyanasundaram, K., and Gratzel, M. (1982) *Struct. Bonding*, **49**, 37–125.
154 Detellier, C., and Villemure, G. (1984) *Inorgan. Chim. Acta*, **86**, L19–L20.
155 Villemure, G., Kodama, H., and Detellier, C. (1985) *Can. J. Chem.*, **63**, 1139–1142.
156 Della Guardia, R.A., and Thomas, J.K. (1983) *J. Phys. Chem.*, **87**, 990–998.
157 Quayle, W.H., and Lunsford, J.H. (1982) *Inorg. Chem.*, **21**, 97–103.
158 Dutta, P.K., and Borja, M. (1993) *Nature*, **362**, 43–45.
159 Kincaid, J.R. (2000) *Chem. Eur. J.*, **6**, 4055–4061.
160 Yamamoto, E.H., Kim, Y.I., Schmehl, R.H., Wallin, J.O., Shoulders, B.A., Richardson, B.R., Haw, J.F., and Mallouk, T.E. (1994) *J. Am. Chem. Soc.*, **116**, 10557–10563.
161 Breck, D.W. (1974) *Zeolite Molecular Sieves*, John Wiley, New York.
162 Jacobs, P.A., and Uytterhoeven, J.B. (1977) *J. Chem. Soc. Chem. Commun.*, 128–129.
163 Kuznicke, S.M., and Eyring, E.M. (1978) *J. Am. Chem. Soc.*, **100**, 6790–6791.
164 Anandan, S., and Yoon, M. (2003) *J. Photochem. Photobiol. A: Chem.*, **160**, 181–184.
165 Yamase, T., Takabayashi, N., and Kaji, M. (1984) *J. Chem. Soc. Dalton Trans.*, 793–799.
166 Anpo, M. (1997) *Catal. Surv. Jpn.*, **1**, 169–179.
167 Moon, S.C., Matsumura, Y., Kitano, M., Matsuoka, M., and Anpo, M. (2003) *Res. Chem. Intermed.*, **29**, 233–256.
168 Breckenridge, R.C., and Hosler, W.R. (1953) *Phys. Rev.*, **91**, 793–802.
169 Cronemeyer, D.C. (1959) *Phys. Rev.*, **113**, 1222–1226.
170 Asahi, R., Morikawa, T., Ohwaki, T., Aoki, K., and Taga, Y. (2001) *Science*, **293**, 269–271.
171 Jansen, M., and Letschert, H.P. (2000) *Nature*, **404**, 980.
172 Khan, S.U.M., Al-shahry, M., and Ingler, W.B., Jr. (2002) *Science*, **297**, 2243–2245.
173 Serpone, N., Borgarello, E., and Gratzel, M. (1984) *Chem. Commun.*, **6**, 342–344.
174 Tambwekar, S.V., Venugopal, D., and Subrahmanyam, M. (1999) *Int. J. Hydrogen Energy*, **24**, 957–963.
175 Jang, J.S., Choi, S.H., Park, H., Choi, W., and Lee, J.S. (2006) *J. Nanosci. Nanotechnol.*, **6**, 3642–3646.
176 Yao, J.Z., Wei, Y., Yan, P.W., and Zhen, H.W. (2008) *Mater. Lett.*, **62**, 3846–3848.
177 Park, J.H., and Park, O.O. (2006) *Appl. Phys. Lett.*, **89**, 1631061–1631063.
178 Susumu, S., Takayuki, H., and Isao, K. (1998) *Chem. Commun.*, 1439–1440.
179 Karn, R.K., Mridula, M., and Srivastava, O.N. (2000) *Int. J. Hydrogen Energy*, **25**, 407–413.

Part Two
Efficient Energy Storage

6
Nanostructured Materials for Hydrogen Storage
Saghar Sepehri and Guozhong Cao

6.1
Introduction

During recent years, significant progress has been made in the development of alternative energy technologies. Hydrogen has the potential to be a good energy carrier candidate in a carbon-free emission cycle. There are three components to the hydrogen economy: production, storage and usage. Storing hydrogen is a challenging step in hydrogen technology and considerable efforts have been made in synthesizing and investigating novel materials for hydrogen storage in the past decade [1]. Ideally, hydrogen should be stored in such a way as to attain a high storage capacity under near ambient conditions, to be safe and economical and to perform a rapid and reversible hydrogenation and dehydrogenation process for practical applications. Although various techniques and materials have been used or studied to store hydrogen, there is neither method nor material that satisfies all the requirements for perceived hydrogen economy [2]. Hydrogen can be stored as gas, liquid or solid. Storing hydrogen as a compressed gas needs high pressure and heavy containers to support such pressure. Liquification of hydrogen needs energy and consumes more than 20% of the recoverable energy. Also cryogenic containers should be used to decrease the hydrogen boil-off.

Storing hydrogen as solid may offer the best option to store hydrogen through two basic mechanisms: physisorption (or physical adsorption) and chemisorption (or chemical adsorption). In physisorption, molecular hydrogen is adsorbed by intermolecular (van der Waals) forces. Examples of physisorption include the storing of hydrogen in carbon structures and organic or inorganic frameworks. In chemisorption, hydrogen molecules and chemical bonding of the hydrogen atoms dissociate by integration in the lattice of a metal or an alloy, or by the formation of a new chemical compound. Metal, chemical and complex hydrides are examples of chemisorption. Each principle has its own prospects and limitations. A given material can exhibit both chemisorption and physisorption. Chemisorption may provide high volumetric and gravimetric storage capacities, but the chemical bonds need to split or recombine to release hydrogen. Storing hydrogen by physisorption is not subject to this constraint, because the hydrogen stays in its

Nanotechnology for the Energy Challenge. Edited by Javier Garcia-Martinez
© 2009 WILEY-VCH Verlag GmbH & Co. KGaA, Weinheim
ISBN: 978-3-527-32401-9

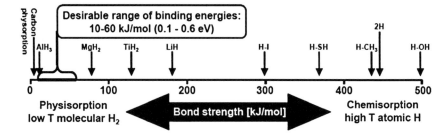

Figure 6.1 Bond strengths for physisorption and chemisorption and the desirable range of binding energies that allow hydrogen release around room temperature [3].

molecular form, but the challenge is to provide materials with a sufficient amount of bonding sites for the hydrogen per volume to achieve a high storage capacity. One of the major differences between physisorption and chemisorption is their binding energies (Figure 6.1). Physisorption bonding is usually too weak (<10 kJ/mol), thus demands cryogenic temperatures for significant storage capacity. Chemisorption shows a stability that is too high (>50 kJ/mol) and demands high desorption temperatures [3].

To achieve an ideal binding energy (in the range of 10–60 kJ/mol), we need to increase the physisorption binding energy or to reduce the chemisorption binding energy. Nanostructured materials with their unique physical, chemical, thermodynamic and kinetics properties can provide effective and sufficient ways to address the challenges involve in hydrogen storage [4, 5]. Nanostructures can offer new opportunities for addressing these challenges. They have the potential for high surface areas and hybrid structures that allow multifunctional performance. This chapter gives an overview of current achievements in developing nanostructured materials and methods for hydrogen storage.

6.2
Hydrogen Storage by Physisorption

Physisorption is a principle where the forces involved are weak intermolecular forces; therefore in general it is associated with fast kinetics and reversibility. But these weak forces also present a challenge for the physisorption of hydrogen. H_2 is the smallest molecule and only has two electrons, hence it is hard to polarize; and in the absence of relatively strong polarizing centers the interaction between the adsorbent and the non-polar hydrogen molecules relies on the weak dispersion forces which are created by temporarily induced dipoles and are typically of the order of 3–6 kJ/mol [6]. Thus, significant hydrogen adsorption often takes place only at a cryogenic temperature. Nanostructured materials may offer advantages for molecular hydrogen storage by providing high surface areas, or encapsulation or trapping hydrogen in microporous media. Using porous nanostructured materials, in general, can reduce the gravimetric and the volumetric storage densities. However, the increased surface area and porosity in nanostructures offers additional binding sites on the surface and in the pores that could increase storage mainly through physisorption. The possibility of storing a significant amount of hydrogen on high surface area density materials has been a key driver in the investigation of the hydrogen sorption properties of nanotubes, graphite sheets, metal–organic frameworks and template ordered porous carbons. Nanostructured carbons, zeolites, metal–organic frameworks, clathrates and polymers with intrinsic microporosity are examples of the investigated physisorption materials.

6.2.1
Nanostructured Carbon

Carbon materials with a high surface area, good chemical stability and low density have received considerable attraction. Nanostructured carbon materials, such as graphitic nanofibers (GNF), multiwalled carbon nanotubes (MWNT), single-walled carbon nanotubes (SWNT), carbon nanorods and carbon aerogels, demonstrate novel but distinct properties that relate to the many possible configurations of the electronic states of carbon atoms. Each carbon atom has six electrons, which occupy $1s^2$, $2s^2$ and $2p^2$ atomic orbitals. The various bonding states are connected with certain structural arrangements, so that sp bonding gives rise to chain structure, sp^2 bonding to planar structures and sp^3 bonding to tetrahedral structures [7]. The structural and practical properties of carbon critically depend on the ratio between the number of sp^2 (graphite-like) and sp^3 (diamond-like) bonds.

Early reports [8, 9] on hydrogen storage in carbon nanotubes and graphitic nanofibers proposed high storage capacities (to 67 wt%) and started an extensive worldwide surge of research. Since then many succeeding experiments have been carried out with different methods, but such high values have not yet been reproduced by other groups [10]. Furthermore, no hypothesis could support the unusually high storage capacities and the high storage capacity results were more related to the faults of experiment [11, 12]. Nevertheless, hydrogen adsorption on carbon materials is still an attractive and improving field. The result of several investigations proposes that the amount of adsorbed hydrogen is proportional to the specific surface area of the carbon material [13, 14]. In the case of activated carbons and activated carbon fibers, the hydrogen absorption of 5 wt% is obtained at low temperature (77 K) and high pressure (30–60 bar; n.b. 30 bar = 3000 kPa) [15]. For GNF, SWNT and MWNT, the reversible hydrogen uptake of 1.5 wt% per 1000 m^2/g under ambient conditions is reported [16]. A hydrogen capacity of 7 wt% is observed for ordered porous carbon with surface area of 3200 m^2/g, prepared by template, at 77 K and 20 bar [17]. Recent studies on carbon aerogels (CAs), another class of amorphous porous carbon structures with high surface area, shows 5 wt% of hydrogen adsorption for surface area of 3200 m^2/g at 77 K and 20–30 bar (Figure 6.2) [18].

Recent research in hydrogen physisorption on carbon nanostructures involves efforts on increasing the surface area of carbon to provide more binding sites and

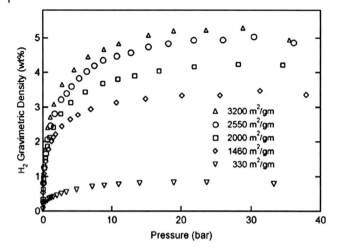

Figure 6.2 Adsorption isotherms at 77 K for the carbon aerogels shows the linear dependency of hydrogen adsorption to the surface area [18].

incorporating functional groups (dopants) in carbon to increase the binding energy between hydrogen and carbon surfaces [19].

6.2.2
Zeolites

Zeolite is an inorganic porous material consisted of a hydrated aluminosilicate mineral with a highly regular structure that exhibits a reversible occlusion of gases. However, because of the high density of the aluminosilicate framework, which contains Si, Al, O and heavy cations, a high gravimetric hydrogen storage density might not be achieved. Yet, zeolites can be ideal choice in studying the hydrogen binding because of their well-known crystal structure and easy ion exchange and those studies may provide a valuable insight for work on other hydrogen adsorbents. The working principle of hydrogen storage in zeolites is that the guest molecules under high temperature and pressure are forced into the cavities of the molecular sieve host. Upon cooling to room temperature or below, hydrogen is trapped inside the cavities; and it can be released again by raising the temperature. The amount of encapsulated hydrogen in zeolite is related to the size of the exchanged cation and a higher storage capacity is observed for zeolites with a high number of small cavities in their structure [20]. It has been showed that the storage capacity of zeolite may be increased at low temperatures; and a hydrogen storage capacity of 1.81 wt% (at 15 bar and 77 K) was obtained for NaY zeolite [21].

The calculated maximum possible hydrogen storage capacity of zeolites is less than 3 wt% [22]. The large mass of the zeolite framework is a limiting factor in the storage capacity; and improvements in the storage capacity of zeolites will therefore include using light elements in the framework and also enhancing the interaction energy between hydrogen and the framework.

6.2.3
Metal–Organic Frameworks

Metal–organic frameworks (MOFs) are crystalline solid compounds consisting of organic ligands connecting metal ions or clusters that form a cage structure. Most MOFs have a three-dimensional interconnected porous framework with uniform pores that provides an ordered network of channels. MOFs can be synthesized using a self-organizational process that allows different combinations of organic linkers and provides a wide range of different functionalities and and pore sizes [23, 24]. MOFs can provide a light porous framework with high surface areas and pore volumes. The surface area of MOFs are usually in the range of 500–3000 m^2/g, while values higher than 5000 m^2/g are also attainable [25]. Large pore volumes such as 1.1 cm^3/g are observed for some MOFs [26]. Similar to carbon nanostructures, the hydrogen storage capacity of MOFs increases with the surface area and microporous volume [5]. At low temperature (77 K) and high pressure (70–90 bar) MOFs with a hydrogen adsorption of 7 wt% are reported [27, 28] but at 298 K and 90 bar the maximum observed hydrogen adsorption is only 1.4 wt% [28].

Increasing the interaction of hydrogen with the organic ligands and metal centers in MOFs can improve the hydrogen adsorption at ambient temperature. Several approaches being investigated include obtaining MOFs with more polarized cations or optimized pore size [29–31].

6.2.4
Clathrates

Clathrates are crystalline structures consisting of a hydrogen-bonded water framework as the "host" lattice providing cavities which hold "guest" molecules. Several natural gases, including methane and carbon dioxide, are well-known to form water clathrates or clathrate hydrates [32] and since the first reports on hydrogen clathrate hydrates [33] possible hydrogen storage in different hydrogen clathrates has promoted an attractive research [34].

Hydrogen-bonded water molecules can produce polyhedral small and large cages around guest molecules to form solid clathrate hydrates. When these cages are empty, they are not stable and may collapse into an ice crystal structure, but the inclusion of gas molecules can stabilize the cages. The formation and stability of H_2/H_2O clathrates need high pressure and low temperatures; therefore research on facilitating the formation of these structures and stabilizing them is necessary.

A binary hydrogen–water clathrate is reported to contain 5.3 wt% hydrogen at 250 K and very high pressure (2 kbar) [35]. To ease the fabrication process and reduce the synthesis pressure a second guest component (such as tetrahydrofuran) can be used to fill the cages of clathrate, but this approach decreases the hydrogen storage capacity of the clathrate to less than 4 wt% [36]. Nonetheless, clathrates have opened up an interesting field for further research.

6.2.5
Polymers with Intrinsic Microporosity

Polymers with intrinsic microporosity (PIMs) are obtained by polymerization of large rigid molecules to form chains and networks that contain interconnected pores and large surface areas (500–1000 m^2/g) [37]. PIMs can be considered as potential hydrogen storage materials because of the low density and large surface area [38]. For network–PIM and hyper-cross-linked polymer a maximum hydrogen adsorption (at 77 K and 10–15 bar) of 3 wt% is observed [39]. In order to improve the hydrogen adsorption on PIMs, they must be optimized further to improve their porosity. However, hydrogen adsorption at near ambient temperature can be much lower than at 77 K due to the weak interaction between the hydrogen and polymer.

6.3
Hydrogen Storage by Chemisorption

Chemisorption is the adsorption of a particle with the formation of a chemical bond. Hydrogen can be stored in hydrides (hydrogen-rich materials) by chemisorption to offer high storage capacity at ambient conditions. Volumetric and gravimetric hydrogen densities of some selected hydrides are compared with other hydrogen storage methods in Figure 6.3 [40].

However, various hydrides suffer from a range of drawbacks, such as poor reversibility, poor thermal conductivity and relatively high dehydrogenation temperature [41]. Developing new hydrides is a very active research topic. Nanostructures can be used to improve the hydrogen storage properties of hydrides. They can change the thermodynamic properties of a hydride which define the theoretical working parameters, that is, the pressure and temperature at which hydrogen can be absorbed and desorbed. They can also change the kinetic properties that determine the rate of hydrogen release. This section discusses recent findings and developments in this field.

6.3.1
Metal and Complex Hydrides

Metal hydrides are solid alloys which are typically composed of metal atoms with a host lattice and hydrogen atoms that are trapped in the interstitial sites, forming a single-phase compound between a metal host and hydrogen. Binary hydrides can essentially be classified into three categories depending on the nature of the bonding between hydrogen and the metal host. Ionic or saline hydrides (e.g., MgH_2, NaH, CaH_2) are formed by alkali and alkaline earth atoms and exhibit ionic

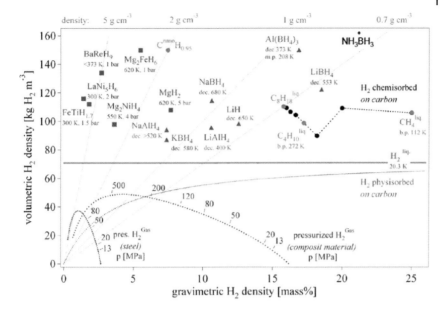

Figure 6.3 The storage density of hydrogen in compressed gas, liquid, adsorbed monolayer (physisorbed), and selected chemical compounds, as a function of the hydrogen mass fraction. The straight lines indicate the total density of the storage medium including, hydrogen and host atoms. Pressurized gas storage is shown for steel (tensile strength 460 MPa, density 6500 kg/m^3) and a hypothetical composite material (tensile strength 1500 MPa, density 3000 kg/m^3) [40].

bonding between the hydrogen and metal atoms. Covalent hydrides are formed by non-metal elements like S, Si, C or B. Metallic hydrides (e.g., LaNi$_5$H$_6$, PdH$_{0.6}$, FeTiH$_2$) originate from the metallic bonding between hydrogen and either a transition metal or a rare earth metal. In addition, group IA, IIA and IIIA light metals form metal–hydrogen complexes (e.g., AlH$_4^-$, BH$_4^-$) which form covalent or ionic bonds with a cation, giving rise to highly stable complex hydrides (e.g., NaAlH$_4$, Mg(AlH$_4$)$_2$). Hydrides have a higher hydrogen storage density than hydrogen in gas or liquid form. For example the hydrogen density of MgH$_2$ is 6.5 H atoms/cm^3, while those of gas and liquid hydrogen are 0.99 and 4.2 H atoms/cm^3, respectively [42].

Some metal hydrides absorb and desorb hydrogen at near ambient temperature and pressure, they demonstrate a very high hydrogen density. However, all the reversible hydrides working around ambient temperature and atmospheric pressure consist of heavy transition metals; therefore, even in nanocrystalline form, the gravimetric hydrogen density of metal hydrides is limited to less than 3 wt% [43]. One way to improve the hydrogen capacity of metal hydrides is to use lightweight materials such as magnesium [41].

Another challenge in hydrogen desorption from metal hydrides is their stability, demanding elevated temperatures for the release of hydrogen. Heat transfer is yet

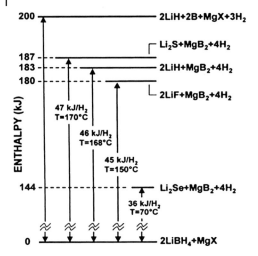

Figure 6.4 Enthalpy diagram for destabilization of LiBH$_4$ by MgX (X = H$_2$, F$_2$, S, Se) [47].

another challenge. In general, the formation of metal hydrides is an exothermic reaction. Efficient heat removal (for absorption) and heat addition (for desorption) has proven extremely difficult to achieve in metal hydride-based systems [3].

The synthesis of new hydrides, particularly complex hydrides, has the potential to develop materials with superior hydrogen storage properties. Complex metal hydrides demonstrate higher gravimetric hydrogen capacities than simple metal hydrides. However, some of them show poor reversibility and once hydrogen is released they need high pressure to adsorb hydrogen again [44]. Moreover, due to the localization of the hydrogen and the slow diffusion rate of the metals in the solid, hydrogen sorption reactions are slow. Also, similar to metal hydrides, most complex hydrides suffer from high thermal stability [45].

Destabilization approaches can be used to improve the hydrogen storage properties of stable hydride materials [46]. Thermodynamic destabilization of light-metal hydrides is achieved by using additives that react with metals to form new compounds (as intermediate states) during dehydrogenation and lower the enthalpy and hydrogen release temperature. Figure 6.4 shows the decreased dehydrogenation enthalpies for LiBH$_4$ by adding destabilization agents [47].

However, the dehydrogenation temperature for destabilized LiBH$_4$ is still higher than 350 °C and the kinetics are slow. Also, it should be mentioned that adding any additive may increase the mass and reduce the hydrogen capacity, for example, the hydrogen capacity decreases from 13.6 wt% in pure LiBH4 to 11.4 wt% after adding MgH$_2$, however, the enthalpy is lowered by 25 kJ/mol H$_2$ [47]. Improving the reaction kinetics by decreasing the size of hydride may offer an interesting approach without increasing the mass. Increased surface area in nanosized hydrides can augment their surface energy and reduce the dehydrogenation enthalpy drastically. The increased surface area of hydrides facilitates the dissociation of hydrogen atoms by offering a larger number of dissociation sites and

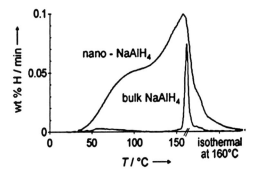

Figure 6.5 The improved hydrogen desorption for sodium alanate (NaAlH$_4$) supported on carbon nanofibers [50].

allowing fast gaseous diffusion [3]. Different methods (including laser ablation, vapor condensation, sputtering, ball milling) can be used to reduce the size of metal hydride particles. Reduced particle and crystallite size is shown to enhance the hydrogen sorption kinetics in aluminum- and magnesium-based hydrides [48, 49]. Figure 6.5 demonstrates the effect of size reduction on the dehydrogenation kinetics of nanosized sodium alanate (NaAlH$_4$), deposited as clusters on carbon nanofibers. The enhanced hydrogen desorption in nano-NaAlH$_4$ can be attributed to the minimization of the solid-state diffusion path length during the loading and desorption of hydrogen [50].

In conclusion, although metal and complex hydrides are considered to be potential candidates for hydrogen storage, significant fundamental research should be performed to obtain hydrides with practical hydrogen storage properties. The dehydrogenation temperatures should be decreased and the kinetics of reaction should be improved. Other issues also need to be addressed, including thermal management, reversibility and durability [1].

6.3.2
Chemical Hydrides

Chemical hydrides store hydrogen as M–H bonds, where M is a light main group element such as C, B, N, or O [51]. They can release hydrogen through a chemical reaction which is typically not easily reversible. Sometimes, metal and complex hydrides are also categorized as chemical hydrides, but those classes often refer to reversible dehydrogenation. Common reactions to release hydrogen from chemical hydrides involve the reaction with water (hydrolysis) or alcohols (alcoholysis) and thermal decomposition (pyrolysis). In all these methods, several issues such as controllability of reaction and regeneration energy should be considered.

A number of chemical hydrides, with both exothermic and endothermic dehydrogenation through different reaction are currently under investigation. Moreover, new chemical hydrides with high hydrogen densities can offer promising approaches for hydrogen storage.

One of the early chemical hydrides studied, ammonia (NH_3) has been used in fuel cells and power plants for more than 40 years [52, 53]. Anhydrous ammonia has a high gravimetric hydrogen density of 17.5 wt% and the byproduct of the hydrogen dissociation process is nitrogen which has no adverse environmental effects. However, decomposition (cracking) of ammonia is an endothermic reaction that happens efficiently at temperatures higher than 500 °C, with an enthalpy of +46 kJ/mol. Therefore, it takes energy to gain hydrogen from ammonia. There are also safety and toxicity issues, such as a propensity for reacting with water, reactivity with container materials and the high toxicity of the vapor if released into the air. These drawbacks should be addressed before using ammonia as a hydrogen storage material; however, because of the high hydrogen density and well-established technology, ammonia is being considered as a means for delivering hydrogen.

Several boron hydrides have a high hydrogen content. Ammonia–borane (AB), also known as borazane or by formula NH_3BH_3, is of great interest as a hydrogen storage material. At ambient temperature and pressure, AB is a stable, white, crystalline solid (orthorhombic at lower temperatures, tetragonal above −50 °C), with a low molecular weight (30.7 g/mol) and a high gravimetric hydrogen capacity (19.6 wt%) [54]. There have been several experimental studies on the multistep thermal decomposition of AB [55–57]. It was found that AB releases one mole of hydrogen (per mole of AB) around 110 °C and a second mole of hydrogen at 150 °C. Different methods, including milling and catalysts, have been used to improve the kinetics of AB dehydrogenation reactions and lower its decomposition temperature [58]. These findings are encouraging; however, future research should address several issues including reducing the dehydrogenation temperature, minimizing the formation of volatile products and developing economically viable methods for the regeneration of AB [59].

6.3.3
Nanocomposites

Nanocomposites refer to materials consisting of at least two phases, with one dispersed in another (called the matrix) to form a three-dimensional network [60]. At the nano-scale, materials can show distinctly different properties than those of their bulk analogs. New fabrication techniques offer new opportunities to design materials with a specific structure to achieve desired properties. In hydrogen storage studies, nanocomposites have observed to significantly improve the thermodynamics and kinetics of hydrogen sorption by providing a high surface area and hybrid structures that offers multifunctional performance. Decreasing the particle size increases the surface/volume ratio, resulting in enhanced surface energies and altering the hydrogen release mechanisms.

Using a nanoporous scaffold as a structure-directing agent to host hydrides can facilitate the formation of nano-size hydrides within the scaffold while reducing hydrogen diffusion distances. It has been shown that infusing AB in a nanoporous silica scaffold lowers the activation barrier for hydrogen release, significantly

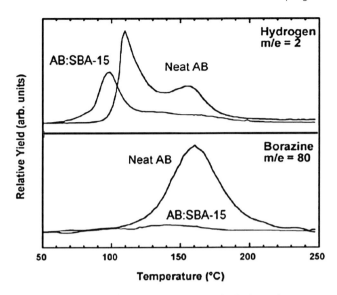

Figure 6.6 Mass spectrometry (at 1 °C/min) of volatile products generated by heating neat AB (solid line) and AB:SBA-15 (dashed line); m/e = 2 (H_2) and m/e = 80 (borazine, c-$(NHBH)_3$) [61].

improves the dehydrogenation kinetics, lowers the dehydrogenation temperature and suppresses unwanted volatile products (Figure 6.6) [61].

Using scaffolds with a high pore volume and low weight can minimize the gravimetric and volumetric penalties associated with this method. Carbon aerogels (CAs) and carbon cryogels (CCs) are nanoporous carbons with a high pore volume and surface area, and tunable densities and pore sizes can serve this purpose. CAs and CCs can be prepared from hydrogels generated by sol–gel polycondensation of organic monomers such as resorcinol and formaldehyde in aqueous solution in the presence of a polymerization catalyst [62]. The precursor hydrogels can be dried by different methods, including supercritical drying and freeze-drying. Supercritically dried hydrogels are called aerogels, while freeze-dried gels are known as cryogels. CAs and CCs are produced by pyrolysis of cryogels and aerogels and consist of an interconnected porous carbon skeleton with high porosity (above 90%) and surface area (above 1000 m^2/g), and pore diameters ranging from <1 nm to 100 nm. Because of the low density and high porosity, CAs and CCs can accommodate a large fraction of hydrides with little addition of weight. Moreover, the extremely high surface area facilitates an intimate contact between hydrides and the carbon network.

Incorporation of $LiBH_4$ into CAs has been shown to enhance the dehydrogenation kinetics and lower the dehydrogenating temperature of $LiBH_4$. Figure 6.7 shows the thermo-gravimetric analysis for hydrogen release from LiBH4 confined in two aerogels with pore sizes of 13 and 26 nm, activated carbon with pore sizes of <2 nm and a non-porous graphite control sample [63]. This study shows that

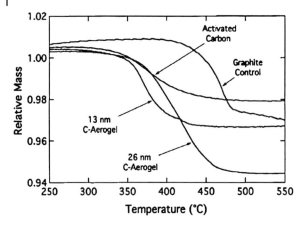

Figure 6.7 Thermogravimetric analysis of LiBH$_4$ dehydrogenation, shows that the reaction temperatures decrease with decreasing scaffold pore size [63].

incorporation of LiBH4 into the CA accelerates the dehydrogenation, reduces the energy barrier to release hydrogen and decreases the hydrogen release temperature, with a lower dehydrogenation temperature observed for CA with a smaller pore size.

Using CC to host AB in CC-AB nanocomposites has been shown to release more than 5 wt% H$_2$ (including material) in one exothermic event at decreased dehydrogenation temperatures while suppressing undesirable volatile products [64, 65]. Moreover, studying the effect of pore size on the dehydrogenation temperature of CC-AB nanocomposites using diffraction scanning calorimetry (DSC) shows that the CC-AB nanocomposites release H$_2$ at temperatures less than that of bulk AB, and the dehydrogenation temperature decreases with the reducing pore size of the CAs (Figure 6.8) [66].

Porous carbon–hydride nanocomposites have shown impressive results in enhancing the dehydrogenation kinetics by reducing the diffusion distance and increasing the reaction surface area. The incorporation of hydrides into functional frameworks or matrices with desired chemical and physical properties must also be investigated more intensively. Increasing the porosity of the nanoscaffold to attain a large accessible pore volume will improve the storage capacity. The tunable structure of nanoporous carbon offers different ways to catalyze the dehydrogenation reaction. The thermal properties of the nanoscaffold play an important role in the reaction kinetics. Fabrication of a light and thermally conductive nanoporous material (such as porous carbon) provides a compelling approach. Furthermore, modifying the structure of the carbon scaffold by adding other elements can increase the heat transfer while catalyzing the dehydrogenation. Using nanocomposites for hydrogen storage opens up the possibility of designing functional systems, where an external matrix could act as a multifunctional destabilization system for hydrides. It can decrease the hydride size, increase the surface energy, increase the heat transfer and chemically catalyze dehydrogenation to achieve the

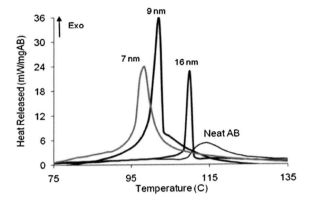

Figure 6.8 DSC exotherms for CC-AB nanocomposites (average pore sizes of 7, 9, 16 nm) and bulk AB shows reduced dehydrogenation temperatures with pore size [66].

desirable thermodynamic and kinetic properties of the hydride. Such research is expected to generate new ideas that can lead to fundamental innovations in hydrogen storage technology.

6.4
Summary

Solid-state storage of hydrogen is a promising and appealing approach as a practical method, but there is still no material or method that meets the requirements for an ideal storage system. Nanostructured materials with their unique properties offer new opportunities to address the challenges involved, including: hydrogen storage capacities, hydrogen sorption near ambient conditions, thermodynamics, and kinetics properties. Recent developments in the nanotechnological approach to address hydrogen storage challenges have shown the improved performance of nanostructured materials and opened new doors to more development.

References

1 Satyapal, S., Petrovic, J., Read, C., Thomas, G., and Ordaz, G. (2007) *Catal. Today*, **120**, 246.
2 Orimo, S., Nakamori, Y., Eliseo, J.R., Zuttel, A., and Jensen, C.M. (2007) *Chem. Rev.*, **107**, 4111.
3 Bérubé, V., Radtke, G., Dresselhaus, M., and Chen, G. (2007) *Int. J. Energy Res.*, **31**, 637.
4 Fichtner, M. (2005) *Adv. Eng. Mater.*, **7**, 443.
5 van den Berg, A.W.C., and Areán, C.O. (2008) *Chem. Commun.*, **2008**, 668–681.
6 Thomas, K.M. (2007) *Catal. Today*, **120**, 389.
7 Dresselhaus, M.S., Dresselhaus, G., and Elkund, P.C. (1996) *Science of Fullerenes and Carbon Nanotubes*, Academic Press, Boston.
8 Chambers, A., Park, C., Baker, R.T.K., and Rodriguez, N.M. (1998) *J. Phys. Chem. B*, **102**, 4253.

9 Dillon, A.C., Jones, K.M., Bekkedahl, T.A., Kiange, C.H., Bethune, D.S., and Heben, M.J. (1997) *Nature*, **386**, 377.
10 Wang, Q., and Johnson, J.K. (1999) *J. Phys. Chem. B*, **103**, 277.
11 Züttel, A. (2004) *Naturwissenschaften*, **91**, 157.
12 Kowalczyk, P., Holyst, R., Terrones, M., and Terrones, H. (2007) *Phys. Chem. Chem. Phys.*, **9**, 1786.
13 Ströbel, R., Garche, J., Moseley, P.T., Jörisen, L., and Wolf, G. (2006) *J. Power Sources*, **159**, 781.
14 Schimmel, H.G., Kearley, G.J., Nijkamp, M.G., Visser, C., de Jong, K., and Mudler, F.M. (2003) *Chem. Eur. J.*, **9**, 4764.
15 Zhou, L., Zhou, Y.P., and Sun, Y. (2004) *Int. J. Hydrogen Energy*, **29**, 319.
16 Zuttel, A., Sudan, P., Mauron, P., Kiyobayashi, T., Emmenegger, C., and Schlapbach, L. (2002) *Int. J. Hydrogen Energy*, **27**, 203.
17 Yang, Z., Xia, Y., and Mokaya, R. (2007) *J. Am. Chem. Soc.*, **129**, 1673.
18 Kabbour, H., Baumann, T.F., Satcher, J.H., Jr., Saulnier, A., and Ahn, C.C. (2006) *Chem. Mater.*, **18**, 6085.
19 Chaung, T.C.M., Jeong, Y., Chen, Q., Kleinhammers, A., and Wu, Y. (2008) *J. Am. Chem. Soc.*, **130**, 6668.
20 Weitkamp, J., Fritz, M., and Ernst, S. (1995) *Int. J. Hydrogen Energy*, **20**, 967.
21 Langmi, H.W., Walton, A., AL-Mamouri, M.M., Johnson, S.R., Book, D., Speight, J.D., Edwards, P.P., Gameson, I., Anderson, P.A., and Harris, I.R. (2003) *J. Alloys Comp.*, **356**, 710.
22 Vitillo, J.G., Ricchiardi, G., Spoto, G., and Zecchina, A. (2005) *Phys. Chem. Chem. Phys.*, **7**, 3948.
23 Eddaoudi, M., Kim, J., Rosi, N., Vodak, D., Wachter, J., Keefee, M.O., and Yaghi, O.M. (2002) *Science*, **295**, 469.
24 Park, H., Britten, J.F., Mueller, U., Lee, J.Y., Li, J., and Parise, J.B. (2007) *Chem. Mater.*, **19**, 1302.
25 Roswell, J.L.C., Millward, A.R., Park, K.S., and Yaghi, O.M. (2004) *J. Am. Chem. Soc.*, **126**, 5666.
26 Ma, S., Sun, D., Ambrogio, M.W., Fillinger, J.A., Parkin, S., and Zhou, H.C. (2007) *J. Am. Chem. Soc.*, **129**, 1858.
27 Wong-Foy, A.G., Matzger, A.J., and Yaghi, O.M. (2006) *J. Am. Chem. Soc.*, **128**, 3494.
28 Collins, D.J., and Zhou, H.C. (2007) *J. Mater. Chem.*, **17**, 3154.
29 Dinca, M., Yu, A.F., and Long, J.R. (2006) *J. Am. Chem. Soc.*, **128**, 8904.
30 Luo, J., Xu, H., Liu, Y., Zhao, Y., Daemen, L.L., Brown, C., Timofeeva, T.V., Ma, S., and Zhou, H.C. (2008) *J. Am. Chem. Soc.*, **130**, 9626.
31 Liu, Y., Kabbour, H., Brown, C.M., Neumann, D.A., and Ahn, C.C. (2008) *Langmuir*, **24**, 4772.
32 Sloan, E.D. (1997) *Clathrate Hydrates of the Natural Gases*, Dekker, New York.
33 Dyadin, Y.A., Larionov, E.G., Aladko, E.Y., Manakov, A.Y., Zhurko, F.V., Mikina, T.V., Komarov, V.Y., and Grachev, E.V. (1999) *J. Struct. Chem.*, **40**, 790.
34 Hu, Y.H., and Ruckenstein, E. (2011) *Angew. Chem. Int. Ed.*, **45**, 2006.
35 Mao, W.L., Mao, H.K., Goncharov, A.F., Struzhkin, V.V., Guo, Q.Z., Hu, J.Z., Shu, J.F., Hemley, R.J., Somayazulu, M., and Zhao, Y.S. (2002) *Science*, **297**, 2247.
36 Kim, D.Y., Park, Y., and Lee, H. (2007) *Catal. Today*, **120**, 257.
37 McKeown, N.B., Ganhem, B., Msayib, K.J., Budd, P.M., Tattershall, C.E., Mahmood, K., Tan, S., Book, D., Langmi, H.W., and Walton, A. (2006) *Angew. Chem. Int. Ed.*, **45**, 1804.
38 Budd, P.M., Butler, A., Slbie, J., Mahmood, K., McKeown, N.B., Ghanem, G., Msayib, K., Book, D., and Walton, A. (2007) *Phys. Chem. Chem. Phys.*, **9**, 1802.
39 Ghanem, G.S., Msayib, K.J., McKewon, N.B., Harris, K.D.M., Pan, Z., Budd, P.M., Butler, A., Selbie, J., Book, D., and Walton, A. (2007) *Chem. Commun.*, **2007**, 67–69.
40 Züttel, A. (2003) *Mater. Today*, **6**, 24.
41 Seayad, A.M., and Antonelli, D.M. (2004) *Adv. Mater.*, **16**, 765.
42 Sakintuna, B., Lamari-Darkrim, F., and Hirscher, M. (2007) *Int. J. Hydrogen Energy*, **32**, 1121.
43 Li, L., and Hurley, J.A. (2007) *Int. J. Hydrogen Energy*, **32**, 6.
44 Nakamori, Y., and Orimo, S.J. (2004) *Alloys Comp.*, **370**, 271.

45 Züttel, A., Wenger, P., Sudan, P., Mauron, P., and Ormio, S. (2004) *Mater. Sci. Eng. B*, **108**, 9.

46 Grochala, W., and Edwards, P.P. (2004) *Chem. Rev.*, **104**, 1283.

47 Vajo, J.J., and Olson, G.L. (2007) *Scripta Mat.*, **56**. 829.

48 Zhu, M., Wang, H., Ouyang, L.A., and Zeng, M.Q. (2006) *Int. J. Hydrogen Energy*, **31**, 251.

49 Dorenheim, M., Eigen, N., Barkhordarian, G., Klassen, T., and Bormann, R. (2006) *Adv. Eng. Mater.*, **8**, 377.

50 Baldé, C.P., Hereijgers, B.P.C., Bitter, J.H., and de Jong, K.P. (2008) *J. Am. Chem. Soc.*, **130**, 6761.

51 Fakioğlu, E., Yurum, Y., and Veziroğlu, T.N. (2004) *Int. J. Hydrogen Energy*, **29**, 1371.

52 Simons, E.L., Cairns, E.J., and Surd, D.J. (1969) *J. Electrochem. Soc.*, **116**, 556.

53 Strickland, G. (1984) *Int. J. Hydrogen Energy*, **9**, 759.

54 Wolf, G., Baumann, J., Baitalowa, F., and Hoffmann, F.P. (2000) *Thermochim. Acta*, **343**, 19–25.

55 Miranda, C.R., and Ceder, G. (2007) *J. Chem. Phys.*, **126**, 184703.

56 Baumann, J., Baitalow, F., and Wolf, G. (2005) *Thermochim. Acta*, **430**, 9.

57 Stowe, A., Shaw, W., Linehan, J., Schmid, B., and Autrey, T. (2007) *Phys. Chem. Chem. Phys.*, **9**, 1831.

58 Benedetto, S.D., Carewska, M., Cento, C., Gislon, P., Pasquali, M., Scaccia, S., and Prosini, P.P. (2006) *Thermochim. Acta*, **441**, 184.

59 Marder, T.B. (2007) *Angew. Chem. Int. Ed.*, **46**, 8116.

60 Cao, G.Z. (2004) *Nanostructures and Nanomaterials, Synthesis, Properties and Applications*, Imperial College Press, London.

61 Gutowska, A., Li, L., Shin, Y., Wang, C.M., Li, X.S., Linehan, J.C., Smith, R.S., Kay, B.D., Schmid, B., Shaw, W., Gutowski, M., and Autrey, T. (2005) *Angew. Chem. Int. Ed.*, **44**, 3578.

62 Al-Muhtaseb, S.A., and Ritter, J.A. (2003) *Adv. Mater.*, **15**, 101.

63 Gross, A.F., Vajo, J.J., Van Atta, S.L., and Olson, G.L. (2008) *J. Phys. Chem. C*, **112**, 5651.

64 Feaver, A.M., Sepehri, S., Shamberger, P., Stowe, A., Autrey, T., and Cao, G.Z. (2007) *J. Phys. Chem. B*, **111**, 7469.

65 Sepehri, S., Feaver, A.M., Shaw, W.J., Howard, C.J., Zhang, Q., Autrey, T., and Cao, G.Z. (2007) *J. Phys. Chem. B*, **111**, 14285.

66 Sepehri, S., Garíca, B.B., and Cao, G.Z. (2008) Tuning dehydrogenation temperature of carbon–ammonia borane nanocomposites. *J. Mater. Chem.*, **18**, 4034.

7
Electrochemical Energy Storage: the Benefits of Nanomaterials
Patrice Simon and Jean-Marie Tarascon

7.1
Introduction

Global warming, finite stocks of fossil fuel and city pollution (transportation is responsible for 30% of CO_2 emission) show, among other things, how important it is to turn towards an intense and efficient use of renewable energies (REN) and to find innovating solutions to ease the progressive transition from thermal to electric vehicles. The intermittency of renewable energies and the need to bring power on board in electrical vehicles, so as to ensure a sufficient autonomy, conspire together to make the invention of new energy storage technologies the greatest challenge for the next 50 years.

Electrode materials are the core of any electrochemical device as the quantity of energy stored by the system and the adequate power depend on them. Whatever the present systems of energy storage (fuel cells, batteries, supercapacitors) they are all dependent on the whims of chemistry, which is frequently reticent in producing the right electrode materials on demand. This penury of materials associated with the problems created by the control of interfaces has slowed down the development of various energy storage technologies and consequently many application sectors. We should recall that the concepts of electric vehicle and cellular phone are no invention of the 20th century, but are more than one century old; therefore their materialization was constantly delayed due to the lack of adequate materials.

The slow progress in this field has often been criticized by the press via harsh comments such as: "Research in the field of batteries is moving at a glacier pace" or "The time scale of micro electronics is much shorter than that of the batteries", where the performances are always limited by … CHEMISTRY. While not looking for excuses, it has to be said that the tuning of new devices for electrochemical storage of energy, whatever they are, is a multifaceted problem, whose complexity (Figure 7.1) resides mainly in: (i) formulating electrodes made of the proper electrochemically active material, binder and electronic conducting additive, (ii) choosing the nature of the electrolyte (liquid, gel or polymer), but more importantly, (iii) mastering the macroscopic/microscopic interfaces, which are the key troublemakers of any electrochemical system.

Nanotechnology for the Energy Challenge. Edited by Javier Garcia-Martinez
© 2009 WILEY-VCH Verlag GmbH & Co. KGaA, Weinheim
ISBN: 978-3-527-32401-9

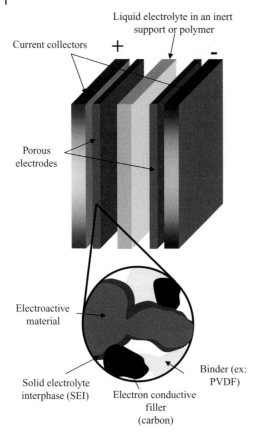

Figure 7.1 Schematic representation of an electrochemical cell with a zoom view of the electrode to illustrate its complex formulation enlisting active material, binders, electron conductive filler and the forming SEI interphase.

Nevertheless, the aforementioned criticism disregards the emergence of new technologies such as nickel metal hydrides (Ni-MH) and lithium metal polymers (LMP) in the 1970s and 1980s and Li-ion technology in the 1990s [1]. The latter is now supplying more than 80% of the portable electronic and constitutes the greatest electrochemical advance of the past century. With the arrival of the Li-ion technology, cell energy density has been boosted by a factor of three and two, respectively, as compared to lead acid 35–40 Wh/kg) and Ni–Cd (50–60 Wh/kg) technologies; and more importantly the energy density of the Li-ion technology has doubled over the past 25 years, passing from 100 Wh/kg to about 200 Wh/kg. However, even the promising Li-ion technology falls short of meeting future application demands linked to the field of renewable energies and automotive transportation in terms of energy density, power and life span. Thus at this juncture a legitimate question is: will the Li-ion technology ever meet the market demands and if so when? This paper, by giving a glimpse of noteworthy recent progress linked to the arrival of new material, will attempt to answer this question.

Today's Li-ion cells use electrode materials functioning according to the classic processus of ions insertion/deinsertion. Among the guidelines prevailing in the search for the ideal insertion material, we find: (i) high electronic and ionic conductions, (ii) the presence of vacant sites in the crystalline structure, (iii) a high redox potential, (iv) a high chemical stability, (v) a low specific surface (e.g., noticeable grain size) and (vi) a low cost and toxicity. These guidelines strongly narrow down the number of today's candidates to three: $LiMn_2O_4$, $LiCoO_2$ and $LiNiO_2$. By playing with the ternary $LiCoO2$, Li_2MnO_3 and $LiMn_{1/2}Ni_{1/2}O_2$ phase diagrams, another material, $LiMn_{0.33}Ni_{0.33}Co_{0.33}O_2$, was unraveled which displays a few advantages in terms of security and performance versus its non substituted parents and this is presently implemented in today Li-ion cells.

In spite of the numerous research efforts made during recent decades, it must be noted that for all these oxides the number of useful available electrons per 3d metal (e.g., Co, Mn, Ni) is always inferior to one [2]. This is why the experts in the field agree this battery technology has reached its limits in terms of energy density. Well aware of this limitation, intrinsic to the Li-ion technology, some chemists/electrochemists decided at the beginning of 2000 to move away from the traditional approach of the intercalation chemistry and to explore new tracks, among which is the use of nanomaterials.

Thus, nanocomposites, nanostructured, nano-architectured, nanoporous electrodes were born, giving the spectacular results we know today. The field of supercapacitors was not spared, as exacerbated capacitances in nanoporous carbons, contrary to the well-established beliefs of these past 20 years, were achieved in 2006. At this stage, one may wonder why it took so long for nanomaterials to penetrate the field of energy storage, while a few decades ago they were present at the origin of the spectacular micro-electronic progress that we are enjoying today. The reason is simple and enlists the catalytic reactions that occur at the interface of electrode materials and electrolytes, as explained below.

Let us keep in mind that Li-ion batteries deliver a 4 V output voltage while using non-aqueous liquid electrolytes in which the thermodynamic stability domain lies between 1.0 and 3.5 V versus Li^+/Li^0. Electrolyte degradation reactions catalyzed by the oxidizing or reducing nature of the electrode materials can occur outside the 1.0–3.5 V potential window. Thus, based on this thermodynamic aspect, the Li-ion batteries should not function. Luckily, the catalytic reactions have slow kinetics. Therefore, they become exaggerated as soon as we reach the nanometric scale because of the surface increase (e.g., interfaces); this is the reason why present Li ion batteries use bulk insertion materials of very low specific surface ($2 m^2/g$). So it must be realized that within an electrode material, depending on its operating voltage, there is possibility of competition between redox electrochemistry generating core redox reactions and catalysis generating surface parasitic reactions. These two worlds, seemingly in total opposition, can converge when nanomaterials, whose redox potential corresponds to the stability domain of electrolytes, are used or, on the contrary, when one masters the nanomaterial/electrolyte interface using protective layers obtained via chemical and physical coating techniques. Hence today's emergence of nanomaterials in Li-ion technologies.

Examples showing the positive attributes of nanometric, nanostructured materials, etc. are so numerous that it is beyond the scope of this paper to cover all of them. In contrast, we decided to be more selective and solely use a few specific ones pertaining to the field of batteries and supercapacitors that could best convey our messages. Thus illustrative examples aiming to show how downsizing particle size combined or not with coating steps and nanostructuring, have enabled us: (i) to transform a once rejected isolating insertion compound $LiFePO_4$ into a strongly attractive electrode material, (ii) to make the creation of metal-based (Si) negative electrodes based on alloy reactions possible and (iii) to promote new reaction mechanisms; these examples are described here. In order to extend the advantages of nano to any electrochemical device, we use examples taken from the field of supercapacitors taking into account the new breakthroughs in the domain of porous materials, with carbide-derived carbons. Finally, to be clear and coherent, examples pertaining to nanometric electrochemistry are taken individually with a reminder of the problematic and the context.

7.2
Nanomaterials for Energy Storage

7.2.1
From Rejected Insertion Materials to Attractive Electrode Materials

In 1997, J.B. Goodenough *et al.* showed that, because they are mixed conductors [3], polyanionic-type structures made from MO_6 octahedra linked to $(XO_4)_n^{n-}$ (X = Mo, W, S, P, As) tetrahedra could be used as host structure for Li-ions. Thus, a mineral, known under the name of triphylite, capable of inserting, at 3.5 V versus Li^+/Li^0, one Li per Fe atom (total capacity of 170 mAh/g) was identified. Yet the use of this mineral was hindered by its low electronic (σ_e) and ionic (σ_i) conductivities since at ambient temperature only 40% of its theoretical capacity could be used. To bypass this issue, the kinetics of this electrode had to be improved meaning improving the electrons and ions transfers. Because the conductivities are intrinsic to the material, strategies aiming at reducing the traveling distances of the ions and the electrons within the electrode had to be developed. The first successful approach, developed by M. Armand [39], has consisted in coating $LiFePO_4$ particles with a thin layer of conductive carbon, following a combined polymer/pyrolysis process. Using this chemical trick, most of the theoretical capacity of the material could be recouped at 70°C. Although carbon coating is very beneficial, it does solely allow a noteworthy improvement at the electronic level leaving the ionic conductivity unaltered. Modifying the ionic conductivity of such a material via chemical doping has been so controversial that we rather decided to simply downsize the particle size to reduce the Li^+ diffusion distances within the particles. Along that line our group set up a novel "Chimie douce" synthesis method ena-

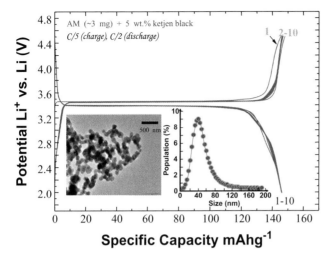

Figure 7.2 Voltage-composition curve for 140 nm LiFePO4 nanoparticles made via a low-temperature solvothermal process (with the courtesy of C. Masquelier).

bling the preparation of monodisperse LiFePO$_4$ powders with sizes nearing 140 nm (Figure 7.2 [4]) with the latter being faulted (e.g., having the presence of Fe and vacancies in the Li crystallographic sites). The performances of these powders, although free of carbon coating, turned out to be excellent in terms of capacity as well as power (Figure 7.2). Downsizing further the particles to 40 nm led to faulted particles (having Fe and vacancies in the Li crystallographic sites) and a profound effect on the electrochemical behavior as a full solid solution domain was found at 298 K upon electrochemical lithium insertion/deinsertion in contrast to a two-phase domain for 140 nm particles.

Overall, moving to the nanometric scale has enabled to turn LiFePO$_4$, an isolating material, into a very attractive electrode material arousing a keen interest from batteries manufacturers; this is due to its abundance and low cost, which are very important factors in the development of sustainable electrodes. It has become important to realize, in the quest for new electrode materials, that particle size in addition to composition and structure added a third dimension to circumvent the intrinsic electronic and ionic conduction limitations, thus giving wide opportunities. From then on, a panel of once rejected isolating electrode materials has been revisited. Among them are (Li$_2$VOSiO$_4$ [5], Li$_2$FeSiO$_4$ [6]) which display interesting electrochemical activities when prepared at the nanometric scale in the presence of carbon coating.

Just as spectacular to witness the advantages of nanometric scale are the recent syntheses of nanoparticles of rutile-type TiO$_2$ via a sol–gel method and the demonstration of their electrochemical activity with reversible and sustainable capacities of 200 mAh/g; thus defying 20 years of mistaken beliefs – namely the existence of an electrochemical activity for the sole anatase-type TiO$_2$. Similarly, we must underline the preparation, via hydrothermal synthesis, of TiO$_2$ (B) nanotubes capable of reversibly and rapidly react versus Li to lead to capacities neighboring 200 mAh/g [7] too. Taking into account the size as an additional variable, besides composition and structure, to tune the electrochemical properties of a compound, another route of research has opened, and currently the number of compounds presenting an interesting activity towards Li is growing exponentially.

7.2.2
The Use of Once Rejected Si-Based Electrodes

Advantages of the nanometric approach are not specific to positive electrode materials, as the ones aforementioned, but do also exist for negative electrodes based on Li-alloying reactions, which stand as serious contenders to replace graphite-type materials and which are widely used in all Li-ion batteries that are commercialized today. It should be recalled that graphite electrodes can reversible intercalate up to one Li atom per six C atoms (LiC_6; e.g., 370 mAh/g) at a potential close to 90 mV versus Li^0 while showing excellent cycle life. However, an inherent drawback to the graphite electrode is low insertion potential; the latter leads to a risk of metal Li deposit on the carbon during the rapid charges of the battery and has a detrimental effect on the battery safety and cycle life.

To eschew this issue, present research aims to look for alternative materials able to insert as much, even more, Li ions than carbon but at slightly higher potentials. This is the "gold rush" and many groups have been working for many years on intermetallic alloys; they are very interesting capacity-wise (up to 3800 mAh/g for Si for instance), but they display a very bad reversibility associated with a loss in electric percolation during cycling owing to humongous volumes changes (e.g., 300% volume increase in going from Si to $Li_{4.4}Si$). To fight this problem, the nanomaterials approach was favoured because nanomaterials are able to absorb constraints and, consequently, to act as an elastic "buffer" to maintain the electrode electric integrity. Besides downsizing the Si particles, a coating approach was also pursued so as to reduce the electrolyte degradation catalyzed by the Si electrode which does react with Li at a potential well below (at least 0.3 V) the thermodynamic stability potential window of the electrolyte. Coated Si particles were obtained as follows. Si/polymer mixtures were made by dispersing nanometric Si powders (obtained separately via a plasma process) in a PVC solution solubilized by propylene oxide so as to obtain a gel [8–10]. Spreading this gel gives a plastic-like layer, which is later on pyrolyzed at 800 °C under a reducing atmosphere to lead to an Si/C composite with very promising electrochemical properties as shown in Figure 7.3. Although the nanometric aspect of Si is important, Raman spectroscopy and EELS studies underlined that the Si/C interface with a few nanometers (2–3 nm SiO_2) between Si and C plays a key role as well. Effectively, Raman spectroscopy (position and shape of the peaks) showed that besides its coating aspect, carbon coating (e.g., a shell around the particles) generated compression constraints at Si surface thus limiting its peeling off during Li insertion, while limiting the direct contact Si-electrolyte. Reassured by these observations, new coating strategies are presently developed. They aim at the *in situ* simultane-

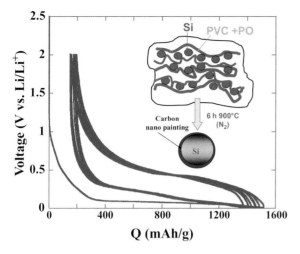

Figure 7.3 Voltage-composition curve for carbon-coated Si nanoparticles. The carbon coating was realized by pyrolysis of polymers (PVC) under reducing ambiance.

ous formation of Si nanoparticles and polymerization reactions, with the aim of having homodisperse nanometric particles topped within a carbon matrix. It goes without saying that some of these processes are not specific to Si and have been successfully applied to other compounds (e.g., Cu_2Sb) [11].

7.2.3
Conversion Reactions

Combined particle downsizing and coating/interface grafting chemistry will enable the use of isolating compounds such as $LiFePO_4$, Li_2VOSiO_4, Li_2FeSiO_4 or $LiFeBO_3$ as positive electrode materials together with the foreseen use of Si as a negative electrode material in the years to come. Another inherent advantage to nanomaterials, linked mainly to their high surface area and hence enhanced chemical reactivity, is their ability to trigger new reaction mechanisms, as described below. Working in that direction, it was to everybody's surprise that our group demonstrated in 2000 [12] that the simple oxides M_xO_y, which do not exhibit the selective criteria for insertion/deinsertion (not possessing an open structure and comprising a metal unable to alloy with Li), could reversibly react towards Li, thus enabling us to reach reversible capacities greater than 700 mAh/g for hundreds of cycles. This new reaction mechanism, which proceeds according to the reaction below and leads to a composite matrix made of an homogenous distribution of metal nanoparticles (e.g., Co from CoO_2) in a "plastic" matrix of Li_2O (Figure 7.4) [12], is presently termed a conversion reaction in the battery community:

$$M_xO_y + 2ye^- + 2yLi^+ \Rightarrow xM^0 + yLi_2O$$

Figure 7.4 Schematic of the conversion reaction mechanism [12].

Thanks to the formation of these nanoparticles the reaction is reversible, so that during the following oxidation these metal nanoparticles are re-oxidized to form oxide nanoparticles. It was rapidly demonstrated, owing to the worldwide contribution of several other groups, that this surprising electrochemical activity, contrary to well-established beliefs, is not specific to oxides but can also include sulfides, nitrides, phosphides and fluorides. Thus a panel of materials electrochemically active versus Li on a scale of potentials ranging from 0.2 to 3.3 V versus Li+/Li has been unraveled, with high potentials for fluorides and low for phosphides because of the strong ionic and covalent nature of M–X bonds, respectively.

Contrary to classic insertion reactions, restricted to a maximum of 1e⁻ per 3d metal atom as previously discussed, these new conversion reactions can imply 2e⁻ (CoO), 3e⁻ (FeF$_3$) [13] up to 6e⁻ (NiP$_2$) [14] per atom of 3d metal, leading then to staggering capacity gains. Yet, any brand new concept generally has its limitations, and conversion electrodes are no exception, as such electrodes suffer not only from reversibility problems but also from polarization issues (e.g., important potential difference between charge and discharge) making their energetic yield lower than that of insertion electrodes. This polarization ΔV was measured to decrease from fluorides ($\Delta V \approx 1.1$ V) to oxides ($\Delta V \approx 0.9$ V), sulfides ($\Delta V \approx 0.7$ V) and phosphides ($\Delta V \approx 0.4$ V), tracking the decrease in the ionic character of the M–X bond from M–F to M–P. Searching further to implement the conversion reaction to other family of compounds, our group recently investigated the electrochemical reactivity of metal hydrides towards Li. Among them MgH$_2$ was shown to display a large reversible capacity of 1120 mAh/g (triple that of conventional Li/C electrodes) at a voltage of 0.44 V versus Li/Li⁺ together with the lowest (0.02 V) polarization ever reported for conversion electrodes. Equally promising was the finding that the reduced composite, made of highly divided Mg particles embedded into a LiH matrix obtained during the electrochemical discharge of a MgH$_2$/Li cell, or the oxidized composite made of divided MgH$_2$ show enhanced hydrogen sorption/desorption kinetics, respectively.

Besides the aforementioned approach, several engineering approaches have been put forward to address the technological problem linked to the conversion of electrode poor kinetics, and more specifically to the charge transfer limitation somewhat linked to the bad ionic/electronic conductivity of the binary oxides. The commonly developed strategy was to improve the electronic/ionic percolation within the electrode, either by altering the morphology of the electrode material (nanostructuring) or by altering the electrode material/current collector interface.

7.3
Nanostructured Electrodes and Interfaces for the Electrochemical Storage of Energy

Mastering the electrodes interfaces for energy storage, via the adjustment of specific properties in terms of developed surface, conductivity, controlled porosity, etc., is today an important area of research which contributes to the set up of the next generation of electrochemical storage devices. In the following, we go through a few examples of interface designs via the elaboration of nanostructured electrodes which concept can equally be applied to batteries and supercapacitors. These examples not only enable us to quantify the contribution of these nanostructures but also to validate a scientific approach that, from the obtained results, deserves special attention.

7.3.1
Nanostructuring of Current Collectors/Active Film Interface

7.3.1.1 Self-Supported Electrodes
Various approaches have been tested by our competitors, like the elaboration of composite electrode from Cr_2O_3 or by ourselves like the temperature-driven growth of metal oxide/phosphide layers on various substrates. For instance, we found the thermal treatment of stainless steel in air to lead, owing to the cations migration, to a nanostructured surface layer made of chromium and iron mixed oxides displaying high electrochemical capacity versus Li^+/Li^0 (850 mAH/g of oxide layer) and good cyclability (more than 100 cycles) [15]. Besides, it must be noted that the use of this type of steel electrode, which is a technological breakthrough, does not require any drastic alteration to the manufacturing method of Li-ion batteries. In addition to improving the energy density, this kind of negative electrode, which has moved towards using stainless steel tissues, enables us to: (i) suppress a copper current collector, (ii) stop relying on special graphites (MCMB) and (iii) stop using organic binders and conductive additives [14, 15].

We applied the same concept to prepare self-supported NiP_2 electrodes. The experiment was conducted in a sealed quartz vessel within which we have placed at the two opposite ends a nickel foam and phosphorus powders. The tube was then placed into an oven to grow NiP_2 on top of the Ni foam. By adjusting the reaction conditions (temperature and heating time) we obtained a conversion electrode, free of ligand and carbon, and which presented cycling and power electrochemical performances clearly superior to those obtained via classic elaboration processes.

7.3.1.2 Nano-Architectured Current Collectors
The nanostructuring of current collectors is another approach developed by our groups within the framework of ALISTORE [16] to create high power rate electrodes. Figure 7.5 represents a sectional view of a planar electrode, with the current collector covered with a film of electrochemically active material so as to highlight the continuous interface between both of them. Assuming a perfectly planar

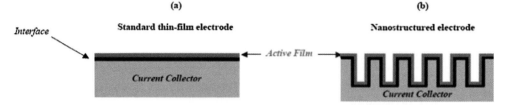

Figure 7.5 Schematic cross-section view of a planar electrode (a) and a nanostructured one (b).

electrode, the contact surface between the active material film and the current collector is equal to the geometrical S_g surface of the electrode (footprint area). Figure 7.5b displays different electrode geometry where the film of active material (same thickness as before) is placed on a nanostructured current collector, namely a current collector with a developed surface S_{dev} which is much larger than the geometrical surface S_g. Creating such an electrode must enable, for the same amount of active material deposited per geometrical surface unit, to improve the electrode power rate capability since the thickness of the film of active material decreases. This approach, although very attractive on paper, needs in first place to elaborate a nanostructured current collector which surface is covered with orderly and evenly spaced pillars [15].

A method derived from the "template synthesis" initially published by C.R. Martin's [17] group was used to prepare such current collector. The choice of copper was based on its use as a negative current collector in Li-ion batteries. Cu was electrochemically deposited into the pores of an alumina membrane placed between a sacrificial Cu anode and a Cu cathode foil acting as the current collector [18]. After Cu deposition by a pulsed current method, the alumina membrane is dissolved and the copper rods remain as columns perpendicular to the copper sheet support. A view of the electrochemical cell used for Cu deposit is represented in Figure 7.6a, b that shows a cross-section view of the prepared electrode. The diameter of the copper rods, which is the size of the pores of the alumina membrane used as template, is about 100 nm.

The second step of the process consists in depositing the active material onto the nano-architectured Cu current collector. Two criteria are essential to achieve an optimum interface between the rods and the active material film: (i) the film must cover the nano-architectured current collector and (ii) it must adhere nicely to guarantee low impedance contact. Iron oxide (Fe_3O_4, magnetite) is a low-cost, non-toxic, electrochemically active material which reacts with lithium according to the following conversion reaction:

$$Fe_3O_4 + 8e^- + 8Li^+ \rightarrow Fe + 4Li_2O \qquad (7.1)$$

Fe_3O_4 can also be electrochemically deposited; accordingly, it was selected as our active electrode material. The Fe_3O_4 deposit shown in Figure 7.7 was carried out

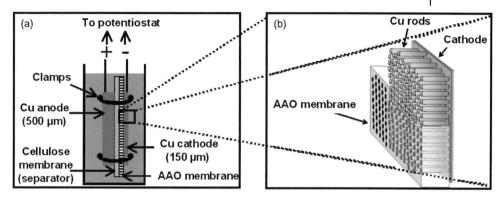

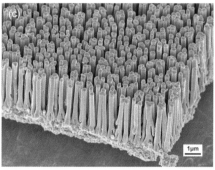

Figure 7.6 (a) Drawing of the electrochemical cell used for the template synthesis of nanostructured Cu current collector is shown (left) with a schematic drawing of the nanostructured current collector expected to be obtained at the end of the electrolysis, before and after removal of the AAO membrane (right). (b) Cross-sectional views of Cu-nanostructured current collector before and after Fe_3O_4 deposits. From [18]. (c) Cross-sectional view of the Cu current collector obtained after electrolysis and membrane removal [18].

by cathodically reducing a Fe(III) chelate in an alkaline medium according to a procedure set up by Switzer's group [19, 20]. These nano-architectured electrodes have been tested in [Li // electrolyte // electrode nano-Fe_3O_4] coin cells using a mixture of ethylene carbonate and dimethyl carbonate in the 1:1 ratio as the electrolyte. Figure 7.8 represents the change of the normalized capacity of the cell with the discharge rate (1C = 1Li^+ exchanged in 1 h) for various deposit times of Fe_3O_4 (e.g., different thicknesses of the Fe_3O_4 layer) together with the characteristic of a classic composite electrode using Fe_3O_4 powder deposited on a Cu planar collector as a reference.

Whereas the capacity of the powder-based planar electrode drops by 80% as the discharge rate is increased from C/32 to 4C, the capacity of the nano-architectured electrodes drops by 20% only. Spectacularly, the nano-architectured electrode can

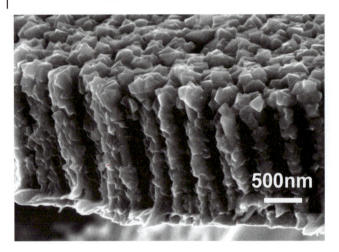

Figure 7.7 Cross-sectional view of Cu-nanostructured current collector after Fe_3O_4 deposits [18].

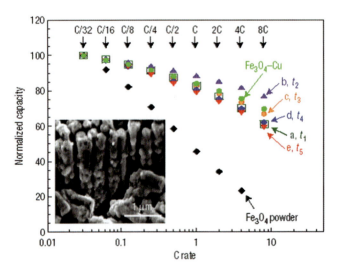

Figure 7.8 Rate capability plots (Peukert's plot) for five different Fe_3O_4 deposits (a–e correspond to increased deposition time) on Cu nanostructured electrodes, as compared to: (i) a Fe_3O_4 deposit denoted "Fe_3O_4-Cu" grown on a planar Cu foil electrode using the same experimental conditions as the nanostructured deposits and (ii) a 1 cm^2 plastic positive electrode film based on commercial Fe_3O_4 powders. Insert: SEM image of a copper-supported Fe_3O_4 deposit cycled galvanostatically 100 times at a high rate (i.e., 1 Li$^+$/0.3 h) showing the good stability of the electrode although a very low amount of active material (lowest deposition time, $t_1 = 120$ s). From [18].

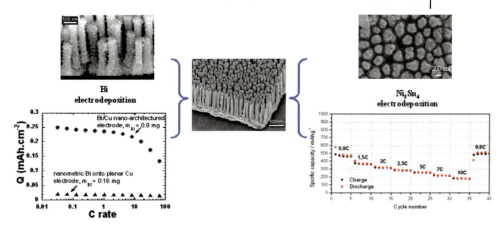

Figure 7.9 Electrochemical performances of nanostructured Cu electrodes electrochemically deposited with Bi (left) and Ni$_3$Sn$_4$ (right). From [21, 22].

sustain a rate of 8C while still delivering 80% of its total energy density, whereas the planar powder-based electrode no longer deliver energy at these rates. Such results unambiguously confirm the benefits of mastering the current collector-active material interface to develop both high capacity and high power rate electrodes.

This approach is obviously not restricted to the sole use of Fe$_3$O$_4$. Other kinds of active materials were also deposited onto nano-architectured current collectors. Figure 7.9 shows the results obtained with nanostructured current collector electrodes covered with either Ni$_3$Sn$_4$ or Bi [21, 22] active materials which react to Li according to the following reactions:

$$Ni_3Sn_4 + 17.6\,Li^+ + 17.6e \rightarrow 4Li_{4.4}Sn + 3Ni \tag{7.2}$$

then

$$xLi^+ + xe + Sn \leftrightarrow Li_xSn \text{ with } x \leq 4.4 \tag{7.3}$$

for Ni$_3$Sn$_4$ and

$$xLi^+ + xe + Bi \leftrightarrow Li_xBi \text{ with } x \leq 3 \tag{7.4}$$

for Bi.

The rate capability curves for either Ni$_3$Sn$_4$ or Bi-based nano-architectured electrodes indicate remarkably high power rate capabilities, further confirming the general trend deduced from the previous Fe$_3$O$_4$ study.

Overall, the aforementioned results demonstrated, regardless of the active materials that we have considered, the benefits of using a nanostructured current collector as opposed to a planar one to reach outstanding power rate densities with conversion reaction electrodes. Moreover, besides the performances achievements, such work opens new paths on the way towards the design 3-D microbatteries that are eagerly needed to power the next generation of miniaturized power sources (MEMS) to satisfy recent microelectronic developments.

7.3.2
NanoStructuring of Active Material/Electrolyte Interfaces

Whatever the electrochemical system considered, we have to keep in mind that the first event taking place prior to the onset of any redox reaction is the formation of the double layer capacitance at the electrode/electrolyte interface. Thus, it is necessary to design this interface in accordance with the nature of the storage mechanism considered (capacitive vs faradic) and the storage location within the electrode (volume vs surface of the active material for batteries and supercapacitors, respectively). Functionalizing this interface with the aim of optimizing the charge storage is a key, and here we give two examples related to the field of batteries and supercapacitors to illustrate this approach.

7.3.2.1 Application to Li-Ion Batteries: Mesoporous Chromium Oxides

The staggering electrochemical reactivity of vanadium oxides aerogels (6 Li per V_2O_5 between 3.5 and 1.5 V vs Li^+/Li^0), as reported by Smyrl et al. in 1995 [23], had the merit, although very controversial, of rejuvenating works on the elaboration of mesoporous electrodes. The benefits of these electrodes are that they can be totally flooded with the electrolyte via its internal pores, thus guaranteeing a large surface contact with the electrolyte and therefore an important Li flow through the interface.

The main difficulty in elaborating mesoporous electrodes lies in the control of two types of porosity, one intrinsic to the particles of the studied material and the other depending on how the particles are put together. "Soft chemistry" synthesis ways, besides a few auto-combustion methods, are among the most used to prepare large porosity but fragile materials (e.g., aerogels). It turns out that such type of porosity is much altered even lost during the electrode elaboration, hence the need to create and control micro-porosity at the grain level. To circumvent this issue, the template approach has been preferred to the sol-gel, and a few binary oxides such as MnO_2, Co_3O_4 and $Cr2O3$ were recently prepared with orderly mesoporosity. This elaboration implies the use of mesoporous silica, with a 3-D pore structure like KIT-6 or SBA-15, as a starting template. The experimental protocol is quite universal and first consists of dissolving in water the 3d metal precursor salt of the oxide we want to make and then using the obtained solution to fill in the pores of the mesoporous silica. The loaded template is then annealed to favor the formation of the binary oxide in the pores, and finally the silica mold is dissolved in NaOH medium to recover the oxide; the latter possesses a replica structure of the initial mesoporous silica, as seen in the TEM shots of the mesoporous Cr_2O_3 compound (Figure 7.10) [24], which is the result of such a elaboration. The pores are three-dimensionally ordered with a 10 nm pore size and a 10 nm wall thickness.

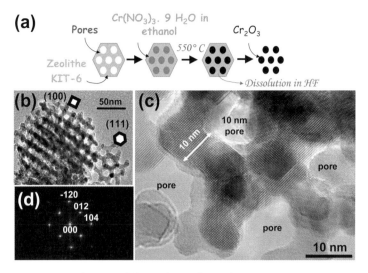

Figure 7.10 Schematic of the synthesis of mesoporous powders of Cr_2O_3 from a silica KIT-6 template (a). Height resolution electron microscopy images show clearly the porosity of the Cr_2O_3 phase obtained (b), together with both the pore and wall dimensions (c) SAED confirms the existence of Cr_2O_3 (d). (Courtesy of L. Dupont [24]).

Such mesoporous Cr_2O_3 electrodes were shown to present electrochemical performances, in terms of power and cycling behavior, superior to those of electrodes made of nanometric powder of same specific surface. This improvement in power rate capabilities is the result of a better contact between the nanometric particles inside the walls, but mainly of short diffusion paths for Li^+ and e^- inside the walls. In contrast the excellent cycling behavior of such electrodes is believed to be rooted in their mechanical strength as supported by HRTEM measurements, which showed that the electrode local structure was not altered after 100 cycles. Another advantage of these powders made of mesoporous grains is that they can be submitted to the various manufacturing processing steps (mixing, induction, compaction, lamination) involved in the electrode confection without losing their porosity, in contrast to the high porosity materials such as aerogels, for example. This approach to electrode nanostructuring/nanotexturing is currently successfully being extended to the elaboration of $LiCoO_2$- and $LiMn_2O_4$-based positive electrodes. Nevertheless, it should be kept in mind that the electrodes manufactured from mesoporous particles possess a relatively low final density and consequently a lower energy density than the standard electrodes.

7.3.2.2 Application to Electrochemical Double-Layer Capacitors

Electrochemical double-layer capacitors (EDLCs), also called supercapacitors or ultracapacitors, are energy storage systems that stand between batteries and dielectric capacitors in terms of energy and power density (Table 7.1). While batteries are able to store a higher energy density than supercapacitors, they deliver less

Table 7.1 Comparison of average performances of different energy storage systems (ESS).

	Dielectric capacitors	Supercapacitors	Batteries
Gravimetric energy (Wh/kg)	<0.1	2–5	20–150
Gravimetric power (kW/kg)	>20	5–15	<2
Charging time (s)	10^{-3}–10^{-6}	1–30	0.2–10.0 h
Cyclability (cycles)	∞	$\geq 10^6$	300–10 000
Life span (years)	≥30	≥20	5
Energy efficiency (%)	≈100	92–98	75–90

power; as compared to dielectric capacitors, supercapacitors can store a higher energy density with less delivered power [25]. Present performances of supercapacitors reach about 5 Wh/kg and 5–10 kW/kg.

These differences in behavior originate from the energy storage mechanism [26]. Unlike batteries, they store energy at the electrolyte/electrode interface through reversible ion adsorption onto the active material surface, thus charging the so called "double-layer capacitance" [27]; no Faradic (redox) reaction is involved in the charge storage mechanism. Charge storage is thus based only on electrostatic interactions between the electrolyte ions and the surface charges of the electrodes, with no charge transfer reaction; the active material/electrolyte interface works as a dielectric capacitor.

Classic active materials used as active materials in supercapacitors are carbons with a large specific surface area (SSA; from 1000 to 25 000 m^2/g). To reach such values, carbon is first ground into particles of a few microns in diameter, then activated (oxidized) via either physical treatment (dry route, high temperature) or chemical treatments (in solution) enabling us to develop the porosity inside the particles, leading to the so-called activated carbon (AC). Activated carbon specific capacitance ranges from 100 to 200 F/g in aqueous medium and from 80 to 100 F/g in organic medium. The very large majority of supercapacitors commercialized today use organic electrolytes based on fluorinated salts dissolved in carbonate- or acetonitrile-based solvents; note that the use of acetonitrile might be limited in some specific applications because of the low flash point (2 °C). In these electrolytes, the operating voltage of carbon/carbon systems reaches 2.5 V.

Initial research on activated carbon was directed towards increasing the pore volume by developing high SSA and refining the activation process. However, the capacitance increase was limited even for the most porous samples. Realizing the importance of how this specific surface was created rapidly changed the deal, and research was then directed to understanding the relationship between the electrolyte ion size and the carbon pore size, to try to answer the basic question: what is the optimal pore size needed in the volume of activated carbon grains to optimize the charge storage at the carbon/electrolyte interface? In other words, how do we optimize the charge storage capacity while controlling the carbon/electrolyte interface?

The pores, which are at the origin of the increase in the developed surface area, are created inside carbon particles during the activation treatment. Three different pore groups are defined [28]: micro, meso and macro pores with diameters, respectively, <2 nm, 2 nm ≤ d < 50 nm and >50 nm. These pores must be neither too big to lead to a significant increase in the specific area (m^2/g) nor too small to host, during the capacitor charge/discharge, the solvated ions from the electrolyte (average size between 1 and 2 nm in organic electrolytes [29]). The first progress in this area was to elaborate carbon materials with a pore structure adapted to the size of the solvated ions in order to optimize charge storage. This is how porous activated carbons were developed with a pore size centered on small mesoporosity (3 to 5 nm), that is, twice of the size of the solvated ions, to allow ion adsorption on both pore walls. The physical or chemical activation of carbons needed to prepare the AC does not allow the fine control of the pore size distribution (PSD); therefore new routes were created to prepare these fine-tuned pore size materials [30]. Among them, the "template" route revealed itself as particularly efficient [31, 32]. It consists of first impregnating a mesoporous silica-type substrate with a gaseous, liquid or solid carbon precursor [31–33]; this step is followed by a pyrolysis and finally dissolution of the Si mold in HCl medium. As a result, the obtained carbon has a pore size that is a replica of the template silica structure. Thanks to this technique, more than 20% capacity gains were obtained (up to 115 °F/g in organic medium).

If the abovementioned mesoporous carbons can be an alternative in terms of improving charges storage capacity, a major breakthrough is still to be found to drastically increase this capacity. It is within this context that carbons appeared with a sub-nanometric pore size, known as carbide-derived carbons (CDC) [34]. The CDC are obtained from metal carbide chlorination at temperatures ranging from 400 to 1000 °C according to the following reaction:

$$MC + nCl_2 \rightarrow MCl_{2n} + C_s \tag{7.5}$$

The point of this method is to allow the synthesis of carbons with a very controlled pore size, since porosity is controlled by the leaching out of metal atoms M. By playing with temperature and thermal treatment duration, we have prepared TiC carbons which porosity ranges from 0.6 to 1.2 nm. These carbons have been tested in an acetonitrile-based electrolyte containing 1 M NEt_4, BF_4 electrolytes, for which the size of the solvated ions is 1.7 nm and 1.5 nm for NEt_4^+ and BF_4^-, respectively.

The change of the gravimetric capacity (F/g) and the volumetric capacity (F/cm^3) versus the CDC pore size (Figure 7.11) brings out two remarkable points. First, when the size of the pores diminishes to be smaller than the size of the solvated ions, the capacity increases; this result is in obvious contradiction with the traditional beliefs which have prevailed for more than 20 years. They anticipated that only mesopores with a size close to twice that of the solvated ions enabled us to reach maximum specific capacities. Besides, the values of the CDC gravimetric and volumetric capacitances are notably higher (140 F/g, 84 F/cm^3) than those of commercial activated carbons, measured under the same conditions (100 F/g,

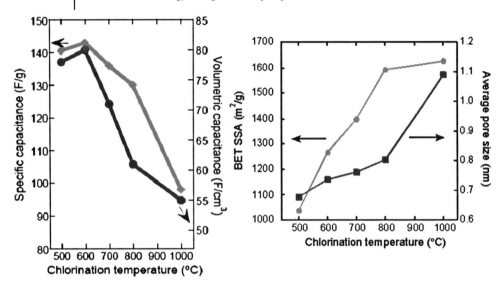

Figure 7.11 Specific capacitance and volumetric capacitance of TiC-CDC synthesized in the 500–1000 °C range (left). Specific surface area as well as average pore size change versus synthesis temperature (right). From [34].

45 F/cm^3). To understand this unexpected behavior, the variation of CDCs normalized capacity (µF/cm^2) was plotted versus the pores size (Figure 7.12). The normalized capacity was obtained by dividing the gravimetric capacity (F/g) by the SSA (m^2/g) for each of the tested CDCs. Up to a pore size of 1 nm, the normalized capacity decreased simultaneously with the decrease in pore size, following the traditional behavior reported in the literature (dotted line). For pore sizes less than 1 nm, the normalized capacity dramatically increased, contrary to what was expected. This figure clearly shows the very significant contribution of the subnanometric pores to the charge storage mechanism in CDCs. The proposed hypothesis to explain this unexpected result is the distortion of the ion solvatation shell which thus enables the ions to access pores in spite of their size. The ion approaching distance (d) to the carbon surface (1 nm, Figure 7.12d) being reduced, the capacitance (C) increases according to:

$$C = \varepsilon A/d \tag{7.6}$$

where ε is the electrolyte dielectric constant, A is the surface area accessible to the ions and d the distance between the center of the ion and the carbon surface.

Recent results obtained in organic liquids as well as in ionic liquid electrolytes confirm the hypothesis of the partial desolvation of the ions when entering the small pores [29, 35, 36].

Beyond the obtained result, the evidence of a strong contribution of the subnanometer pores (<1 nm) to the capacitive storage is questionable for several reasons.

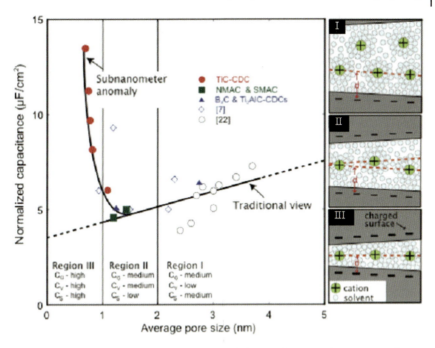

Figure 7.12 Plot of specific capacitance normalized by specific surface area for the carbons in the study and two other studies with identical electrolytes. Illustrations showing solvated ions residing in pores with distance between adjacent pore walls (Region I) >2 nm, (Region II) 1–2 nm and (Region III) <1 nm illustrate this behavior schematically [34].

First, it shows a lack of fundamental understanding in the field of ionic transportation in nanoporous materials. Today, a theory is needed to understand the physics of the ionic transport and the charge storage in confined pores that forbids any existence of a diffuse layer.

Also, these results have a direct consequence for applications since they demonstrate the interest in going from today's mainly porous carbons to nanostructured porous carbons. A new generation of high-energy density supercapacitor could result, thus paving the way to applications in the field of transportation in general and hybrid vehicles in particular, in association with Li-ion batteries. Decorating these nanoporous carbons with nanometric oxides could lead to the development of pseudo-capacitance or hybrid supercapacitors to offer promising perspectives to the future of supercapacitors too. These particles using nanometric oxide particles, notably ruthenium and manganese [37, 38] using fast faradic reactions, could result in long- or medium-term upheavals in rapid energy storage/use.

Finally, these sub-nanometer porous materials can also be used as model materials for a deeper investigation of the influence of the pore size/solvated pore size ratio on the capacity of the carbon double layers, to improve our knowledge of the fundamental understanding of the ion transport in a confined environment for supercapacitors as well as other applications, such as water desalinization or biomedicine.

7.4
Conclusion

The field of electrochemical energy storage, leading to new opportunities, has been conquered in these past few years by nano-objects. We have proved the importance of acting at the particle size, manipulating the electrode material nanotexture/nanostructure and elaborating nanocomposite electrodes, even mesoporous, or nano-architectured current collectors, in order to prepare battery electrodes as well as supercapacitors with intensified properties. The field of electrochemical energy storage makes a new start. These possibilities being clearly demonstrated, our research must stop making nano fit every occasion (as shown by the number of scientific papers) and start integrating these nano-objects in batteries so that they do not remain a laboratory curiosity. The first attempts were fruitful; in particular A123 Systems and Sony companies commercialized new battery systems based on the use of nanometric $LiFePO_4$ and divided Sn, respectively. The obtained product displays attractive performances but always at the expense of another performance. This is how in A123 batteries the gain in power density is obtained to the expense of the energy density; and regarding NexeLion–Sony the energy density gain is obtained to the expense of the life span on cycling. It is therefore important to realize that an intrinsic aspect associated with the change into nano resides in the difficulty of compacting nano powders in order to obtain electrodes of more important packed densities than those obtained from a bulk material. The hope aroused by nanomaterials is based on their ability to induce new reaction mechanisms, such as: (i) conversion reactions which have increased from one to six the theoretical limit of the number of electrons exchanged per 3d metal according to Li insertion/deinsertion mechanism governing today's batteries and (ii) the exacerbated storage capacity of nanoporous carbons, which has been underestimated for several decades, and which can now foresee reaching capacities increased by 150% according to the classic carbon electrodes of high specific area. Other advantages resulting from our work on current collectors' nano-architecture and self-supported electrodes concern the shaping/assembling of electrodes, which predicts the conception of 3-D micro-batteries with interpenetrated network. All this will benefit the microsystems (MEMS, etc.) presently fed by energy-limited 2-D batteries. The possibilities are numerous, but as usual, any new concept has its limits, especially the polarization issue for conversion electrodes or the fine control of carbon porosity, as well as the elaboration of adequate electrolytes for these nanoporous carbons. This leads us to think that the commercialization will be an exciting and long-term adventure which will require, besides an enormous amount of work, a pluridisciplinary approach involving electrochemists, metallurgists, materials scientists and organic chemists. In view of the increasing multidisciplinary joint efforts that are developing, the odds are that new batteries and supercapacitors will be born in the next decade.

Acknowledgments

The authors wish to thank M. Nelson for her great help in editing this paper and their colleagues from the European network ALISTORE who have been involved with the research into nanoporous materials, as described in this book. They also wish to thank the European Commission for their trust in our setting up such a network, whose evolution has been ratified as a European Research Institute, as well as for funding through the Erasmus Mundus program of the "Materials for Energy Storage and Conversion" Masters degree (http://www.u-picardie.fr/mundus_MESC/) which was attended by many of our students.

References

1. Nagaura, T., and Tozawa, K. (1990) Lithium ion rechargeable battery. *Prog. Batt. Solar Cells*, **9**, 209.
2. Tarascon, J.M., and Armand, M. (2001) *Nature*, **414**, 359–367.
3. Padhi, A.K., Nanjundaswamy, K.S., Masquelier, C., Okada, S., and Goodenough, J.B. (1997) *J. Electrochem. Soc.*, **144** (5), 1609–1613.
4. Delacourt, C., Poizot, P., Levasseur, S., and Masquelier, C. (2006) *Electrochem. Solid State Lett.*, **9** (7), A352–A355.
5. Prakash, A., Rozier, P., Dupont, L., Vezin, H., Sauvage, F., and Tarascon, J-M. (2006) *Chem. Mater.*, **18** (2), 407–412.
6. Nytten, A., Abouimane, A., Armand, M., Gustafsson, T., and Thomas, J.O. (2005) *Electrochem. Commun.*, **7** (2), 156–160.
7. Armstrong, A.R., Armstrong, G., Canales, J., and Bruce, P. (2004) *Agew. Chem. Int. Ed.*, **43**, 2286.
8. Yang, J., Wang, B.F., Wang, K., Liu, Y., Xie, J.Y., and Wen, Z.S. (2003) *Electrochem. Solid State Lett.*, **6** (8), A154.
9. Larcher, D., Beattie, S., Morcrette, M., Edström, K., Jumas, J.-C., and Tarascon, J.-M. (2007) *J. Mater. Chem.*, **17**, 3759–3772.
10. Saint, J., Morcrette, M., Larcher, D., Laffont, L., Beattie, S., Pérès, J.-P., Talaga, D., Couzi, M., and Tarascon, J.-M. (2007) *Adv. Func. Mat.*, **17** (11), 1765–1774.
11. Morcrette, M., Larcher, D., Tarascon, J-M., Edstrom, K., Vaughey, J.T., and Thackeray, M.M. (2007) *Electrochim. Acta*, **52** (17), 5339–5345.
12. Poizot, P., Laruelle, S., Grugeon, S., Dupont, L., and Tarascon, J-M. (2000) *Nature*, **407**, 496.
13. Amatucci, G.G., and Pereira, N. (2007) *J. Fluor. Chem.*, **128** (4), 243–262.
14. Gillot, F., Boyanov, S., Dupont, L., Doublet, M.L., Morcrette, M., Monconduit, L., and Tarascon, J-M., (2005) *Chem. Mater.*, **17** (25), 6327–6337.
15. Grugeon, S., Laruelle, S., Dupont, L., Chevallier, F., Gireaud, L., Taberna, P.L., Simon, P., and Tarascon, J.M. (2005) *Chem. Mater.*, **17**, 5041–5047.
16. Tarascon, J.M. (2008) The Alistore Network of Excellence, see http://www.u-picardie.fr/alistore/
17. Martin, C.R. (1996) *Chem. Mat.*, **8**, 1739–1746.
18. Taberna, P.L., Mitra, S., Poizot, P., Simon, P., and Tarascon, J.M., (2006) *Nat. Mater.*, **5**, 567–573.
19. Kothari, H.M., et al. (2006) *J. Mater. Res.*, **21**, 293–301.

20 Mitra, S., Poizot, P., Finke, A., and Tarascon, J.M. (2006) *Adv. Funct. Mater.*, **16**, 2281.

21 Hassoun, J., Panero, S., Taberna, P.L., Simon, P., and Scrosati, B. (2007) *Adv. Mater.*, **19**, 1632–1635.

22 Finke, A., Poizot, P., Guéry, C., Dupont, L., Taberna, P-L., Simon, P., and Tarascon, J-M. (2008) *Electrochem. Solid State Lett.*, **11**, E5–E9.

23 Le, D.B., Passerini, S., Guo, J., Ressler, J., Owens, B.B., and Smyrl, W.H. (1996) High surface area V_2O_5 aerogel intercalation electrodes. *J. Electrochem. Soc.*, **143** (7), 2099–2104.

24 Dupont, L., Laruelle, S., Grugeon, S., Dickinson, C., Zhou, W., and Tarascon, J.M. (2008) *J. Power Sources*, **175**, 502–509.

25 Miller, J.R., and Simon, P. (2008) *Science*, **321**, 651–652.

26 Simon, P., and Gogotsi, Y. (2008) *Nat. Mater.*, **7**, 845–854.

27 Helmholtz, H.V. (1879) *Ann. Phys.*, 337.29

28 Sing, K.S.W., Everett, D.H., Haul, R.A.W., Moscou, L., Pierotti, P.A., and Rouquerol, J., (1985) *Pure Appl. Chem.*, **57**, 603.

29 Lin, R., Taberna, P.L., Chmiola, J., Guay, D., Gogotsi, Y., and Simon, P. (2009) *J. Electrochem. Soc.*, **156** (1), A7–A12.

30 Pandolfo, A.G., and Hollenkamp, A.F. (2006) *J. Power Sources*, **157**, 11–27.

31 Fuertes, A.B., Lota, G., Centeno, T.A., and Frackowiak, E. (2005) *Electrochim. Acta*, **50**, 2799–2805.

32 Vix-Guterl, C., Saadallah, S., Jurewicz, K., Frackowiak, E., Reda, M., Parmetier, J., Patarin, J., and Beguin, F. (2004) *Mater. Sci. Eng.*, **B108**, 148–155.

33 Fuertes, A.B. (2003) *J. Mater. Chem.*, **13**, 3085.

34 Chmiola, J., Yushin, G., Gogotsi, Y., Portet, C., Simon, P., and Taberna, P.L. (2006) *Science*, **313**, 1760–1763.

35 Chmiola, J., Largeot, C., Taberna, P.-L., Simon, P., and Gogotsi, Y. (2008) *Angew. Chem.*, **47**, 3392–3395.

36 Largeot, C., Portet, C., Chmiola, J., Taberna, P.-L., Gogotsi Y., and Simon, P. (2008) *J. Am. Chem. Soc.*, **130** (9), 2730–2731.

37 Naoi, K. (2004) "RuO_2 nano-dot encapsulated KB", Japanese patent number 2004-173452.

38 Cottineau, T., Toupin, M., Brousse, T., and Bélanger, D. (2006) *Appl. Phys. A*, **82**, 599–606.

39 Ravet, N., *et al.* (1999) Abstract 127, *196th Meeting of the Electrochemical Society, Hawaï*. Armand, M., Gauthier, M., Magnan, J.-F., and Ravet, N. (1999) Method for synthesis of carbon-coated redox materials with controlled size. World Patent WO 02/27823 A1.

8
Carbon-Based Nanomaterials for Electrochemical Energy Storage
Elzbieta Frackowiak and François Béguin

8.1
Introduction

Recently, improving the energy efficiency of mobile and stationary systems has become an important challenge in order to reduce fossil fuel consumption and CO_2 emissions. In this context, electrochemical storage systems are an undeniable part of the energy management strategy. For example, they can be used as buffer elements for storing intermittent renewable energies, but also for controlling the energy flows which are related with the fluctuations of electricity consumption. In the transportation systems, energy can also be recovered during deceleration or braking and further used for the acceleration steps.

The most effective electrochemical energy storage systems are Li-ion batteries and supercapacitors, also called electrochemical capacitors. Lithium batteries are characterized by a relatively important energy density, but a moderate power density, whereas supercapacitors display a high power density, but their energy density is relatively low. Although both devices can be used in combination, their performance must be dramatically improved to reach the objectives imposed by energy management in stationary and transport applications.

Among the parameters which control the performance of lithium batteries and supercapacitors, electrode materials, particularly nanocarbons, are undoubtedly the most important. This chapter shows the influence of nanotexture and surface functionality of carbons on their electrochemical properties in the two systems. The main objectives are to correlate the electrochemical behavior with the physico-chemical parameters of carbons and, from this knowledge, to define routes for optimizing the performance.

8.2
Nanotexture and Surface Functionality of sp² Carbons

Apart from hexagonal graphite, which is a crystalline form of carbon, the other forms of sp^2 carbons are in general poorly organized. The most usual way for

Nanotechnology for the Energy Challenge. Edited by Javier Garcia-Martinez
© 2009 WILEY-VCH Verlag GmbH & Co. KGaA, Weinheim
ISBN: 978-3-527-32401-9

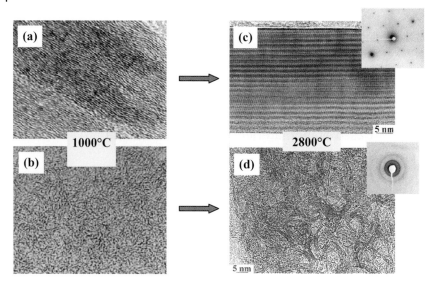

Figure 8.1 002 lattice fringe HRTEM images of nanocarbons obtained by (a, b) pyrolysis of anthracene and saccharose at 1000 °C, (c, d) treatment of anthracene and saccharose cokes at 2800 °C. The insets show electron diffraction patterns (Courtesy of J.N. Rouzaud, Ecole Normale Supérieure, Paris).

obtaining sp^2 carbons is the pyrolysis of polymers and fossil (coal, etc.) or biomass precursors at temperatures ranging from 500 to 1000 °C (for more information on the synthesis of carbons and their structural/textural properties see [1]). They can be further physically (CO_2, H_2O) or chemically (KOH, H_3PO_4) activated in order to develop porosity.

Figure 8.1a, b shows HRTEM 002 lattice fringe images of cokes obtained by the pyrolysis of anthracene and saccharose at 1000 °C under a neutral atmosphere. For anthracene coke, one easily distinguishes fringes which are more or less parallel to a given direction, whereas they are fully misoriented in the case of saccharose coke. The former is transformed into graphite by heat treatment at 2800 °C, as shown by the perfectly parallel and long fringes together with characteristic diffraction spots in Figure 8.1c, while the later gives nanoporous turbostractic carbon (or glassy carbon) characterized by diffraction rings (Figure 8.1d). Anthracene coke is named graphitizable or soft carbon, and saccharose coke is non-graphitizable or hard carbon. In between these extremes, all kinds of nanotextural/structural organizations are possible, depending on the precursor and the pyrolysis conditions. From the foregoing, low-temperature sp^2 carbons appear to be constituted of more or less disoriented nanometer-size polyaromatic units, as shown by the schematic model of a hard carbon in Figure 8.2. A careful observation of Figure 8.1b reveals some stacks of two or three layers, indicating that the model of Figure 8.2 is only an approximation. In particular, it is noteworthy that the aromatic units are interconnected or even continuous with some curvatures. Otherwise, such a

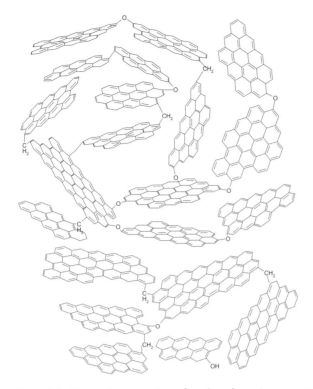

Figure 8.2 Schematic presentation of graphene layers in nanoporous carbon.

carbon with independent units would be soluble in most aromatic solvents. Notwithstanding the imperfections of such model, it is important to observe that the disorientation of the aromatic layers creates some porosity, which can strongly affect the electrochemical properties.

Low-temperature carbons are never exclusively containing the element carbon. They also include heteroatoms, such as hydrogen, oxygen, nitrogen, etc. The latter are in part remains from the precursor, but they can be also included by post-treatments, for example activation. Most heteroatoms are located as functional groups at the edge of the aromatic layers. Nitrogen is either substituted to carbon (*lattice nitrogen*) or included in the form of functional groups (*chemical nitrogen*) at the periphery of polyaromatic structural units [2, 3] (see Figure 8.3). Figure 8.4 shows the usual oxygenated functionalities present in carbons [4]. The electrochemical properties of carbons are strongly influenced by these functionalities and differ with the kind of electrolyte (aqueous, non-aqueous). Beside the oxygenated groups, the model in Figure 8.4 shows also free edge sites and unpaired electrons on the basal plane. As demonstrated later, the latter might participate in the decomposition of organic electrolytes, influencing the extent of *solid electrolyte interphase* formation in lithium batteries and the cycle life of supercapacitors.

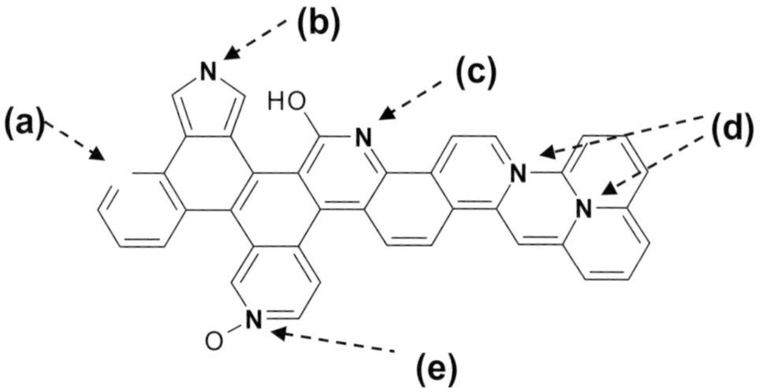

Figure 8.3 Nitrogen functionalities in a carbon material. (a) Pyridinic (N-6), (b) pyrrolic, (c) pyridonic (N-5), (d) quaternary (N-Q), (e) oxidized nitrogen (N-X).

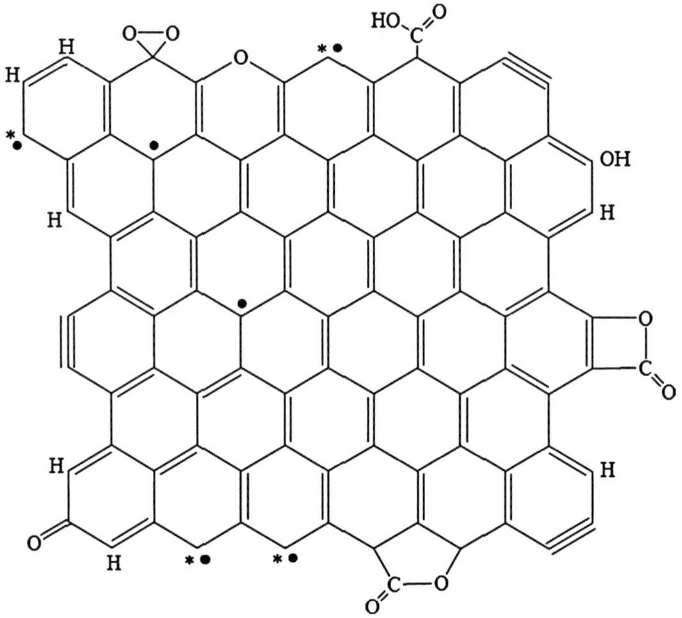

Figure 8.4 Oxygen functionality and free radicals in the carbon network [4].

8.3
Supercapacitors

8.3.1
Principle of a Supercapacitor

An electrochemical capacitor, also called a supercapacitor, golden capacitor or ultracapacitor, is a device able to store charges in the electrode/electrolyte interface [5–9]. In principle, the energy storage is based on the separation of charges (ions)

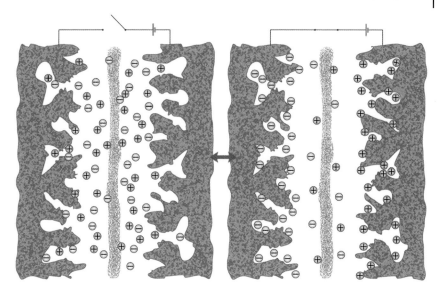

Figure 8.5 Principle of an electrochemical capacitor. On the left, the system is discharged, while it is charged on the right.

in an electric double layer (EDL). As shown in Figure 8.5, an electrochemical capacitor is constructed with two porous electrodes (positive, negative) immersed in an electrolytic solution. During charging, the negative plate attracts the positively charged cations, while the anions are accumulated in the pores of the positive electrode. According to Figure 8.5, each electrode can be treated as a single capacitor of capacitance C_1 or C_2. Since the two electrodes are in series, the total capacitance is controlled by the electrode with the smallest capacitance value according to equation (8.1):

$$\frac{1}{C} = \frac{1}{C_1} + \frac{1}{C_2} \tag{8.1}$$

For each electrode the EDL capacitance, C_{dl}, can be expressed by equation (8.2). It is proportional to the surface area s of the electrode/electrolyte interface and to the electrolyte permittivity ε. The d value represents the EDL thickness which is strongly determined by the type of pores.

$$C_{dl} = \varepsilon \cdot \frac{s}{d} \tag{8.2}$$

An increase of the capacitor voltage U causes a significant enhancement of power P and energy E because of the following dependences in equations (8.3) and (8.4), where R_S stands for the equivalent series resistance:

$$E = \frac{1}{2}CU^2 \tag{8.3}$$

and

$$P = \frac{U^2}{4R_S} \tag{8.4}$$

8.3.2
Carbons for Electric Double Layer Capacitors

It is well known that micropores (diameter <2 nm) play an essential adsorption role for the formation of the electric double layer. However, these micropores must be electrochemically accessible for the ions, therefore the presence of mesopores (diameter 2–50 nm) is necessary for efficient charge propagation to the bulk of the electrode material, allowing the so-called frequency response to be fulfilled, that is, the energy to be extracted at higher frequencies, for example, 1–10 Hz [10]. Hence, the availability and wettability of pores, with dimensions adapted to the size of solvated anions and cations which have to be transported from the electrolytic solution, is crucial for high capacitor performance.

For capacitor applications, activated carbons from different precursors and prepared by different activation processes have been widely used [11–21]. It can often be found in the literature that the higher the BET specific surface area, the higher the capacitance values. However, for a wide variety of activated carbons, this trend is not perfectly followed [17]. In fact, the narrow micropores may not contribute to the total double layer capacitance due to a sieving effect [22–24], which in part explains the absence of proportionality with the BET surface area. This is confirmed by a study which shows that capacitance values as high as 175 F g^{-1} in aqueous medium can be reached using a carbon with a surface area of only 1300 m^2 g^{-1}, when the pore size is optimized by chemical activation [25]. Usually, the capacitance values of activated carbons range from 100 to 200 F g^{-1} in aqueous medium and from 50 to 150 F g^{-1} in organic medium. The larger values in aqueous electrolyte are essentially justified by a smaller size of solvated ions and a higher dielectric constant than in organic media. However, organic electrolytes are generally preferred for applications, due to their high voltage window which allows more energy to be stored than in aqueous solution.

The effect of pore size in aqueous (6 mol L^{-1} KOH, 1 mol L^{-1} H$_2$SO$_4$) and organic (1 mol L^{-1} TEABF$_4$ in acetonitrile) electrolytes has been studied on a series of porous carbons prepared by KOH activation of bituminous coal chars [26]. Since the materials were all prepared from the same precursor, by simply changing the carbonization temperature (520–1000 °C), one can reasonably assume that they are nanotexturally identical (same pore shape and pore interconnection) and that they only differ by their average pore width. As the pyrolysis temperature of coal increases, the BET specific surface area (S_{BET}) decreases from ca. 3000 to 1000 m^2 g^{-1} and the average micropore width (L_{0N2}) from 1.4 to 0.8 nm. Figure 8.6 shows that the volumetric capacitance [gravimetric capacitance (F g^{-1}) divided by the Dubinin–Raduskevich micropore volume (cm^3 g^{-1})] or the normalized capacitance [gravimetric capacitance (F g^{-1}) divided by the BET specific surface area (m^2 g^{-1})] of these

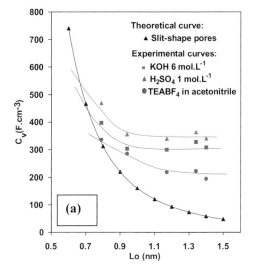 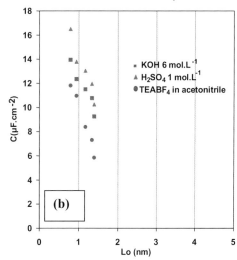

Figure 8.6 Capacitance in various electrolytes (1 mol L^{-1} H$_2$SO$_4$, 6 mol L^{-1} KOH and 1 mol L^{-1} TEABF$_4$ in acetonitrile) versus average pore size L_0 for a series of bituminous coal derived carbons activated by KOH at 800 °C. (a) Experimental volumetric capacitance (gravimetric capacitance divided by Dubinin-Raduskevich micropore volume) and theoretical volumetric capacitance (calculated for one ion occupying one pore of volume L_0^3). (b) Normalized capacitance (gravimetric capacitance divided by the BET specific surface area; adapted from [26]).

carbons increases as the average micropore size, L_0, decreases [26]. These data fit well with formula (8.2), showing that capacitance increases when the distance d between pore walls and ions decreases.

A significant increase in carbon capacitance has also been shown for average pore sizes <1 nm using carbide-derived carbons (CDCs) obtained by chlorination of carbides at various temperatures [27]. A distortion of the solvation shells in these pores was suggested that allows close approach of the ion center to the electrode surface. However, further detailed investigations showed that such behavior is typical for many carbons and there is no anomalous behavior for carbide-derived carbons [28].

Hence, an adequate pore size is more important than a high surface area for obtaining high values of capacitance. The saturation of gravimetric capacitance, observed in all the electrolytic media for the high surface area materials, is related to the increase of the average pore width L_0 (Figure 8.6). Although the pore volume of these carbons is very high, their pores are too wide for an effective participation in the formation of the double layer. In two-electrode cells, pore filling seems to be optimal when the pore size is close to 0.7 nm and 0.8 nm in an aqueous and organic medium, respectively [26]. By using a three-electrode configuration, it is possible to demonstrate that the optimal pore size differs with each ion, being slightly higher at the negative electrode than at the positive one in TEABF$_4$ in acetonitrile [29].

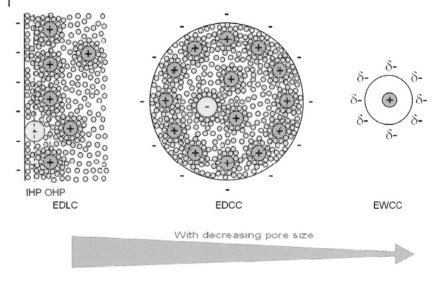

Figure 8.7 Schematic diagrams of the different types of electric double-layer capacitors depending on the pore dimensions [30].

The simplification of the carbon/electrolyte interface to an EDLC (Figure 8.7) does not consider the curvature of each pore. Accordingly, equation (8.2) for the simple parallel-plate capacitor model can be a good approximation for macropores, but it does not reflect the actual scenario for micropores and mesopores, where the close interaction with the pore walls should be considered. Different models of EDL capacitors depending on the size and shape of pores, and the corresponding expression for capacitance, have lately been proposed by Huang [30] and are shown in Figure 8.7. EDCC states for an electric double cylinder capacitor are formed by a negatively charged mesopore and solvated cations approaching the pore wall. A negatively charged micropore and solvated or desolvated cations lining up form an electric wire-in-cylinder capacitor (EWCC).

Apart from the kinetic parameters connected with the pore size distribution, the electrical conductivity of the carbon electrode is another limiting factor of capacitor performance, especially for power density [31]. Generally, higher porosity leads to poorer electrical conductivity. The conductivity of activated carbons varies from a few tens of mS cm^{-1} to 100 mS cm^{-1}, depending on the kind of material [11]. Hence, the addition of a conducting agent to the electrode material has crucial consequences on the electrochemical performance.

In conclusion, the micropores play an essential role for ions adsorption, whereas the small mesopores are necessary for their rapid transportation to the bulk of the material. An optimal performance is expected with a carbon of high surface area with preferably narrow pores below 1 nm and well-balanced micro/mesoporosity. Hence, a strict control of the carbonization and activation processes is necessary for preparing optimized carbon materials. Even if the conventional carbons are characterized by a low cost and a large specific surface area, their EDLC application is rather limited, because their pore size has a wide distribution and the pores are randomly connected, limiting both charge storage capacity and rate capability. Closed or isolated pores may not be wetted by the electrolyte and an irregular pore connection makes ionic motion difficult. Hence, porous carbons should be designed with micropores of narrow pore size (to supply high capacitance values) interconnected with a regular network of small mesopores to assure a good charge propagation.

8.3.3
Carbon-Based Materials for Pseudo-Capacitors

In general, two modes of energy storage are combined in electrochemical capacitors: (i) the electrostatic attraction between the surface charges and the ions of opposite charge (electric double layer); (ii) a pseudo-capacitive contribution which is related with quick faradic charge transfer reactions between the electrolyte and the electrode [5, 8, 9]. Whereas the redox process occurs at an almost constant potential in an accumulator, the electrode potential varies proportionally to the charge utilized dq in a pseudo-capacitor, summarized by equation (8.5):

$$dq = C * dU \tag{8.5}$$

The electrical response of such a system is comparable to that of a capacitor. Being of faradic origin and non-electrostatic, this capacity is distinguished from the double-layer one and is called pseudo-capacity. The electric double-layer formation is a universal property of a polarized material surface, while pseudo-capacity is an additional property which depends both on the type of material and electrolyte. Compared to the double-layer specific capacity ($\sim 10\,\mu F\,cm^{-2}$), it has generally a high value (100–400 $\mu F\,cm^{-2}$), because it involves the bulk of the electrode and not only the surface. From a practical point of view, pseudo-capacity contributes to enhancing the capacity of materials and their energy density.

8.3.3.1 Pseudocapacitance Effects Related with Hydrogen Electrosorbed in Carbon
The reversible electrosorption of hydrogen on the surface of a carbon network by cathodic decomposition of water is a good example of pseudo-capacitance [32–35]. In a unique step, hydrogen is produced by water electrolysis and simultaneously stored in the carbon substrate. The hydrogen produced *in situ* easily penetrates into the nanopores of carbon, where it is adsorbed due to the driving force of the negative polarization. By this process, higher pressures (calculated approximately from the Nernst law) than in the gas phase can be reached [32]. In alkaline solution, water is reduced according to the following reaction:

$$H_2O + e^- \rightarrow H + OH^- \tag{8.6}$$

and the formed nascent hydrogen is adsorbed on the surface of nanopores:

$$\langle C \rangle + xH \rightarrow \langle CHx \rangle \tag{8.7}$$

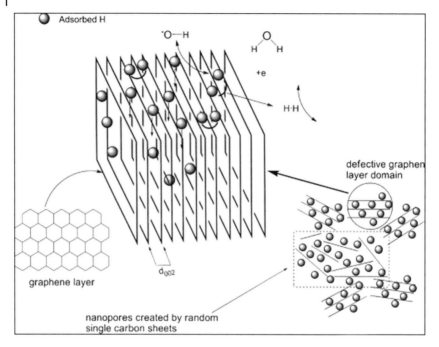

Figure 8.8 Schematic presentation of hydrogen electrosorption into nanoporous carbon [33].

where <C> represents the carbon host, and <CHx> is the hydrogen adsorbed on the latter.

The overall charge/discharge phenomenon is summarized as follows:

$$\langle C \rangle + xH_2O + xe^- \leftrightarrow \langle CHx \rangle + xOH^- \tag{8.8}$$

It is now generally accepted that a nanoporous carbon is required for hydrogen storage although the size and structure of the pores is open to debate. It seems likely that a combination of micropores (where the storage occurs) and mesopores is necessary, to facilitate the transport that could be required for high charge/discharge rates. It is also possible that the mesopores could facilitate the initial water decomposition [35].

The process of hydrogen insertion into the carbon network is schematically presented in Figure 8.8. It seems that especially defective domains can adsorb hydrogen more preferably [33]. However, the filling of intervals between graphene layers through an intercalation phenomenon is still controversial. For understanding the mechanism of hydrogen electrosorption, various techniques must be used.

Cyclic voltammetry is a well-adapted electrochemical technique to elucidate the mechanism and kinetics of reversible hydrogen storage. An example of voltammetry characteristics using a microporous activated carbon cloth from viscose (AC;

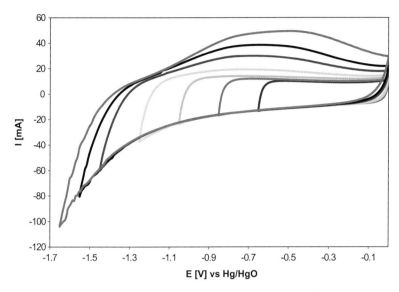

Figure 8.9 Cyclic voltammograms (5 mV s^{-1}) of AC cloth in 6 mol L^{-1} KOH with a gradual shift to more negative values of potential.

S_{BET} = 1390 m^2 g^{-1}) in 6 mol L^{-1} KOH, is shown in Figure 8.9. The minimum potential is gradually shifted for each cycle towards more negative values. When the electrode potential is higher than the thermodynamic value which corresponds to water decomposition, the voltammetry curves have the rectangular shape typical of charging the electric double layer. From the third loop, as the potential becomes lower than this value, both the double layers are charged and hydrogen is adsorbed in the carbon pores. During the anodic scan, the reactions run in opposite directions and a peak corresponding to the electro-oxidation of adsorbed hydrogen is observed. This pseudo-capacity contributes to the total capacity in addition to the electric double layer capacity. When the negative potential limit decreases, the anodic current increases, and the corresponding hump shifts towards more positive values of potential. The noticeable polarization between the cathodic and anodic processes indicates that hydrogen electrochemically stored is trapped more energetically than for a classic physisorption and/or that there are important diffusion limitations.

The fact that hydrogen is strongly trapped in the carbon matrix is also confirmed by the galvanostatic intermittent titration technique (GITT), using a meso/microporous carbon prepared by the template method. By selecting a carbon with an interconnected network of mesoporous canals and micropores, one can get rid of an important part of the diffusion effects. For the GITT experiment presented in Figure 8.10, the galvanostatic cycle includes periods of 1 h at a constant current of ±25 mA g^{-1} interrupted by relaxation periods of 2 h at open circuit. A curve close to equilibrium would be obtained by including all the potential values at the end of each relaxation period (Figure 8.10). The important polarization (overpotential)

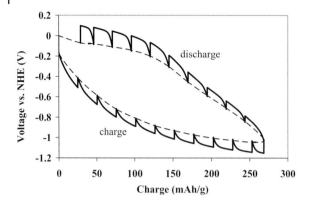

Figure 8.10 Galvanostatic intermittent titration technique (GITT) for a template mesoporous carbon in 6 mol L^{-1} KOH [34].

between charge/discharge indicates a strong interaction of adsorbed hydrogen with the carbon substrate. Since it can be still desorbed in ambient conditions, one must conclude there is a weak chemical bonding of hydrogen.

Such type of carbon–hydrogen interactions have been proved by thermo-programmed desorption (TPD) coupled with mass spectrometry. A peak ($m/e = 2$) located at 200 °C in the TPD curve confirmed this state of hydrogen [34]. An activation energy of 110 kJ mol^{-1} for hydrogen desorption was estimated from this peak; such a value is characteristic of a weakly chemisorbed state. The desorption temperature is smaller than the value which is generally observed for covalently bonded hydrogen, for example, 700 °C.

The particularity of electrochemical storage, compared to the typical adsorption under gas pressure, is due to the production of nascent hydrogen through water electroreduction. This very reactive form of hydrogen interacts with the active sites of carbon, being very energetically trapped. From a practical point of view, such a state of hydrogen is particularly attractive, as it is stabilized in the carbon substrate (e.g., the self-discharge is not important and can be desorbed in ambient conditions). Taking into account that the capacity is quite high (up to 500 mAh g^{-1}) and that this reversible process occurs in the negative range of potentials, it is clear that nanoporous carbons are very interesting as negative electrode of asymmetric supercapacitors.

8.3.3.2 Pseudocapacitive Oxides and Conducting Polymers

Electrically conducting polymers (ECPs) [36–38], for example, polyaniline (PANI), polypyrrole (PPy), poly-ethylenedioxythiophene (PEDOT) and transition metal oxides [39–43], for example, MnO$_2$, RuO$_2$, Ni(OH)$_2$, are attractive materials with pseudocapacitance properties because of the following reactions:

$$MnO_x(OH)_y + \delta e^- + \delta H^+ \leftrightarrow MnO_{x-\delta}(OH)_{y+\delta} \quad (8.9)$$

$$[PPy^+ A^-] + e^- \leftrightarrow [PPy] + A^- \quad (8.10)$$

They might be a promising alternative for the development of high performance supercapacitors. However, in most reports, a very thin layer of the active material coats a metallic current collector – very often platinum – that is far from the requirements for an industrial application. Moreover, swelling and shrinkage of ECPs may occur during doping/dedoping of the active film, leading to mechanical degradation of the electrode and fading of the electrochemical performance during cycling. Adding carbon, especially carbon nanotubes, is the most effective solution proposed to improve the resiliency and electrical conductivity of electrodes based on pseudo-capacitive materials [36, 38, 42]. The good performance of composites based on electrically conducting polymers and hydrous oxides is generally limited to a narrow range of potential. Oxides are preferable materials for the positive electrode, whereas ECPs, for example, PPy or PEDOT, can serve as materials for both polarities.

For the preparation of ECP composites with multi-walled carbon nanotubes (MWNTs), the chemical or electrochemical polymerization of the monomer (e.g., pyrrole, aniline) on the nanotubular materials can be applied. The capacitance of an electrochemically obtained MWNTs/polypyrrole (PPy) composite reaches the value of $170\,F\,g^{-1}$ with a good cyclic performance over 2000 cycles. Comparing the results of the two coating techniques, the non-homogeneous PPy layer deposited chemically is more porous and less compact than that electrochemically deposited. Consequently, the diffusion of ions proceeds more easily in the chemically formed composites, giving a better efficiency for charge storage. It has also been shown that the values of capacitance for the composites with polyaniline (PANI) and PPy strongly depend on the cell construction [38]. The high capacitance values found with the MWNT/ECP composites are due to the unique property of the entangled nanotubes which supply a perfect three-dimensional volumetric charge distribution and a highly accessible electrode/electrolyte interface.

As mentioned previously, carbon nanotubes can also be a perfect support for cheap transition metal oxides of poor electrical conductivity, such as amorphous manganese oxide ($a\text{-}MnO_2 \cdot nH_2O$) [42]. The $a\text{-}MnO_2$/MWNT composite can be prepared by precipitation of $a\text{-}MnO_2$ from a $KMnO_4 + Mn(OAc)_2 \cdot 4H_2O$ mixture which contains a predetermined amount of carbon nanotubes. The SEM image presented in Figure 8.11 for the $a\text{-}MnO_2$/MWNT composite containing 15 wt% of nanotubes shows a perfect adherence of MnO_2 when the nanotubes are oxidized by $KMnO_4$. Consequently, the composite electrodes have a good resiliency, and their porosity is high enough to favor the access of ions to the bulk of the active material.

The $a\text{-}MnO_2$/MWNT composite has a capacitance of $140\,F\,g^{-1}$ with a good cyclability and high dynamic of charge propagation. However, the voltage window of MnO_2-based capacitors is limited to 0.6 V, due to the irreversible conversion of Mn(IV) to Mn(II) at the negative electrode and Mn(IV) to Mn(VII) at the positive one [43]. In order to circumvent this drawback, an asymmetric configuration has been proposed, where the positive electrode consists of $a\text{-}MnO_2$ and the negative is built from activated carbon [43–45]. When the $a\text{-}MnO_2$/MWNT composite is used as the positive electrode, the supercapacitor demonstrates perfectly rectan-

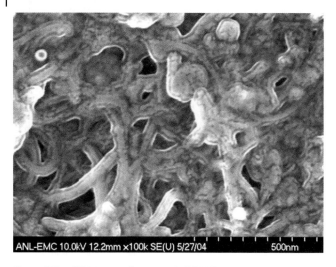

Figure 8.11 SEM image of an a-MnO$_2$/MWNT composite containing 15 wt% of nanotubes (Courtesy of M. Thackeray, Argonne National Laboratory, USA).

gular voltammograms and it can be operated up to 2 V in aqueous medium with an extremely good cycleability.

An asymmetric or hybrid configuration, with two electrodes of different nature, for example, conducting polymer, transition metal oxide and activated carbon [43, 44], taking into account the optimal potential range of each electrode, is an excellent way to reach a wide operating voltage. The most promising material for use as a negative electrode is activated carbon, whereas conducting polymers and transition metal oxides can easily operate as positive or negative electrode. The advantage of the asymmetric combination over the symmetric one is obvious. A significant increase of energy and power can easily be obtained especially due to the extension of the supercapacitor voltage.

8.3.3.3 Pseudo-Capacitive Effects Originated from Heteroatoms in the Carbon Network

Beside the resistive transition metal oxides and expensive and not always environment-friendly conducting polymers (ECPs), more simple pseudo-capacitive materials based on carbons were proposed for supercapacitor application. Instead of developing the specific surface area of carbons, an alternative is to introduce heteroatoms (nitrogen, oxygen, etc.) which provide pseudo-faradaic reactions. In the case of nitrogen, the variety of surface functionality (Figure 8.3) results both from the position occupied in the ring system and from the extent of association with oxygen, which hardly can be avoided during synthesis. The surroundings of the nitrogen atom in a graphene layer obviously affect its charge, electron donor/acceptor properties and the contribution to the delocalized π electron system. Generally, the presence of nitrogen in a carbon material can enhance capacitance

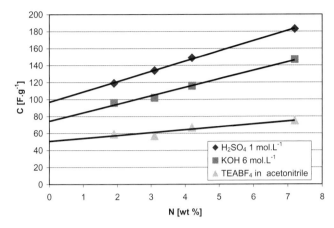

Figure 8.12 Evolution of capacitance values for a series of activated carbon materials with a similar nanotexture and surface area but various nitrogen content [56].

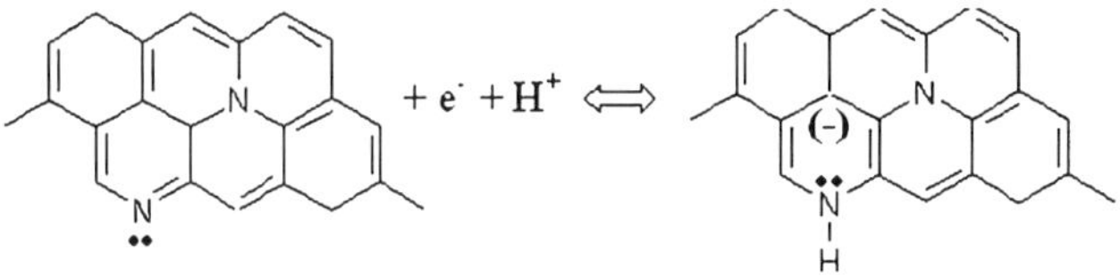

Figure 8.13 Possible pseudo-faradaic reaction of the pyridinic group in aqueous medium.

due to faradaic redox reactions but also because of the modification of its electronic character as well as improved wettability [18, 46–56].

The effect of various amounts of substitutional nitrogen in the carbon network on the pseudo-capacitive properties has been studied on a series of nitrogen doped carbons of comparable BET surface area ($S_{BET} \approx 800\,m^2\,g^{-1}$) and micropore volume [48, 49]. A careful analysis of the results obtained shows no remarkable differences in organic medium, whereas capacitance is proportional to the N content (Figure 8.12) in alkaline and especially in acidic medium. It is noteworthy that the basic character of carbons is also proportional to the nitrogen content. The enhancement of the capacitance values in H_2SO_4 medium is interpreted by pseudo-faradaic charge transfer reactions due to the nitrogenated functionality, as illustrated in Figure 8.13 or by equations (8.11) and (8.12):

$$C^*=NH + 2e^- + 2H^+ \leftrightarrow C^*H\text{-}NH_2 \tag{8.11}$$

$$C^*\text{-}NHOH + 2e^- + 2H^+ \leftrightarrow C^*\text{-}NH_2 + H_2O \tag{8.12}$$

where C* stands for the carbon network.

The beneficial effect of nitrogen in a composite with an incorporated nanotubular backbone has been proved using melamine as nitrogen-rich carbon precursor

Table 8.1 Physicochemical and electrochemical characteristics of nanocomposites rich in nitrogen. C stands for the capacitance estimated at 5 A g^{-1} current load [55].

Sample	S_{BET} [m^2 g^{-1}]	V_{total} [cm^3 g^{-1}]	V_{micro} [cm^3 g^{-1}]	C [F g^{-1}]	C [μF cm^{-2}]	Nitrogen [wt%]	Nitrogen [atm%]
M+F	329	0.162	0.152	4	1.2	21.7	24.8
Nt+3M+F	403	0.291	0.174	100	24.8	14.0	13.6
Nt+2M+F	393	0.321	0.167	126	32.1	11.7	12.7
Nt+M+F	381	0.424	0.156	83	21.8	7.4	7.9

(45 wt%). Composites with different proportions of nitrogen have been prepared at 750 °C from polymerized blends of melamine with formaldehyde, in the presence of a controlled amount of multiwalled carbon nanotubes [55]. The final carbonization products were named M+F (i.e., melamine and formaldehyde without carbon nanotubes), Nt+M+F, Nt+2M+F Nt+3M+F (composites with carbon nanotubes and one-, two- and threefold melamine proportion in the blend, respectively). The results of elemental analysis show that the nitrogen content varied in the final product from 7.4 to 21.7 wt% (Table 8.1). The oxygen content calculated by difference is comparable in all the samples, varying from 5.9 to 7.8 wt%. The BET specific surface area is quite similar for all the samples, ranging from 329 to 403 m^2 g^{-1}, being most developed for the Nt+3M+F composite.

TEM images of the composites show that the entangled morphology of the nanotubes with the presence of open mesopores is well-preserved [55]. Yet, the simple carbonization product of the melamine/formaldehyde (M+F) blend – without nanotubes – gives an amorphous texture. In the case of the Nt+M+F composite with ~7 wt% nitrogen, the carbon nanotubes are hardly covered by the carbonization product. An ideal good adhesion of carbon homogeneously distributed on nanotubes is demonstrated in Nt+2M+F, whereas increasing the melamine content in the case of Nt+3M+F, the composite morphology gradually changes with a compact texture together with some agglomerates.

The values of capacitance of the nitrogen enriched composites in sulfuric acid medium are given in Table 8.1. The capacitance of the non-mesoporous M+F carbon is negligible, whereas it reaches 126 F g^{-1} for the Nt+2M+F composite. Although the BET specific surface area is comparable for all the materials, the improvement in the case of the composites is due to the presence of open mesopores which enhance the access of the electrolyte to the active nitrogenated functionalities. Figure 8.14 shows the dependence of capacitance versus the current load, from 50 mA g^{-1} to 50 A g^{-1}, for all the composites. It is well visible how the ability of charge accumulation diminishes with the load, but the Nt+2M+F sample is still able to supply a capacitance of 60 F g^{-1} at 50 A g^{-1} current density. The high charge propagation can be explained, both by an exceptional electronic transport in carbon nanotubes which are still preserved after the carbonization

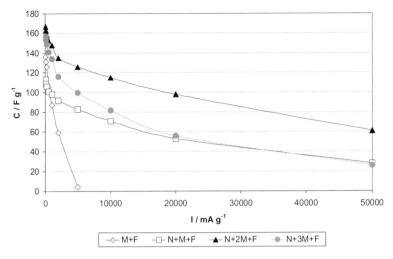

Figure 8.14 Capacitance vs current load for the nitrogen-enriched composites in 1 mol L^{-1} H$_2$SO$_4$ electrolytic solution (adapted from [55]).

process, and by their opened mesoporous network which enhances the mobility of ions.

The form in which N participates in the carbon network is especially important. It seems that the effect of NH$_2$- groups outside of the matrix, with N chemically bound to the carbon network ("chemical nitrogen"), is of less importance and most probably such groups could block the entrance to the pores. Nitrogen substituted to carbon ("lattice nitrogen") in the periphery, in the form of pyridinic groups, could play some useful role according to the reversible redox reaction shown in Figure 8.13. The donor properties of nitrogen are responsible for filling the conduction band by electrons, and in turn, more ions can be sorbed in the electric double-layer, especially for composites where carbon is optimally substituted by nitrogen. However, it was confirmed experimentally that a higher amount of nitrogen (>15%) could lead to a significant aggravation of conductivity. It should be also stressed that these composites have been prepared without any activation process, hence, their moderate surface area gives profitable values of specific surface capacitance as well as attractive volumetric capacity.

Attractive materials of high density can also be obtained by one-step carbonization of biopolymers, for example, sodium alginate at 600 °C under argon flow without any additional activation process [57]. The carbonized material is moderately microporous (S_{BET} = 273 m^2 g^{-1}) but it contains 15 atm% oxygen in the carbon framework. Even if the specific surface area of this carbon is low, the capacitance in 1 mol L^{-1} H$_2$SO$_4$ electrolyte reaches 200 F g^{-1}. Voltammetry curves of this material in a three electrode cell show reversible humps at around −0.1 to 0.0 V versus

Hg/Hg$_2$SO$_4$ [57]. It is well-known that peaks at such positions are connected with electrochemical reactions of oxygen surface functionalities such as the quinone/hydroquinone pair [58]. Lately, it has been shown that pyrone-like structures (combinations of non-neighboring carbonyl and ether oxygen atoms at the edge of the graphene layers) can effectively accept two protons and two electrons in the same range of electrochemical potential as the quinone/hydroquinone pair [59]. Consequently in the carbon obtained from sodium alginate or other biopolymers, for example, sea weeds [60], the high value of capacitance is related with charge transfer reactions on the quinone, phenol and ether groups. Apart from an attractive gravimetric capacitance, this weakly porous material has a high density and a high electrical conductivity, that allows a volumetric capacitance higher than that of typical activated carbons, and the capacitors can be charged at high regimes without any conductivity additive in the electrodes [57, 60].

8.4
Lithium-Ion Batteries

Lithium-ion batteries are generally based on the reversible transfer of lithium ions between two intercalation/insertion materials through an aprotic electrolyte. The cathodic (positive electrode) materials are lithium transition metal oxides represented by the general formula Li$_y$MO$_2$ ($y \approx 1$), whereas graphite is the most frequently used negative electrode (anode) material. The following equations represent the reversible redox reactions in the cell:

$$6C + xLi^+ + xe^- \Leftrightarrow Li_xC_6 \tag{8.13}$$

$$Li_yMO_2 \Leftrightarrow Li_{y-x}MO_2 + xLi^+ + xe^- \tag{8.14}$$

With graphite, the battery operates at about 3.5 V during discharge, which makes the system very attractive for its high energy density [61]. During intercalation (reduction of graphite), stages 3, 2 and 1 intercalation compounds are formed with charge transfer to carbon, while the opposite sequence occurs during deintercalation [62]. The capacity C_C of cathodic materials varies from 140 mAh g^{-1} for LiCoO$_2$ to about 200 mAh g^{-1} for LiMnO$_2$ and its derivatives [63]. For the above C_C range, the estimated capacity of a Li-ion cell noticeably increases when the anode specific capacity C_A is increased up to about 1200 mAh g^{-1} [63]. With graphite, the theoretical reversible capacity reaches 372 mAh g^{-1} (saturation composition of one lithium for six carbon atoms – LiC$_6$). Therefore, for new energy applications, it is worth developing anodic materials that display a higher reversible capacity than graphite, for example, in the range of 1000 mAh g^{-1}.

This presentation introduces attempts for enhancing the reversible capacity of lithium batteries by using other anodic materials as disordered carbons and silicon/carbon composites. The physico-chemical parameters controlling the irreversible capacity are also discussed in order to define strategies for improving the performance of lithium-ion batteries.

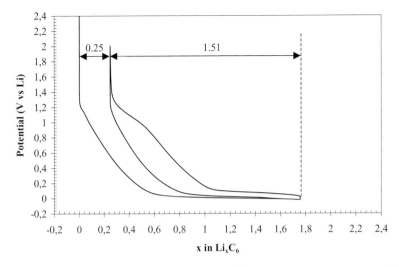

Figure 8.15 Galvanostatic insertion/deinsertion of lithium in a composite constituted from viscose carbon coated by a pyrocarbon film [65].

8.4.1
Anodes Based on Nanostructured Carbons

Many authors have presented exceptional properties of soft and hard carbons. A documented literature review may be found in reference [64]. However, most of these materials have not been applied industrially. Figure 8.15 shows a typical galvanostatic charge/discharge curve for these kinds of carbons [65]. Although the reversible capacity of the viscose carbon coated by pyrolytic carbon is 1.51 times higher than that of graphite, the real application of such a material is very questionable because the main part of lithium is inserted at potentials close to 0 V versus Li/Li^+. Low values of insertion potentials are not recommended, because of the risks of lithium metal plating and subsequent short-circuit and battery explosion.

Ex situ ^{7}Li nuclear magnetic resonance (^{7}Li-NMR) has been used for confirming the state of lithium at different steps of reduction or oxidation of disordered carbons [66, 67]. However, because of the possible alteration of the samples after their extraction from the cell, either by washing or by the impurities of the glovebox atmosphere, it seemed preferable to perform *in situ* ^{7}Li NMR during galvanostatic cycling of a supple ultrathin plastic carbon/lithium cell [68, 69]. The carbon electrode was the viscose carbon/pyrolytic carbon composite which galvanostatic characteristics are presented in Figure 8.15. During the reduction, two lines related with lithium insertion were observed: (i) the minor one at 18 ppm was attributed to intercalated lithium (with a charge transfer to carbon), (ii) the most important line was down-field shifted during lithium insertion to reach values *ca.* 120 ppm at full reduction of carbon. Such values are characteristic of quasi-metallic lithium

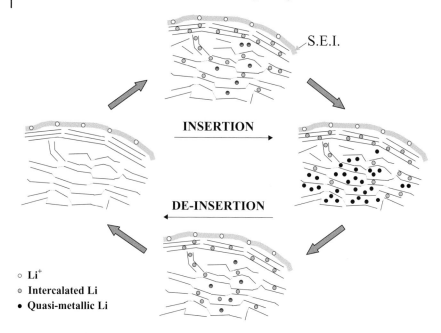

Figure 8.16 Schematic model based on HRTEM observations and *in situ* ^{7}Li-NMR during galvanostatic reduction/oxidation of a carbon fiber/pyrolytic carbon composite. During reduction, once the SEI is formed, ionic lithium penetrates at first in the smallest intervals. Then it diffuses to the largest intervals where quasi-metallic clusters are formed. During oxidation, the process runs in opposite direction (adapted from [68]).

in the nanopores. These two lines completely disappeared during the oxidation step. Taking into account the ^{7}Li-NMR data and the structural/nanotextural data provided by HRTEM observation of the host carbon, a model of lithium insertion/extraction has been proposed (Figure 8.16). Lithium is first intercalated in the small intervals between nanometer-size graphitic type layers and it then penetrates into the nanopores where growing quasi-metallic clusters are formed [68].

From the foregoing, it is now clear that the main part of lithium inserted in disordered carbon is in a quasi-metallic state. Although it seems to be essentially located in the porosity, its state is too close to metallic lithium for taking the risk of implementing these carbons in high-energy-density lithium batteries.

8.4.2
Anodes Based on Si/C Composites

In the search of high capacity anodes for batteries, several metals which can be electrochemically alloyed with Li (including Sn, Sb, Pb, Al, Zn, Cd) have been investigated. Among them, silicon with a theoretical capacity around 4200 mAh g^{-1}, for example, 4.2 Li atoms per silicon atom, appears to be very interesting [70]. Moreover, the insertion/deinsertion potential plateau of lithium into/from silicon is located close to 0.5 V versus Li$^+$/Li, that is, at a high enough value to avoid any risk of lithium plating. Unfortunately, the high volume changes (more than 300%) during lithium storage induce a complete disintegration of the silicon electrode, with a subsequent loss of electrical contact and rapid capacity fade [71].

It has been proposed to realize electrodes from silicon films or silicon dispersed in a matrix in order to alleviate the mechanical strains during alloying. In the case of films, the effect of volume expansion is limited by reducing the film thickness [72] or by using binary Si alloy films [73]. Although such electrodes were reported to keep a high reversible capacity for hundreds of cycles [74], the films preparation is complicated and expensive and the mass of electrode material is too low to allow devices to be developed.

In composite anodes, the silicon particles are dispersed in a matrix which can be electrochemically active (C, Ag, Mg) or inactive (e.g., TiN, SiC, TiC) [75]. The carbonaceous matrix is highly preferred in the research developments because of numerous advantages, for example, softness, good electrical conductivity, low density, small volume expansion and an additional lithium uptake. Even if an interesting electrochemical performance is sometimes reported in the literature for Si/C composites, it is important to mention that the results have been obtained by using specific electrochemical protocols [76, 77] which are not acceptable for practical applications of Li-ion batteries. Therefore, it is important to study the influence of charging/discharging conditions in order to state on the interest of these composites.

The cycleability of Si/C composites has been investigated as a function of current density and low cut-off potential [78]. The composites were prepared by mixing silicon nanoparticles (~50 nm) with pitch and further pyrolyzing the blend at 1000 °C under a neutral atmosphere. Figure 8.17 shows the first galvanostatic cycle and the second discharge of the Si (10 wt%)/C composite with 5 mV cut-off

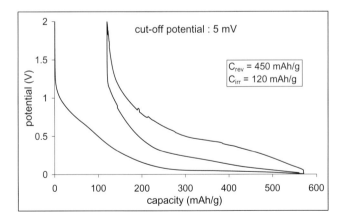

Figure 8.17 First galvanostatic discharge/charge cycle and second discharge at C/10 (a capacity of 372 mAh g^{-1} in 10 h) of a Si (10%)/C composite treated for 1 h at 1000 °C. The upper value of potential is 2 V and the discharge cut-off potential is 5 mV.

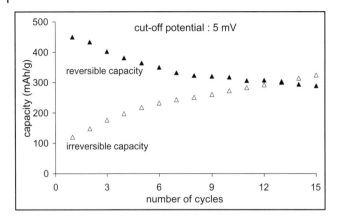

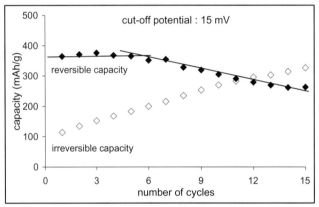

Figure 8.18 Reversible and irreversible capacities of the Si (10%)/C composite during cycling at C/10 with 5 and 15 mV cut-off potentials.

potential. Lithium is mostly inserted/deinserted at a potential around 0.3–0.5 V versus Li/Li$^+$, and the small hysteresis between charge and discharge is typical for silicon. Upon fixing the cut-off potential at values of 5, 10 and 15 mV, the irreversible capacity was found to be constant (115–120 mAh g^{-1}); it might be attributed to the solid electrolyte interface (SEI) formation during the first cycle. In contrast, the reversible capacity increases from 364 to 465 mAh g^{-1} when the cut-off limit changes from 15 to 10 mV. The highest value of capacity (465 mAh g^{-1}) is lower than the maximum theoretical capacity of the composite estimated on the basis of the proportions of the two components and their respective maximum capacity, for example, 646 mAh g^{-1}. In other words, the silicon particles are not fully saturated by lithium after the first discharge.

The dependence of the irreversible and reversible capacities of the Si (10 wt%)/C composite during cycling has been studied at different values of cut-off potential from 5 to 15 mV, and Figure 8.18 shows the particular case of 5 and 15 mV cut-off. Whatever the value of cut-off potential, the irreversible capacity continuously increases with the number of cycles. The reversible capacity keeps constant over a few cycles for the highest values of cut-off potential, for example, 10 and 15 mV. Then it decreases with a rate comparable to the irreversible capacity increase, suggesting that the reversible capacity loss of the Si/C composite is due to lithium unextracted from the silicon nanoparticles. At 15 mV cut-off potential, the reversible capacity keeps around 360 mAh g^{-1} during the first cycles, and the irreversible capacity increases from 115 to 216 mAh g^{-1} (Figure 8.18). On the seventh and next cycles, the sum of the reversible and irreversible capacities is $C_{rev} + C_{irr} = 354 + 216 = 570$ mAh g^{-1}. A comparable value of the sum is found at the first cycle for a cut-off potential of 5 mV, for example, $C_{rev} + C_{irr} = 450 + 120 = 570$ mAh g^{-1}. Such a result suggests that the silicon particles are saturated with lithium (reversible and irreversible lithium) when the reversible capacity drop starts to occur. Similar conclusions could be drawn from galvanostatic cycling performed on the same composite at current densities from C/20 to C/2.5 with a cut-off potential of 10 mV [78].

In summary, the performance of Si/C composites, and particularly their cycleability, is strongly depending on the electrochemical conditions. Si/C composites are able to demonstrate a higher reversible capacity than graphite, but for a good cycle life the real discharge cut-off potential must be kept enough high in order to avoid saturation of particles by lithium [78].

8.4.3
Origins of Irreversible Capacity of Carbon Anodes

The possible origins of irreversible capacity C_{irr} have been carefully studied by a number of authors. Some papers claim that lithium could be irreversibly trapped through electrostatic forces by the surface functional groups of carbon [79], or that it could react with di-oxygen or water molecules adsorbed on the carbon surface [80]. A linear dependence has been found between the irreversible capacity of a series of carbons and their micropore volume [81], and ^{7}Li-NMR experiments lead to the conclusion that metallic lithium could be irreversibly trapped in the micropores of the materials [82]. However, the main contribution to the irreversible capacity seems to be the formation of the SEI during the first reduction cycle at *ca.* 0.8 V versus Li/Li$^+$. Some papers claim that the extent of the decomposition reaction is directly related to the BET specific surface area of the carbon host measured by nitrogen adsorption at 77 K [83–85]. In fact, this relationship between C_{irr} and the BET surface area is not always verified, as shown for example in Figure 8.19 (top graph) for graphite ball-milled in different atmospheres and for the same samples coated by pyrolytic carbon after milling [86], confirming that other parameters control the value of C_{irr}. This is not surprising, because the BET specific surface area is essentially a geometric parameter based on nitrogen physisorption on the basal carbon planes, whereas the SEI formation involves higher energies. Therefore, it has been suggested [65] to correlate the irreversible capacity to the active surface area (ASA) [87] which depends on the number of active sites on the carbon surface.

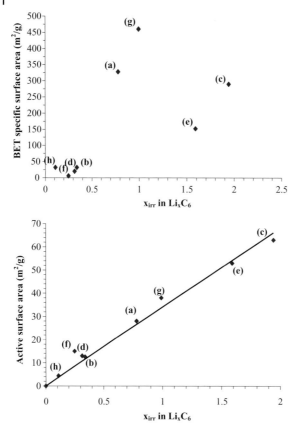

Figure 8.19 Relation between the BET specific surface area (top graph) or the active surface area (bottom graph) and the irreversible capacity x_{irr} of graphite samples ball-milled in different conditions, or ball-milled and subsequently coated by pyrolytic carbon. (a) 10 h in vacuum, (b) 10 h in vacuum + pyrolytic carbon deposition, (c) 10 h under H_2, (d) 10 h under H_2 + pyrolytic carbon deposition, (e) 10 h under O_2, (f) 10 h under O_2 + pyrolytic carbon deposition, (g) 20 h in vacuum, (h) 20 h in vacuum + pyrolytic carbon deposition [86].

The ASA of carbon materials corresponds to the cumulated surface area of the different types of defects present on the carbon surface (stacking faults, single and multiple vacancies, dislocations) [87, 88]. Figure 8.19 (bottom graph) shows a perfect linear relationship between the irreversible capacity C_{irr} and the value of ASA for the different graphite samples, whereas C_{irr} could not be correlated with the BET specific surface area [86]. Moreover, all the samples coated with a thin carbon layer by pyrolytic decomposition of propylene demonstrate the lowest values of irreversible capacity and ASA. A similar improvement was mentioned after coating of graphite and subsequent carbonization, either with pitch and resin blend [89] or with coal tar pitch [90]. Due to its oriented nanotexture with only few edge planes accessible for adsorbed species, the thin carbon coating forms a barrier preventing from the diffusion of the large solvated lithium ions to the active sites of the fibres. For the same reasons, in the ASA determination experiment, di-oxygen cannot easily diffuse to reach the active sites through the pyrolytic carbon layer, therefore the ASA is apparently low in the coated samples.

Hence, the active sites present on the surface of carbonaceous materials are highly responsible of electrolyte decomposition and extent of SEI formation. Their blocking is highly desirable in order to control the amount of SEI formed.

8.5 Conclusions

The high potentialities of sp^2 nanocarbons for energy storage have been shown through few examples. Beside the electrochemical performance, the interest of these materials is their high versatility, availability and relatively low cost. The electrochemical characteristics are fully controlled by the surface functionality and nanotexture. In particular, the presence of functional groups or active sites might be harmful in organic electrolyte, because it favors electrolyte decomposition and it impacts the cycleability of the systems. In contrast, surface groups turn out to be highly desirable in aqueous medium where they can be involved in pseudo-fararadaic redox reactions which enhance the capacitance properties of the materials. The porosity, and in particular the pore size, has a strong influence on the electric double-layer capacitance. Designing highly porous carbons with a narrow pore size distribution in the range of 1 nm represents a very high challenge for improving the energy density of supercapacitors in organic medium. The electrochemical hydrogen storage in water medium is a very promising direction for designing new energy storage systems operating at high voltage in an environment friendly electrolyte. Fundamental work is necessary for elucidating the sites responsible of hydrogen trapping; *in situ* observation by spectroscopic techniques during electrochemical storage of hydrogen could be extremely helpful.

Beside their role of active material, carbons are also very interesting as a support or a coating material. Carbon nanotubes are extremely interesting as a conductive support of pseudo-capacitive materials, for example, transition metal oxides and conducting polymers as well as carbons doped with heteroatoms; the open mesoporosity of nanotubes favors very positively the access of electrolyte to the active surface. In the case of silicon, carbon can also constrain the expansion while insuring a conducting and protective role.

In the future, composites based on multifunctional nanocarbons appear to be the key for developing high performance energy storage systems.

References

1. Béguin, F., and Frackowiak, E. (2009) *Carbons for Electrochemical Energy Storage and Conversion Systems*, CRC Press/Taylor&Francis, Boca Raton.
2. Pels, J.R., Kapteijn, F., Moulijn, J.A., Zhu, Q., and Thomas, K.M. (1995) *Carbon*, **33**, 1641.
3. Kapteijn, F., Moulijn, J.A., Matzner, S., and Boehm, H.P. (1999) *Carbon*, **37**, 1143.
4. Radovic, L.R., and Bockrath, B. (2005) *J. Am. Chem. Soc.*, **127**, 5917.
5. Conway, B.E. (1999) *Electrochemical Supercapacitors – Scientific Fundamentals and Technological Applications*, Kluwer Academic/Plenum, New York.
6. Burke, A. (2000) *J. Power Sources*, **91**, 37.
7. Kötz, R., and Carlen, M. (2000) *Electrochim. Acta*, **45**, 2483.
8. Frackowiak, E., and Béguin, F. (2001) *Carbon*, **39**, 937–950.
9. Frackowiak, E. (2007) *Phys. Chem. Chem. Phys.*, **9**, 1774.
10. Pandolfo, A.G., and Hollenkamp, A.F. (2006) *J. Power Sources*, **157**, 11.
11. Gambly, J., Taberna, P.L., Simon, P., Fauvarque, J.F., and Chesneau, M. (2001) *J. Power Sources*, **101**, 109.
12. Lozano-Castello, D., Cazorla-Amoros, D., Linares-Solano, A., Shiraishi, S., Kurihara, H., and Oya, A. (2003) *Carbon*, **41**, 1765.
13. Kierzek, K., Frackowiak, E., Lota, G., Gryglewicz, G., and Machnikowski, J. (2004) *Electrochim. Acta*, **49**, 515.
14. Qu, D., and Shi, H. (1998) *J. Power Sources*, **74**, 99.
15. Endo, M., Kim, Y.J., Ohta, H., Ishii, K., Inoue, T., Hayashi, T., Nishimura, Y., Maeda, T., and Dresselhaus, M.S. (2002) *Carbon*, **40**, 2613.
16. Shiraishi, S., Kurihara, H., Tsubota, H., Oya, A., Soneda, Y., and Yamada, Y. (2001) *Electrochem. Solid State Lett.*, **4**, A5.
17. Barbieri, O., Hahn, M., Herzog, A., and Kötz, R. (2005) *Carbon*, **43**, 1303.
18. Jurewicz, K., Babel, K., Ziolkowski, A., and Wachowska, H. (2003) *Electrochim. Acta*, **48**, 1491.
19. Endo, M., Kim, Y.J., Osawa, K., Ishii, K., Inoue, T., Nomura, T., Miyashita, N., and Dresselhaus, M.S. (2003) *Electrochem. Solid State Lett.*, **6**, A23.
20. Jiang, Q., Qu, M.Z., Zhou, G.M., Zhang, B.L., and Yu, Z.L. (2002) *Mat. Lett.*, **57**, 988.
21. Pell, W.G., Conway, B.E., and Marincic, N. (2000) *J. Electroanal. Chem.*, **491**, 9.
22. Salitra, G., Soffer, A., Eliad, L., Cohen, Y., and Aurbach, D. (2000) *J. Electrochem. Soc.*, **147**, 2486.
23. Eliad, L., Salitra, G., Soffer, A., and Aurbach, D. (2001) *J. Phys. Chem. B*, **105**, 6880.
24. Eliad, L., Salitra, G., Soffer, A., and Aurbach, D. (2002) *J. Phys. Chem. B*, **106**, 10128.
25. Guo, Y., Qi, J., Jiang, Y., Yang, S., Wang, Z., and Xu, H. (2003) *Mat. Chem. Phys.*, **80**, 704.
26. Raymundo, E., Kierzek, K., Machnikowski, J., and Béguin, F. (2006) *Carbon*, **44**, 2498.
27. Chmiola, J., Yushin, G., Gogotsi, Y., Portet, C., Simon, P., and Taberna, P.L. (2006) *Science*, **313**, 1760.
28. Fernandez, J.A., Arulepp, M., Leis, J., Stoeckli, F., and Centeno, T.A. (2008) *Electrochim. Acta*, **53**, 7111.
29. Chmiola, J., Largeot, C., Taberna, P.L., Simon, P., and Gogotsi, Y. (2008) *Angew. Chem.*, **120**, 3440.
30. Huang, X.J. (2008) *Angew. Chem.*, **47**, 520.
31. Shi, H. (1996) *Electrochim. Acta*, **41**, 1633.
32. Jurewicz, K., Frackowiak, E., and Béguin, F. (2004) *Appl. Phys. A*, **78**, 981.
33. Qu, D. (2008) *J. Power Sources*, **179**, 310.
34. Béguin, F., Friebe, M., Jurewicz, K., Vix-Guterl, C., Dentzer, J., and Frackowiak, E. (2006) *Carbon*, **44**, 2392.
35. Fang, B., Zhou, H., and Honma, I. (2006) *J. Phys. Chem. B*, **110**, 4875.
36. Jurewicz, K., Delpeux, S., Bertagna, V., Béguin, F., and Frackowiak, E. (2001) *Chem. Phys. Lett.*, **347**, 36.
37. Arbizzani, C., Mastragostino, M., and Soavi, F. (2001) *J. Power Sources*, **100**, 164.
38. Khomenko, V., Frackowiak, E., and Béguin, F. (2005) *Electrochim. Acta*, **50**, 2499.

39 Miller, J.M., Dunn, B., Tran, T.D., and Pekala, R.W. (1999) *Langmuir*, **15**, 799.
40 Toupin, M., Brousse, T., and Bélanger, D. (2002) *Chem. Mater.*, **14**, 3946.
41 Wu, N.L. (2002) *Mater. Chem. Phys.*, **75**, 6.
42 Raymundo-Piñero, E., Khomenko, V., Frackowiak, E., and Béguin, F. (2005) *J. Electrochem. Soc.*, **152**, A229.
43 Khomenko, V., Raymundo-Piñero, E., and Béguin, F. (2005) *J. Power Sources*, **153**, 183.
44 Khomenko, V., Raymundo-Piñero, E., Frackowiak, E., and Béguin, F. (2005) *Appl. Phys. A*, **82**, 567.
45 Brousse, T., Toupin, M., and Bélanger, D. (2004) *J. Electrochem. Soc.*, **151**, A614.
46 Jurewicz, K., Babel, K., Ziolkowski, A., Wachowska, H., and Kozlowski, M. (2002) *Fuel Process. Technol.*, **77–78**, 191.
47 Hulicova, D., Yamashita, J., Soneda, Y., Hatori, H., and Kodama, M. (2005) *Chem. Mater.*, **17**, 1241.
48 Lota, G., Grzyb, B., Machnikowska, H., Machnikowski, J., and Frackowiak, E. (2005) *Chem. Phys. Lett.*, **404**, 53.
49 Frackowiak, E., Lota, G., Machnikowski, J., Vix, C., and Béguin, F. (2006) *Electrochim. Acta*, **51**, 2209.
50 Béguin, F., Szostak, K., Lota, G., and Frackowiak, E. (2005) *Adv. Mater.*, **17**, 2380.
51 Hulicova, D., Kodama, M., and Hatori, H. (2006) *Chem. Mater.*, **18**, 2318.
52 Leitner, K., Lerf, A., Winter, M., Besenhard, J.O., Villar-Rodil, S., Suarez-Garcia, F., Martinez-Alonso, A., and Tascon, J.M.D. (2006) *J. Power Sources*, **153**, 419.
53 Li, W., Chen, D., Li, Z., Shi, Y., Wan, Y., Huang, J., Yang, J., Zhao, D., and Jiang, Z. (2007) *Electrochem. Comm.*, **9**, 569.
54 Kodama, M., Yamashita, J., Soneda, Y., Hatori, H., Nishimura, S., and Kamegawa, K. (2004) *Mater. Sci. Engineer B.*, **108**, 156.
55 Lota, G., Lota, K., and Frackowiak, E. (2007) *Electrochem. Commun.*, **9**, 1828.
56 Frackowiak, E., and Béguin, F. (2006) *Recent Advances in Supercapacitors* (ed. V. Gupta), Transworld Research Network, Kerala, India, pp. 79–114.
57 Raymundo-Pinero, E., Leroux, F., and Béguin, F. (2006) *Adv. Mater.*, **18**, 1877.
58 Biniak, S., Swiatkowski, A., and Makula, M. (2001) Electrochemical studies of phenomena at active carbon – electrolyte solution interfaces, Ch. 3, in *Chemistry and Physics of Carbon* (ed. L.R. Radovic), Marcel Dekker, New York, pp. 125–225.
59 Montes-Moran, M.A., Suarez, D., Menendez, J.A., and Fuente, E. (2004) *Carbon*, **42**, 1219.
60 Raymundo-Pinero, E., Cadek, M., and Béguin, F. (2009) *Adv. Funct. Mat*, **19**, 1.
61 Sawai, K., Iwakoshi, Y., and Ohzuku, T. (1994) *Solid State Ionics*, **69**, 273.
62 Guérard, D., and Hérold, A. (1975) *Carbon*, **13**, 337.
63 Yoshio, M., Tsumura, T., and Dimov, N. (2005) *J. Power Sources*, **146**, 10.
64 Zheng, T., and Dahn, J.R. (1999) Applications of carbon in Lithium-ion batteries, in *Carbon Materials for Advanced Technologies* (ed. T.D. Burchell), Elsevier, Oxford, pp. 341–388.
65 Béguin, F., Chevallier, F., Vix, C., Saadallah, S., Rouzaud, J.N., and Frackowiak, E. (2004) *J. Phys. Chem. Solids*, **65**, 211.
66 Gautier, S., Leroux, F., Frackowiak, E., Faugère, A.M., Rouzaud, J.N., and Béguin, F. (2003) *J. Phys. Chem. A*, **105**, 5794.
67 Tatsumi, K., Akai, T., Imamura, T., Zaghib, K., Iwashita, N., Higuchi, S., and Sawada, Y. (1996) *J. Electrochem. Soc.*, **143**, 1923.
68 Chevallier, F., Letellier, M., Morcrette, M., Tarascon, J.M., Frackowiak, E., Rouzaud, J.N., and Béguin, F. (2004) *Electrochem. Solid State Lett.*, **6**, A225.
69 Letellier, M., Chevallier, F., Clinard, C., Frackowiak, E., Rouzaud, J.N., Béguin, F., Morcrette, M., and Tarascon, J.M. (2003) *J. Chem. Phys.*, **118**, 6038.
70 Wang, G.X., Ahn, J.H., Yao, J., Bewlay, S., and Liu, H.K. (2004) *Electrochem. Commun.*, **6**, 689.
71 Ryu, J.H., Kim, J.W., Sung, Y.E., and Oh, S.M. (2004) *Electrochem. Solid State Lett.*, **7**, A306.
72 Yoshimura, K., Suzuki, J., Sekine, K., and Takamura, T. (2005) *J. Power Sources*, **146**, 445.

73 Fleischauer, M.D., Topple, J.M., and Dahn, J.R. (2005) *Electrochem. Solid State Lett.*, **8**, A137.
74 Takamura, T., Ohara, S., Uehara, M., Suzuki, J., and Sekine, K. (2004) *J. Power Sources*, **129**, 96.
75 Kasavajjula, U., Wang, C., and Appleby, A.J. (2007) *J. Power Sources*, **163**, 1003.
76 Liu, W.R., Wang, J.H., Wu, H.C., Shieh, D.T., Yang, M.H., and Wu, N.L. (2005) *J. Electrochem. Soc.*, **152**, A1719.
77 Holzapfel, M., Buqa, H., Krumelch, F., Novak, P., Petrat, F.M., and Velt, C. (2005) *Electrochem. Solid State Lett.*, **8**, A516.
78 Eker, Y., Kierzek, K., Raymundo, E., Machnikowski, J., and Béguin, F. (2009) Electrochim. Acta, doi:10.1016/j.electacta.2009.09.011.
79 Larcher, D., Mudalige, C., Gharghouri, M., and Dahn, J.R. (1999) *Electrochim. Acta*, **44**, 4069.
80 Xing, W., and Dahn, J.R. (1997) *J. Electrochem. Soc.*, **144**, 1195.
81 Guérin, K., Fevrier-Bouvier, A., Flandrois, S., Simon, B., and Biensan, P. (2000) *Electrochim. Acta*, **45**, 1607.
82 Guérin, K., Ménétrier, M., Février-Bouvier, A., Flandrois, S., Simon, B., and Biensan, P. (2000) *Solid State Ionics*, **127**, 187.
83 Winter, M., Novak, P., and Monnier, A. (1998) *J. Electrochem. Soc.*, **145**, 428.
84 Fong, R., Von Sacken, U. and Dahn, J.R. (1990) *J. Electrochem. Soc.*, **137**, 2009.
85 Simon, B., Flandrois, S., Fevrier-Bouvier, A., and Biensan, P. (1998) *Mol. Cryst. Liq. Cryst.*, **310**, 333.
86 Béguin, F., Chevallier, F., Vix-Guterl, C., Saadallah, S., Bertagna, V., Rouzaud, J.N., and Frackowiak, E. (2005) *Carbon*, **43**, 2160.
87 Lahaye, J., Dentzer, J., Soulard, P., and Ehrburger, P. (1991) Carbon gasification: the active site concept, in *Fundamental Issues of Control of Carbon Gasification Reactivity* (eds J. Lahaye and P. Ehrburger), Academic Publishers, pp. 143–158.
88 Laine, N.R., Vastola, F.J., and Walker, P., Jr. (1963) *J. Phys. Chem.*, **67**, 2030.
89 Kuribayashi, I., Yokoyama, M., and Yamashita, M. (1995) *J. Power Sources*, **54**, 1.
90 Yoon, S., Kim, H., and Oh, S.M. (2001) *J. Power Sources*, **94**, 68.

9
Nanomaterials for Superconductors from the Energy Perspective
Claudia Cantoni and Amit Goyal

9.1
Overcoming Limitations to Superconductors' Performance

Efficient transport and storage of energy are two key factors in the formidable quest to meet the world's growing energy needs. Because electricity is the energy form of choice for most uses, and the most effective way to transport energy from and to different locations, addressing the energy problem means necessarily addressing the shortcomings of the present electric power grid. The capacity and reliability of the present grid are entirely inadequate to accommodate an expected growth in energy demand of 50% by the year 2030, most of which will be concentrated in urbanized areas whose infrastructures are already severely congested. In addition, 7–10% of the electricity generated is lost in the grid by heat dissipation, and transmission limitations have already caused black-outs throughout the world. Superconducting cables made with high critical temperature (T_c) oxide superconductors have the potential to transform the grid to meet the 21st century demands. Such cables can carry five times more power then copper cables with the same cross-section and exhibit considerable lower losses (zero dc resistance). Because lower losses enable longer transmission lengths, superconductors provide the best choice for building a new super grid that will enable massive long-distance power transmission, interconnect entire continents, and provide widespread, local energy storage. In such a grid, power plants in the most remote areas can deliver power anywhere in the continent according to real-time demands, and episodic surpluses are readily stored for future use. High-temperature superconducting (HTS) cables can in fact be used for building underground transmission cables, oil-free transformers, superconducting magnetic-energy storage (SMES) units, fault current limiters, motors and compact generators, all with much improved efficiency and reliability and more environmentally friendly as compared to the present infrastructure. All these applications require that the high T_c superconductor of choice, $YBa_2Cu_3O_{7-x}$ (YBCO), be formed into long and flexible conductors that can replace copper wires. By itself, this is already a formidable obstacle because all high T_c superconductors are brittle ceramics, and the large anisotropy of their physical and transport properties requires that all the crystallites in the conductors be very

Nanotechnology for the Energy Challenge. Edited by Javier Garcia-Martinez
© 2009 WILEY-VCH Verlag GmbH & Co. KGaA, Weinheim
ISBN: 978-3-527-32401-9

closely aligned along the three principle crystal axes, with the current flowing in the basal (*a,b*) plane. Only ten years ago, there were still serious doubts about whether YBCO could be shaped in the form of kilometers-long wires with the required performance for commercial applications. This doubt has now been lifted thanks to tremendous advances in nanotechnology and the consequent ability to control and engineer oxide/metal interfaces and multilayers of various thicknesses. Successful technology partnerships between research institutions and industrial companies has allowed for the newest scientific achievements to be promptly transferred from the laboratory to large-scale production. Currently, HTS wires are fabricated using the second generation (2G), also called coated conductors (CCs) wire technology. They utilize YBCO or similar cuprate oxides with the chemical formula $RBa_2Cu_3O_{7-x}$ where R is a rare earth or a mixture of rare-earth ions (Nd, Sm, Gd, Dy, etc.). The superconductor is processed in the form of an epitaxial film deposited by different methods on a near single-crystal flexible template comprising a metal substrate and epitaxial oxide buffer layers. The template is manufactured using two major technologies: the rolling assisted biaxially textured substrate (RABiTS) technology [1–3] and the ion beam assisted deposition (IBAD) technology [4–6]. As shown in Figure 9.1, the RABiTS approach relies on producing a biaxially textured Ni-5%W alloy template on which oxide buffer layers and the superconductor are epitaxially deposited. In the IBAD approach, the metal

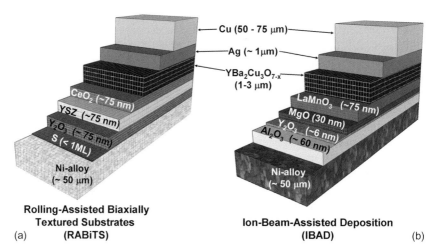

Figure 9.1 Schematic of the two major technologies for HTS wires. The RABiTS architecture (a) involves a sharply textured, single-crystal-like Ni–W alloy substrate on which Y_2O_3, YSZ and CeO_2 are epitaxial deposited to provide chemical and structural compatibility for the superconductor (YBCO) deposition. In the IBAD architecture (b) the metal tape is polycrystalline Hastelloy. The texture is developed in the MgO film and transferred to the YBCO through a $LaMnO_3$ layer. The Al_2O_3 layer serves as a diffusion barrier between oxides and metal. In both techniques the superconductor is capped with a ~1 μm Ag layer followed by a ~50 μm Cu layer for electrical, thermal and environmental stability.

alloy substrate is polycrystalline and the biaxial texture is developed in an MgO thin film by the assisted ion beam deposition and then transferred by epitaxy to the subsequent layers. Wires of 100–500 m in length carrying currents of over 300 A cm^{-1} width at 77 K (LN$_2$ boiling temperature) can now be made [7]. Several pilot projects have been successfully carried out in the United States, and companies are now moving to the pre-commercial development phase.

Despite the tremendous progress made in this area, in order for superconducting cables to become reality, their cost/performance ratio needs further improvement. The critical objective today is to increase the amount of current the superconductor can carry without dissipation in a range of magnetic fields required by the specific application. This can be achieved by carefully playing on two different and related fronts – trying to push the dissipation-free current density (J_c) as close as possible to the theoretical limit, and at the same time, increasing the thickness of the superconductive layer as much as possible without degrading performance. It is in this final challenge that the newest achievements in nanotechnology come into play as J_c enhancements in thick superconducting films can only be obtained by introducing and engineering extrinsic nano-sized defects within the film.

9.2
Flux Pinning by Nanoscale Defects

The importance of nanoscale defects derives from the concepts of magnetic flux lines and flux pinning in the superconductor. The current carried by a type II superconducting film, such as a YBCO film, generates a magnetic field (self field) that penetrates the superconductor in the form of tubular structures called vortices or flux lines, each of them carrying one unit of magnetic field, the flux quantum ϕ_0. The current interacts with the magnetic flux lines generating a Lorenz-type force (equal to $J \times \phi_0$) that pushes the vortices across the conductor [8–12]. Because the material within the vortex core is in the normal state, this motion generates dissipation and the superconductor ends up resembling a low-resistance conductor. Fortunately, naturally occurring nano-sized defects and secondary phases provide spatial variations in the thermodynamic free energy that act like potential wells for the flux lines. A flux line has a lower free energy when it is positioned in the energy well of the defect than it does in the bulk matrix. This energy difference acts as a pinning force constraining the vortex to remain within the well [13–16]. Figure 9.2 is a schematic that illustrates the interaction between flux lines and different types of defects in a superconducting film. As the current increases, so does the Lorenz force, and when this exceeds the pinning force (at $J = J_c$) dissipation begins. Therefore, the higher the pinning force, the higher the current the superconductor can sustain without dissipation.

The picture gets more complicated in presence of an external magnetic field, which is the case for several power applications. In this case, as more flux lines are generated within the superconductor, the density of pinning sites, in addition

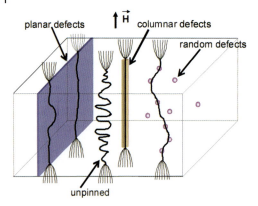

Figure 9.2 Schematic diagram illustrating unpinned and pinned magnetic flux lines and random and extended defects. Columnar defects are capable of pinning the flux lines over the full length and, therefore, provide strong pinning effects.

to their strength, becomes important. The most effective pinning sites are defects whose size is close to the vortex's normal core diameter 2ξ where ξ, the superconducting coherence length, is a few nanometers for the superconductors of interest. This is why artificially introduced nanostructures play a dramatic role in optimizing pinning without destroying an unnecessary volume fraction of the superconductor, thereby allowing for a significant increase in J_c.

9.3
The Grain Boundary Problem

Although flux pinning has been the object of numerous studies for many years, efforts aimed at increasing flux pinning in coated conductors by means of extrinsically added defects and nanostructures started only around 2004. The reason for this late interest is that early efforts to optimize coated conductors performance had to face more pressing issues. One of the greatest quandaries was that, even after more ordinary materials issues, such as interdiffusion of chemical species between the different layers and non-epitaxial growth were addressed, early CCs still showed reduced J_c values as compared to the J_c of YBCO films of the same thickness on single crystal substrates (e.g., $SrTiO_3$). The reason for this discrepancy resided in the imperfect texture of the metal-based template, which showed grain boundaries as high as 10°. From early studies of YBCO films on bi-crystals, scientists learned that grain boundaries of this magnitude dramatically suppress J_c across the grain boundary region (intergrain J_c) [17]. Therefore, strategic research on CCs towards the end of the 1990s focused mainly in studying J_c across YBCO grain boundaries in the low angle regime (2–10°) to establish the critical angle at which the intergrain J_c started departing from the J_c within the grain (intragrain J_c), and in improving the CC texture. The grain boundary problem, and the related

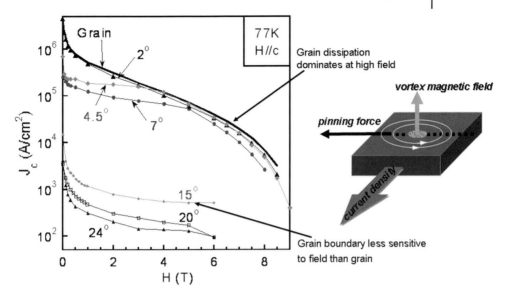

Figure 9.3 J_c dependence on magnetic field for YBCO thin films on a single crystal, representing the intragranular J_c (black line), and SrTiO$_3$ bicrystals with [001] tilt or in-plane tilt angles of 2°, 4.5°, 7°, 15°, 20° and 24°. YBCO films grown on single crystal SrTiO$_3$ are expected to show a mosaic spread of at least 1.8° as a consequence of twinning domains [20]. The schematic to the right of the figure illustrates the direction of magnetic field, current and pinning force in the measurements.

role of weak links in YBCO, cast a doubt on the technological application of CCs for many years but was finally resolved for both major routes of fabricating CCs through the improvement of the superconducting film texture [18, 19]. Today both techniques are able to produce long wires with a true in-plane texture full width half maximum (FWHM) in the YBCO film of ~4°. J_c values comparable to those obtained on single crystals are obtained for wires with such sharp textures. While FWHMs less than 4° are routinely obtained in IBAD-coated conductors, RABiTS conductors still show some margin for improving the in-plane texture in the YBCO film. However, comparable J_c values in IBAD and RABiTS conductors suggest that grain boundary meandering during YBCO growth on RABiTS might reduce the current-suppressing grain boundary effect in these conductors. Grain boundaries of less than 2° do not essentially affect the performance of YBCO films and actually might be beneficial since their widely spaced dislocation cores provide additional pinning in magnetic fields. As a result of the study of low angle grain boundaries, it was found that the intergrain $J_c(H)$ curve, although lower than the intragrain $J_c(H)$ curve, always joined the latter at a certain field value (see Figure 9.3), which increased as the grain boundary angle increased [20]. For fields larger than this crossover field the grain boundary can in principle sustain a higher current density than the grain itself. This finding implies that, since all applications of CCs involve some magnetic field, a substrate with a texture of 3–4° is equivalent to a single crystal substrate in terms of current performance. In addition, because for fields larger than the crossover field, J_c is limited by the intragrain component, further J_c improvements in this regime can only be achieved by optimizing bulk flux pinning. Additional fixes on the grain boundaries, such as, for example, Ca doping [21] do not produce any observable benefit. Once the limitations imposed by the grain boundaries were reduced, the efforts turned to improving flux pinning by introducing atomic-scale and nano-scale defect structures in addition to those naturally occurring during YBCO processing.

It is worth noting that, while the performance of YBCO films on single crystal was considered an upper limit in early CCs days, such performance lies well below the theoretical limit. The critical current density at which superconductivity viewed as pairing of electrons is destroyed in the absence of any extrinsic material issue is named the depairing current and is roughly three or four times higher than the best J_c values obtained today on any YBCO film. Therefore, CCs with performances well above those of thin YBCO films on single crystal are in principle possible. The questions to be addressed at this point are: (i) what kind of nanostructures offer the best pinning properties and (ii) how can these nanostructures be implemented in various YBCO processing techniques?

9.4
Anisotropic Current Properties

Since the deposition of the first YBCO films on single crystals, many studies have been conducted to investigate the origin for the high J_c of epitaxial films as compared to melt textured and single crystal YBCO samples. The discovery of a well defined type of defect that would dramatically increase flux pinning throughout the H,T phase space has however eluded scientists and the picture that gradually emerged was that of a multitude of point-like and extended defects that acted synergistically producing different outcomes in different regimes. The reason for this complex behavior is that the pinning force depends on many characteristics of a single defect (size, shape, composition) and is not additive for different types of defects. In addition, pinning is affected by vortex–vortex interactions, which become more prominent at higher magnetic fields (the vortices are more numerous and closer together), by thermal activation processes, which can induce depinning of vortices, and by the electronic mass anisotropy, which is due to the layered crystal structure. As a result, the objective is not to increase pinning in general but to target the enhancement in particular H,T regions of interest and for specific orientations of the magnetic field relative to the sample. Often, the pinning enhancement might be very significant in some H,T regions and completely negligible or even deleterious in others. For example, a highly crystalline YBCO film with very few but strong columnar defects along the c axis might have a record high J_c in the self field [22] because all the vortices produced by the small self magnetic field can be accommodated in deep narrow-potential wells, which take away very little superconducting material from the nearly perfect surrounding matrix. However, once the vortices exceed the number of columnar defects, they

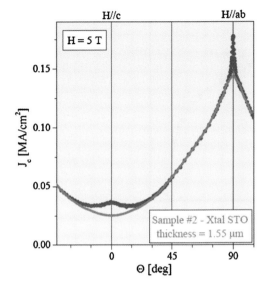
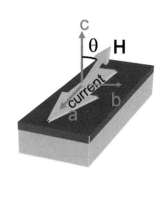

Figure 9.4 Angular dependence of J_c for a 1.55 μm YBCO film on SrTiO$_3$ at 5 T. Note that the largest current density is obtained when the field lies in the a,b plane. The solid line represents the random pinning contribution. The various angular ranges of pinning are indicated. Reprinted with permission from L. Civale et al. Physica C **412–414** 976–982 (2004) Copyright 2004 Elsevier B. V. The schematic on the right illustrates the geometry used in the measurements.

cannot be effectively pinned and J_c decreases so rapidly with field that the film performance becomes completely inadequate for any real application.

As a consequence of the anisotropic crystal structure and the presence of many point-like random defects, J_c in as grown YBCO films obtained by various techniques is always maximal when the magnetic field is directed parallel to the a,b planes, a situation which is not preferred in all power applications. Figure 9.4 shows a typical curve for J_c as function of magnetic field orientation for a YBCO film grown by pulsed laser deposition (PLD) on SrTiO$_3$. Superimposed to the data points is a curve that represents the contribution to J_c from uncorrelated pinning, produced by randomly distributed localized defects. The random pinning contribution is obtained by fitting the $J_c(\theta)$ data to the scaling curve predicted for electronic mass anisotropy effects alone, according to the analysis of Civale et al. [23]. The difference in J_c between the data and the random pinning component is due to correlated pinning provided by extended, parallel pinning centers, either linear or planar, which naturally arise during growth. These defects are usually oriented both along the c axis (especially for PLD films) and parallel to the a,b planes. A final contribution to the peak at $H//a,b$ is given by intrinsic pinning, also due to the layered crystal structure.

From the point of view of meeting power applications criteria, the goal would be to introduce artificial pinning sites that enhance J_c for orientations close to $H//c$ near to the level of J_c at $H//a,b$, giving rise to a more isotropic angular dependence.

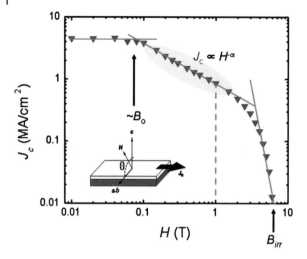

Figure 9.5 Magnetic field dependence of J_c. Three regions are distinguishable: a plateau at very small fields, a power low dependence at intermediate fields and a rapid decay for fields larger than 1–2 T.

Another way to quantify desirable pinning effects is to analyze the $J_c(H)$ curve for the least favorable field orientation ($H//c$). Typically, in YBCO with only naturally occurring defects, $J_c(H)$ shows a plateau at small fields of the order of 0.1 T, where the defects outnumber the vortices (see Figure 9.5). Generally, added pinning sites produce a small effect in this regime because there are already enough defects to pin all the vortices produced by the small field. Although some added defects (such as impurity nanophases) may produce stronger pinning forces than other naturally occurring defects, they are also made of non-superconducting material and necessarily reduce the superconducting areal section and often T_c as well. At intermediate fields, J_c decreases like $H^{-\alpha}$ where α is an exponent ranging between 0.5 and 0.7. However, in the case of strong pinning, such as that produced by nanoparticles/nanorods additions, the α exponent decreases, reflecting a smaller J_c degradation with increasing magnetic field. This is the regime were the introduction of artificial pinning sites plays the largest effect. For fields larger that 2.0 T, a rapid decay of J_c toward zero at the irreversibility field H_{irr} is observed because of depressed vortex line tension and strong thermal activation. Improved performance in this regime can also be observed with added pinning sites, as observed for strong linear defects with high areal density.

9.5
Enhancing Naturally Occurring Nanoscale Defects

It is well known from flux pinning studies that only defects that are to some degree extended or correlated along the c direction can give rise to enhancement of J_c

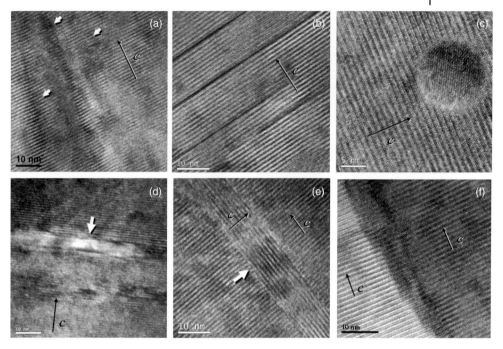

Figure 9.6 Nanosize defects typically observed in YBCO films for coated conductors: (a) antiphase boundaries (the crystal cells of two adjacent grains are out-of-registry along the c axis), (b) stacking faults, (c) second-phase spherical nanoparticles, (d) elongated, platelet-like intergrowths, (e) a axis grains (the YBCO grows with the [100] perpendicular to the substrate), (f) low angle grain boundaries formed when slightly misoriented grains coalesce during growth.

when $H//c$. Similar to earlier heavy ion irradiation studies on single crystals [24], the best performance results from columnar structures of non-superconducting material that accommodate the vortex core over the entire thickness of the film with a certain amount of splay [25, 26]. In contrast, randomly distributed localized defects (uncorrelated pinning centers) can produce very strong pinning, but do not give rise to a peak for $H//c$ in the J_c angular dependence at 77 K. One of the first attempts of correlating J_c performance for $H//c$ with the density of extended pinning centers along the c axis was carried out in 1999 [27]. The defects targeted by this study were threading edge and screw dislocations normally occurring during the YBCO growth process, especially in PLD films with their columnar growth mode. Subsequently, another type of strong, naturally occurring pinning defects, known as antiphase boundaries, were studied in detail using controlled vicinal substrate surfaces to tailor their density [28, 29]. Figure 9.6 shows a collection of various defects usually occurring in YBCO films as imaged by high resolution transmission electron microscopy (HRTEM).

One approach to optimize flux pinning is to try to tune deposition conditions and/or the substrate morphology in order to increase the most effective naturally

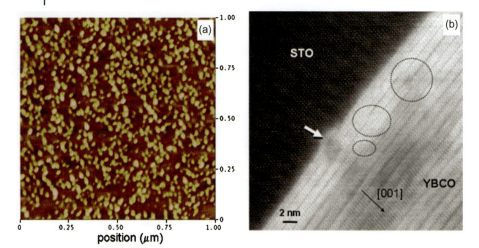

Figure 9.7 (a) SrTiO$_3$ substrate decorated with ZrO$_2$ nanodots deposited by PLD. (b) Antiphase boundaries, stacking faults and plane buckling developed around a surface nanoparticle in the YBCO film grown on the decorated substrate shown in (a). In this particular sample J_c was enhanced by a factor of two at H = 1 T and T = 40 K with respect to a control YBCO film on bare SrtiO$_3$.

occurring crystal defects. For example, by controlling the size of the YBCO growth islands and the mosaic spread within a single substrate grain, additional threading dislocations along the c axis can be introduced. However, the increase in defect density obtained through this route is modest and limited to low magnetic fields. Moreover, such an approach is highly dependent on the YBCO deposition method.

By decorating the substrate surface with chemically compatible nanoparticles, one can try to disrupt the subsequent YBCO growth just enough to introduce low-angle grain boundaries or dislocations of another nature that might tread to the film surface. This approach has been pursued by various research groups in the past few years using different materials deposited in the form of nanoparticles on the substrate (Ir, Ag, Y$_2$O$_3$, ZrO$_2$, MgO, BaZrO$_3$) or processing the oxide template so that it develops outgrowths of nanoscale size [30–35]. However, the pinning induced by the defects and strain fields stemming from the surface particles is generally of a random nature and does not provide the desirable enhancement of J_c for H//c. Figure 9.7 shows an example of surface decoration on a SrTiO$_3$ (001) substrate by PLD deposited ZrO$_2$ nanodots. The YBCO growth is disturbed by the presence of the nanodots, and antiphase boundaries terminating in stacking faults are produced in the growing film. In two cases c-axis-correlated defects with consequent significant J_c enhancement for H//c were reported using Y$_2$O$_3$ nanoparticles and nano-sized SrTiO$_3$ outgrowths [33, 36]. Nevertheless, an understanding of the correlation between nanoparticles or outgrowths and the type of defects they generate is lacking and similar nanostructures on the surface of the substrate give rise to different pinning mechanisms within the superconductor. However, a recent result obtained on *ex situ* YBCO films that did not contain artificially introduced pinning sites shows that the relation between defects and resulting pinning landscape is still not well understood and *c*-axis-correlated defects might not be the only route for enhanced performances [37]. In that study, crystal defects (not unambiguously identified) gave rise to very strong pinning, yielding an almost isotropic $J_c(\theta)$ curve at magnetic fields as high as 3 T.

9.6
Artificial Introduction of Flux Pinning Nanostructures

A different approach, that has produced overall better results, is to introduce nano-sized particles of chemically compatible, non-superconducting material throughout the superconductor matrix. This can be accomplished by deposition from multiple sources, in the case of *in situ* methods, or by adding excess elements to the superconductor precursor, in the case of *ex situ* methods. *In situ* ablation from a superconductor pellet already containing the nanoparticle material in a disperse form is another option. Although the uncorrelated pinning from randomly distributed point-like defects is generally weak, randomly dispersed finite-size particles can provide strong pinning with significant enhancements throughout the H,T phase space. As several research groups worked on this approach, several examples of impurity additions in the form of nanoparticles and/or nanodots are available. In the first attempts a YBCO non-superconducting phase Y_2BaCuO_5, also known as 211 was added by intermittently depositing YBCO layers and discontinuous 211 layers by PLD using two deposition targets [38, 39]. This approach resulted in pancake-like 211 particles with an average size of 15 nm distributed in the YBCO film, which produced some enhancement in pinning along *c* and much higher improvements for H//*a,b*. Correlated pinning along the *c*-axis was also observed when seemingly randomly distributed 10-nm $BaZrO_3$ (BZO) particles were inserted in the YBCO film by ablating from a PLD target synthesized using Ba excess and added Zr [40]. In the last study, the observed correlated pinning along the *c*-axis was related to misfit dislocations between the nanoparticles and the superconducting matrix, which aligned along the *c* direction, with an areal density of 400 μm^{-2} compared to 80 μm^{-2} for undoped YBCO films. Figure 9.8 shows high resolution Z-contrast scanning transmission electron microscope (STEM) images of a YBCO film with incorporated Y_2O_3 nanoparticles obtained with the multilayer approach (depositing from a YBCO and a Y_2O_3 target alternately) [41]. In this case, the Y_2O_3 nanodots (3–10 nm in size) grew epitaxially aligned with the YBCO in a coherent fashion, without obvious signs of lattice distortions, stacking faults or strain fields. Therefore, the observed pinning enhancement can be assumed to derive from the nanodots themselves and not from defects induced by the nanodots. The films in this study showed strong flux pinning characterized by a better J_c retention in magnetic field compared to pure YBCO films, as indicated by a lower value (0.3 vs 0.5) for the α exponent. However, the dependence of J_c on the

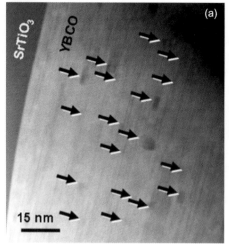 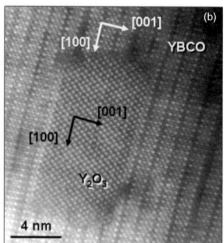

Figure 9.8 Cross-sectional high-resolution STEM of a YBCO/Y$_2$O$_3$ composite film deposited with the multilayer approach. (a) Nanoparticles (indicated by arrows) are clearly discernible, with sizes ranging from 3 to 10 nm and about 5–20 nm apart. (b) High-resolution image of a Y$_2$O$_3$ nanoparticle, coherent with the YBCO matrix.

orientation of the magnetic field was typical of random pinning and showed no correlation or preferential direction for the pinning.

Nanoparticle addition has the great advantage of being applicable to all YBCO deposition methods, including solution-based *ex situ* processes. For example, nanocomposite YBCO films containing randomly oriented, non-coherent BZO nanoparticles in their matrix were recently processed from complex metal–organic solutions [42]. These films showed strong isotropic flux pinning generated by stacking faults and other lattice defects emanating in the YBCO from the embedded nanoparticles.

9.7
Self-Assembled Nanostructures

When BZO nanoparticles are epitaxially oriented within the YBCO matrix a significant strain field is expected in the growing film due to the large mismatch between the two lattices, which is about 9% when BZO(001)//YBCO(001) and BZO(100)//YBCO(100). Such a strain, if properly tuned, can induce a remarkable self-assembly of the nanoparticles resulting in the growth of almost continuous and uniform BZO nanorods in the YBCO matrix. The mechanism responsible for self-assembly has been explored for semiconducting heteroepitaxial structures and is quite complex. Tersoff *et al.* [43, 44] demonstrated that, in multilayer structures with large mismatch, buried dots can influence the nucleation in subsequent layers, leading to self-organization of a more ordered and uniform array. Formation of an island on the surface is most favorable where the strain at the surface reduces the mismatch between the surface and the island. Because the nucleation rate is an exponential function of the nucleation barrier, which in turn depends sensitively upon the strain, islands nucleate wherever this strain gives a local minimum in the misfit. Strain mediated self-assembly of nanodots is therefore more prominent for large mismatched heteroepitaxial systems. Analogous to semiconductor heterostructures, a multilayer approach in which continuous YBCO layers and BZO nanodots are deposited sequentially is expected to produce a near ideal regular array of linear defects embedded within the HTS film matrix, when the layers thickness is properly controlled.

Goyal et al. [45, 46] were able to produce such an ordered array of BZO nanodots in YBCO by simply performing laser ablation from a single target comprising a mixture of YBCO powder and BZO nanoparticles. During simultaneous deposition of YBCO and BZO, the two phases were found to separate in the growing film and interact through the strain field similar to consecutively deposited phases. Goyal et al. were able to produce similar nanostructures using different oxides in addition to BZO ($CaZrO_3$, YSZ, MgO, $Ba_xSr_{1-x}TiO_3$) and therefore demonstrate the general validity of this approach [47]. Theoretical formulations to explain how self-assembly occurs in the YBCO-BZO system via both energetic and kinetic arguments have been developed [48].

Figure 9.9 shows a HRTEM image obtained from YBCO/BZO composite films deposited from a YBCO target doped with 2 vol% BZO nanoparticles. The BZO columns produced by stacking of nanodots and continuous nanorods are clearly visible in the TEM cross-section. The contrast of the BZO rods is generally enhanced in the TEM images by tilting the sample along the g = 001 vector of YBCO. The nanocolumns extend throughout the thickness of the film, have a diameter of 2–3 nm and are aligned along the c axis of YBCO.

Figure 9.10 shows a plan view TEM image in low magnification with the electron beam directed parallel to the YBCO c axis. It is interesting to note that the areal distribution of nanocolumns as shown in this image is not random and the spacing between nearest neighboring columns is fairly uniform, suggesting an interaction-mediated self-organization in the basal plane of the film. The degree of order in the areal distribution of columns can be probed by plotting the autocorrelation function for a bidimensional plot of the nanocolumn positions in Figure 9.10a. The autocorrelation function of an image I is defined as the inverse Fourier transform of $F(I) \times F^*(I)$, where $F(I)$ indicates the Fourier transform of the Image and $F^*(I)$ is the complex conjugate. As the autocorrelation determines the self-similarity of an image, the result always shows a peak in the center (each image point is correlated to itself) and an additional structure that ranges from an amorphous background, for a random distributions of dots, to an ordered peak array, for an ordered lattice of dots. As shown in Figure 9.10b, several distinct peaks are observed for the BZO columns distribution around the central maximum and their symmetry resembles that of a square lattice with constant $a \cong 15$ nm. From this average spacing we can estimate a maximum pinning force scenario for an applied magnetic field $B_\phi = \phi_0/a^2$ of 8–10 T, known as the matching field, corresponding to the ideal situation of each nanocolumn accommodating one flux line. According

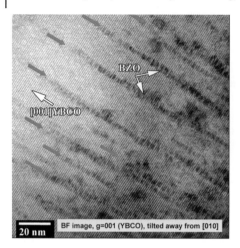

Figure 9.9 Cross-sectional TEM image of a YBCO film on RABiTS with self-assembled BaZrO$_3$ (BZO) nanodots and nanorods. The image shows that the BZO nanodots are aligned along the c axis of YBCO and are about 2–3 nm in diameter.

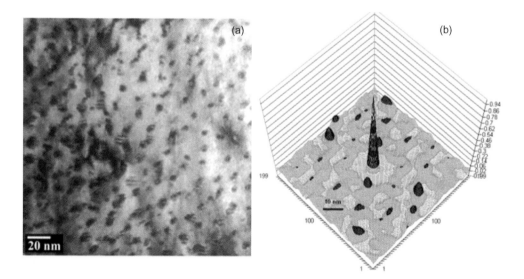

Figure 9.10 Low-magnification, plan view TEM image of a YBCO film on RABiTS with self-assembled nanodots of BaZrO$_3$ (BZO) showing some degree of order in the areal distribution of the nanocolumns (a). Corresponding autocorrelation function plot for an image derived from (a) by highlighting the BZO columns (b). The appearance of peaks around the central (auto-correlation) peak indicates a degree of ordering among the BZO columns.

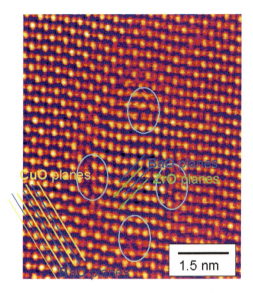

Figure 9.11 Z-contrast STEM image of a single BZO nanodot. Four misfit dislocations can be seen around the BZO nanodot. The extra semiplane in the edge dislocation cores are marked in blue. Different sets of planes in the YBCO and BZO are indicated.

to this value, significant J_c enhancement is expected in these films for all field regimes that relevant for certain power applications.

The high-resolution Z-contrast STEM image of Figure 9.11 is a plan view of a single BZO nanorod. Different atomic columns are distinguishable thanks to a dependence of the image intensity on the atomic number. In the YBCO matrix, the Ba/Y columns appear brighter than the Cu/O columns, and within the BZO, BaO columns appear brighter than ZrO. The image shows that the BZO is epitaxially oriented and coherent with the YBCO matrix, and misfit dislocations form to accommodate mismatch at the corners of the BZO nanodot. Since the nanodots are aligned along the YBCO c direction, the strain from the misfit dislocations is also aligned and extended along this direction. Both the arrays of dislocations and the nanodots themselves are therefore expected to form ideal flux pinning columns, similar to damage tracks by heavy ion irradiation [49, 50].

The pinning enhancement is evident in Figure 9.12, which compares J_c dependencies on field magnitude and orientation for an undoped YBCO film on RABiTS and a BZO-doped YBCO film on the same substrate. Both films have a thickness of 0.2 μm. In particular, we notice that the correlated c axis peak in the angular dependence of J_c is very pronounced for the YBCO film containing the BZO nanodots, indicative of strong pinning defects along the c axis. Although the J_c values shown in Figure 9.12 are large, for practical application of HTS it is also essential to increase the overall current carrying capacity of the coated conductor. The figure of performance of a coated conductor is not the J_c of the YBCO film but the engineering current density J_E of the whole conductor, which is the total current carried

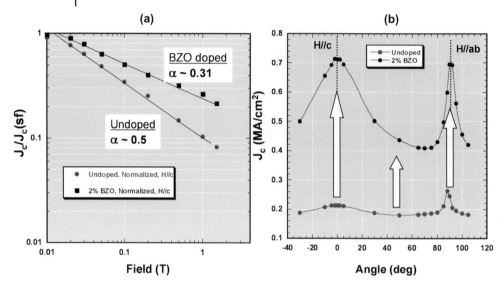

Figure 9.12 (a) Log-log plot of normalized J_c as a function of applied field showing that the power low exponent in the relation $J_c \sim H^{-\alpha}$ is lower in the YBCO film with self-assembled nanodots than for the control, undoped YBCO film. (b) Angular dependence of J_c at 77 K, 1 T for an undoped YBCO film on RABiTS compared to a YBCO film with 2 vol% BZO nanodots. Significant enhancement of J_c is seen at all angles, particularly at $H//c$ (angle = 0°). Both superconducting films have a thickness of 0.2 μm.

by the superconductor, I_c, divided by the total cross-section of the conductor. Because the thickness of the superconducting film is only a small fraction (2–3%) of the total conductor thickness, including the substrate and the stabilizing metal overlayer (see Figure 9.1), YBCO films with a thickness of a few microns, and therefore larger I_c values are in fact preferred, provided their J_c is similar to that of 0.2-μm films. Unfortunately, gradual deterioration of the critical current density occurs with increasing superconductor thickness, and in some cases a dead layer, which provides no contribution to the current carrying ability, forms after a critical thickness of ~1.5 μm.

This effect has been attributed both to extrinsic material issues, such as roughening of the film/vapor interface, coarsening of precipitates, secondary outgrowths, etc. and the intrinsic healing with thickness of treading misfit dislocations originating at the substrate interface. A first solution to this problem was found by fabricating multilayered films with alternating YBCO layers less than 1.5 μm thick and intervening CeO_2 interlayers [51]. More recently, several group were able to overcome the extrinsic issues involved with the J_c decay in thick films by optimizing PLD YBCO deposition conditions, and/or using a mixture of Y and Eu instead of pure Y to obtain a smoother and denser film microstructure [52]. Goyal et al. have recently demonstrated record-high critical current values by incorporating 3D self-assembled BZO nanodots, like the ones shown above, in films as thick as 4 μm. The achievement of 3D self-assembly throughout such large thickness was obtained by careful control of the PLD target concentration, homogeneity and

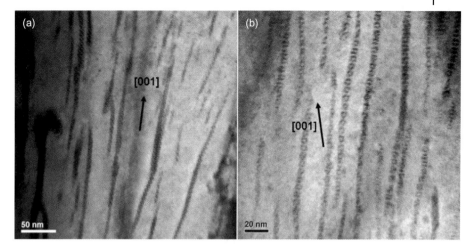

Figure 9.13 Cross-sectional HRTEM images of BZO nanocolumns in a 4-μm YBCO film. The columns extend through the entire thickness of the film and show small deviations in their direction from the YBCO c direction.

surface roughness, and by optimization of deposition repetition rate and substrate temperature [53]. Figure 9.13 shows cross-sectional TEM images of the 4-μm film, illustrating the BZO nanorod structure in the film. It appears that the columns of self-assembled BZO nanodots are not perfectly aligned along the c-axis of YBCO at the high deposition rate used to grow the film. The columns show a range of angles with respect to the c-axis of YBCO and resemble a "splayed" defect microstructure. Such a "splayed" microstructure appears to be more favorable for flux-pinning at all field orientations [26]. The performance of this 4-μm film on a short section of CC template (I_c of 353 A cm^{-1} at T = 65 K, H = 3T) is higher than record values previously reported and exceed minimum requirements for transmission cables, electric ship propulsion systems, large-scale motors and generators. This result is even more impressive when considering that, as reported recently at the 2008 peer review on the HTS program in United States, the BZO nanodots technology was successfully transferred from ORNL to SuperPower, one of the leading United States companies in 2G-coated conductors manufacture [54]. Interestingly, the self-assembly of BZO dots is not confined to the PLD process but also occurs, with proper processing tuning, during YBCO growth by MOCVD, a highly scalable technique used by SuperPower to produce kilometer-long superconducting cables.

9.8
Control of Epitaxy-Enabling Atomic Sulfur Superstructure

In addition to the nanostructures added to the superconductor to enhance its performance, the successful processing of long lengths of 2G HTS wires is based on the employment of several other nanotechnologies. A great example of nanotechnology use in coated conductors is given by the atomic surface treatment of the Ni–W surface prior to the deposition of oxide buffer and superconductor layers in RABiTS coated conductors. Before 2001, it seemed natural to assume that a deposited oxide layer nucleated on a clean and pure Ni or Ni–W surface. Before deposition, the native metal oxide was removed by an *in situ* reducing heat treatment, and unwanted re-oxidation of the Ni surface was prevented by choosing an oxide layer thermodynamically more stable than NiO (e.g., CeO_2 or Y_2O_3) and by depositing it in reducing conditions for Ni. At this point, the lattice match between metal and oxide buffer layer was supposed to be the main factor influencing epitaxial growth. This picture, however, ultimately proved to be inadequate, particularly after the introduction of continuous reel-to-reel processes for fabricating kilometers-long buffered tapes. In fact, irreproducibilities in the oxide layer texture and unwanted crystal orientations threatened to stop further development of coated conductors in the early stage of industrial scale-up. Careful structural and chemical investigations subsequently revealed the existence of a $c(2 \times 2)$ sulfur superstructure on the textured Ni surface that formed after diffusion and surface segregation of sulfur contained in the metal bulk at a level less than 30 wt ppm [55]. It was found that a fully developed atomic sulfur superstructure on the Ni surface, corresponding to a half monolayer of S atoms, enabled the epitaxial nucleation of the oxide buffer layer and consequently the right crystal orientation for the superconductor. In cases where the bulk S contents of particular batches of Ni or Ni–W were much lower than 30 wt ppm, and therefore inadequate to spontaneously produce the surface superstructure, the buffer layer deposition process produced films with degraded texture or mixed crystal orientations. However, annealing the Ni–W substrate at 800 °C in the presence of a small amount of H_2S was sufficient to produce a stable $c(2 \times 2)$-S with the desired coverage of 0.5 mL [56]. This simple step allowed creating a complete S template independently of the coverage obtained through segregation. Implementing this step in a continuous RABiTS fabrication process provided a solution to the irreproducibility issues and enabled subsequent stages of development. The role played by the S superstructure can be partially explained on the basis of structural and chemical considerations. The S layer behaves like a template that matches and mimics the arrangement of the oxygen atoms of particular (001) sublattice planes of YSZ, CeO_2, or STO. Sulfur belongs to the VI group and is chemically very similar to oxygen, often exhibiting the same electronic valence. Therefore, it is plausible that during the oxide buffer layer deposition the cations easily bond to the S atoms already present on the substrate surface, giving rise to the desired (001) epitaxial growth of the film, which otherwise would not take place. Such structural argument for the effect of the S template on buffer layer nucleation is supported by the observation that other oxides with crystal structures very similar to perovskites and fluorites, like $LaMnO_3$, and Gd_2O_3 or Y_2O_3 also grow epitaxially on (100) Ni (or Ni–W) only in presence of an intervening $c(2 \times 2)$-S superstructure. Figure 9.14 illustrates the above-proposed model of epitaxial growth on a S-terminated (001) Ni surface for $SrTiO_3$ and CeO_2. Today, Y_2O_3 is used instead of CeO_2 or $SrTiO_3$ as a first oxide buffer layer for its improved

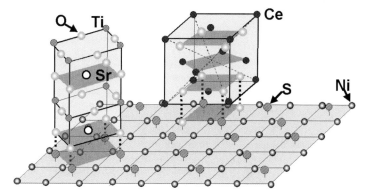

Figure 9.14 Schematic model for the nucleation of CeO_2 and STO on a (001) Ni surface with chemisorbed $c(2 \times 2)$ sulfur superstructure. The dashed lines indicate correspondence between oxygen sites in (001) planes of the seed layer and sulfur sites on the Ni surface. The seed layer cations impinging on the Ni surface bond easily to the sulfur atoms present on the metal surface, promoting the (001) orientation of the growing film. In the STO case, there is a 1:1 correspondence between oxygen atoms in the SrO plane and sulfur atoms on the Ni surface. Therefore, it is plausible that Ti ions initially bond to the S surface atoms to form the first TiO_2 plane of the STO structure. In the CeO_2 case, two of four oxygen ions/unit cell match the sulfur atoms of the $c(2 \times 2)$. During nucleation of CeO_2, oxygen atoms may fill in the empty fourfold Ni hollows and the Ce cations subsequently bond to the template formed by S and O.

structural and diffusion properties. However, the Y_2O_3 crystal structure is very similar to that of CeO_2 and the model applies to Y_2O_3 as well.

Acknowledgments

The authors would like to thank Prof. L. Civale, Dr. D. Christen and Dr. M. Varela for providing some of the figures. Research was sponsored by the United States Department of Energy, Office of Electricity Delivery and Energy Reliability-Superconductivity Program, under contract DE-AC05-00OR22725 with UT-Battelle, LLC managing contractor for Oak Ridge National Laboratory.

References

1 Goyal, A., Norton, D.P., Budai, J.D., Paranthaman, M., Specht, E.D., Kroeger, D.M., Christen, D.K., He, Q., Saffian, B., List, F.A., Lee, D.F., Martin, P.M., Klabunde, C.E., Hatfield, E., and Sikka, V.K. (1996) High critical current density superconducting tapes by epitaxial deposition of yba$_2$cu$_3$ox thick films on biaxially textured metals. *Appl. Phys. Lett.*, **69**, 1795–1797.

2 Goyal, A., Paranthaman, M., and Schoop, U. (2004) The RABiTS approach: Using

3 Goyal, A. (2005) Epitaxial superconductors on rolling-assisted-biaxially-textured-substrates (rabiTS), in *Second Generation HTS Conductor* (ed. A. Goyal), Kluwer, Norwell, MA, pp. 29–46.

4 Iijima, Y., Tanabe, N., Ikeno, Y., and Kohno, O. (1991) Biaxially aligned YBa$_2$Cu$_3$Ox thin film tapes. *Physica C*, 185, 1959–1960.

5 Wu, X.D., Foltyn, S.R., Arendt, P.N., Townsend, J., Adams, C., Tiwari, P., Coulter, J.Y., and Peterson, D.E. (1994) High current YBa$_2$Cu$_3$Ox thick films on flexible nickel substrates with textured buffer layers. *Appl. Phys. Lett.*, 65, 1961–1963.

6 Arendt, P. (2005) IBAD template for HTS coated conductors, in *Second Generation HTS Conductors* (ed. A. Goyal), Kluwer, Norwell, MA, pp. 1–28.

7 See 2007 and 2008 presentations for DOE Annual Peer Review on Superconductivity available at http://www.energetics.com/supercon08/agenda.html. For further info, see also http://apps1.eere.energy.gov/news/news_detail.cfm/news_id=11609

8 Lee, P.J. (2001) *Engineering Superconductivity*, John Wiley & Sons, Inc., New York.

9 Campbell, A.M., and Evetts, J.E. (1972) Flux vortices and transport current in type-II superconductors. *Adv. Phys.*, 21, 194–428.

10 Larkin, A.I., and Ovchinnikov, Yu.N. (1979) Pinning in type II superconductors. *J. Low Temp. Phys.*, 34, 409–428.

11 Blatter, G., Feigel'man, M.V., Geshkenbein, V.B., Larkin, A.I., and Vinokur, V.M. (1994) Vortices in high-temperature superconductors. *Rev. Mod. Phys.*, 66, 1125–1388.

12 Tinkham, M. (1975) *Introduction to Superconductivity*, McGraw Hill, New York.

13 Nelson, D.R. (1996) Points, lines and planes: vortex pinning in high-temperature superconductors. *Phys. C*, 263, 12–16.

14 Agassi, D., and Cullen, J.R. (1999) New vortex state in the presence of a long Josephson junction. *Phys. C*, 316, 1–12.

15 Agassi, D., Christen, D.K., and Pennycook, S.J. (2002) Flux pinning and critical currents at low-angle grain boundaries in high-temperature superconductors. *Appl. Phys. Lett.*, 81, 2803–2805.

16 Kumar, R., Malik, S.K., Pai, S.P., Pinto, R., and Kumar, D. (1992) Self-field-induced flux creep in YBa$_2$Cu$_3$O$_{7-y}$ thin films. *Phys. Rev. B*, 46, 5766–5768.

17 Dimos, D., Chaudhari, P., and Mannhart, J. (1990) Superconducting transport properties of grain boundaries in YBa$_2$Cu$_3$O$_7$ bicrystals. *Phys. Rev. B*, 41, 4038–4049.

18 Foltyn, S.R., Arendt, P.N., Jia, Q.X., Wang, H., MacManus-Driscoll, J.L., Kreiskott, S., DePaula, R.F., Stan, L., Groves, J.R., and Dowden, P.C. (2003) Strongly coupled critical current density values achieved in YBa2Cu3O$_{7-\delta}$ coated conductors with near-single-crystal texture. *Appl. Phys. Lett.*, 82, 4519–4521.

19 Goyal, A., Rutter, N., Cantoni, C., and Lee, D.F. (2005) Long-range current flow and percolation in RABiTS-type conductors and the relative importance of out-of-plane and in-plane misorientations in determining J(c). *Phys. C*, 426, 1083–1090.

20 Verebelyi, D.T., Christen, D.K., Feenstra, R., Cantoni, C., Goyal, A., Lee, D.F., Paranthaman, M., Arendt, P.N., DePaula, R.F., Groves, J.R., and Prouteau, C. (2000) Low angle grain boundary transport in YBa2Cu3O$_{7-\delta}$ coated conductors. *Appl. Phys. Lett.*, 76, 1756–1758.

21 Schmehl, A., Goetz, B., Schulz, R.R., Schneider, C.W., Bielefeldt, H., Hilgenkamp, H., and Mannhart, J. (1999) Doping-induced enhancement of the critical currents of grain boundaries in YBa$_2$Cu$_3$O$_{7-\delta}$. *Europhys. Lett.*, 47, 110–115.

22 Wu, J.Z., Emergo, R.L.S., Wang, X., Xu, G., Haugan, T.J., and Barnes, P.N. (2008) Strong nanopore pinning enhances Jc in YBa$_2$Cu$_3$O$_{7-\delta}$ films. *Appl. Phys. Lett.*, 93, 062506-1–062506-3.

23 Maiorov, B., and Civale, L. (2007) Identification of vortex pinning centers and regimes in coated conductors, in *Flux Pinning and AC Loss Studies on YBCO Coated Conductors* (eds M.P. Paranthaman and V. Selvamanickam), Nova Science Publishers, Inc., New York, pp. 35–58.

24 Civale, L., Marwick, A.D., Worthington, T.K., Kirk, M.A., Thompson, J.R., Krusin-Elbaum, L., Sun, Y., Clem, J.R., and Holtzberg, F. (1991) Vortex confinement by columnar defects in $YBa_2Cu_3O_7$ crystals: Enhanced pinning at high fields and temperatures. *Phys. Rev. Lett.*, **67**, 648–651.

25 Hwa, T., Le Doussal, P., Nelson, D.R., and Vinokur, V.M. (1993) Flux pinning and forced vortex entanglement by splayed columnar defects. *Phys. Rev. Lett.*, **71**, 3545–3548.

26 Civale, L., Krusin-Elbaum, L., Thompson, J.R., Weeler, R., Marwick, A.D., Kirk, M.A., Sun, Y.R., Holtzberg, F., and Feild, C. (1994) Reducing vortex motion in $YBa_2Cu_3O_7$ crystals with splay in columnar defects. *Phys. Rev. B*, **50**, 4102–4105.

27 Dam, B., Huijbregtse, J.M., Klaassen, F.C., van der Geest, R.C.F., Doornbos, G., Rector, J.H., Testa, A.M., Freisem, S., Martinezk, J.C., Staüble-Pümpin, B., and Griessen, R. (1999) Origin of high critical currents in $YBa_2Cu_3O_{7-\delta}$ superconducting thin films. *Nature*, **399**, 439–442.

28 Haage, T., Zegenhagen, J., Li, J.Q., Habermeier, H.-U., Cardona, M., Jooss, C., Warthmann, R., Forkl, A., and Kronmüller, H. (1997) Transport properties and flux pinning by self-organization in $YBa_2Cu_3O_7\delta$ films on vicinal $SrTiO_3(001)$. *Phys. Rev. B*, **56**, 8404–8418.

29 Cantoni, C., Verebelyi, D.T., Specht, E.D., Budai, J., and Christen, D.K. (2005) Anisotropic nonmonotonic behavior of the superconducting critical current in thin $YBa_2Cu_3O_{7-\delta}$ films on vicinal $SrTiO_3$ surfaces. *Phys. Rev. B*, **71**, 054509-1–054509-9.

30 Crisan, A., Fujiwara, S., Nie, J.C., Sundaresan, A., and Ihara, H. (2001) Sputtered nanodots: a costless method for inducing effective pinning centers in superconducting films. *Appl. Phys. Lett.*, **79**, 4547–4549.

31 Aytug, T., Paranthaman, M., Leonard, K.J., Kang, S., Martin, P.M., Heatherly, L., Goyal, A., Ijaduola, A.O., Thompson, J.R., Christen, D.K., Meng, R., Rusakova, I., and Chu, C.W. (2006) Analysis of flux pinning in $YBa2Cu_3O_{7-\delta}$ films by nanoparticle-modified substrate surfaces. *Phys. Rev. B*, **74**, 184505-1–184505-8.

32 Aytug, T., Paranthaman, M., Gapud, A.A., Kang, S., Christen, H.M., Leonard, K.J., Martin, P.M., Thompson, J.R., Meng, D.K. Christen, R., Rusakova, I., Chu, C.W., and Johansen, T.H. (2005) Enhancement of flux pinning and critical currents in $YBa2Cu_3O_{7-\delta}$ films by nanoscale iridium pretreatment of substrate surfaces. *J. Appl. Phys.*, **98**, 114309.

33 Matusmoto, K., Horide, T., Osamura, K., Mukaida, M., Yoshida, Y., Ichinose, A., and Horii, S. (2004) Enhancement of critical current density of YBCO films by introduction of artificial pinning centers due to the distributed nano-scaled $Y2O3$ islands on substrates. *Phys. C*, **412–414**, 1267–1271.

34 Mele, P., Matsumoto, K., Horide, T., Miura, O., Ichinose, A., Mukaida, M., Yoshida, Y., and Horii, S. (2006) Critical current enhancement in PLD $YBa2Cu3O7-x$ films using artificial pinning centers. *Phys. C*, **445–448**, 648–651.

35 Nie, J.C., Yamasaki, H., Yamada, H., Nakagawa, Y., Develos-Bagarinao, K., and Mawatari, Y. (2004) Evidence for c-axis correlated vortex pinning in $YBa2Cu_3O_{7-\delta}$ films on sapphire buffered with an atomically flat $CeO2$ layer having a high density of nanodots. *Supercond. Sci. Technol.*, **17**, 845–852.

36 Maiorov, B., Wang, H., Foltyn, S.R., Li, Y., DePaula, R., Stan, L., Arendt, P.N., and Civale, L. (2006) Influence of naturally grown nanoparticles at the buffer layer in the flux pinning in $YBa_2Cu_3O_7$ coated conductors. *Supercond. Sci. Technol.*, **19**, 891–895.

37 Solovyov, V.F., JWiesmann, H., Wu, L, Li, Q, Cooley, L.D., Suenaga, M., Maiorov, B., and Civale, L. (2007) High

critical currents by isotropic magnetic-flux-pinning centres in a $_3$ μm-thick YBa2Cu3O7 superconducting coated conductor. *Supercond. Sci. Technol.*, **20**, L20–L23.

38 Haugan, T., Barnes, P.N., Wheeler, R., Meisenkothen, F., and Sumption, M. (2004) Addition of nanoparticle dispersions to enhance flux pinning of the YBa2Cu3O7-x superconductor. *Nature*, **430**, 867–870.

39 Haugan, T., et al. (2005) Flux pinning strengths and mechanisms of YBCO with nanoparticle addition, in *Epitaxial Growth of Functional Oxides* (eds A. Goyal, Y. Kuo, O. Leonte, and W. Wong-Ng), The Electrochemical Society Inc., Pennington, NJ, pp. 359–366.

40 MacManus-Driscoll, J.L., Foltyn, S.R., Jia, Q.X., Wang, H., Serquis, A., Civale, L., Maiorov, B., Hawley, M.E., Maley, M.P., and Peterson, D.E. (2004) Strongly enhanced current densities in superconducting coated conductors of YBa2Cu3O7-x + BaZrO3. *Nat. Mater.*, **3**, 439–443.

41 Gapud, A.A., Kumar, D., Viswanathan, S.K., Cantoni, C., Varela, M., Abiade, J., Pennycook, S.J., and Christen, D.K. (2005) Enhancement of flux pinning in YBa2Cu3O7–δ thin films embedded with epitaxially grown Y2O3 nanostructures using a multi-layering process. *Supercond. Sci. Technol.*, **18**, 1502–1505.

42 Gutiérrez, J., Llordés, A., Gázquez, J., Gibert, M., Romá, N., Ricart, S., Pomar, A., Sandiumenge, F., Mestres, N., Puig, T., and Obradors, X. (2007) Strong isotropic flux pinning in solution-derived YBa2Cu3O7-x nanocomposite superconductor films. *Nat. Mater.*, **6**, 367–373.

43 Tersoff, J. (1998) Self-organized epitaxial growth of low-dimensional structures. *Phys. E*, **3**, 89–91.

44 Teichert, C., Legally, M.G., Peticolas, L.J., Bean, J.C., and Tersoff, J. (1996) Stress-induced self-organization of nanoscale structures in SiGe/Si multilayer films. *Phys. Rev. B*, **53**, 16334–16337.

45 Goyal, A., Kang, S., Leonard, K.J., Martin, P.M., Gapud, A.A., Varela, M., Paranthaman, M., Ijaduola, A.O., Specht, E.D., Thompson, J.R., Christen, D.K., Pennycook, S.J., and List, F.A. (2005) Irradiation-free, columnar defects comprised of self-assembled nanodots and nanorods resulting in strongly enhanced flux-pinning in YBa2Cu3O7–δ films. *Supercon. Sci. Technol.*, **18**, 1533–1538.

46 Kang, S., Goyal, A., Li, J., Gapud, A.A., Martin, P.M., Heatherly, L., Thompson, J.R., Christen, D.K., List, F.A., Paranthaman, M., and Lee, D.F. (2006) High-performance high-Tc superconducting wires. *Science*, **311**, 1911–1914.

47 Goyal, A. (2008) Engineered Columner Defects for Coated Conductors. Presented at the 2008 DOE Annual Peer Review on Superconductivity, available at http://www.energetics.com/supercon08/agenda.html

48 Goyal, A., et al. (2009) Manuscript in preparation.

49 Weinstein, R., Sawh, R., Gandini, A., and Parks, D. (2005) Improved pinning by multiple in-line damage. *Supercond. Sci. Technol.*, **18**, S188–S193.

50 Li, Q., Suenaga, M., Foltyn, S.R., and Wang, H. (2005) Jc(H) crossover in YBCO thick films and Bi2223/Ag tapes with columnar defects. *IEEE Trans. Appl. Supercon.*, **15**, 2787–2789.

51 Foltyn, S.R., Wang, H., Civale, L., Jia, Q.X., Arendt, P.N., Maiorov, B., Li, Y., Maley, M.P., and MacManus-Driscoll, J.L. (2005) Overcoming the barrier to 1000 A/cm-width superconducting coatings. *Appl. Phys. Lett.*, **87**, 162505.

52 Zhou, H., Maiorov, B., Wang, H., MacManus-Driscoll, J.L., Holesinger, T.G., Civale, L., Jia, Q.X., and Foltyn, S.R. (2008) Improved microstructure and enhanced low-field Jc in (Y0.67Eu0.33)Ba2Cu3O7–δ films. *Supercond. Sci. Technol.*, **21**, 025001–025006.

53 Wee, S.H., Goyal, A., Zuev, Y.L., and Cantoni, C. (2008) High performance superconducting wire in high applied magnetic fields via nanoscale defect engineering. *Supercond. Sci. Technol.*, **21**, 092001.

54 Goyal, A., Selvamanickam, V., Paranthaman, M., and Aytug, T. (2008) ORNL/SuperPower CRADA: Development of MOCVD-based,

IBAD-2G wire. Presented at the 2008 DOE Annual Peer Review on Superconductivity, available at http://www.energetics.com/supercon08/agenda.html

55 Cantoni, C., Christen, D.K., Feenstra, R., Norton, D.P., Goyal, A., Ownby, G.W., and Zehner, D.M. (2001) Reflection high-energy electron diffraction studies of epitaxial oxide seed-layer growth on rolling-assisted biaxially textured substrate Ni(001): The role of surface structure and chemistry. *Appl. Phys. Lett.*, **79**, 3077.

56 Cantoni, C., Christen, D.K., Heatherly, L., Kowalewski, M.M., List, F.A., Goyal, A., Ownby, G.W., Zehner, D.M., Kang, B.W., and Kroeger, D.M. (2002) Quantification and control of the sulfur c (2 x 2) superstructure on {100}<100> Ni for optimization of YSZ, CeO2 and SrTiO3 seed layers texture. *J. Mater. Res.*, **17**, 2549–2554.

Part Three
Energy Sustainability

10
Green Nanofabrication: Unconventional Approaches for the Conservative Use of Energy
Darren J. Lipomi, Emily A. Weiss, and George M. Whitesides

10.1
Introduction

Many of the chapters in this book describe the ways that nanostructured materials contribute to the conversion and storage of energy. The goal of this chapter is to highlight strategies that make efficient use of energy[1)] for the production of the nanostructures themselves ("green nanofabrication"). We define nanofabrication as the collection of methods that generates structures with sizes of less than 100 nm in at least one dimension [1]. By this definition, nanostructures have been fabricated as components of microelectronic devices for about a decade. Microprocessors are fabricated by high-performance techniques in facilities with rigorous environmental controls. These techniques – which we refer to as "conventional techniques" – make information technology possible and will not be replaced in the foreseeable future. They are, however, intensive consumers of energy and natural resources. Many emerging applications of nanotechnology – in optics, sensing, separations, biology and photovoltaics – tend to be more defect-tolerant than are microchips. Conventional techniques often are capable of producing structures for such defect-tolerant applications, but the sacrifice of efficiency (and cost) for precision made by conventional methods is, at best, unnecessary and, at worst, unacceptable.

Real and potential applications of nanotechnology pervade almost every branch of science and engineering. This book is concerned with applications of nanotechnology for energy, but comparably important applications exist in biology, chemistry, medicine, materials and the field in which it already enjoys the most success – information technology. As industrial and governmental spending increases in these and other areas, nanofabrication itself is becoming a greater proportion of total energy consumption in the developed world. The costs of nanomanufacturing (defined very broadly) are becoming significant contributors

---

1) By "efficient use of energy" we imply also the efficient use of materials and water, with a minimal production of toxic waste.

Nanotechnology for the Energy Challenge. Edited by Javier Garcia-Martinez
© 2009 WILEY-VCH Verlag GmbH & Co. KGaA, Weinheim
ISBN: 978-3-527-32401-9

of the costs of running the society. It is important that nanoscience and nanotechnology grow in environmentally responsible ways – not only developing nanomaterials with low toxicity, but also minimizing the energy intensity of nanofabrication. Green techniques of nanofabrication and nanomanufacturing can make a significant contribution to energy sustainability.

In the past several years, many unconventional approaches to nanofabrication have emerged; many of these approaches share the characteristic – intentionally designed or not – that they are conservative with respect to energy. The emphasis on the fabrication of nanostructures is distinct from the important functions that nanotechnology is likely to fulfill in the energy economy: efficient microprocessors, heterogeneous catalysts for the production and use of fuels, systems for global carbon management [2], systems for solar energy harvesting (photovoltaics, lenses, mirrors), membranes for purification and separation (O_2, water, CH_4, CO_2, H_2 CO, small particles) [3], devices for communications (optical fiber and microwave technology), displays, systems for lighting, materials to reduce friction, lightweight nanocomposite materials for the transportation industry [4] and many others.

10.1.1
Motivation

All strategies for meeting future energy demands require both improved and alternative sources of power and improvements in the efficiency with which energy is used [5]. Any potential solution to the energy challenge may be severely constrained in its form by environmental concerns (especially global warming) and issues of national security (e.g., nuclear power). The scientific community is justifiably interested in technologies that have a large, long-term potential for the clean production of energy (e.g., photovoltaics, carbon sequestration). Progress in these areas has come in notoriously small increments, however. Correspondingly small increases in efficiencies of sufficiently large consumers of energy (e.g., water purification, indoor lighting) or small reductions in losses (e.g., due to electrical transmission or friction) would make an immediate impact on the conservation of energy.

Nanofabrication, in all of its forms, is already a significant consumer of energy and thus can be regarded as a target for strategies to reduce energy consumption [6]. We justify this assertion by briefly characterizing the energetic aspects of nanofabrication in the context of the semiconductor industry and in academic laboratories. We find high energetic costs associated with fabricating highly ordered nanostructures that resemble the forms – but not necessarily the functions – of microelectronic components. We do not focus on the fabrication of random bulk nanostructured materials in which the "nano" may be incidental (e.g., carbon fiber tennis racquets, cosmetics, simple porous membranes). We avoid these materials in part because it is difficult to estimate the energetic costs associated with manufacturing them and also because the criteria for materials that do or do not qualify as "nano" are unclear.

10.1.2
Energetic Costs of Nanofabrication

The manufacturing sector of the United States economy consumes 24% of all energy distributed. The computers and electronics industry accounts for 12% of the manufacturing sector, by economic value [7]. Electronic products are driven by the semiconductor industry, which is growing at 15% per year with concomitant increases in inputs of energy [8]. The reduction in size of semiconductor devices has come at a price of increased complexity of both the devices and the techniques used to manufacture them. Wilson *et al.* conducted a rigorous analysis of the inputs into the production of a 32 MB DRAM chip (weighing 2 g, measuring 1.24 cm^2) and concluded that 1.6 kg of chemicals were consumed during the process [9]. The value includes estimates of processing chemicals and fossil fuels consumed at every stage of device production, starting from quartz and including its conversion into silicon wafers, but excludes a large contribution from water combined with purified elemental gases, which have a substantial energy of production as well.

In semiconductor fabrication, the ratio of the weight of input materials to the weight of finished products (microchips) is several orders of magnitude higher than that for other manufactured goods. In fact, an average desktop computer consumes four times more energy in production than in use, over a three-year life cycle [10]. The total carbon footprint (in production plus use) of a desktop computer is 1.3× that of a refrigerator, even if the computer lasts 3 years and the refrigerator lasts 13 years [10]. While the production energy of a manufactured product should not be entangled with its value to society [8, 11], the minimization of the energy intensity of nanomanufacturing can be addressed most efficiently if divorced from its (difficult to estimate) societal costs and benefits. The notably high production energy associated with semiconductor manufacturing arises from the purity of the materials and chemicals, the high-precision processing tools, and the environmental control within the fabrication facility (fab).

Semiconductor fabs and academic cleanrooms are significant consumers of energy [12]. Hu *et al.* found an average power consumption value of 2 kW m^{-2} for fabs in Taiwan (this figure is typical for fabs in other countries as well) [13]. For comparison, a typical residential energy use is <10 W m^{-2} in the United States. A survey of an anonymous research-oriented nanofabrication facility, a modern class-100 cleanroom – where the class number refers to the number of particles >500 nm allowed per cubic foot of air – revealed a consumption rate of 3 kW m^{-2}. Figure 10.1 summarizes the power distribution within the cleanroom. Processing tools (evaporators, furnaces, etc.) consume two-thirds of the energy, while climate controls and air polishing (filtration) consume the balance. The proportions are reversed in industrial fabs, where tools consume one-third and climate controls consume two-thirds [13].

The energetic requirement *per unit output* of an academic cleanroom is significantly greater than that of an industrial fab, because the energy consumed *per area* by the two types of facilities is comparable. For example, the anonymous

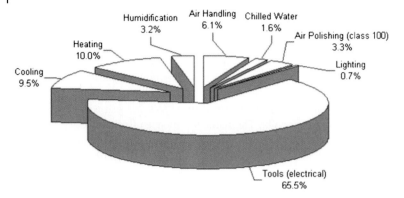

Figure 10.1 The power distribution of environmental controls, tools, and peripheral devices in a research-oriented, class-100 cleanroom. The cleanroom is 975 m² in area and in total consumes 3.0 MW. Electricity, including both lighting and processing tools, accounts for nearly two-thirds of the power.

cleanroom averages 100 users per day; liberally assuming that each user produces one finished wafer (or "experiment") per visit, the energy required for a single experiment is 2.7 GJ, which excludes the costs associated with chemicals and starting materials. This value is about 100 times greater than the analogous energy required by a fab to produce a 32 MB DRAM chip [9]. Conservation of energy usually is not a constraint when designing an experiment in a research setting, but economics usually is. The significant energetic costs of available techniques (which are almost always linked to economic costs) limit the extent to which fundamental studies of nanostructures may be carried out in academic laboratories.

We note that almost any process could be improved if performed under precise control over the environment and in the absence of dust. The extents to which processes and applications tolerate variations in climate and contamination can, however, vary greatly. A single defect might ruin an entire microprocessor, whereas a damaged feature of a nanostructured solar cell would be inconsequential. This chapter is not an attempt to label the semiconductor industry as inefficient or to dismiss high-precision instruments and cleanrooms as unnecessary. Such an interpretation would be wholly incorrect. In the event that a green approach produces a functionally equivalent structure to that of a conventional approach, the green approach should, however, be chosen. Economics – for better or worse – usually determine the choice in industry, unless sustainability is required by regulation.

10.1.3
Use of Tools

Processing tools are central to fabrication, and consume a significant percentage of the energy per experiment. Additional energy related to instruments goes into

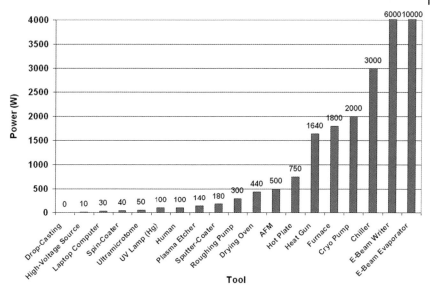

Figure 10.2 Power consumption values for common laboratory tools. All tools <1.8 kW were measured with a P3 "Kill A Watt" model P4400 commercial AC power meter except where otherwise noted. E-beam evaporator: maximum power of filament only. E-beam writer: Raith-150 maximum power. Chiller: recirculating water chiller maximum power. Cryo pump: CTI-Cryogenics compressor power rating. Furnace: Thermolyne model 48010 at 1200 °C max from manufacturer's specifications. Heat gun: Varitemp model VT 750-C at maximum setting. AFM: Veeco Nanoman manufacturer's maximum power rating. Hot plate: Cole-Parmer Instruments model 4658 with heating element on; add 19 W for maximum stirring speed. Drying oven: VWR Scientific Products 1300 U. Roughing pump: Alcatel 2010 SD at base pressure. Sputter-Coater: Cressington sputter coater equipped with a Pfeiffer vacuum pump, model MVP 015-2, idling at base pressure 134 W, 180 W at 60 mA. Plasma etcher: SPI Plasma Prep II, idled at 47 W, vacuum pump not included. Ultramicrotome: Leica Ultracut UCT in normal operation with lights off; it consumes 32 W idle and 92 W with all lamps on. Spin-coater: Headway Research model PMW 32 at 3 krpm, does not include energy of house vacuum system. Laptop Computer: IBM ThinkPad T43p. High-voltage DC power supply: Spellman CZE 1000R at 10 kV.

a chilled-water system, which is required for heat management. Sophisticated tools can also have substantial large production energies, which are paid even if they are never used. Figure 10.2 shows typical values of power consumption for several key tools and peripheral devices discussed in this chapter. The power ratings were measured by us or taken from the manufacturer's specifications (see figure caption). Unsurprisingly, instruments that heat and cool are the largest consumers of energy, which include the filament of an e-beam evaporator, furnaces, heat guns, hot plates and chillers. Tools that produce a plasma or convert electrical energy into mechanical energy tend to consume less energy. Most instruments consume energy only when activated (e.g., heat gun), but some idle at their maximum power (e.g., drying oven) or a large fraction of it (e.g., because it is wasteful and time-consuming to turn off and regenerate a cryogenic pump between uses of an e-beam evaporator).

Conventional tools of nanofabrication include electron-beam (e-beam) lithography (for creating lithographic masters) and photolithography (for replication). E-beam lithography is the dominant technology for the masterless creation of nanoscale features. It has the advantage of making arbitrary patterns at high resolution, but at the disadvantages of being serial, slow and energy intensive. Photolithography typically uses a photomask defined by e-beam lithography to transfer a pattern to a photoresist film. The technique is parallel and is the workhorse of replication for microelectronics. Lithography combines with processes such as physical vapor deposition (including sputter-coating, e-beam, thermal evaporation), reactive ion etching (RIE) and ion implantation to generate structures and to build complexity.

One of the strategies of unconventional approaches to nanofabrication is the adaptive re-use of analytical instruments for the purposes of fabrication. For example, instruments designed for analysis at the nanoscale can also be used as fabrication techniques, such as dip-pen nanolithography [14] (DPN, which uses an atomic force microscope) and nanoskiving [15] (which uses an ultramicrotome). Another strategy is to extract nanoscale information from unconventional sources. Thin films [16], shadows [17], edges [18], cross-sections [19], cracks [20] and lattice steps [21] can all be exploited for any nanoscale information they contain. The spontaneous organization of molecules into nanostructures (self-assembly) also has significant potential in nanofabrication [22]. The creative use of tools, the extraction of nanometric information from unconventional sources, and the discovery of new phenomena in self-assembly are thus central to this chapter.

10.1.4
Nontraditional Materials

The use of nontraditional materials is expected to reduce significantly the costs associated with nanofabrication. High-purity silicon wafers for microelectronics are produced in a sophisticated and energy-intensive process, which requires ~10 GJ kg^{-1} of electricity (~3000 kWh, ~2 t of CO_2) and has a yield of silicon of 9.5% from quartz, including all steps of manufacturing [9]. The formation of nanostructures from semiconductor nanocrystals [23], π-conjugated polymers and small molecules [24], materials derived from sol-gel precursors [25], and other nontraditional materials often cannot be accomplished with conventional techniques. The fabrication of devices based on these materials thus provides additional motivation for the development of new ideas in nanofabrication.

10.1.5
Scope

The concept of reducing the use of energy and raw materials is well established in the field of chemistry. A manifestation of this trend is in the impact factor of the journal *Green Chemistry*, which has risen steadily in the decade since the journal's inception [26]. Some of the tenets of green chemistry are also applicable to green nanofabrication, including the substitution of "harsh" processes for "mild" ones (temperatures, pressures, etc.), the use of water [27] and supercritical CO_2 [6] as solvents and the exploitation of self-assembly wherever possible.

Previously, we reviewed several unconventional approaches for nanofabrication – embossing, molding, printing, scanning-probe lithography, edge lithography, self-assembly [1, 28]. Unconventional approaches to nanofabrication tend to be low-cost alternatives to e-beam and photolithography and are simple enough to be accessible to users in fields far removed from solid-state physics and electrical engineering, such as biology and chemistry. They were invented, in part, to expand the scope of applications of nanostructures; while the conservation of energy was incidental in the invention of these techniques, it is the focus of this review.

We consider green nanofabrication a subset of both green chemistry and unconventional nanofabrication. Green techniques share many (but not necessarily all) of several key characteristics. They tend to: (i) be parallel, rather than serial, (ii) produce little waste by being additive, rather than subtractive (i.e., they form structures by successively adding or assembling materials rather than etching or discarding large fractions of material), (iii) benefit marginally or not at all by the use of a cleanroom or another form of strict environmental control, (iv) have the fewest number of steps possible, (v) replace materials that are costly to mine and purify, such as high-quality silicon, with widely available or synthetic materials, (vi) use simple, low-energy instruments and (vii) have the potential for large-area ($>cm^2$) fabrication.

This chapter is organized in terms of general approaches for green nanofabrication. Each section describes the general strategy and then gives examples of relevant techniques. Potential applications are emphasized throughout, although we pay particular attention to techniques for emerging applications of nanostructured materials – rather than techniques that are only intended for integration into semiconductor fabrication. The summary of each section compares the technique with others that are capable of producing similar structures. Many approaches in the literature can be regarded as green, but we must restrict our attention, in the interest of space, to a relatively small number of them.

The unconventional approaches to nanofabrication discussed in this chapter occupy a range of developmental stages. For example, block copolymer lithography is already incorporated into commercial processes in semiconductor fabrication [29], while step-and-flash imprint lithography (SFIL) is regarded as a practical technology for its cost, whole-wafer printing capability and potential to surpass the limits of resolution of photolithography imposed by optical diffraction [30, 31] (currently 40 nm; prospectively 15–20 nm in commercial production). Other techniques, such as particle replication in nonwetting templates (PRINT) [32] and nanoskiving [15], have promising capabilities but are in their exploratory stages. Our selection of techniques is idiosyncratic and emphasizes techniques in which pattern transfer – from one material or dimension to another – is a key step. The techniques produce or have the potential to generate long-range, laterally ordered or multicomponent nanomaterials, which are required for electronics, optics, photovoltaics, fuel cells, sensing and other applications.

10.2
Green Approaches to Nanofabrication

10.2.1
Molding and Embossing

Molding and embossing are perhaps the most successful strategies for unconventional nanofabrication [1]. The resolution of molding and embossing is limited only by the graininess of matter and thus has the potential of circumventing the theoretical limits of photolithography set by diffraction. These techniques are divided into two categories depending on whether the pattern transfer element (the mold or stamp) is hard or soft. Techniques that require a hard mold include nanoimprint lithography (NIL) [33–35] and step-and-flash imprint lithography (SFIL) [36, 37], while those that use a soft mold include replica molding (REM) [38, 39], solvent-assisted micromolding (SAMIM) [40] and particle replication and nonwetting templates (PRINT) [32]. A molding or embossing technique, *per se*, cannot generate nanoscale information; it relies on other patterning techniques, usually e-beam- or photolithography, to pattern a hard mold directly or to pattern a master for a soft mold. The extent to which these techniques can be used to conserve energy, therefore, depends on: (i) the techniques used to prepare the mold and (ii) the number of times the mold can be reused.

10.2.1.1 Hard Pattern Transfer Elements
Embossing using hard molds in various forms has been used commercially on the sub-micron scale for decades in the production of compact discs (CDs), DVDs, diffraction gratings and holograms [1]. The most common hard molds are made from quartz, silicon and metals. They are prepared by first patterning a resist film on the flat substrate by e-beam or photolithography and then modifying the exposed regions using reactive-ion etching, wet etching or electroplating. Quartz molds have the advantage of optical transparency, which enables registration with preexisting features on the substrate. The smallest lateral features transferred into quartz and silicon molds are 20 and 10 nm, using e-beam lithography [1].

Nanoimprint lithography is the use of a hard mold to emboss a polymer heated above its glass-transition temperature. This technique is an important element in the toolkit of nanofabrication, but has been reviewed extensively elsewhere [1]. SFIL is a variation of nanoimprint lithography [41]: it replaces the photomask in traditional photolithography with a quartz mold, which renders complex optical systems unnecessary. It is performed at ambient temperature, low applied pressures (<1 lb in^{-2}; n.b., equivalent to <7 kPa) and avoids the baking and solvent-processing steps of photolithography. An additional benefit is that the transparent mold facilitates alignment and registration. The accuracy of alignment in SFIL is

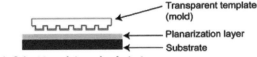
Step 1: Orient template and substrate

Step 2: Dispense drops of liquid imprint resist

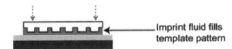
Step 3: Lower template and fill pattern

Step 4: Polymerize imprint fluid with UV exposure

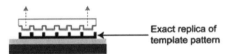
Step 5: Separate template from substrate

Figure 10.3 Schematic illustration of the SFIL process. A transparent mold (quartz) is pressed into a layer of imprint resist, a photocurable prepolymer. UV irradiation cures the imprint resist, after which the mold is removed. The resist contains a relief structure of the original mold, which can direct further modification (etching, evaporation, ion implantation, etc.) of the substrate. Reprinted with permission from [42]. Copyright 2008, American Chemical Society.

as high as ±10 nm (3σ) [36]. SFIL thus has potential for incorporation into microprocessor fabrication [30].

SFIL works by molding a photocurable prepolymer (imprint resist) with a quartz template against a flat substrate. The prepolymer is irradiated and crosslinked through the transparent mold. Release of the mold reveals the relief structure in the crosslinked polymer. A residual (scum) layer connects the molded features in the imprint resist to one another. A breakthrough etch removes the scum layer and completes the pattern transfer step. The unconnected features of imprint resist can then direct further elaboration of the substrate [30]. Figure 10.3 summarizes the process.

The use of hard molds can save energy in nanofabrication as long as the molds are reusable. Increasing the lifetime of a hard mold is a significant challenge and an object of research for the adoption of molding techniques in industrial nanofabrication. Deposition of crosslinked polymer on the mold (fouling) can cause irreversible adhesion of the mold to the substrate. Treatment of the mold using fluorinated silanes lowers the surface energy of the mold and extends the lifetime, but fewer than 100 uses per mold is typical [1].

A new approach has been described in which the imprint resist material is crosslinked with a linear molecule containing a selectively cleavable functional group along its backbone [41, 42]. The most successful group used was an acetal, which is labile in acidic solution. A mold contaminated with this material could be cleaned by de-crosslinking the polymer adhered to the features of the mold with a simple acidic wash. Figure 10.4 shows the cross-section of a SiO_2 stamp filled with a reversibly crosslinked imprint resist (top) and a different stamp after dissolution of the resist (bottom). In this study, all traces of crosslinked imprint resist could be removed by a simple acidic wash. While a statistical analysis of the lifetime of molds using this technique was not reported, the development of imprint resists that can be washed from the molds is an important line of research in SFIL.

10.2.1.2 Soft Pattern Transfer Elements

Soft molds are prepared by casting a prepolymer against a photolithographically patterned master, in a collection of techniques known as soft lithography [43]. The most common material used for this purpose, by far, is poly(dimethylsiloxane) (PDMS). PDMS is an elastomer with several useful characteristics, including chemical inertness, durability, transparency above 280 nm and the ability to pattern curved substrates [43]. The polymer is available in several commercial formulations; our laboratory uses Dow Corning Sylgard 184.

In the process of replica molding (REM), a photolithographically patterned master structure templates a PDMS mold, which then molds another polymer. Individual PDMS molds have been used more than 20 times in the context of the replica molding process, although a practical maximum for the number of replications has not been established. PDMS can replicate features in a topographically patterned master down to the atomic scale (0.2 nm), as demonstrated by Xu and coworkers by casting the polymer against a crack in a Si wafer [20].

In solvent-assisted micromolding (SAMIM), a solvent softens or dissolves a polymer film. A PDMS stamp is placed in contact with the polymer. During evaporation of the solvent, the polymer film conforms to the features of the PDMS mold. SAMIM has been used to pattern lines with widths of 60 nm in a Novolac photoresist [40]. This process is also amenable to several other polymers, including polystyrene (PS) [40], poly(methylmethacrylate) (PMMA) [40] and cellulose acetate (CA) [40]. A related technique, micromolding in capillaries (MIMIC), has been used to fabricated conjugated polymer field-effect transistors [44]. Like SFIL, SAMIM leaves behind a scum layer in the patterned polymer film. Figure 10.5 summarizes the SAMIM process.

Nonlithographic master structures may also be used to template the formation of a mold. McGehee and coworkers used a nanoporous anodic aluminum oxide

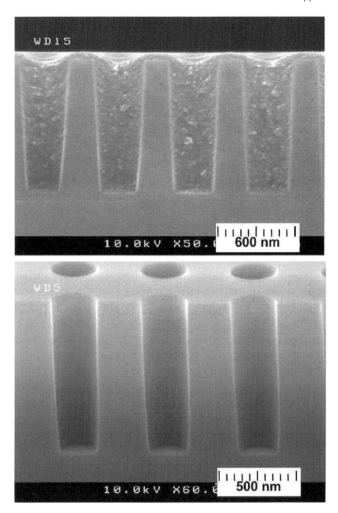

Figure 10.4 Scanning electron micrograph of a SiO$_2$ substrate patterned with pores with reversibly crosslinked imprint resist (top) and a different sample after de-crosslinking and dissolving the imprint resist (bottom). Reprinted with permission from [41]. Copyright 2007, American Chemical Society.

(AAO) template to direct the formation of a PMMA mold, which contained a relief structure of the template – nanopillars of 35–65 nm diameter with aspect ratios up to 3 [25]. This mold, in turn, was used to stamp a titania sol-gel precursor film, which was then calcined. In this way, the morphology of the nanostructured AAO template could be transferred to semiconducting metal oxide films. The choice of PMMA was necessary because of the high compression modulus (~2–3 GPa) compared to that of PDMS (~2 MPa), which was unable to keep the high-aspect-ratio features rigid. While both the AAO template and the PMMA mold

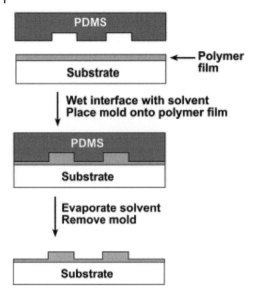

Figure 10.5 Schematic illustration of SAMIM. A PDMS mold is wet with solvent and placed in contact with a polymer film. After evaporation of the solvent, the mold is released. A relief pattern of features remains in the polymer film. Reprinted with permission from [1]. Copyright 2005, American Chemical Society.

are consumed by dissolution during the process, both materials are inexpensive and readily available. High-surface-area structures are believed to facilitate charge collection in excitonic solar cells (including polymer, small molecule, semiconductor nanocrystal-based devices) [45–47].

The ability to tailor the surface chemistry of molds is paramount to the success of molding techniques. SFIL, NIL, RM and SAMIM usually yield a residual layer (scum) between embossed features, which must be removed with a breakthrough etch step. The use of photocurable perfluoropolyether (PFPE) – a nonswellable, nonwettable elastomer (8–10 dynes cm^{-1}; n.b., 10 dynes = 10^{-4} N) – as both a mold and a flat template do not form a scum layer when embossing many types of materials. In a technique called *particle replication in nonwetting templates* (PRINT, Figure 10.6), DeSimone and coworkers showed that embossing films with PFPE templates produces discrete, unconnected particles, which can be easily detached from the substrate and harvested [32]. Examples of polymers molded into discrete nanoparticles include poly(ethylene glycol diacrylate), triacrylate resin, poly(lactic acid), poly(pyrrole) [32] and proteins [48]. Potential applications for these monodisperse nanoparticles include electronics and drug delivery [49].

The PRINT technique was recently extended to the preparation of discrete particles of inorganic oxides derived from sol-gel precursors [50]. First, a liquid sol was filled into the cavities of a mold. The mold was placed face-down on a substrate and held at an elevated temperature, at which the material underwent the sol-gel

Figure 10.6 Schematic drawing of the PRINT process. A solvated precursor is compressed between a nonwetting substrate and a PFPE mold. The procedure produces isolated particles, which is a unique capability among soft lithographic techniques. Reprinted with permission from [32]. Copyright 2005, American Chemical Society.

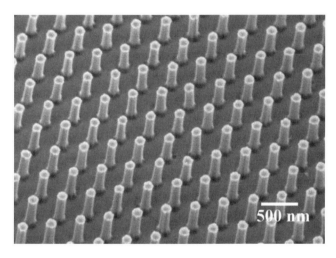

Figure 10.7 SEM image of discrete anatase TiO$_2$ nanopillars fabricated by PRINT. Reprinted with permission from [50]. Copyright 2008, Wiley-VCH Verlag GmbH & Co. KGaA.

transition. Removal of the mold revealed isolated oxide particles in their xerogel state. Calcination provided the crystalline phase. This strategy was able to produce either isolated features or embossed films of TiO$_2$, SnO$_2$, ZnO, ITO and BaTiO$_3$. Figure 10.7 is a SEM of isolated TiO$_2$ nanopillars, which could be useful, for example, as high-surface-area nanoelectrodes for excitonic solar cells. The sheer range of materials amenable to the PRINT technique makes it one of the most promising soft lithographic techniques for molding and embossing.

10.2.1.3 Outlook

The energetic costs of molding and embossing are tied to the fabrication and lifetime of the molds. The creation of arbitrary patterns with maximum resolution still requires the use of e-beam lithography and other conventional techniques;

nonlithographic patterns, such as porous block copolymer films or AAO membranes, avoid these limitations, but are only available in a limited number of patterns.

Once a mold is fabricated it can be reused several times, much in the way that a mask can be reused in photolithography, with the important difference that molds are subject to surface fouling and deterioration with repeated use (particularly hard molds). New materials, surface treatments and imprint resists are required to decrease the cost of molding and embossing. Soft molding techniques overcome some of the difficulties associated with surface fouling of the mold, because the low surface energy and mechanical flexibility facilitates detaching the mold from the substrate. The native surface energy of PDMS is 21.6 dynes cm^{-1}, which may be lowered by passivation with a fluorosilane to a the value to ~12 dynes cm^{-1}, which is similar to the value for poly(tetrafluoroethylene) (Teflon) [1]. Each photolithographic master patterned with SU-8 negative photoresist on a Si wafer and fluorosilanized is capable of producing >50 PDMS stamps. The lifetime of PDMS molds has not been established rigorously, but in our experience the features do not deteriorate if used at room temperature over a period of several months for >50 uses each. The number of end uses that can be derived from one step of photolithography (one trip to the cleanroom) is therefore >2500. The PRINT technology is a particularly exciting technology in the toolbox of soft lithography, as it has the advantage of making discrete nanoparticles in addition to two-dimensional surface patterns.

10.2.2
Printing

Nanofabrication by printing is the transfer of material from a topographically patterned stamp or movable tip to a surface. In a general sense, the advantages of printing are its potentially high throughput and the fact that it accommodates many types of materials. In terms of the conservation of energy, printing is green because it is additive: material is deposited only where it is required. (Conventional lithography is subtractive: large amounts of photoresist and other materials are usually discarded during spin-coating, development and any lift-off processes.) This section describes recent developments in two different types of printing: (i) microcontact printing (µCP) [43], which is a soft-lithographic technique that uses a stamp and (ii) dip-pen nanolithography (DPN) [14], which is a direct-write technique that uses an AFM tip.

10.2.2.1 Microcontact Printing
The most common incarnation of printing strategy is microcontact printing (µCP), which can be used to pattern alkane thiolates, silanes, biomolecules, colloidal particles, and polymers on metal and semiconductor surfaces [43]. Microcontact printing has been reviewed elsewhere [43], but we note that its key element is the generation of a topographically patterned stamp, usually PDMS, from a master that is typically patterned by photolithography. Printing self-assembled monolay-

ers (SAMs) of alkane thiolates on metal surfaces is the most prevalent form of μCP. SAMs can alter the wetting properties of surfaces and template the formation of secondary structures of another material. SAMs can also be used as resists for a variety of wet-etching processes. The lateral resolution of μCP can be <100 nm; it depends on the properties of the stamp and the extent to which the "ink" diffuses across the substrate. The smallest features transferred into a metal film by selective etching were 35 nm trenches separated by 350 nm [51].

A series of strategies of μCP can also be used to pattern features from a stamp that are smaller than those on the original master [52]. For example, mechanical deformation of the PDMS stamp can cause the features to converge; the lateral spreading of the contact area with the surface decreases the spacing between features. A single stamp can thus generate submicron linewidths by compression using a stamp patterned with lower resolution. Swelling the stamp with a solvent can be used to achieve a similar effect [52]. An extension of μCP, electrical-μCP, uses a PDMS stamp coated with a metallic film to print charge into a film of a dielectric polymer, PMMA [53]. Patterned charge can potentially be used as a high-density storage system for digital information. The resolution obtained in this study was 100 nm, but the theoretical limit has yet to be established. The energetic costs associated with μCP are similar to those of any soft-lithographic technique and depend largely on the number of prints a single stamp can make. Our laboratory has found that stamps may be reused over the course of months without noticeable deterioration in quality.

10.2.2.2 Dip-Pen Nanolithography

The Mirkin laboratory has developed a direct-write process using an AFM tip known as dip-pen nanolithography (DPN) [14, 54]. In DPN, the AFM tip is inked in a solution placed into contact with a surface. Direct transfer of the ink to the surface can create nanoscale patterns in arbitrary geometries. DPN accommodates a range of inks, including alkane thiols, proteins, conjugated polymers, DNA, sol-gel precursors, metal salts and monomers for polymerization [14]. Like μCP, DPN is an additive technique – material is deposited only where it is required, so it avoids blanket exposure steps to chemicals and other processes. Figure 10.8 shows the way in which an AFM tip deposits material on a substrate: a water meniscus forms between the AFM tip and the substrate. The meniscus serves as a carrier for molecular transport. Like μCP, the first demonstration of DPN used alkane thiolate SAMs on Au [55]. An additional useful characteristic of DPN is that multiple inks can be deposited on the same substrate[56].

The use of a single AFM tip only allows the serial reproduction of patterns. Recently, however, massively parallel DPN was demonstrated, in which 55 000 AFM tips were connected in a microfabricated, two-dimensional array. This array was able to produce copies of a dot matrix map of Au features over a $1 \times 1 \, cm^2$ area (Figure 10.9). The ability to produce arbitrary patterns should have applications in nanoscale bioassays and other areas [54].

Parallel DPN is a maskless technique that, in principle, combines the resolution of a mastering technique [58] like e-beam lithography with the throughput of

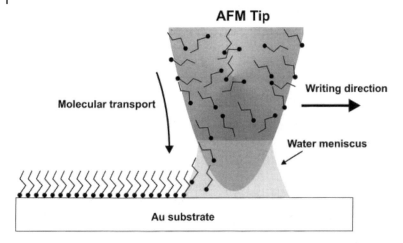

Figure 10.8 Schematic illustration of the DPN process. The AFM tip is "inked" with small molecules, which transfer to the surface of a solid substrate. Reprinted with permission from [55]. Copyright 1999, American Association for the Advancement of Science.

photolithography. It can serve simultaneously as a mastering step or a replication step. In the massively parallel process, each tip is passive and thus the whole array generates multiple copies of the same structure. There is no reason why the technique cannot be used to translate each tip independently, but this capability represents a significant challenge in engineering. As it stands, the DPN process is slow, but it is a promising new technology with unique capabilities. As a technique capable of generating masters, it may combine well with techniques requiring soft, patterned stamps.

10.2.2.3 Outlook

Nanofabrication by printing is now ubiquitous in research laboratories, and both µCP and DPN have significant potential for manufacturing due to their parallel capabilities. The additive nature of printing attenuates the production of chemical waste, which is produced in photoresist processing steps used in photolithography. The energy associated with the µCP process is in its use of conventional photolithographic mastering techniques, while the printing step itself can be done manually or with simple tools. DPN uses an AFM, a low-wattage instrument. There is evidence, however, that the transport of ink in the DPN process is a function of humidity and temperature of the ambient air [59]. Some control over the environment is necessary for the reproducibility of DPN.

10.2.3
Edge Lithography by Nanoskiving

Conventional lithography and thin film deposition are typically regarded as two-dimensional techniques. In many cases, the edges of structures (including the

10.2 Green Approaches to Nanofabrication | 247

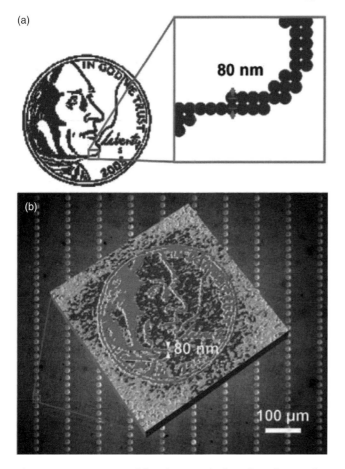

Figure 10.9 (a) An array of dots depicting the front face of a United States five-cent coin. (b) This image was reproduced in an array using the parallel array of AFM tips. Reprinted with permission from [57]. Copyright 2007, Wiley-VCH Verlag GmbH & Co. KGaA.

sidewalls of photolithographic features, heights of thin films and step edges of crystalline lattices) contain nanoscale information. Edge lithographic techniques extract such information in two general ways: (i) pattern transfer directed by the edge of a feature and (ii) conversion of a feature that is thin in the vertical dimension to a feature that is thin in the lateral dimension [1]. Since our initial reviews of unconventional techniques of nanofabrication [1, 28], our laboratory has expanded significantly the types of structures accessible by the technique of nanoskiving – fabrication by sectioning with an ultramicrotome – which falls under (ii) [15]. We have applied nanoskiving to generate nanostructures and heterostructures of a range of organic and inorganic materials. Applications of devices include frequency-selective surfaces (FSS) [60], single-crystalline plasmon resonators [61], chemical sensors [19] and organic photovoltaics (OPVs) [62]. The technique is operationally simple, can access structures with lateral dimensions as small as 10 nm and can be completely nonphotolithographic: it requires only an ultramicrotome (which is accessible in almost any research setting) and a method for the deposition of thin films or chemically synthesized crystals. It is inherently conservative with respect to energy (particularly in concert with spin-coating) and raw materials, and a cleanroom is not required.

10.2.3.1 The Ultramicrotome

The ultramicrotome is a mechanical cutting tool that was originally developed for thin sectioning of biological specimens for electron microscopy. The instrument is low-energy and is capable of producing sections of 30-nm thickness with routine use. An ultramicrotome consists of a sample arm, a diamond knife attached to a water-filled trough, a movable stage/knife holder and a stereomicroscope (Figure 10.10). The steps of the procedure are: (i) embedding the master structure in a hard polymer matrix (Young's modulus >1500 MPa, typically epoxies or PMMA) to make the "block", (ii) trimming and aligning the block, (iii) sectioning the block into thin slabs and (iv) collecting the slabs, which slide from the edge of the knife to the surface of the water trough. The surface tension of the water prevents compression of the slabs, which tend to self-assemble into ribbons while floating on the surface of the water. After the slabs are collected, optional processing steps include wet or dry etching. The rest of this section illustrates examples of materials and structures that are amenable to nanoskiving to form functional devices.

10.2.3.2 Nanowires with Controlled Dimensions

The simplest application of nanoskiving is the sectioning of strips of metal to form nanowires with precise control of every dimension. Figure 10.11 illustrates the procedure. A silicon wafer is patterned with Au stripes (micron width), which are transferred to the surface of an epoxy substrate. The Au stripes are embedded in epoxy and sectioned with the ultramicrotome, which yields nanowires. These nanowires displayed clear, size-dependent plasmon resonances [63].

10.2.3.3 Open- and Closed-Loop Structures

Sectioning parallel to the surface of a topographically patterned metal film can generate arrays of open- or closed-loop structures. Figure 10.12 illustrates this variation of the procedure. A topographically patterned epoxy substrate, with posts patterned by replica molding, is coated on one or more sides of each post with a metal film, using shadow evaporation. Embedding these posts in additional epoxy and then sectioning it parallel to the plane of the topography generates arrays of open-loop structures. Closed-loop structures can be generated using a single step of metal deposition by using sputtering (a noncollimated source) to coat the sides of the posts uniformly. A grating with micron-width trenches is used as the master [64]. These metallic structures behave as frequency-selective surfaces [60] and can also serve as resists for further elaboration of the substrate [65].

10.2 Green Approaches to Nanofabrication | 249

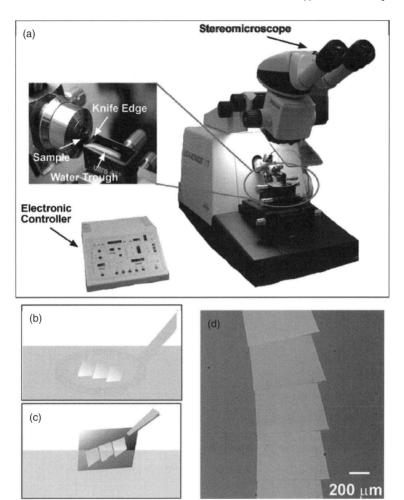

Figure 10.10 (a) Photograph of the model of ultramicrotome (Leica Ultracut UCT) used by our laboratory and close-up of the diamond knife and water trough. (b, c) Schematic illustration of collection of epoxy slabs with the use of a metal loop, which supports the slabs in a film of water, or by direct-capture with a substrate (a piece of a silicon wafer is depicted in 8F). (d) Micrograph of self-assembled epoxy slabs on the surface of water. Reprinted with permission from [15]. Copyright 2007, American Chemical Society.

10.2.3.4 Linear Arrays of Single-Crystalline Nanowires

Physical vapor deposition typically produces thin films that are polycrystalline. Sectioning these films, in turn, yields polycrystalline nanostructures, which are unsuitable for many applications. Chemically synthesized Ag nanowires have been shown to behave as low-loss plasmon resonators, with potential for use as waveguides in photonic integrated circuits and as biological sensors [66]. Silver

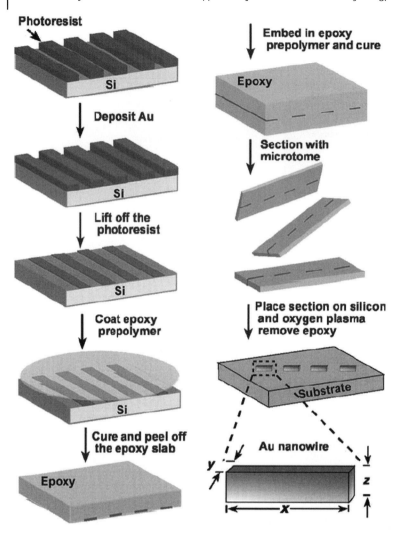

Figure 10.11 Schematic of the procedure used to fabricate Au nanowires. Reprinted with permission from [63]. Copyright 2006, Wiley-VCH Verlag GmbH & Co. KGaA.

structures, however, have limited potential for real applications because they are unstable due to oxidation under ambient conditions, oriented randomly on a substrate and polydisperse. Our laboratory used nanoskiving to fabricate single-crystalline nanowires of Au by sectioning chemically synthesized Au microplates [67]. The Au nanowires produced by nanoskiving also behave as plasmon resonators, but have the characteristics that they are stable to the atmosphere, oriented co-linearly within an epoxy slab, and reproducible from slab to slab. Figure 10.13 shows five quasi copies of a Au nanowire that were derived from a single microplate. The ability to produce monodisperse nanowires of this type has not been demonstrated using chemical synthesis alone.

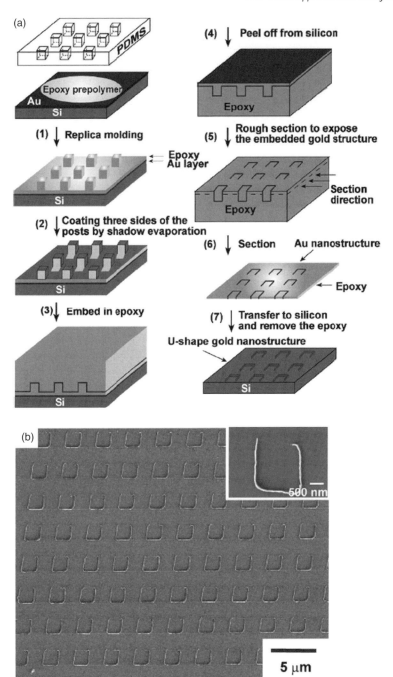

Figure 10.12 (a) Schematic illustration of the procedure used to generate complex metallic nanostructures from sectioning a topographically patterned surface (containing posts). (b) SEM images of structures obtained by sectioning parallel to the surface. Arrays of these structures serve as frequency-selective surfaces. Reprinted with permission from [15]. Copyright 2007, American Chemical Society.

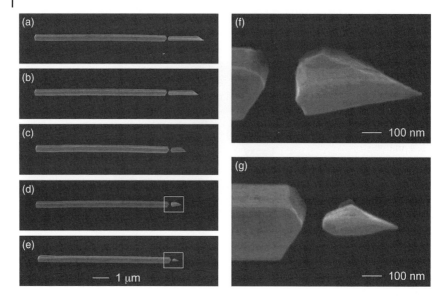

Figure 10.13 (a–e) Five quasi copies of a Au nanowire that were cut consecutively from a single hexagonal microplate. There is a section missing between (b) and (c). The structures on the right of (a–e) are nanowires derived from a smaller, triangular microplate (close-up shown in f, g). Reprinted with permission from [61]. Copyright 2008, American Chemical Society.

10.2.3.5 Conjugated Polymer Nanowires

Conjugated polymers possess many of the useful properties of crystalline semiconductors: electroluminescence [68], photovoltaic response [69] and modulation of conductivity by gate voltage [70] or by doping [71]. Polymers tend to be more mechanically flexible and less expensive to produce and process than crystalline semiconductors. As a result of this potential utility, there are many strategies for the fabrication and patterning of functional conjugated polymer structures [24]. One of the simplest (but most useful) structures is the nanowire [72, 73].

Potential applications of conjugated polymer nanowires include: chemical and biological sensors, field-effect transistors [74], tools for studying one-dimensional charge transport [75], actuators [76] and interconnects [77]. Nanowires composed of conjugated polymers are well-suited for sensing because they have a high ratio of surface area to volume, which permits rapid diffusion of an analyte to and from a wire [78, 79]. Electrical response and recovery rates are thus higher for nanowires than they are for thin films or fibrous networks. Electrochemical biosensors based on polymer nanowires have the potential for label-free detection of biological analytes. Incorporation of molecular recognition elements into conjugated polymers is relatively straightforward by synthesis, while analogous modifications of carbon nanotubes and inorganic nanowires require surface reaction(s) carried out postfabrication [80].

There is not yet a truly general technique for the fabrication of conjugated polymer nanowires, despite interest in the structures. In principle, any film-forming material can be formed into wires by nanoskiving, although the extent to which materials are damaged by the process depends on many properties. Many conjugated polymers are processible from solution and form stable films. We found that nanoskiving easily accommodates conjugated polymers. Figure 10.14 summarizes our strategy for the formation of two conjugated polymers with different physical and electronic properties, poly(2-methoxy-5-(2´-ethylhexyloxy)-1,4-phenylenevinylene) (MEH-PPV) and poly(benzimidazobenzophenanthroline ladder) (BBL).

The polymers can be identified in the SEM by etching. Figure 10.15a shows the transition between the composite MEH-PPV/BBL film and the free MEH-PPV nanowires. We obtained the image by covering a portion of the epoxy section with a conformal slab of poly(dimethylsiloxane) (PDMS) and treating the uncovered portion with a drop of MSA (as shown schematically on the left-hand side of Figure 10.15a) for ~5 s. We rinsed the MSA off the substrate with ethanol and removed the slab of PDMS. Figure 10.15b shows the transition between the intact composite film (right-hand side) and the free BBL nanowires after dry etching of the MEH-PPV and epoxy matrix (left-hand side).

When exposed to vapor from an I_2 crystal, the conductivity of the MEH-PPV nanowires increased by several orders of magnitude. Upon removal of the crystal, the conductivity rapidly decreased by the evaporative loss of I_2, which was concomitant with a decrease in free charge carriers. These results suggest that conjugated polymers formed by nanoskiving could function as high-surface-area sensors.

10.2.3.6 Nanostructured Polymer Heterojunctions

Given the fact that nanoskiving easily accommodates conjugated polymers, we sought to extend the technique to a different type of nanostructured architecture – the ordered bulk heterojunction for organic photovoltaics (OPVs). OPV devices are promising low-cost alternatives to solar cells based on crystalline semiconductors for the efficient production of electrical energy. Devices based on conjugated polymers, in particular, are ideally suited for coverage of many surfaces because they are processible from solution and mechanically flexible. One of the primary disadvantages of conjugated polymers (along with small molecules and semiconductor nanocrystals) is that they require a large amount of interfacial area between electron donating ("p-type") and electron accepting ("n-type") phases in order to separate photoexcited states (excitons) into free charge carriers (by photoinduced charge transfer). Once created, the electrons (e^-) drift toward a reflective, low-work-function electrode (LWFE), and the holes (h^+) drift toward a transparent, high-work-function electrode (HWFE). Additionally, the distance an exciton can travel before it decays (the exciton diffusion length, or L_D) is about ten times shorter than the thickness of material required for efficient absorption of photons (100–200 nm). The architecture that satisfies the requirements of both L_D and the thickness for optimal absorption of light is known as the ordered bulk heterojunc-

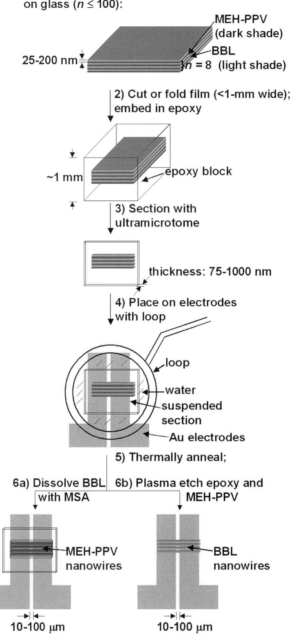

Figure 10.14 Summary of the procedure used for fabrication of multiple nanowires of MEH-PPV or BBL. We started by spin-coating a composite film of MEH-PPV alternating with BBL (step 1). Next, we cut the film into strips and embedded it in epoxy resin (the "block", step 2). We sectioned the block with an ultramicrotome (step 3), which yielded slices of the conjugated polymer film, framed by epoxy. Using a metal loop, we transferred manually the sections from the water boat of the ultramicrotome to photolithographically patterned Au electrodes on a SiO_2 substrate (step 4). Thermal annealing (step 5) improved adhesion between the thin sections and the electrodes. Selective dissolution of BBL with methanesulfonic acid (MSA; step 6a) gave free MEH-PPV nanowires or selective etching of MEH-PPV and epoxy with oxygen plasma (step 6b) gave free MEH-PPV or BBL nanowires. Reprinted with permission from [19]. Copyright 2008, American Chemical Society.

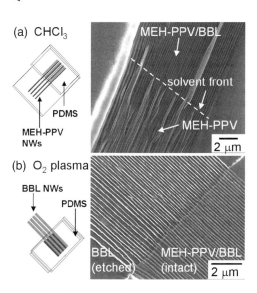

Figure 10.15 Scanning electron micrographs (SEM) showing the transition between the composite MEH-PPV/BBL film and free MEH-PPV nanowires (NWs). Reprinted with permission from [19]. Copyright 2008, American Chemical Society.

tion (Figure 10.16). It has a cross-section of p-type and n-type phases that is interdigitated on the length scale of L_D and is 100–200 nm thick.

We used nanoskiving to fabricate an ordered bulk heterojunction of two conjugated polymers in a three-step process: (i) spin-coating a composite film with 100 alternating layers of BBL (n-type) and MEH-PPV (p-type), (ii) rolling this multilayer film into a cylinder (a "jelly roll") and (iii) nanoskiving the jelly roll (Figure 10.17). The cross-section of a slab of the jelly roll has an interdigitated arrangement of the two polymers. The thickness of the slab is determined by the ultramicrotome and the spacing between the two materials is determined by spin-coating. Figure 10.18 shows a top-down image of a jelly roll slab embedded in an epoxy membrane, as well as a cross-section, which shows the interdigitated arrangement of MEH-PPV and BBL.

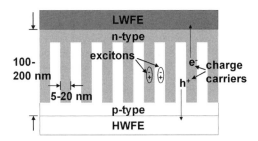

Figure 10.16 Schematic illustration of the ordered bulk heterojunction. Reprinted with permission from [62]. Copyright 2008, Wiley-VCH Verlag GmbH & Co. KGaA.

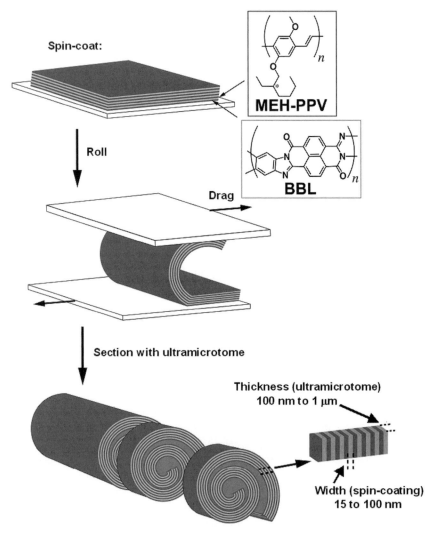

Figure 10.17 Schematic drawing of the procedure used to make nanostructured heterojunctions of MEH-PPV and BBL by nanoskiving a rolled, free-standing thin film (a jelly roll). Reprinted with permission from [62]. Copyright 2008, Wiley-VCH Verlag GmbH & Co. KGaA.

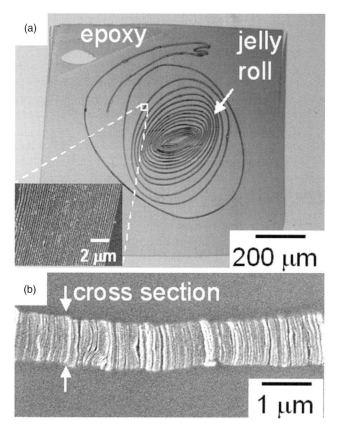

Figure 10.18 (a) Optical image of a slab of a jelly roll embedded in epoxy. The inset is an SEM close-up image of the exposed surface of the multilayer film of BBL (light gray stripes) and MEH-PPV (dark gray stripes). (b) Cross-sectional SEM of a slab of a jelly roll that shows the interdigitated structure of BBL alternating with MEH-PPV. Reprinted with permission from [62]. Copyright 2008, Wiley-VCH Verlag GmbH & Co. KGaA.

When placed in a junction between two electrodes with asymmetric work functions, the heterostructures exhibit a photovoltaic response under white light (although the efficiency of conversion of optical to electrical energy is low). Selective excitation of BBL with red light confirms that the photovoltaic effect is the result of photoinduced charge transfer between BBL and MEH-PPV. An initial experiment to determine the effect of n-type and p-type buffer layers between the jelly roll and the electrodes on the photovoltaic performance showed a marked increase in efficiency. Although the performance of the structure as a device is currently limited, we suggest that this approach to donor/acceptor heterojunctions could be useful in photophysical studies and might ultimately suggest new approaches to OPV devices.

10.2.3.7 Outlook

Nanoskiving is a simple and inexpensive technique in which nanoscale features are derived from the edges of thin films, not from two-dimensional photolithographic features. While some of the applications we described required the use of a photolithographically derived master, a cleanroom was not necessarily required for the micron-scale resolution reported here (only insofar as mask aligners are typically found in cleanrooms). In most applications of nanoskiving, the energetic costs associated with cleanrooms may be avoided entirely.

The structures produced by nanoskiving are embedded and stabilized in a thin epoxy slab, which is a macroscopic object that can be manipulated. These slabs may be placed on arbitrary surfaces, including curved substrates. If compatible with the nanostructured material contained in the slab, the epoxy can be removed by dry etching with an oxygen plasma.

Nanoskiving can access features with lateral widths as thin as 10 nm in several classes of film-forming materials. The technique shares this capability with only a few other methods, including scanning-beam and scanning-probe lithographies. These techniques, however, are not capable of producing large heights of features, which nanoskiving accomplishes easily. In this respect, nanoskiving has some analogy with nanoimprint lithography and deep reactive-ion etching, although the aspect ratios obtainable by nanoskiving can be significantly greater. Sectioning thin films into slabs with thickness ranging from 30 nm to 10 μm can produce structures with height-width aspect ratio of 10^3 (depending on the properties of the materials), which is not possible by any conventional technique.

Nanoskiving can function as a technique of both mastering and replication. As a mastering technique, it is somewhat like electron-beam or scanning-probe lithography; as a replication technique, it is analogous to photolithography or soft lithography. As potential manufacturing technique, nanoskiving is conservative with respect to raw materials because of the large number of quasi-copies that can be derived from a single block. For example, a 1-cm epoxy block in principle contains thousands of 100-nm sections. (Note that fewer sections may be produced when sectioning parallel to the surface of a topographically patterned film.)

The energetic costs of nanoskiving are low, particularly when sectioning spin-coated polymer films as the active material. An ultramicrotome consumes about 35 W when idling and 92 W during operation. Of these 92 W, 45 W are used by the lighting system, which may be turned off during operation of the instrument. A typical production energy value for epoxy, which is inexpensive, is 140 MJ kg^{-1} [9]. Diamond knives with a sharp wedge angle (35°) give the best results, and, while they are expensive (US$1000–5000, depending on size and quality), may be re-sharpened after ~1 year of ordinary wear (at about half the cost of a new knife). Disposable glass knives are significantly less expensive, but are typically discarded after a single use.

As with any technique, nanoskiving has limitations. It is currently limited to structures with unconnected line segments, because the lateral features of the desired structure must be encoded in the cross-section of the block. Multilayer patterning would, in principle, allow connectivity between features within a slab, but it would also erode the primary strength of nanoskiving – its simplicity. There is not yet a precise way to orient slabs on a substrate or to transfer them from one substrate to another. Some orientation is possible by hand using the perfect loop or direct transfer of slabs from the water trough to the substrate, but stacking sections to produce layered structures is difficult to accomplish with precise registration. The ultramicrotome also tends to compress samples somewhat along the axis perpendicular to the edge of the knife. This effect necessitates some flexibility in addressing the structures after sectioning and potentially limits the extent to which the slabs can be incorporated into precisely registered, multilayer structures.

10.2.4
Shadow Evaporation

Shadow evaporation is a method for forming nanoscale features by evaporating materials onto a substrate that is positioned at an oblique angle to the evaporation source, through a shadow mask, which determines the shape of projection of the beam of evaporated material onto the substrate [17]. Using different angles of evaporation for each material can form overlays of different materials.

10.2.4.1 Hollow Inorganic Tubes

Our group has used shadow evaporation to fabricate electrically continuous arrays of nanotubes of metals and indium-doped tin oxide (ITO) with controllable dimensions (height, outer diameter) [17]. We have shown these arrays to be useful as substrates for detecting small concentrations of organic molecules using surface-enhanced Raman spectroscopy. Potential applications exist for three-dimensional, nanostructured versions of devices, such as solar cells, light-emitting diodes (LEDs), electrochromics and batteries. For these devices, the nanostructured arrays provide: (i) a high ratio of surface area to volume for interfacial charge collection/separation, ion transport across liquid/solid interfaces (i.e., for mass-transport-limited processes in general) and (ii) a template for depositing nanostructured films of small molecules or polymers that serve as optically and electronically active layers for these devices. In addition, while there have been many efforts to fabricate nanostructures of ITO (the most popular transparent conducting oxide for use in such devices) [81–89], templated shadow evaporation is the first to produce uniform, free-standing ordered arrays of electrically connected ITO nanotubes with controllable dimensions.

As a nanostructured shadow mask, we used a nonlithographic template: a commercially available anodized alumina (AAO) membrane (Whatman Anodisc; 60 μm thick, 200 nm diam. pores). These membranes dissolve easily in aqueous base and have a high density ($\sim 10^9$–10^{11} cm^{-2}) of straight pores with monodisperse diameters. Alumina membranes with regular, cylindrical pores with diameters from 10 nm to microns can also be fabricated through established anodic processing techniques.

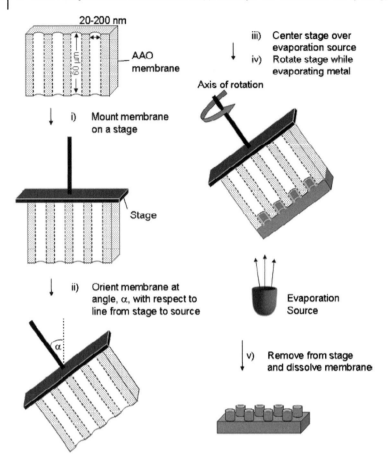

Figure 10.19 Procedure for fabrication of nanotube arrays: We (i) mounted an anodized alumina (AAO) membrane onto a stage, (ii, iii) centered the stage directly above the evaporation source at an incident angle α, defined as the angle between the axis of rotation and a line drawn from the source to the AAO membrane) and (iv) evaporated onto the membrane. The edges of the pores of the membrane cast shadows into the pores. (v) We immersed the coated membrane in 1 M NaOH to completely etch the AAO membrane and yield an array of nanotubes connected by a continuous backing of the same material. Reprinted with permission from [17]. Copyright 2007, American Chemical Society.

Figure 10.19 shows the general procedure for fabricating arrays of nanotubes. We mounted the membrane flush against a rotating plate driven by a battery-powered DC motor and e-beam evaporated ~150 nm of metal (Au or Pt) or ITO onto the membrane. We fixed the evaporation incident angle (α) between 0 and 90° to control the height of the nanotubes. After deposition, we removed the AAO membrane from the rotating support. The alumina dissolved when the AAO–metal (or ITO) composite was immersed in aqueous base. This procedure yielded an array of nanotubes supported by an electrically continuous backing of the same material.

In principle, the technique we describe is compatible with any material that can be vapor-deposited, either through thermal or e-beam evaporation. Electron-beam evaporation is the preferred method for this particular technique because, in contrast to other evaporation techniques (e.g., sputtering), the evaporating material produced by an electron beam has a long mean free path due to the low pressure in the chamber ($\sim 1 \times 10^{-6}$ torr; n.b., 1 torr = 133 Pa) and, thus, is approximately collimated after traveling from the source to the substrate ($\sim$45 cm total distance). The collimated beam is critical for this application because the shadows cast by the AAO membrane into its pores defines the geometry of the tubes.

The heights of the structures depended on the angle of evaporation (α, the angle between the axis of rotation and a line between the source and the AAO membrane). Figure 10.19 shows micrographic evidence of the dependence of height of Au tubes on incident angle of evaporation for the two most extreme angles (0°, 90°), and an intermediate angle (45°). At a glancing angle (Figure 10.20a, $\alpha \sim 90°$), the Au structures formed a loosely connected array of rings, because the collimated beam did not penetrate the pores, and only a small amount of deposited material adhered to the membrane. The diameter of the rings was approximately 200 nm (the diameter of the pores of the AAO membrane). At $\alpha \sim 45°$ (Figure 10.20b), the tubes formed structures whose height was approximately the same as their diameter ($\sim$200 nm). At a normal angle (Figure 10.20c, $\alpha \sim 0°$), the structures formed an array of long nanotubes ($\sim$1.5 μm tall). These are the highest aspect ratio nanotubes formed using a one-step physical vapor deposition process on AAO membranes. The walls of the tubes, however, were thin since the collimated beam projected parallel to the walls of the pores. Many of the tubes in Figure 10.20c therefore mechanically separated from the substrate or collapsed during sample preparation.

10.2.4.2 Outlook

In photolithography (and most other methods of pattern transfer), the fabricated structure has a one-to-one correspondence to the master structure. Shadow evaporation is unique because the information is encoded in the three-dimensional topography of the master. This "line-of-sight" deposition allows for an infinite number of possible structures or patterns to be fabricated using the same master simply by changing the angle of the substrate with respect to the source. Shadow evaporation, therefore, increases the effective complexity of a topographically patterned master by introducing the angle as another variable that is capable of rendering different product structures.

The technique is potentially conservative with respect to energy because the same master structure can be used with different angles and source materials to generate multicomponent structures. While photolithography is typically used for pattern transfer in two dimensions, the heights of the individual photoresist features may be exploited in shadow evaporation. In fact, electronic components such as resistors, capacitors, and metal oxide semiconductor field-effect transistors (MOSFETs) can be formed using only a single layer of photolithography on a silicon wafer [90].

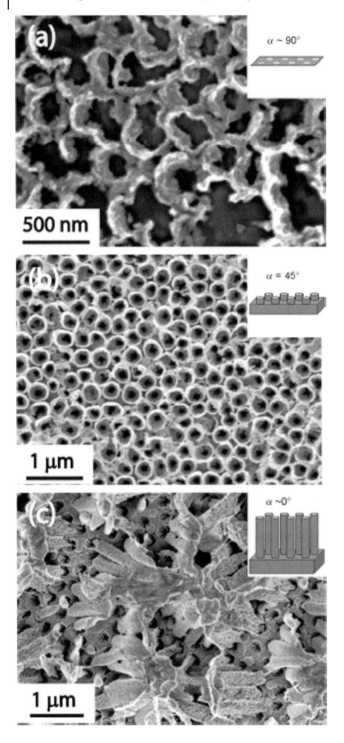

Figure 10.20 SEM images (top view) of Au nanostructures formed at various incident angles α (insets: the corresponding oblique-angle schematic diagrams illustrating the dependence of the ideal geometry of the nanotube on α for a given amount of deposited material). (a) α ~ 90° (glancing): an array of small rings formed with a minimal backing layer. The collimated evaporation source resulted in negligible penetration into the pores and a thin backing layer. (b) α ~ 45°: an array of short tubes formed. (c) α ~ 0° (normal): an array of tall, thin-walled tubes formed with a thick backing layer. The penetration was deep into the pores, but the walls of the tube were thin because collimation of the vaporized material was parallel to the pores. Reprinted with permission from [17]. Copyright 2007, American Chemical Society.

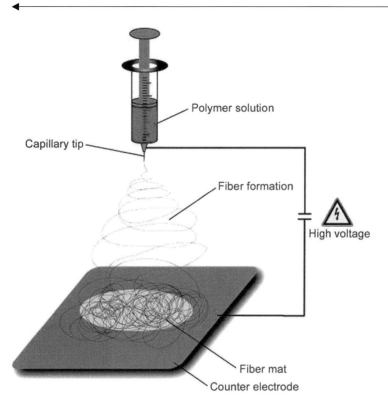

Figure 10.21 Schematic illustration of the electrospinning process. Reprinted with permission from [92]. Copyright 2007, Wiley-VCH Verlag GmbH & Co. KGaA.

10.2.5
Electrospinning

Electrospinning is a technique for making nanofibers of a range of organic and inorganic materials by the electrostatic extraction of a solid fiber from a solution or melt [91]. While it has been reviewed before [92], we focus on the newest developments, which include methods to produce aligned and multicomponent nanofibers. The typical setup is depicted in Figure 10.21. A syringe containing a

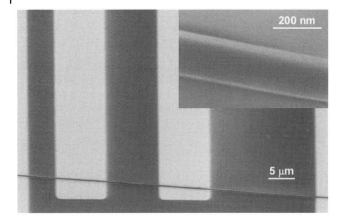

Figure 10.22 A single electrospun fiber of P3HT spanning microfabricated electrodes on an Si/SiO$_2$ substrate. The inset is a closeup of the fiber. The nanofiber behaves as a field-effect transistor. Reprinted with permission from [74]. Copyright 2005, American Institute of Physics.

solution-based precursor is held at a high voltage (10–50 kV). A drop of solution is extruded from the needle or capillary from which a jet is electrostatically ejected toward a grounded substrate platform. As the jet of material reaches the substrate, the solvent evaporates or the material otherwise hardens to form a continuous fiber. In the typical case, the deposition of the fiber is not controlled, so it forms a disordered mat on the substrate. We restrict our attention to modifications of the electrospinning technique that generate aligned nanofibers and those that generate multicomponent core-shell or hollow nanofibers.

10.2.5.1 Scanned Electrospinning

Craighead and coworkers have developed a procedure that produces single nanofibers of polymers by a scanned electrospinning process [93]. In this procedure, a microfabricated spinneret is dipped in a solution and held in close proximity (*ca.* 2 cm) from the grounded substrate. The jet originates from the sessile drop attached to the spinneret. The key to scanned electrospinning is a motor that translates the substrate during the spinning process. The relative motion of the substrate with respect to the spinneret allows fibers to be deposited in a predictable orientation. The use of photolithographically patterned electrodes on the substrate platform enables the characterization of single-nanowire devices. For example, a chemical sensor based on polypyrrole [78] and an organic field-effect transistors (OFETs) based on poly(3-hexylthiophene) (P3HT) have been demonstrated (Figure 10.22) [74]. In the case of the P3HT nanowire OFET, the measured charge carrier mobility was comparable to those found in devices based on thin films. This result suggests that electrical properties of the P3HT were intact after the electrospinning process. With greater control, scanned electrospinning might become useful for nanoelectronic applications; it is already useful as a low-energy method for making electronic test structures in academic laboratories.

10.2 Green Approaches to Nanofabrication | 265

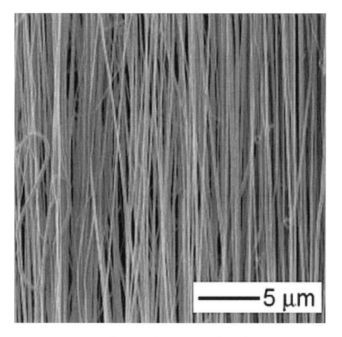

Figure 10.23 Uniaxially oriented BaTiO$_3$ nanofibers fabricated by electrospinning over an insulating gap between a metal electrodes. Reprinted with permission from [98]. Copyright 2006, Elsevier.

10.2.5.2 Uniaxial Electrospinning

Self-assembly of nanofibers as they are extruded from the spinneret is another route that is being investigated for control over the electrospinning process. For example, Xia and coworkers are developing a model for the ways in which electric fields influence the orientation of nanofibers on a substrate [94–97]. They have found that the presence of insulating shapes patterned on an otherwise conductive surface direct the formation of parallel nanofibers. Figure 10.23 shows uniaxial collections of barium titanate nanofibers.

10.2.5.3 Core/Shell and Hollow Nanofibers

Multicomponent nanofibers in a core/shell geometry have also been demonstrated for several combinations of materials (Figures 10.24 and 10.25) [100, 101]. This process is performed with a compound spinneret, in which a smaller needle is placed inside a larger needle. Electrospinning from this arrangement causes the jet to have a core/shell of two different polymers. For example, conjugated polymers have been placed in the interior of fibers made of poly(vinylpyrrolidone) (PVP) [102]. Two conjugated polymers together may also serve as the core and shell at the same time [101]. Such heterostructures could find use in optoelectronic applications. The inner spinneret can also contain a sacrificial material, such as mineral oil, which upon rinsing may be removed to form hollow nanofibers [102].

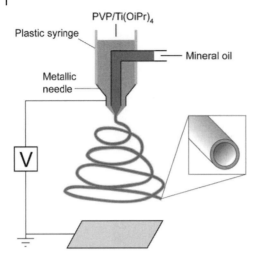

Figure 10.24 Schematic illustration of the compound spinneret used to generate core-shell multicomponent polymer nanofibers. Reprinted with permission from [99]. Copyright 2005, Royal Society of Chemistry.

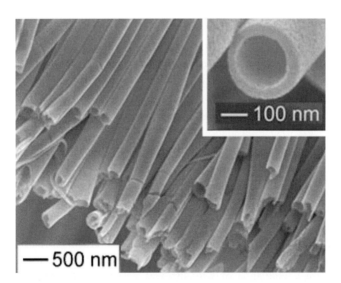

Figure 10.25 Hollow nanofibers of aligned anatase TiO_2 collected across an insulating gap on a conductive substrate. The cross-sections of the fibers were exposed using a razor blade. Reprinted with permission from [99]. Copyright 2004, American Chemical Society.

10.2.5.4 Outlook

The electrospinning phenomenon is remarkable in its applicability to most materials that can be processed from melt or solution. The basic function of electrospinning is to deposit randomly oriented mats of fibers. The challenge is to deposit single nanowires or oriented collections of nanofibers. Some combination of stage translation and electrostatic self-assembly might enable the eventual integration of electrospun fibers into nanoscale devices.

The energetic costs associated with electrospinning are low. A typical high-voltage DC power supply drains about 10 W in normal operation. Scanned electrospinning also requires an electric motor to translate the stage, which typically draw <100 W. The in-flight, uniaxial assembly of nanofibers as described by Xia and coworkers requires conductive and insulating zones patterned on a substrate [95].

10.2.6
Self-Assembly

Self-assembly (SA) is a ubiquitous strategy in nanofabrication [22, 103]. It is the spontaneous organization of one or more components using covalent or noncovalent interactions into ordered superstructures. There are elements of SA in the chemical synthesis of nanocrystals and their organization into superlattices, the formation of Langmuir–Blodget films and of course the tertiary structure of proteins. An entire organism is an example of dynamic SA [22]. The operation of a cell relies on precisely evolved nanostructures, which are much more complex than any manmade system. Thus, biology provides proof-of-principle of the potential utility and efficiency of SA. In the context of green nanofabrication, SA is potentially an elegant strategy, if the assembly step occurs under sufficiently mild conditions.

Two types of SA are important to fabrication: nontemplated and templated [1]. Nontemplated self-assembly is the purest form. It relies only on the interactions of the individual components with each other, without external forces and spatial constraints. Materials that are assembled this way include nanoparticles, SAMs and structures that assemble from liquid crystals and block copolymers. The most appealing aspect of nontemplated self-assembly is that, ideally, the only energetic inputs are those of the synthesis of the individual components. While nontemplated SA is presently only able to generate a limited set of structures, it has significant potential. Templated SA relies on external forces and spatial constraints of a rationally designed master structure, which is usually derived from a top-down process. An example of a combination of templated and nontemplated strategies is the use of patterned substrate to increase the length of ordering in a block copolymer film [104].

The ubiquity of SA in nanofabrication precludes a thorough treatment of the subject here. We direct the reader to the book by Ozin and Arsenault for an excellent review of SA and other chemical approaches to nanotechnology [105]. The rest of this section will focus on two examples of SA: stacked films of semiconductor nanocrystal superlattices and block copolymer lithography. These two examples also have potential applications in the conversion and storage of energy.

10.2.6.1 Hierarchical Assembly of Nanocrystals

One of the simplest routes to manufacturing nanomaterials on a large scale is solution-phase, bulk chemical synthesis. The energetic costs associated with these "wet" processes are the same as those associated with ordinary chemical synthesis: chemical precursors, solvents, temperature control and ventilation. A large number of syntheses are available for metallic and semiconducting nanocrystals and carbon nanotubes [106]. These processes are successful in bulk applications where they are used as ordinary chemical reagents. For academic purposes, randomly oriented, individual nanostructures are selected from a collection and characterized (electronically or optically). The extent to which structures from wet chemical synthesis can be integrated into devices depends on the ease with which single nanostructures can be deposited on predefined locations with a high density. Chemical synthesis produces nanostructures in the solution phase, so they can be deposited on any substrate they will stick to, including low-cost substrates like plastic and metallic foil, by a variety of bench-top techniques (spin-coating, dip-coating, printing, etc.) and in a variety of thicknesses and interparticle spacings.

Wet chemical synthesis can produce macroscopic quantities of semiconductor quantum dots with narrow polydispersities. The average size of the dots can be selected by slightly altering the conditions of the synthesis. For instance, many groups [97, 107–113] now routinely synthesize monodisperse ($\sigma < 4\%$ rms) colloidal CdSe quantum dots (QDs) at temperatures less than 400 °C using wet-chemical procedures. The QDs have diameters ranging from 1.2 to 15.0 nm (the bulk exciton radius of CdSe is ~5 nm [114]), good electronic passivation and uniform shape [107, 115, 116]. These synthetic methods make CdSe QDs useful and highly developed building blocks for the fabrication of superlattices ordered over hundreds of microns [117–120], with controllable nearest neighbor distances [117, 121]. Furthermore, CdSe QDs have a finely tuned profile of absorption versus size with good coverage of the visible spectrum: for $d = 1.2$ nm to 15 nm, the band gap (E_g) ranges from 2.9 eV (~425 nm) to 1.75 eV (~710 nm) [107, 117, 121–125].

Our group produced three different sizes of CdSe QDs in order to construct junctions of three-dimensional arrays containing CdSe QDs of multiple sizes (Figure 10.26). The arrays of QDs are hierarchically self-assembled: first, atoms into the dots themselves and then the dots into polycrystalline superlattices. These superlattices are stacked using spin-coating, which introduces an additional level of complexity to the composite film. These multi-size arrays enabled us to infer some of the electronic consequences of quantum confinement that have been largely unexplored and unexploited in devices based on QDs: the importance of energetic alignment of the orbitals of the QDs and the work functions of the electrodes [127, 128] in determining the electrical characteristics of the junctions.

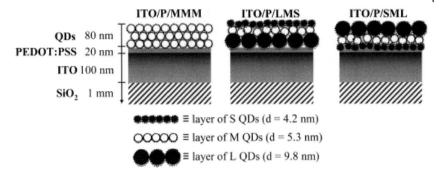

Figure 10.26 Schematic diagrams of selected films of QDs on ITO/PEDOT:PSS: ITO/P/MMM, ITO/P/LMS, and ITO/P/SML. "S," "M," and "L" indicate small (d = 4.2 nm), medium (d = 5.3 nm), and large (d = 9.8 nm) CdSe QDs, respectively, and "P" indicates a ~20 nm layer of polymer. Each monolayer of QDs in the diagram represents a multilayer (25–30 nm thick) in the actual film. Reprinted with permission from [126]. Copyright 2008, American Chemical Society.

10.2.6.2 Block Copolymers

Recent advances in polymer synthesis and an increased understanding of the physics of block copolymers have led to their incorporation in microelectronics applications [104]. Block copolymer chains contain two or more covalently linked, chemically distinct segments. These compounds form complex structures in the solid phase because of preferential association of chemically similar blocks minimizes the free energy of the system. Covalent linkage of the two blocks in a copolymer prevents long-range phase segregation of the two components, which occurs in blends of two or more homopolymers. Instead, block copolymers self-assemble into microdomains. These microdomains are nanometric features that can be used directly or as a template to direct further elaboration of the substrate [129].

One of the most useful morphologies of a block copolymer thin film has cylindrical microdomains of the minor block with their axes perpendicular to the substrate. Thin films of polystyrene-*block*-poly(ethylene oxide) (PS-*b*-PEO) spin-coated on a silicon wafer and annealed in a vapor of benzene can produce laterally ordered arrays of hexagonally packed cylinders with grain sizes of tens of square microns (Figure 10.27) [130]. This technique is known as solvent-induced ordering, which is a key to long-range self-assembly of block copolymer films [104].

Block copolymer lithography is the transfer of the thin film morphology of a block copolymer to another surface. This process is usually accomplished by selective etching of the minor component to reveal a nanoporous array. The nanoporous film then may be used directly, or serve as a pattern transfer element (as a mask or a template). In some sense, block copolymer templates are analogous to anodic aluminum oxide templates, though block copolymer films can have a higher density of features [104].

Figure 10.27 Scanning force micrograph of a highly ordered array of hexagonally packed cylindrical microdomains in a thin film of PS-b-PEO. Reprinted with permission from [130]. Copyright 2004, Wiley-VCH Verlag GmbH & Co. KGaA.

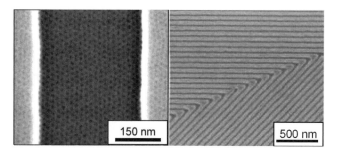

Figure 10.28 Two types of epitaxial self-assembly of block copolymer films. (Left) A trench etched in a SiO$_2$ substrate directs the orientation of the grains in a film of poly(α-methylstyrene)-*block*-poly(4-hydroxystyrene) film. Reprinted with permission from [131]. Copyright 2008, American Chemical Society. (Right) A chemically patterned surface directs the orientation of domains in a ternary blend of PS-*b*-PMMA, PS, and PMMA. The chemically patterned lines are commensurate with the natural lamellar period of the block copolymer film. Homopolymers are recruited to the points of discontinuity (the bends) in the lines. Reprinted with permission from [133]. Copyright 2005, American Association for the Advancement of Science.

For many applications, it is not sufficient that a block copolymer film display simple periodicity. Electronic devices require the ability to produce arbitrary patterns. In various techniques of epitaxial deposition, a lithographically patterned surface can direct the formation of self-assembled structures. This underlying pattern can be of a topographic [131] or chemical nature [132]. Topographic patterns can affect the long-range periodicity and the morphology of microdomains perpendicular to the substrate (nanodots) within trenches. Chemical patterning can direct the formation of patterns of block copolymer films in which the microdomains lie parallel to a chemically patterned surface. Figure 10.28 shows examples of these two types of hierarchical ordering. Chemical patterning of surfaces with a periodicity close to that of the bulk lamellar spacing of a block copolymer film can cause the lamellae to follow angular discontinuities. While the lamellae can become strained and discontinuous when coerced to follow angles on a patterned substrate, added homopolymers can be recruited to the sites of discontinuities and alleviate the strain. In this way, Nealey and coworkers were able to overlay a ternary blend of the diblock copolymer polystyrene-*block*-poly(methylmethacrylate) (PS-*b*-PMMA) and homopolymers of PS and PMMA atop a chemically patterned surface in which the lamellae were physically continuous over bends in the substrate.

10.2.6.3 Outlook

Self-assembly is progressing as an approach to nanofabrication, though in its nontemplated form it does not yet possess the capability of producing structures as complex as those available to top-down techniques [1]. Templated systems mix "top-down" with "bottom-up" methods of nanofabrication. Epitaxial self-assembly makes it possible to integrate block copolymer structures with conventional lithographic patterns. We are optimistic that the role of self-assembly for nanofabrication will increase in the future, especially because of the potential for fabrication in three dimensions, the opportunity for reversible [134] self-assembly and the possibility that self-assembled structures can self-repair or self-replicate [135].

10.3
Future Directions: Toward "Zero-Cost" Fabrication

This section describes three examples of extremely low-cost fabrication: the use of Scotch tape to fabricate islands of graphene monolayers [136] paper as an extremely low-cost substrate for microfabrication [137, 138] and Shrinky-Dinks as a photoresist substitute for soft lithography mastering [139, 140]. While not necessarily "nano", these techniques are salient examples of simplicity in fabrication and are perfectly suitable for low-cost (and low-energy) academic research. These methods demonstrate that even the simplest of techniques can yield potentially important advances: the Scotch-tape method led to the experimental discovery of graphene [136]; patterning paper as a substrate for microfabrication has potential for manufacturing portable bioassays for developing countries [138]; and printing on biaxially prestressed polymeric substrates could lead to an inexpensive, non-photolithographic method of miniaturizing printed features [139, 140].

10.3.1
Scotch-Tape Method for the Preparation of Graphene Films

Graphene is a two-dimensional monolayer of carbon atoms packed into an infinite array of fused benzene rings. Fullerenes, carbon nanotubes, and graphite can be regarded as zero-, one- and three-dimensional arrangements of graphene sheets

[141]. For half a decade, graphene was considered too unstable to exist, and was only studied theoretically. In 2004, Novoselov et al. discovered the first atomically thin carbon films using an astonishing combination of simple tools: scotch tape and an optical microscope [136]. Scotch tape was used to cleave individual graphene planes off of bulk graphite, and the characteristic optical interference patterns of graphene layers on an Si/SiO_2 wafer of the appropriate oxide thickness allowed the flakes of atomic thickness and microscopic width to be visualized (scanning-probe microscopy and SEM techniques were unable to find the specimens). Since that time, there has been an explosion of reports in the literature on experimental studies of graphene that have verified theoretical models and demonstrated the potential utility of the material for applications in nanoscale electronics.

10.3.2
Patterned Paper as a Low-Cost Substrate

Our group has published a method for patterning paper to create well-defined, millimeter-sized channels, comprising hydrophilic paper bounded by hydrophobic polymer (see Figure 10.29, left panel) for details of the fabrication process) [137, 138]. This type of patterned paper could be used for low-cost, portable and technically simple multiplexed bioassays (Figure 10.29, right panel). We made assay devices based on paper by patterning photoresist onto chromatography paper to form defined areas of hydrophilic paper separated by hydrophobic lines or walls; these patterns provide spatial control of biological fluids and enable fluid transport owing to capillary action in the millimeter-sized channels. In a fully developed technology, patterned photoresist would be replaced by an appropriate printing technology.

10.3.3
Shrinky-Dinks for Soft Lithography

While the various forms of soft lithography have been adopted by a large number of research laboratories, most incarnations of the technique require access to modern photolithographic equipment (mask aligners, etc.) and materials (photoresists, wafers, etc.). Khine and coworkers have recently developed tools for soft lithography based on the childrens' toys "Shrinky-Dinks." Shrinky-Dinks are biaxially prestressed polystyrene thermoplastic sheets. When heated, the sheets contract laterally by approximately 63% with a concomitant increase in height of features printed on the substrate by 500% [140]. Recessed features also become deeper. These characteristics were used in the mastering [140] and direct fabrication [139] of microfluidic devices.

In a first demonstration, a laser printer was used to pattern lines on sheets of prestressed polystyrene [140]. Upon heating, the substrate contracted while the printed features of ink were raised. PDMS was then poured over the polystyrene/ink master and cured (Figure 10.30a). The PDMS stamp bonded easily to a glass

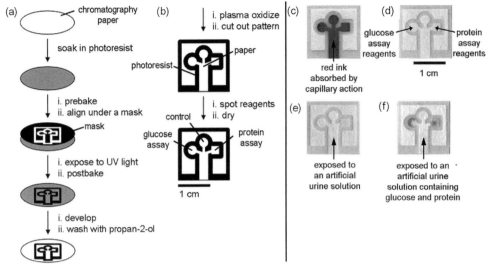

Figure 10.29 (Left) Diagram depicting the method for patterning paper into millimeter-sized channels: (a) Photolithography was used to pattern SU-8 photoresist embedded into paper, (b) the patterned paper was modified for bioassays. (Right) Chromatography paper patterned with photoresist. Darker lines are cured photoresist, whereas lighter areas are unexposed paper. (c) Patterned paper after absorbing Waterman red ink (5 μL) by capillary action. The central channel absorbs the sample by capillary action and the pattern directs the sample into three separate test areas. (d) Complete assay after spotting the reagents. The square region on the right is the protein test and the circular region on the left is the glucose test. The circular region on the top was used as a control well. (e) Negative control for glucose (left) and protein (right) by using an artificial urine solution (5 μL). (f) Positive assay for glucose (left) and protein (right) by using a solution that contained 550 mM glucose and 75 μM BSA in an artificial urine solution (5 μL). The control well was spotted with potassium iodide solution, but not with enzyme solution. Reprinted with permission from [137]. Copyright 2007, Wiley-VCH Verlag GmbH & Co. KGaA.

slide to form complex patterns of microchannels (with a maximum height of 80 μm) through which cells could flow. Iterative printing could increase the heights of features. Multiple heights of features—which are difficult to obtain with conventional photolithographic mastering—could be obtained by printing over selected features of the master. The minimum linewidth was 63 μm, but the authors suggest that a higher-resolution printer could achieve a smaller value.

In a second demonstration, the polystyrene itself was scribed manually with lines using a hypodermic needle [139]. Several complementary designs were stacked and shrunk together to form an interconnected, three-dimensional microfluidic device. In a manufacturing setting, the manual scribing step would be replaced by an automated indentation tool. A photograph of a test structures appears in Figure 10.30b. Microfluidic devices based on PDMS are amenable to

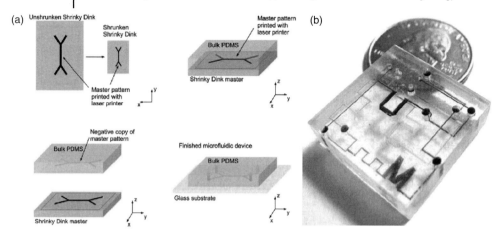

Figure 10.30 (a) Schematic illustration of the use of printed Shrinky-Dinks as masters for soft lithography. (b) Photograph of a three-dimensional, all-polystyrene microfluidic device, which contains three layers (separate networks of channels). Reprinted with permission from [139] and [140]. Copyright 2008, Royal Society of Chemistry.

rapid prototyping in a research setting, but PDMS is not favored industrially because of issues of swelling and compression [139]. A benefit of the Shrinky-Dink method is that it is amenable to rapid prototyping, while polystyrene is a rigid, industrially viable material.

10.4
Conclusions

Nanostructured materials may contribute to the efficient storage and production of energy, but only if they are produced in environmentally responsible ways. We have described several of the energetic aspects of conventional and unconventional approaches to nanofabrication and several techniques that satisfy many of the criteria for energy conservation. These techniques possess significant potential for test structures in academic laboratories and some offer manufacturing capabilities.

We have two goals in pursuing this topic:

The first, immediate goal is to increase awareness among researchers of the design constraints for green nanofabrication. Major universities are already taking the lead in reduction of greenhouse gas emissions. In 2008, an anonymous university taskforce reported that the university emits $282\,000\,\text{t}\,\text{year}^{-1}$ of carbon dioxide equivalents. The electricity bill of the nanofabrication facility, alone, accounts for about 2% of the energy budget of the entire university. We hope that, by characterizing some of the successes of unconventional approaches to nanofabrication, we have shown that simple tools and materials can often yield functionally equivalent structures to those made by conventional techniques. In many cases, combining the re-use of an existing tool with the exploitation of nanoscale information in unconventional sources can yield totally unique structures. We expect that green nanofabrication could find immediate use for the fabrication of test structures in a research setting, as well as forestall regulations on nanoscience based on the costs of performing experiments in this often energy-intensive area of research.

The second, long-term goal is to promote the development of green approaches to nanofabrication that have some potential for manufacturing. It is understood that integrating a new approach into an established field, such as microelectronics, is hindered by the risks of tampering with an existing infrastructure. We expect that green approaches to semiconductor manufacturing, such as the use of supercritical CO_2 and block copolymer lithography, will continue to decrease the energy intensity of the industry. Green techniques, however, have the characteristic that their products are perhaps the most relevant to emerging fields – areas outside of microelectronics that do not yet posses an infrastructure for manufacturing.

Acknowledgments

This research was supported by the United States Department of Energy under DE-FG02-00ER45852 and the National Science Foundation under CHE-0518055. D.J.L. acknowledges a graduate fellowship from the American Chemical Society, Division of Organic Chemistry, sponsored by Novartis. E.A.W. thanks the Petroleum Research Fund of the American Chemical Society fellowship (PRF no. 43083-AEF). The authors thank Noah Clay and Jiangdong Deng (Harvard) for helpful discussions.

References

1 Gates, B.D., Xu, Q.B., Stewart, M., Ryan, D., Willson, C.G., and Whitesides, G.M. (2005) *Chem. Rev.*, **105**, 1171.

2 Arakawa, H., Aresta, M., Armor, J.N., Barteau, M.A., Beckman, E.J., Bell, A.T., Bercaw, J.E., Creutz, C., Dinjus, E., Dixon, D.A., Domen, K., DuBois, D.L., Eckert, J., Fujita, E., Gibson, D.H., Goddard, W.A., Goodman, D.W., Keller, J., Kubas, G.J., Kung, H.H., Lyons, J.E., Manzer, L.E., Marks, T.J., Morokuma, K., Nicholas, K.M., Periana, R., Que, L., Rostrup-Nielson, J., Sachtler, W.M.H., Schmidt, L.D., Sen, A., Somorjai, G.A., Stair, P.C., Stults, B.R., and Tumas, W. (2001) *Chem. Rev.*, **101**, 953.

3 Yang, S.Y., Ryu, I., Kim, H.Y., Kim, J.K., Jang, S.K., and Russell, T.P. (2006) *Adv. Mater.*, **18**, 709.

4 Garcia, E.J., Hart, A.J., Wardle, B.L., and Slocum, A.H. (2007) *Adv. Mater.*, **19**, 2151.

5 Whitesides, G.M., and Crabtree, G.W. (2007) *Science*, **315**, 796.

6 O'Neil, A., and Watkins, J.J. (2004) *Green Chem.*, **6**, 363.

7 National Association of Manufacturers (2008) Facts About U.S. Manufacturing, http://www.nam.org/s_nam/bin.asp?CID=202325&DID=233605&DOC=FILE.PDF (accessed 13 July 2008).

8 Williams, E.D., Ayres, R.U., and Heller, M. (2004) *Environ. Sci. Technol.*, **38**, 1916.

9 Williams, E.D., Ayres, R.U., and Heller, M. (2002) *Environ. Sci. Technol.*, **36**, 5504.
10 Williams, E. (2004) *Environ. Sci. Technol.*, **38**, 6166.
11 Shadman, F., and McManus, T.J. (2004) *Environ. Sci. Technol.*, **38**, 1915.
12 Hu, S.C., Wu, J.S., Chan, D.Y.L., Hsu, R.T.C., and Lee, J.C.C. (2008) *Energy Build.*, **40**, 1765.
13 Hu, S.C., and Chuah, Y.K. (2003) *Energy*, **28**, 895.
14 Ginger, D.S., Zhang, H., and Mirkin, C.A. (2004) *Angew. Chem. Int. Ed.*, **43**, 30.
15 Xu, Q., Rioux, R.M., and Whitesides, G.M. (2007) *ACS Nano*, **1**, 215.
16 Xu, Q.B., Gates, B.D., and Whitesides, G.M. (2004) *J. Am. Chem. Soc.*, **126**, 1332.
17 Dickey, M.D., Weiss, E.A., Smythe, E.J., Chiechi, R.C., Capasso, F., and Whitesides, G.M. (2008) *ACS Nano*, **2**, 800.
18 Aizenberg, J., Black, A.J., and Whitesides, G.M. (1998) *Nature*, **394**, 868.
19 Lipomi, D.J., Chiechi, R.C., Dickey, M.D., Whitesides, G.M., and Nano, L. (2008) *Nano Lett.*, **8**, 2100.
20 Xu, Q.B., Mayers, B.T., Lahav, M., Vezenov, D.V., and Whitesides, G.M. (2005) *J. Am. Chem. Soc.*, **127**, 854.
21 Zach, M.P., Ng, K.H., and Penner, R.M. (2000) *Science*, **290**, 2120.
22 Whitesides, G.M., and Grzybowski, B. (2002) *Science*, **295**, 2418.
23 Gur, I., Fromer, N.A., Geier, M.L., and Alivisatos, A.P. (2005) *Science*, **310**, 462.
24 Menard, E., Meitl, M.A., Sun, Y.G., Park, J.U., Shir, D.J.L., Nam, Y.S., Jeon, S., and Rogers, J.A. (2007) *Chem. Rev.*, **107**, 1117.
25 Goh, C., Coakley, K.M., and McGehee, M.D. (2005) *Nano Lett.*, **5**, 1545.
26 Royal Society of Chemistry (2008) *Green Chemistry*'s Impact Factor Continues to Rise. From http://www.rsc.org/Publishing/Journals/gc/News/2008/ImpactFactor.asp (accessed 13 July 2008).
27 Linder, V., Gates, B.D., Ryan, D., Parviz, B.A., and Whitesides, G.M. (2005) *Small*, **1**, 730.
28 Gates, B.D., Xu, Q.B., Love, J.C., Wolfe, D.B., and Whitesides, G.M. (2004) *Annu. Rev. Mater. Res.*, **34**, 339.
29 Johnson, R.C. (2007) *EE Times*, http://www.eetimes.com/showArticle.jhtml?articleID=199203911 (accessed 8 November 2008).
30 Bratton, D., Yang, D., Dai, J.Y., and Ober, C.K. (2006) *Polym. Adv. Technol.*, **17**, 94.
31 Willson, C.G., and Roman, B.J. (2008) *ACS Nano*, **2**, 1323.
32 Rolland, J.P., Maynor, B.W., Euliss, L.E., Exner, A.E., Denison, G.M., and DeSimone, J.M. (2005) *J. Am. Chem. Soc.*, **127**, 10096.
33 Chou, S.Y., Krauss, P.R., and Renstrom, P.J. (1996) *Science*, **272**, 85.
34 Chou, S.Y., Krauss, P.R., and Renstrom, P.J. (1996) *J. Vac. Sci. Technol., B*, **14**, 4129.
35 Chou, S.Y., Krauss, P.R., and Renstrom, P.J. (1995) *Appl. Phys. Lett.*, **67**, 3114.
36 Stewart, M.D., Johnson, S.C., Sreenivasan, S.V., Resnick, D.J., and Willson, C.G. (2005) *J. Microlith. Microfab. Microsys.*, **4**, 011002.
37 Bailey, T.C., Johnson, S.C., Sreenivasan, S.V., Ekerdt, J.G., Willson, C.G., and Resnick, D.J. (2002) *J. Photopolym. Sci. Technol.*, **15**, 481.
38 Xia, Y.N., McClelland, J.J., Gupta, R., Qin, D., Zhao, X.M., Sohn, L.L., Celotta, R.J., and Whitesides, G.M. (1997) *Adv. Mater.*, **9**, 147.
39 Xia, Y.N., Kim, E., Zhao, X.M., Rogers, J.A., Prentiss, M., and Whitesides, G.M. (1996) *Science*, **273**, 347.
40 Kim, E., Xia, Y.N., Zhao, X.M., and Whitesides, G.M. (1997) *Adv. Mater.*, **9**, 651.
41 Palmieri, F., Adams, J., Long, B., Heath, W., Tsiartas, P., and Willson, C.G. (2007) *ACS Nano*, **1**, 307.
42 Heath, W.H., Palmieri, F., Adams, J.R., Long, B.K., Chute, J., Holcombe, T.W., Zieren, S., Truitt, M.J., White, J.L., and Willson, C.G. (2008) *Macromolecules*, **41**, 719.
43 Xia, Y.N., and Whitesides, G.M. (1998) *Annu. Rev. Mater. Sci.*, **28**, 153.
44 Beh, W.S., Kim, I.T., Qin, D., Xia, Y.N., and Whitesides, G.M. (1999) *Adv. Mater.*, **11**, 1038.

45 Coakley, K.M., and McGehee, M.D. (2004) *Chem. Mater.*, **16**, 4533.
46 Yang, F., and Forrest, S.R. (2008) *ACS Nano*, **2**, 1022.
47 Gur, I., Fromer, N.A., and Alivisatos, A.P. (2006) *J. Phys. Chem. B*, **110**, 25543.
48 Kelly, J.Y., and DeSimone, J.M. (2008) *J. Am. Chem. Soc.*, **130**, 5438.
49 Petros, R.A., Ropp, P.A., and DeSimone, J.M. (2008) *J. Am. Chem. Soc.*, **130**, 5008.
50 Hampton, M.J., Williams, S.S., Zhou, Z., Nunes, J., Ko, D.H., Templeton, J.L., Samulski, E.T., and DeSimone, J.M. (2008) *Adv. Mater.*, **20**, 2667.
51 Biebuyck, H.A., Larsen, N.B., Delamarche, E., and Michel, B. (1997) *IBM J. Res. Dev.*, **41**, 159.
52 Xia, Y.N., and Whitesides, G.M. (1997) *Langmuir*, **13**, 2059.
53 Jacobs, H.O., and Whitesides, G.M. (2001) *Science*, **291**, 1763.
54 Salaita, K., Wang, Y.H., and Mirkin, C.A. (2007) *Nat. Nanotechnol.*, **2**, 145.
55 Piner, R.D., Zhu, J., Xu, F., Hong, S.H., and Mirkin, C.A. (1999) *Science*, **283**, 661.
56 Hong, S.H., Zhu, J., and Mirkin, C.A. (1999) *Science*, **286**, 523.
57 Salaita, K., Wang, Y.H., Fragala, J., Vega, R.A., Liu, C., and Mirkin, C.A. (2006) *Angew. Chem. Int. Ed.*, **45**, 7220.
58 Salaita, K.S., Lee, S.W., Ginger, D.S., and Mirkin, C.A. (2006) *Nano Lett.*, **6**, 2493.
59 Rozhok, S., Piner, R., and Mirkin, C.A. (2003) *J. Phys. Chem. B*, **107**, 751.
60 Xu, Q.B., Bao, J.M., Rioux, R.M., Perez-Castillejos, R., Capasso, F., and Whitesides, G.M. (2007) *Nano Lett.*, **7**, 2800.
61 Wiley, B.J., Lipomi, D.J., Bao, J., Capasso, F., Whitesides, G.M., and (2008) *Nano Lett.*, **8**, 3023–3028.
62 Lipomi, D.J., Chiechi, R.C., Reus, W.F., and Whitesides, G.M. (2008) *Adv. Funct. Mater.*, **18**, 3469.
63 Xu, Q.B., Bao, J.M., Capasso, F., and Whitesides, G.M. (2006) *Angew. Chem. Int. Ed.*, **45**, 3631.
64 Xu, Q.B., Perez-Castillejos, R., Li, Z.F., and Whitesides, G.M. (2006) *Nano Lett.*, **6**, 2163.
65 Xu, Q.B., Rioux, R.M., Dickey, M.D., and Whitesides, G.M. (2008) *Acc. Chem. Res.*, **41**, 1566–1577.
66 Ditlbacher, H., Hohenau, A., Wagner, D., Kreibig, U., Rogers, M., Hofer, F., Aussenegg, F.R., and Krenn, J.R. (2005) *Phys. Rev. Lett.*, **95**, 257403.
67 Kan, C.X., Zhu, X.G., and Wang, G.H. (2006) *J. Phys. Chem. B*, **110**, 4651.
68 Sirringhaus, H., Tessler, N., and Friend, R.H. (1998) *Science*, **280**, 1741.
69 Yu, G., Gao, J., Hummelen, J.C., Wudl, F., and Heeger, A.J. (1995) *Science*, **270**, 1789.
70 Sirringhaus, H., Brown, P.J., Friend, R.H., Nielsen, M.M., Bechgaard, K., Langeveld-Voss, B.M.W., Spiering, A.J.H., Janssen, R.A.J., Meijer, E.W., Herwig, P., and de Leeuw, D.M. (1999) *Nature*, **401**, 685.
71 Chiang, C.K., Fincher, C.R., Park, Y.W., Heeger, A.J., Shirakawa, H., Louis, E.J., Gau, S.C., and Macdiarmid, A.G. (1977) *Phys. Rev. Lett.*, **39**, 1098.
72 Wanekaya, A.K., Chen, W., Myung, N.V., and Mulchandani, A. (2006) *Electroanalysis*, **18**, 533.
73 Cui, Y., Wei, Q.Q., Park, H.K., and Lieber, C.M. (2001) *Science*, **293**, 1289.
74 Liu, H.Q., Reccius, C.H., and Craighead, H.G. (2005) *Appl. Phys. Lett.*, **87**, 253106.
75 Duvail, J.L., Retho, P., Fernandez, V., Louarn, G., Molinie, P., and Chauvet, O. (2004) *J. Phys. Chem. B*, **108**, 18552.
76 Smela, E. (2003) *Adv. Mater.*, **15**, 481.
77 Samitsu, S., Shimomura, T., Ito, K., Fujimori, M., Heike, S., and Hashizume, T. (2005) *Appl. Phys. Lett.*, **86**, 233103.
78 Liu, H.Q., Kameoka, J., Czaplewski, D.A., and Craighead, H.G. (2004) *Nano Lett.*, **4**, 671.
79 McQuade, D.T., Pullen, A.E., and Swager, T.M. (2000) *Chem. Rev.*, **100**, 2537.
80 Ramanathan, K., Bangar, M.A., Yun, M., Chen, W., Myung, N.V., and Mulchandani, A. (2005) *J. Am. Chem. Soc.*, **127**, 496.
81 Aoki, Y., Huang, J., and Kunitake, T. (2006) *J. Mater. Chem.*, **16**, 292.

82 Wan, Q., Song, Z.T., Feng, S.L., and Wang, T.H. (2004) *Appl. Phys. Lett.*, **85**, 4759.
83 Limmer, S.J., Cruz, S.V., and Cao, G.Z. (2004) *Appl. Phys. A: Mater. Sci. Process.*, **79**, 421.
84 Wan, Q., Feng, P., and Wang, T.H. (2006) *Appl. Phys. Lett.*, **89**, 123102/1.
85 Cheng, Z.-X., Dong, X.-B., Pan, Q.-Y., Dong, J.-C., and Zhang, X.-W. (2006) *Mater. Lett.*, **60**, 3137.
86 Xue, X.Y., Chen, Y.J., Liu, Y.G., Shi, S.L., Wang, Y.G., and Wang, T.H. (2006) *Appl. Phys. Lett.*, **88**, 201907/1.
87 Jang, H.S., Kim, D.-H., Lee, H.-R., and Lee, S.-Y. (2005) *Mater. Lett.*, **59**, 1526.
88 Yu, D., Wang, D., Yu, W., and Qian, Y. (2003) *Mater. Lett.*, **58**, 84.
89 Orlandi, M.O., Aguiar, R., Lanfredi, A.J.C., Longo, E., Varela, J.A., and Leite, E.R. (2004) *Appl. Phys. A: Mater. Sci. Process.*, **80**, 23.
90 Dickey, M.D., Russell, J.K., Lipomi, D.J., and Whitesides, G.M. (2009) Manuscript in preparation.
91 Reneker, D.H., and Chun, I. (1996) *Nanotechnology*, **7**, 216.
92 Greiner, A., and Wendorff, J.H. (2007) *Angew. Chem. Int. Ed.*, **46**, 5670.
93 Kameoka, J., Czaplewski, D., Liu, H.Q., and Craighead, H.G. (2004) *J. Mater. Chem.*, **14**, 1503.
94 Ostermann, R., Li, D., Yin, Y.D., McCann, J.T., and Xia, Y.N. (2006) *Nano Lett.*, **6**, 1297.
95 Li, D., Ouyang, G., McCann, J.T., and Xia, Y.N. (2005) *Nano Lett.*, **5**, 913.
96 Li, D., Wang, Y.L., and Xia, Y.N. (2003) *Nano Lett.*, **3**, 1167.
97 Li, D., and Xia, Y.N. (2003) *Nano Lett.*, **3**, 555.
98 McCann, J.T., Chen, J.I.L., Li, D., Ye, Z.G., and Xia, Y.A. (2006) *Chem. Phys. Lett.*, **424**, 162.
99 Li, D., and Xia, Y.N. (2004) *Nano Lett.*, **4**, 933.
100 Babel, A., Li, D., Xia, Y.N., and Jenekhe, S.A. (2005) *Macromolecules*, **38**, 4705.
101 Li, D., Babel, A., Jenekhe, S.A., and Xia, Y.N. (2004) *Adv. Mater.*, **16**, 2062.
102 McCann, J.T., Li, D., and Xia, Y.N. (2005) *J. Mater. Chem.*, **15**, 735.
103 Love, J.C., Estroff, L.A., Kriebel, J.K., Nuzzo, R.G., and Whitesides, G.M. (2005) *Chem. Rev.*, **105**, 1103.
104 Hawker, C.J., and Russell, T.P. (2005) *MRS Bull.*, **30**, 952.
105 Ozin, G.A., and Arsenault, A.C. (2005) *Nanochemistry*, RSC Publishing, Cambridge, UK.
106 Burda, C., Chen, X.B., Narayanan, R., and El-Sayed, M.A. (2005) *Chem. Rev.*, **105**, 1025.
107 Murray, C.B., Norris, D.J., and Bawendi, M.G. (1993) *J. Am. Chem. Soc.*, **115**, 8706.
108 Snee, P.T., Chan, Y., Nocera, D.G., and Bawendi, M.G. (2005) *Adv. Mater.*, **17**, 1131.
109 Boatman, E., Lisensky, G.C., and Nordell, K.J. (2005) *J. Chem. Educ.*, **82**, 1697.
110 Peng, X., Wickham, J., and Alivisatos, A.P. (1998) *J. Am. Chem. Soc.*, **120**, 5343.
111 Shim, M., Wang, C., and Guyot-Sionnest, P. (2001) *J. Phys. Chem.*, **105**, 2369.
112 Talapin, D.V., Schevchenko, E.V., Kornowski, A., Gaponik, N., Haase, M., Rogach, A.L., and Weller, H. (2001) *Adv. Mater.*, **13**, 1868.
113 Munro, A.M., Plante, I.J.-L., Ng, M.S., and Ginger, D.S. (2007) *J. Phys. Chem. C*, **111**, 6220.
114 Kittel, C. (1996) *Introduction to Solid State Physics*, 7th edn, John Wiley & Sons, Inc., New York.
115 Bowen Katari, J.E., Colvin, V.L., and Alivisatos, A.P. (1994) *J. Phys. Chem.*, **98**, 4109.
116 Peng, Z.A., and Peng, X. (2001) *J. Am. Chem. Soc.*, **123**, 183.
117 Murray, C.B., Kagan, C.R., and Bawendi, M.G. (2000) *Annu. Rev. Mater. Sci.*, **30**, 545.
118 Sun, B., Marx, E., and Greenham, N.C. (2003) *Nano Lett.*, **3**, 961.
119 Huynh, W.U., Dittmer, J.J., Teclemariam, N., Milliron, D.J., and Alivisatos, A.P. (2003) *Phys. Rev. B*, **67**, 115326.
120 Yu, G., Gao, J., Hummelen, J.C., Wudi, F., and Heeger, A.J. (1995) *Science*, **270**, 1789.

121 Kagan, C.R., Murray, C.B., and Bawendi, M.G. (1996) *Phys. Rev. B*, **54**, 8633.
122 Kagan, C.R. (2003) *Proceedings of the NSF-Conicet Quilmes Nanoscience Workshop* (eds E. Calvo, J. Michl, and C. Kubiak), Quilmes, Provincia de Tucuman, Argentina.
123 Norris, D.J., Bawendi, M.G., and Brus, L.E. (1997) *Molecular Electronics* (eds J. Jortner and M.A. Ratner), Blackwell Science Ltd., Malden, MA, p. 281.
124 Greenham, N.C., Peng, X., and Alivisatos, A.P. (1996) *Phys. Rev. B*, **54**, 17628.
125 Landsberg, P.T., Nussbaumer, H., and Willeke, G. (1993) *J. Appl. Phys.*, **74**, 1451.
126 Weiss, E.A., Chiechi, R.C., Geyer, S.M., Porter, V.J., Bell, D.C., Bawendi, M.G., and Whitesides, G.M. (2008) *J. Am. Chem. Soc.*, **130**, 74.
127 Selmarten, D., Jones, M., Rumbles, G., Yu, P., Nedeljkovic, J., and Shaheen, S. (2005) *J. Phys. Chem. B*, **109**, 15927.
128 Ginger, D.S., and Greenham, N.C. (1999) *Phys. Rev. B*, **59**, 10622.
129 Gowrishankar, V., Miller, N., McGehee, M.D., Misner, M.J., Ryu, D.Y., Russell, T.P., Drockenmuller, E., and Hawker, C.J. (2006) *Thin Solid Films*, **513**, 289.
130 Kim, S.H., Misner, M.J., and Russell, T.P. (2004) *Adv. Mater.*, **16**, 2119.
131 Bosworth, J.K., Paik, M.Y., Ruiz, R., Schwartz, E.L., Huang, J.Q., Ko, A.W., Smilgies, D.M., Black, C.T., and Ober, C.K. (2008) *ACS Nano*, **2**, 1396–1402.
132 Stoykovich, M.P., and Nealey, P.F., (2006) *Mater. Today*, **9**, 20.
133 Stoykovich, M.P., Muller, M., Kim, S.O., Solak, H.H., Edwards, E.W., de Pablo, J.J., and Nealey, P.F. (2005) *Science*, **308**, 1442.
134 Anfinsen, C.B. (1973) *Science*, **181**, 223.
135 Grzybowski, B.A., Radkowski, M., Campbell, C.J., Lee, J.N., and Whitesides, G.M. (2004) *Appl. Phys. Lett.*, **84**, 1798.
136 Novoselov, K.S., Geim, A.K., Morozov, S.V., Jiang, D., Zhang, Y., Dubonos, S.V., Grigorieva, I.V., and Firsov, A.A. (2004) *Science*, **306**, 666.
137 Martinez, A.W., Phillips, S.T., Butte, M.J., and Whitesides, G.M. (2007) *Angew. Chem. Int. Ed.*, **46**, 1318.
138 Martinez, A.W., Phillips, S.T., Carrilho, E., Thomas, S.W., Sindi, H., and Whitesides, G.M. (2008) *Anal. Chem.*, **80**, 3699.
139 Chen, C.S., Breslauer, D.N., Luna, J.I., Grimes, A., Chin, W.C., Leeb, L.P., and Khine, M. (2008) *Lab Chip*, **8**, 622.
140 Grimes, A., Breslauer, D.N., Long, M., Pegan, J., Lee, L.P., and Khine, M. (2008) *Lab Chip*, **8**, 170.
141 Geim, A.K., and Novoselov, K.S. (2007) *Nat. Mater.*, **6**, 183.

11
Nanocatalysis for Fuel Production
Burtron H. Davis

11.1
Introduction

The concept of catalysis was coined by Berzelius in the early 1800s to describe a number of phenomena that had been practiced prior to his definition. Thus, catalysis was practiced rather widely before it was officially recognized and defined. This is also the case of nanocatalysis in the energy field. Because of the high cost of many catalytic materials, it was required that they be highly dispersed if they were to have an impact at the commercial scale. Nanocatalysis is usually considered to involve those particles in the 1–100 nm (10–1000 Å) scale. For cost effectiveness, many of the catalysts utilized for energy applications are dispersed to fall in the nanocatalyst range. In this review we limit the coverage to the use of nanocatalysis in the production of energy, and this primarily limits us to the production of transportation fuels. Astruc, for example, reviews the transition-metal nanoparticles in catalysis and includes many examples but does not include those for the production of transportation fuels [1]. This is the general situation with reviews of nanocatalysis. This in spite of the fact that the production and use of transportation fuels account for a major fraction of catalyst production and usage in the United States and the world today.

One of the features that greatly advanced understanding of the field of nanocatalysis for energy applications has not been the introduction of the catalyst but rather the great advances in the techniques utilized for their preparation and for the characterization of the structure of the finished and working catalysts. The *in situ* infrared studies by Eischens and coworkers at Texaco from the late 1950s to the early 1960s may be viewed as the beginning of *in situ* characterization of heterogeneous catalysis. This was followed by the pioneering work on high-vacuum characterization by such pioneers as Gabor Somorjai and Gehard Ertl, with Ertl receiving the Nobel Prize for his work in this area in 2007. Needless to say, this characterization work added much to our understanding the structure and working mechanisms of the application of nanocatalysis in the energy area.

Nanotechnology for the Energy Challenge. Edited by Javier Garcia-Martinez
© 2009 WILEY-VCH Verlag GmbH & Co. KGaA, Weinheim
ISBN: 978-3-527-32401-9

11.2
Petroleum Refining

For petroleum refining, there are four major processes that utilize catalysts that fall in the nano-scale: naphtha reforming, cracking, hydrocracking and hydrotreating. Because of the volume of the feedstock that is processed, product/catalyst separation is a major concern. Thus, in petroleum processing the nanocatalyst is almost always present on a high surface area support that may function only as a carrier for the nanocatalyst or may provide some catalytic functions. The introduction of naphtha reforming with the $Pt-Al_2O_3$ catalyst introduced nanocatalysis on a large commercial scale.

Another type of nanocatalysis in the energy area involved the hydrotreating catalysts that were required to provide a nearly sulfur-free feedstock for the naphtha reforming catalyst. However, it remained for the advent of sophisticated catalyst characterization techniques for the recognition of the nano-characteristics of these catalysts.

The introduction of catalytic cracking involved large catalyst particles; however, with the introduction of fluid bed catalytic cracking (FCC) the catalyst sphere approached the nano-scale and with today's catalyst the zeolite crystals imbedded in the silica–alumina amorphous material moved the catalyst into the nano-size range. The introduction of hydrocracking expanded the use of nanocatalysis since either nano-size metal particles were incorporated in the zeolite or the supported nickel– or cobalt–molybdena or tungsten oxide on an alumina support were used.

The scale of the use of catalysis in the energy industry is very large compared to other industries. Because of the size, processes dealing with the production of transportation fuels are ones that operate continuously. For this reason, the ability to separate the product from the catalyst is an important factor. To accomplish this for the early use of nanocatalysts, one used larger support materials; however, today processes can utilize relatively small catalyst particles and still effect separations. In spite of this, separation of product from the catalyst is still an important aspect when considering nanocatalysis.

11.3
Naphtha Reforming

One, if not the first, application of nanocatalysis in the production of energy was the use of the naphtha reforming catalyst to produce high octane gasoline. The catalyst developed by Vladimir Haensel and coworkers at UOP involved adding platinum to an acidic alumina support. The high cost of platinum required that maximum efficiency be obtained with the metal function and this was accomplished with a high dispersion of Pt. Today it is difficult to recognize how revolutionary the introduction of these low-loading noble metal catalysts was. During an interview trip to Texaco, one of the people related the difficulty the Texaco researchers had in preparing such a catalyst. To that time, most of the supported catalysts utilized a relatively high loading (10–50 wt%) of the active component on an inert support. The early Pt-containing naphtha reforming catalysts prepared at Texaco therefore had this high Pt loading and the resulting catalyst made a product that was predominantly gaseous and not the desired high-octane gasoline. This was because they were loading the alumina support with about 10 wt% of Pt as they would have done with earlier naphtha reforming catalysts such as chromia–alumina, rather than the lower loadings of about 0.5 wt% where the nanocatalyst particles would be present. When they finally reduced the Pt loading to the low level to produce nano-size Pt particles they were successful.

As indicated above, the need for this process became evident by 1950 and the introduction of the Pt-alumina reforming catalyst was a revolutionary advance. At that time catalyst characterization techniques, by today's standards, were rather meager. Even so, techniques such as hydrogen and carbon monoxide chemisorption allowed the workers to verify that the supported Pt particles were in the nano-size range, even if at the upper end of the scale. Within five years of its introduction, the process had saturated the refinery operations. One of the disadvantages of the Pt-alumina catalyst was that it had to be regenerated frequently; normally the reactor had to be taken off-line after about three months of operation to regenerate the catalyst. The regeneration of this catalyst by oxidizing the carbon, re-reduction of the Pt and re-adding the halide took many days.

The introduction of this naphtha reforming catalyst also led to significant advances in the understanding of catalysis. Mills and co-workers introduced the concept of bi-functional catalysis to explain how the catalyst operated [2] (Figure 11.1). In this process the C_6-ring is dehydrogenated rapidly at the metal site in a typical monofunctional mechanism. However, since the C_5-ring cannot be directly dehydrogenated to an aromatic structure, the dehydrogenation stops at the alkyl cyclopentene stage. The alkyl cyclopentene then desorbs from the metal site and is transported through the gas phase to an acid site where it is adsorbed. The acid-catalyzed ring expansion/contraction reaction converts the cyclopentene to the cyclohexene structure which desorbs and is transported in the gas phase to again adsorb on the metal site. Reaction on the metal site converts the cyclohexane to the aromatic, which then desorbs. The aromatic product is the desired end product since it has a very high octane number. This mechanism requires that the gas phase transport be over a short distance in order to obtain a high reaction rate, and this is only accomplished when there are a large number of metal particles dispersed upon the alumina surface. The requirements of the reaction pathway and the high cost of platinum mandated that the platinum be present in the nano-size range. Shortly after Mills and coworkers advanced the bifunctional mechanism, Weisz and Prater [3] provided a sophisticated mathematical description of bifunctional catalysis that remains a classic even today.

The regeneration of the catalyst involves burning off carbon in an oxygen containing atmosphere at a high temperature (greater than 500 °C). As a result of the oxidation–reduction steps the Pt undergoes growth and redispersion steps.

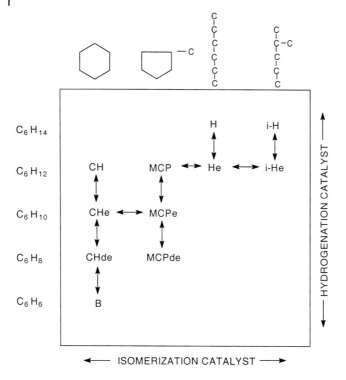

Figure 11.1 Reforming C_6 hydrocarbons with dual function catalyst [2].

Sintering/redispersion of Pt-alumina reforming catalysts is well documented and early work is summarized in reference [4]. During use, the Pt nano-size particle grows through sintering; there is still debate whether the dominate pathway is by coalescence of two Pt particles or the migration of single Pt atoms (Ostwald ripening). In any event, the aged catalyst has much larger Pt crystals than the fresh catalyst. When the catalyst is heated to oxidize the carbon Pt can undergo oxidation to form Pt oxide. Reduction of the Pt oxide then results in a number of much smaller Pt metal particles. Thus, during the oxidation/reduction cycle much smaller Pt particles are formed from the large Pt particle.

Goeke and Datye [5] utilized model oxide supports to study the sintering of nanoparticles at elevated temperatures. These model supports made it possible to observe the same area of the sample, before and after sintering. They found, for example, that Ostwald ripening is the dominant mechanism for Pd sintering on silica at 900 °C, and they found evidence for migration of particles as large as 50 nm for an alumina surface.

The supported catalysts were soon recognized to have important properties other than a high metallic surface area. It was recognized that the activity of the metal depends upon the size of the particle to an extent that exceeded the surface area [6–11]. van Hardeveld and Hartog [12] calculated the dependence of surface atoms of different nearest neighbors on the size and exposed surface of the supported metal crystallites. On the basis of their calculations, they indicated that the sorptive and catalytic behavior of metal catalysts was dependent upon the number of these different sites.

Chevron Oil introduced an improved naphtha reforming catalyst in the late 1960s that included rhenium addition to the Pt-alumina [13]. This catalyst aged at a much slower rate than the initial Pt-alumina catalyst and therefore increased the length of time between regeneration by a factor of two or more. The need to bring this catalyst on-line more slowly than the Pt-alumina catalyst was a minor disadvantage compared to the major advance of the much longer on-line time between regeneration periods. A surprising feature of the PtRe-alumina catalyst was that during or following reduction a small amount of sulfur was added to the catalyst; the role of the sulfur was to moderate the catalytic activity during the initial contact of the catalyst by the naphtha. The surprising feature of this is that sulfur was a severe poison for the Pt-alumina catalyst, and major efforts were made to prevent contact of the catalyst with sulfur.

For these bimetallic catalysts, a number of structures are possible. One that was identified early was for the NiCu alloy. This alloy catalyst was used to hydrogenate alkenes and it was found that the catalytic activity was inversely related to the number of d-band holes: as Cu was added to Ni, electrons were added to the d-band and the number of d-band holes decreased, and the conversion also decreased [14]. Thus, this became a classic example of the catalytic activity being directly related to the number of d-band holes. It was later shown that this was not the case; the alloy had two eutectics and these were responsible for the activity [15, 16].

The structure of the PtRe in the finished catalyst was difficult to define since the two metals do not make a uniform alloy at the compositions used in the commercial catalysts. Initially there was debate over whether the Re was present in the oxide form or the metallic form but there is currently general agreement that at least part of the Re is in the metallic form. The structure of the PtRe-alumina catalyst was reviewed recently [17]. Meitzner et al. [18] reported that essentially all of the Pt and Re were in the zero valence state after hydrogen reduction at 775 K. Pt and Re were coordinated in bimetallic clusters that were not characterized by a single interatomic distance. Rather than one of the metals segregating on the surface of the other metal particle, the data suggested that there was Pt–Re atomic variation for the clusters. These authors indicated that the Pt–Re bonds in the cluster were stronger than would be anticipated based on interatomic distances for the unlike pair of atoms.

Bazin et al. [19] found that the average environment of a monometallic Pt-alumina catalyst was six metal atoms at 2.75 Å and 0.7 oxygen atoms at 2.04 Å, whereas the Pt–Re bimetallic catalyst had five metal atoms around Pt with hardly detectable oxygen but that the Re was surrounded by six to eight metal atoms and two oxygen atoms. They considered their data to support a structure with a rhenium oxide interface between Pt and the alumina support. This raft-like structure of Re bonded to the support stabilizes the Pt nanocluster in smaller crystallites

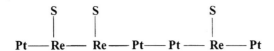

Figure 11.2 Biloen's model [17].

and also inhibits sintering of the Pt crystals. Both of these factors provide enhanced stability to the bimetallic Pt–Re catalyst compared to the Pt-only catalyst.

The EXAFS data [20] provided support for the general model proposed by Biloen et al. [21] and extended by Sachtler [22] where the mixed metal cluster consists of "chains" of Pt–Pt and Re–Re which are connected by Pt–Re bonds (Figure 11.2). The EXAFS data indicate that after sulfidation both Re and Pt atoms are coordinated to about two sulfur atoms. Following a second reduction the Re environment is unchanged but the Pt atoms are coordinated by about only one sulfur atom. Since most of the sulfur is lost gradually during use as a reforming catalyst, it appears that the role of sulfur is to break in the catalyst while maintaining the structure of the initial catalyst for a longer time than if sulfur is not present.

The Pt–Sn alloy catalyst was discovered at about the same time as the Pt–Re catalyst [23] but only became viable as a commercial catalyst when low-pressure reforming with continuous regeneration processing was practiced [24] There were more advances in developing the nature of this catalyst than the PtRe-alumina catalyst. Part of the reason was the utility of Mössbauer spectroscopy in identifying the states of tin in the finished catalyst.

An early preparation of the Pt–Sn catalyst utilized the $Pt_3Sn_6Cl_{20}^{2-}$ complex, a structure that falls at the small end of the nano-scale catalyst [23]. The characterization techniques then available would not allow us to define whether the complex remained after the activation in hydrogen at high temperatures; however, later workers were able to add such complexes to a zeolite structure and show that the complex remained intact, at least for lower temperature reactions [24].

Pt and Sn can form a number of alloys ranging from Pt_3Sn to $PtSn_4$ [25]. In the loading range of metals in catalysts used for naphtha reforming, the PtSn alloy dominates even though it will be possible to find a few particles with another alloy composition (Figure 11.3). The fraction of Pt that is present as an alloy increases for a constant Pt content as the Sn concentration increases (Figure 11.4). While the fraction of Pt present as an alloy does not appear to depend upon the surface area and porosity of the alumina support, the fraction does vary for the alumina and silica support. The alumina supported catalyst retains essentially all of the chloride that is added during impregnation with $PtCl_6^{2-}$ and $SnCl_2$. XPS and Mössbauer data suggest that the Cl is likely to be present as $SnCl_4^{2-}$, $SnCl_6^{2-}$, AlOCl or $AlCl_4^-$-type complexes. Presumably these chloride complexes would be dispersed on both the surface and bulk of the finished catalyst. The XPS and SEM data clearly indicate that at least a portion of the chloride is mobile during calcination and reduction steps.

Somorjai et al. [26] made extensive studies of the conversion of hydrocarbons over single crystals of single or two metals. For the Pt(111) surface containing

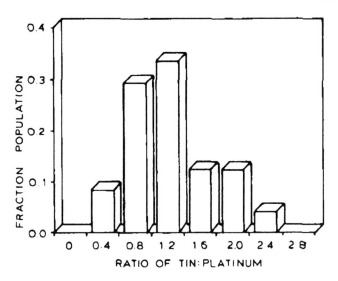

Figure 11.3 Frequency distribution of EDX intensity of Sn:Pt ratios for increasing Sn:Pt concentrations [25].

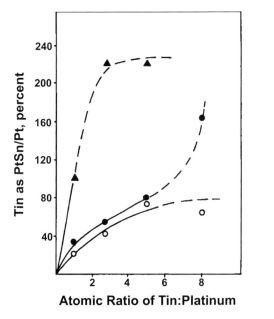
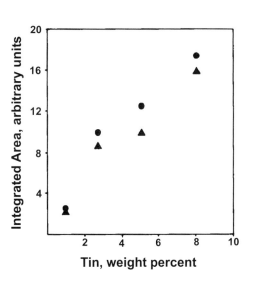

Figure 11.4 Left: The amount of tin calculated to be present as Pt–Sn alloy versus Sn:Pt ratio (▲, silica support; ●, low surface area alumina support; ○, high surface area alumina support). Right: The integrated intensity measured from the 100% Pt–Sn (102) profile versus Sn content (▲ on United Catalysts, Inc., alumina support, and ● on Degussa alumina support) [25].

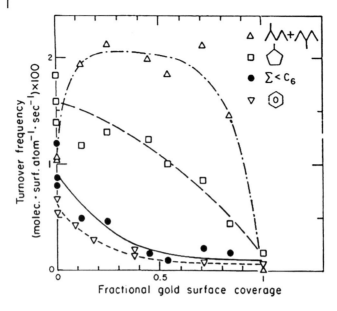

Figure 11.5 Turnover frequencies of isomerization (△), cyclization (□), hydrogenolysis (●) and aromatization (▽) as a function of Au coverage in Pt–Au alloy formed on Pt (111) single crystal [26].

various amounts of gold, they found that the turnover frequency number (TON) did not vary in the same manner for different surface concentrations of Au for the isomerization, cyclization, hydrogenolysis and aromatization reactions (Figure 11.5).

The discovery of the selective aromatization of hexane using Pt-KL zeolite was another example of the utility of nanocatalysis in petroleum refining [27] and was commercialized by Chevron workers [28]. Tauster and Steger [29] offered a geometric explanation where the structure of the KL zeolite oriented the alkane molecule in the linear channels to favor terminal C-adsorption and the $C_{1,6}$ ring closure on the Pt sites. Some authors considered the selective properties of this catalyst as being due to the electron-rich nano-Pt crystals, induced by the interactions of the metal with the basic framework of the zeolite (e.g., [30–32]). McVicker et al. [33] claimed that the agglomeration of the Pt-nanocrystals was hindered by the KL-zeolite structure. Iglesia and Baumgartner [34] claimed that the Pt particle selectivity did not depend on the support; rather the KL catalyst had the channel structure to inhibit the bimolecular reactions that would lead to coke formation. Jacobs et al. [35], among others, supported the view of Iglesia and Baumgartner. Derouane and Vanderveken [36] considered that the cavities of the zeolite would lead to a preorganization of the hydrocarbon chain that adsorbed at C_1 to adapt a structure that resembled a transition state for $C_{1,6}$-ring cyclization (confinement model). Triantafillou et al. [37, 38] considered the selectivity for aromatization to result from a combination of effects such as the size of the Pt-cluster, basicity of the zeolite and constraints imposed by the support channels.

The metal ions used with the zeolite has been considered to impact the aromatization selectivity and probably the cyclization pathway. The initial Chevron work emphasized using the Ba cation. Grau *et al.* [39] found that partially exchanging K^+ with Ba^{2+} did not modify n-heptane aromatization but that exchanging with La^{3+} decreased the dehydrocyclization and promoted acid catalysis. Further, impregnating KL with barium nitrate followed by calcination at 873 K led to improved dehydrocyclization activity, compared to just Pt-KL.

An even earlier example of a zeolite naphtha reforming catalyst was the tellurium-loaded zeolite for dehydrocyclization [40]. It appeared that the highly dispersed Te phase was coordinated to the cations of the zeolite under the reducing atmosphere [41, 42]. The hydrogen partial pressure dependence of the dehydrocyclization reaction led Silvestri and Smith [43] to conclude that the aromatics were formed by consecutive dehydrogenation to hexene, hexadiene and hexatriene with the triene undergoing cyclization. Iglesia *et al.* [44] studied the dehydrocyclization of ^{13}C-labeled heptane using a TeNa-XC zeolite catalyst and concluded that the reaction pathway followed a mechanism involving a consecutive dehydrogenation mechanism to a heptatriene which desorbs and undergoes cyclization in the gas phase. While these Te-zeolite catalysts are interesting, their value to date has been limited because of the rapid loss of Te, together with the loss of catalytic activity.

The mechanism for dehydrocyclization with monofunctional catalysts has been reviewed [45] and the general conclusion is that aromatics are formed by a direct $C_{1,6}$-ring closure mechanism. Arcoya *et al.* [46] consider that two pathways can lead to aromatics with the Pt-KL catalyst: (i) a direct $C_{1,6}$-ring closure followed by dehydrogenation and (ii) successive dehydrogenation to heptatriene and then cyclization. These cyclization steps were considered to occur in the adsorbed phase and on the same active site. This conclusion is consistent with one advanced earlier based on the conversion of n-octane labeled with deuterium in the 2- and 7-carbon positions [47, 48].

11.4
Hydrotreating

While hydrotreating has been practiced for many years [49], it is only recently that sophisticated catalyst characterization techniques have been able to provide solid evidence for the nature of the active catalyst. There are four metals that are used in some combination in nearly all hydrotreating catalysts: Co with Mo or W and Ni with Mo or W. While the catalyst is prepared in the oxide form, at least the Mo and W component are converted to the sulfide before use; the extent that Co or Ni will be sulfided depends upon many factors. The recent emphasis on producing ultra-low sulfur transportation fuels has led to renewed interest in understanding the structure of the active catalyst and how one can improve it.

The nano aspects of the Co–Mo catalyst has received much attention during the past few years [50]. Results reported by Haldor-Topsøe personnel and their co-workers have greatly extended our understanding of the potential of

Figure 11.6 Atomically resolved STM image ($V_t = 5.2$ mV, $I_t = 1.28$ nA) of a triangular single-layer MoS_2 nanocluster on Au(111). The size of the image is 41 Å × 42 Å [50].

nanotechnology in the preparation and use of these catalysts. Much of their outstanding results have come from studies of single layer MoS_2 nanoparticles that were formed by sulfiding molybdenun deposited on a gold surface under sulfiding conditions. Under strong sulfiding conditions, triangular-shaped clusters are formed (Figure 11.6). There are data to suggest that certain sizes of these structures are exceptionally stable and form electron shells that have stability such as found within the atom periodicity of the Periodic Table. For the MoS_2 edges, it was concluded that the electronic structure is predominantly metallic one-dimensional edge states. These edge sites are described as "brim sites" and are considered to have a significant role in catalysis. Under sulfiding conditions, the bare Mo can form two almost equal stability structures (Figure 11.7). Hydrogen treatment of the triangular structure caused a much stronger chemisorbed state of thiophene. This is considered to result at vacancies where sulfur atoms are removed from the edge, leaving sulfur vacancies, and it is presumed that the remaining edge sulfur atoms are present as SH. Adsorption of thiophene leads to hydrogenation of one or both double bonds by the SH groups which undergo subsequent reaction with loss of H_2S. DFT calculations indicate that the structure of the adsorbed intermediate is probably an adsorbed cis-but-2-ene-thiolate compound coordinated through the terminal sulfur atom to the edge sulfur atoms near the brim.

Particles in the nanometer regime may adapt new structural and electronic properties that may vary significantly from those found in the bulk material [51]. This is the case for the triangular MoS_2, where a series of structures with increasing numbers of Mo along an edge has been observed [52] (Figure 11.8). The triangular structures with less than six Mo atoms on an edge of the triangle appears

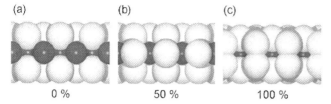

Figure 11.7 Side view of ball models of the molybdenum edge of MoS$_2$ with various coverages (0, 50, 100%). (a) Molybdenum edge exposing a row of under-coordinated molybdenum atoms (dark). Under sulfiding conditions, the most stable structures are saturated with sulfur atoms (light) adsorbed on the edge. This occurs in two configurations that have almost equal stability, but with different total coverages of sulfur. (b) Molybdenum edge with sulfur monomers, corresponding to a coverage of 50%. (c) Molybdenum edge with sulfur dimmers, with 100% coverage. In both configurations (b) and (c) the edge molybdenum atoms keep the coordination to six sulfur atoms, but structurally the edges are reconstructed relative to the bulk configuration [50].

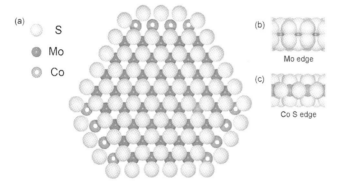

Figure 11.8 (a) Ball model of the proposed CoMoS structure. The CoMoS cluster is shown in top view, exposing the unpromoted molybdenum edge and a cobalt-promoted sulfur edge (dark, molybdenum; bright, sulfur; dark with white spot, cobalt). Also shown on the basal plane is a single cobalt inclusion. (b) The molybdenum edge (shown in side view) is unaffected by cobalt and is covered with sulfur dimers. (c) Cobalt fully substitutes for molybdenum on the sulfur edge. Sulfur monomers adsorbed on the edge produce a tetrahedral coordination of each cobalt atom [50].

different and the appearance of the interior of the cluster in the STM picture is brighter and different from that of normal basal planes. It was suggested that the structural changes were caused by a rearrangement of the cluster edges because of an increase of the S:Mo ratio for the smallest nanoclusters.

A different picture is obtained for the more typical catalyst that contains both Co and Mo. Here the MoS$_2$-like nanoclusters with Mo substituted by Co at edge sites is considered to be the model. Rather than the triangular structure of the MoS$_2$, a hexagonal structure is formed. This change in structure is considered to be due to the formation of the CoMoS phase, as originally proposed by Topsøe

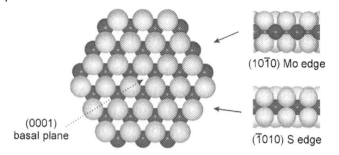

Figure 11.9 Ball model of a hexagonal MoS$_2$ cluster, obtained from a simple truncation of a bulk MoS$_2$(0001) slab. Left: The cluster is oriented with the (0001) basal plane in parallel with the paper and it exposes the two different fundamental low-index edge terminations, the ($10\bar{1}0$) molybdenum edge and the ($\bar{1}010$) sulfur edge. Right: Side view of the edges. The molybdenum atoms (dark) are coordinated to six sulfur atoms (bright) at the sulfur edge, whereas at the molybdenum edge the coordination to sulfur is only four [50].

workers [53]. The nature of the STM picture suggests that the promotion by Co may be explained by a change in the electronic structure of the sulfur edge as well as a change of the interaction strength of sulfur with the cobalt. The replacement of Mo by Co at the edges results in a different edge structure (Figure 11.9).

As a result of these studies and knowledge of the interactions of the cluster with the support, new catalysts were developed by Haldor-Topsøe; these are employed in commercial hydrotreating reactors [54]. This is an excellent example of the utilization of fundamental studies leading to ideas that help the development of better commercial catalysts [55]. The company introduced TK-558 BRIM (CoMo) and TK-559 BRIM (NiMo) for pretreatment of FCC service in 2004 and then a new CoMo catalyst (TK-576 BRIM) for ultra-low sulfur diesel production. These catalysts are now in service in more than 50 hydrotreating units.

While the models developed based on work with nanoparticles supported on gold by the Topsøe workers led to many scientific accomplishments, there may be differences between them and the structures in the nanoparticles present on the high surface area support of real hydrotreating catalysts. This is recognized by the Denmark group [56] as well as by others. For example, de Jong et al. [57] reported high resolution tomography pictures of industrial NiMo-γ-Al$_2$O$_3$ catalysts that show that MoS$_2$ particles form a complex interconnected structure within the alumina support (Figure 11.10). The MoS$_2$ phase occupies a substantial fraction of the mesopore volume. The individual planes of MoS$_2$ were imaged and the shapes were found to differ from the model structures described above. Similar structures have been reported by Breysse et al. [58].

Cortés-Jácome et al. [59] supported Mo and Co–Mo (Co/(Co + Mo) = 0.3) on nanotubular titania. They found that the CoMo catalyst had 1.3–1.7 times the activity of a commercial reference sample. However, one must always keep in mind that commercial catalysts are designed to operate for months or years before

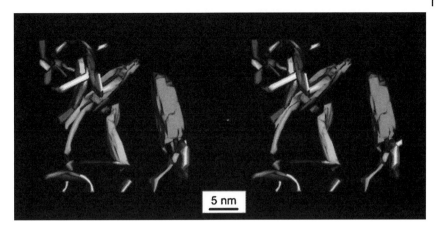

Figure 11.10 Stereoview of interconnected MoS$_2$ particles within a 44 × 44 × 44 nm^3 volume of the 3D reconstruction [57].

regeneration or replacement occurs so that some initial activity may be sacrificed in the commercial catalyst in order to attain a long catalyst life.

In summary, the characterization of nanoparticles has led to a much greater understanding of the hydrotreating catalysts. In this instance, that understanding has been converted to commercial catalysts.

11.5
Cracking

The introduction of the zeolite cracking catalyst utilized nanocrystals of zeolite dispersed in an amorphous silica–alumina matrix [60, 61]. Since their introduction as cracking catalysts, the study of zeolites has become a major branch of scientific investigations. By manipulating the template agent and the reaction conditions a wide range of zeolites can be synthesized. The variety of channel openings and arrangements is great as well as the cavities arranged along the channels. The introduction of VPI-5 [62] and MCM-41 [63] led to a variety of materials with channel openings that were much larger than for those for the zeolites (Figure 11.11).

In addition to the silica–alumina zeolites, the introduction of other elements, such as Ti for the selective oxidation using hydrogen peroxide [64], opened another major area of research [65].

Within a given zeolite structure, a number of factors may impact the catalytic cracking activity. Defining these has involved many workers. The distribution of Al, or other atoms such as Ti, within a unit cell of the zeolite may vary with the dealumination technique. Extra framework aluminum may play a dominant role in the catalytic cracking activity [66] and may even lead to superacidity [67].

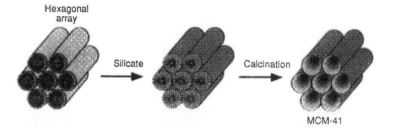

Figure 11.11 Schematic drawing of the liquid-crystal templating mechanism. Hexagonal arrays of cylindrical micelles form (possibly mediated by the presence of silicate ions), with the polar groups of the surfactants (light gray) to the outside. Silicate species (dark gray) then occupy the spaces between the cylinders. The final calcination step burns off the original organic material, leaving hollow cylinders of inorganic material [63].

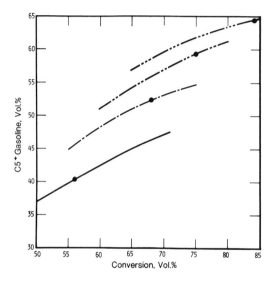

Figure 11.12 Improvements in zeolite cracking catalyst selectivity (—, standard silica-alumina gel; — · —, early zeolite catalysts (REHX); — · · —, improved zeolite catalyst (REHY and copromoter); ●, point where 4% coke (on charge) occurs [60].

A detailed examination of this topic is beyond the scope of this chapter. A series of symposia on fluid catalytic cracking has been held at several ACS meetings and seven volumes have been published with papers describing many aspects of FCC [68]. The recognition of the role of these nano-zeolite catalysts was emphasized by one of the persons responsible for the introduction of zeolite catalytic cracking, Charlie Plank [60] who showed that for a given level of catalytic cracking, much less of the feed was converted to carbon (Figure 11.12). Plank estimated that about

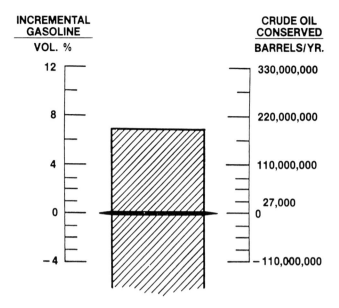

Figure 11.13 Economic value of incremental gasoline [60].

180 million barrels/year of crude oil were conserved due to the increased gasoline yield that was possible by use of zeolite catalytic cracking (Figure 11.13); the savings today in the United States would be at least three times greater than when Plank made his estimate in 1966. Since catalytic cracking is the largest volume process in the refinery, the volume of cracking catalysts is large and is responsible for the largest dollar amount invested in catalysts.

11.6
Hydrocracking

Hydrocracking resembles hydrotreating except that it involves a reduction in molecular weight of the feed, usually defined as >50% of the molecules undergo molecular weight reduction. Like naphtha reforming, hydrocracking involves bifunctional catalysis [69]. Two general catalyst systems are used for hydrocracking: (i) a catalyst that is, or is closely related to, the hydrotreating catalyst described above and (ii) a noble metal-loaded zeolite [70]. The metal zeolite catalyst is a combination of a supported noble metal, such as in the naphtha reforming catalyst, and the nano-zeolite that is similar to the catalytic cracking catalyst. Just as with naphtha reforming, the bifunctional catalyst ages much more slowly than the cracking catalyst so that long on-stream times are possible.

11.7
Conversion of Syngas

There are four major reactions involving the conversion of syngas for energy applications: water-gas shift (WGS), methanation, methanol synthesis and Fischer–Tropsch synthesis. All four of these reactions are mature and are practiced at the commercial scale. Recent work in these four areas are included in this review.

11.8
Water-Gas Shift

The initial work in this area involved the high-temperature WGS and at these temperatures the thermodynamic equilibrium limits the extent of conversion to CO_2 and H_2. To overcome this, the process frequently involves both high- and low-temperature stages to increase the fraction of CO converted and the H_2 produced. Rugged, low-surface area catalysts are utilized for the high-temperature stage but nanocatalysis may be involved in the low-temperature stage; thus, the following only considers the low-temperature stage.

Today's commercial catalyst for the low-temperature stage involves Cu–ZnO, sometimes promoted with other elements and supported on a high surface area material. An extensive review of the low-temperature WGS was reported recently [71]. However, this catalyst is used at a high-enough temperature where a few percent of CO remain, even at the equilibrium composition. Therefore, a search is underway to obtain a catalyst that will operate at lower temperatures than is required for the Cu–ZnO catalyst and one that has received significant attention recently has been based on ceria. The nano-structure of the Cu–ZnO catalyst is described in Section 11.9.

Two major reaction mechanisms are advanced for the WGS reaction: the redox and the formate mechanisms. A major difference in the two mechanisms is whether the $Ce^{4+} \leftrightarrow Ce^{3+}$ reaction occurs and whether formate is a reaction intermediate or a spectator ion. For the redox mechanism, nanoparticles of a metal (Pt, Cu, etc.) are present on the surface and CO adsorbs on the metal and is then oxidized to CO_2 utilizing an oxygen from the ceria at the boundary of the metal–oxide interface, reducing the Ce^{4+} and leaving an anion vacancy. The adsorption of water at the anion vacancy leads to the formation of hydrogen and filling the vacancy while oxidizing the Ce^{3+} to Ce^{4+}. For the formate mechanism, the metal nanoparticle serves to catalyze the reduction of only the surface Ce^{4+} to Ce^{3+} and then serving as a catalyst for the formation of hydrogen from the formate. At this date, the mechanism that operates is being widely debated. For either mechanism, nanocatalysis is involved at least with the supported metal and, depending upon the surface area, for the ceria.

The redox mechanism may be illustrated by the following reactions:

$$Ce^{4+}O + M\text{-}CO \rightarrow Ce^{3+} + M + CO_2$$

$$M + CO \rightarrow M\text{-}CO$$

$$Ce^{3+} + H_2O \rightarrow Ce^{4+}O + H_2$$

The formate mechanism is illustrated by the following reactions:

$$M\text{-}OH + CO \rightarrow M\text{-}O_2CH$$

$$M\text{-}O_2CH + H_2O \rightarrow M\text{-}OH + CO_2 + H_2$$

For the formate mechanism it was initially considered that the role of the metal was to assist in reducing the ceria surface but it has become apparent that the metal must function in some other role since the rate of reaction increases with the loading of metal nanoparticles. Thus, it is proposed that the formate structure in the vicinity of the metal nanoparticle can diffuse to the metal where hydrogen removal is assisted by the metal. This gives a schematic of the surface, as illustrated in Figure 11.14, to show the influence of increasing loading of the metal nanoparticles [72].

During the past ten years, a nanoscale catalyst that has received much attention is the supported gold catalyst. Historically, gold was regarded as an inert catalyst. This view changed dramatically when Haruta *et al.* [73] discovered the amazingly high catalytic activity of Au for low-temperature CO oxidation. Surprisingly, it was soon recognized that only small (<5 nm) gold particles, preferably supported on an oxide of the first transition series (e.g., TiO_2, $\alpha\text{-}Fe_2O_3$), were active. These early observations attracted much attention to gold catalysis. First, gold having catalytic activity was a novel realization. Second, the size of the gold particle is very limited if it is to have high catalytic activity; particles in the 2–4 nm range provide the maximum activity. Third, the controversy over the valence state of the catalytically active form of gold has attracted attention and many investigations. Much of the early work was summarized by Bond and Thompson [74].

Au–Fe_2O_3 and Au–Al_2O_3 catalysts were used for WGS and, at low temperatures, Au–Fe_2O_3 had a higher activity than a commercial Cu–Zn–Al_2O_3 catalyst (Figure 11.15) [75]. The active catalysts had an Au particle size of 3.5 nm. The data indicate that catalysts prepared by deposition–precipitation were more active than those prepared by coprecipitation.

Fu *et al.* [76] found that Au–CeO_2 (5–8% Au) had excellent WGS activity. They report that the presence of Au is crucial for activity below 300 °C. They found that all available surface oxygen was reduced but there was no effect of the bulk oxygen of ceria.

Later it was reported that essentially all of the gold could be removed from the catalyst without decreasing the catalytic activity [77, 78]. The metallic gold was leached from an Au–CeO_2 catalyst using a NaCN solution. More than 90% of the gold loading was removed and the remaining (<10%) gold no longer had a metallic character as determined by XPS. The amazing result was that the leached catalyst was as active for WGS as the parent material. They concluded that the metallic Au nanoparticles were merely spectator species (Figure 11.16). As shown in the figure, the low-loading of Au catalyst (0.44% Au) was about as active as the parent catalyst

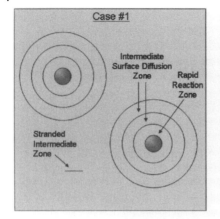
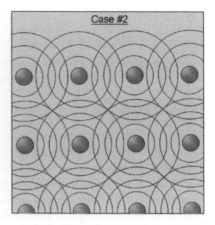
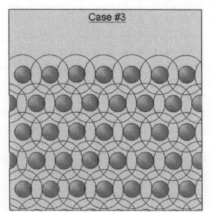

Figure 11.14 Consideration for the interpretation of SSITKA data. Case 1: three formates can exist, including: (a) rapid reaction zone (RRZ) – those reacting rapidly at the metal-oxide interface, (b) intermediate surface diffusion (SDZ) – those at pathlengths sufficient to eventually diffuse to the metal and contribute to overall activity and (c) stranded intermediate zone (SIZ) – intermediates are essentially locked onto surface due to excessive diffusional path lengths to the metal-oxide interface. Case 2: metal particle population sufficient to overcome excessive surface diffusional restrictions. Case 3: all formate intermediates within rapid reaction zone [72].

(4.7% Au); both low- and high-Au catalysts had similar activation energies. Both gold-containing catalysts were much more active than the support [Ce(La)O$_x$] alone.

11.9
Methanol Synthesis

Today methanol, apart from methyl-*t*-butyl ether whose use is declining, is not used for transportation fuels. However, it is included because there are many

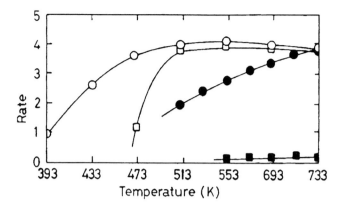

Figure 11.15 Temperature dependence of the catalytic activity of supported gold and base metal oxide catalysts in the water-gas shift. The starting reaction gas mixture was 4.88 vol% carbon monoxide in argon; water vapor partial pressure = 233 Torr; SV = 4000 h^{-1}; 1 atm. ○, Au/α-Fe$_2$O$_3$; □, CuO/ZnO/Al$_2$O$_3$; ●, α-Fe$_2$O$_3$; ■, Au/Al$_2$O$_3$. Rates are expressed in mol m^{-2} h^{-1} × 10^{-2} [75].

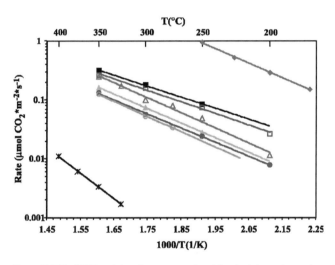

Figure 11.16 WGS activity of as-prepared and leached Au-ceria catalysts in 11% Co-7% Co$_2$-26% H$_2$O–He reformate-type gas [77].

options for its use as a fuel, either directly or as dimethyl ether or indirectly as a hydrogen carrier. Moreover, it is commercially practiced on a large scale today.

It is generally accepted that in the commercial working catalyst copper is present in the metallic state under reaction conditions. The reaction takes place on the copper and the methanol is formed predominantly from CO$_2$. In spite of copper

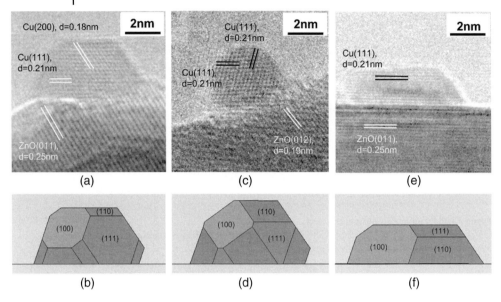

Figure 11.17 In-situ TEM images (a, c, e) of a Cu/ZnO catalyst in various gas environments together with the corresponding Sulff constructions of the Cu nanocrystals (b, d, f). (a) The image was recorded at a pressure of 1.5 mbar of H_2 at 220 °C. The electron beam is parallel to the [011] zone axis of copper. (c) Obtained in a gas mixture of H_2 and H_2O, $H_2:H_2O = 3:1$ at a total pressure of 1.5 mbar at 220 °C. (e) Obtained in a gas mixture of H_2 (95%) and CO (5%) at a total pressure of 5 mbar at 220 °C [81].

being the seat of the synthesis, the ZnO plays an important, though still unsettled, role in achieving the high activity and long catalyst life. While there are reports of a nearly linear relationship between the rate of methanol synthesis and the copper surface area (e.g., [79]), there are also a number of deviations from this (e.g., [80]).

The methanol catalyst may undergo dynamic modification with changes in the atmosphere that the Cu–ZnO catalyst is exposed to [81]. This is illustrated in Figure 11.17 where the crystal is exposed to a hydrogen atmosphere (A), a mixture of H_2 and H_2O (C) and a mixture of H_2 and CO (E); these pictures were obtained under realistic reaction conditions of 1.5–6.0 mbar (150–600 Pa) and 220 °C. The change in crystal shape under these realistic conditions is reversible, as shown in Figure 11.18 where the change induced by the presence of water in the hydrogen (b) is reversed when the water was removed. Thus, one needs to define the structure of the catalyst under realistic reaction conditions.

Initially, the view for the role of ZnO was to stabilize the nano-Cu metal particles. However, that simple picture has been dramatically modified over the years. Thus, Fujitani and Nakamura [82, 83] indicated the role of ZnO is creating the active site for synthesis in addition to the role of dispersing the Cu particles. These authors further assigned the role of Zn in the reaction mechanism to the hydro-

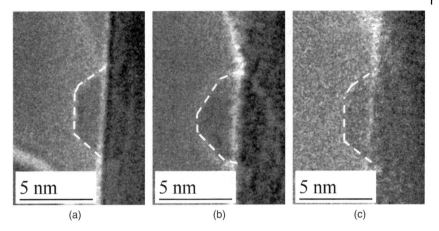

Figure 11.18 TEM images showing the reversible shape change of a Cu nanocrystals. The same Cu nanocrystals is imaged at 220 °C under (a) H_2 at 1.5 mbar, (b) $H_2:H_2O$ (3:1) at a total pressure of 1.5 mbar, and (c) H_2 at 1.5 mbar [81].

genation of the formate to the methoxy species. Their STM results showed the formation of a Cu–Zn surface alloy for Zn/Cu(111) and suggested that this may be the active site for the reaction.

Meitzner and Iglesia [84] investigated the slow approach to steady-state synthesis rate for the SiO_2 and ZnO supports. Since the Cu should not change for the SiO_2 support, they considered that a bifunctional mechanism was involved for methanol synthesis from CO because of the need to form the formate using the hydroxyl groups of the silica support. The bifunctional mechanism was not needed for the conversion of CO_2 since the formate can form directly from CO_2 and H_2 on the Cu surface.

Jansen et al. [85] discussed their results in light of those presented in a recent review on methanol synthesis catalysts [86]. In the review, based on literature and experimental data, they advanced two models: (i) migration of partly reduced ZnO on top of Cu or (ii) reversible formation of flat epitaxial Cu particles upon high-temperature reduction. In spite of arguments against model (i) in the review, Jansen et al. [85] concluded that model (i) seemed the best hypothesis for the formation of active sites of the $Cu/ZnO/SiO_2$ catalyst. Since the solubility of CuO in ZnO is limited, they indicated that Cu is present largely as Cu^{1+} and arrived at the structure in Figure 11.19. When they assumed that the solubility of ZnO in Cu is small, the presence of subsurface metallic Cu would require reduction of most of the subsurface Zn species. Hence, they believed that their model explained the catalytic tests.

Frost [87] proposed that the minute Schottky junctions at the interface between the metal and the oxide in the catalyst affects the surface chemistry of the oxides in a way that correlates with the catalytic behavior. Kasatkin et al. [88] indicated that the abundance of non-equilibrium structures in Cu, such as planar defects

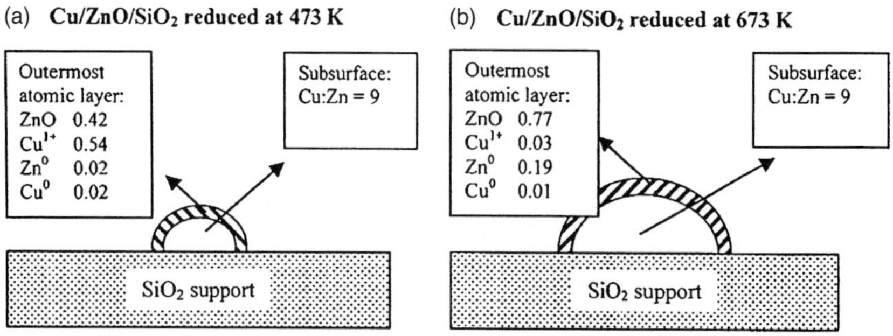

(a) Cu/ZnO/SiO₂ reduced at 473 K

Outermost atomic layer:
ZnO   0.42
Cu^{1+} 0.54
Zn^0 0.02
Cu^0 0.02

Subsurface: Cu:Zn = 9

(b) Cu/ZnO/SiO₂ reduced at 673 K

Outermost atomic layer:
ZnO   0.77
Cu^{1+} 0.03
Zn^0 0.19
Cu^0 0.01

Subsurface: Cu:Zn = 9

Figure 11.19 Model of the surface structure of Cu/ZnO/SiO$_2$ after (a) low-temperature 473 K and (b) high-temperature 673 K reduction as determined with LEIS. One should note that the cluster shape represents an educated guess; it was not determined [85].

and strain clearly correlate with catalytic activity for methanol synthesis. They considered the relevant non-equilibrium structure of Cu is generated during synthesis, either leaving Zn from the mixed metal hydroxycarbonate in the Cu or leaving oxygen from the oxide–suboxide reduction dissolved in the Cu. They concluded that the synthesis of Cu-based catalysts can be redesigned to optimize the exact chemical composition. Kurtz et al. [89] made a similar plea for rational catalyst design with not only the aim of a large surface area of Cu but that the Cu/ZnO interface should also be optimized.

Methanol decomposition to produce hydrogen is one means of producing hydrogen on-site. Nano-catalysts have potential for this reaction. Croy et al. [90] report a size dependent study of methanol decomposition over Pt nanoparticles. Micelle encapsulation was used to prepare Pt nanoparticles that were highly dispersed on anatase-TiO$_2$ supports. They found that the activity increased with decreasing size of the Pt nanoparticles. Samples annealed at higher temperatures had a higher percentage of metallic Pt and higher catalytic activity. The size of the Pt particles was found to be the dominant factor in catalytic activity, and the smaller particles were most active.

In summary, there is a vast literature on the synthesis and decomposition of methanol. For the low temperature synthesis and decomposition, the supported metal is present as a nanoparticle. In spite of the extensive investigation of the topic, there still remains much uncertainty as to the nature of the catalytic site.

11.10
Fischer–Tropsch Synthesis (FTS)

The conversion of syngas (mixture of CO and H$_2$) to hydrocarbons, the Fischer–Tropsch synthesis (FTS), has been practiced for nearly 100 years. Initial work in this area involved low surface area catalysts in fixed bed reactors. The particle size of the catalyst was smaller for the high-temperature FTS conducted in the fixed

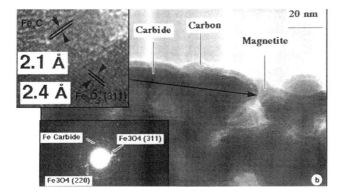

Figure 11.20 TEM view of the catalyst after activation treatments followed by Fischer-Tropsch reaction (H_2/CO reactant ratio = 0.7) for 45 h. Sample initially activated in syngas mixture at 523 K. The carbide phase, the magnetite phase, and the carbon films are shown by markers. Electron diffraction and the high magnification inset show the presence of the carbide and the magnetite phase [91].

fluidized bed reactor at Brownsville (Texas, USA) and the circulating fluidized bed reactor at Sasolburg (South Africa) but they still utilized low surface area catalysts. Thus, it was only with the introduction of the slurry bubble column low-temperature FTS that nanocatalysis was utilized for FTS.

Sasol introduced commercial slurry bubble column reactors for FTS in 1991 utilizing iron catalysts. They are now in the final stages of shake-down runs for a 35 000 bbl/day plant in Qatar that utilizes cobalt nanocatalysis. Actually, at Sasol the low-temperature slurry and fixed bed iron catalysts have common catalyst preparation steps up to the finishing stages, where the fixed bed catalyst is formed into a pellet large enough to ensure a reasonable pressure drop across the bed and the slurry catalyst which is treated to form microspheres in the 30–70 μm range.

The spray dried catalyst is made up of many smaller particles in the 1–3 μm range. The initial catalyst, after calcination is some mixture of the iron oxyhydroxide and Fe_2O_3, the composition depending upon the temperature and length of the calcination step. This catalyst then needs to undergo an activation step and this may involve heating at some temperature for an optimum time period in: (i) hydrogen, (ii) CO, or (iii) synthesis gas. For a commercial operation hydrogen and syngas are readily available for the activation but a pure stream of CO is usually not available.

The steps for the hydrogen activation involves the reduction of Fe_2O_3 to Fe_3O_4 followed by the reduction to the metal (the formation of the intermediate phase FeO is questionable). The activation with CO results in the formation of CO_2 to produce first Fe_3O_4 and then further reduction combined with the Bouduard reaction to produce CO_2 and iron carbide. In actual practice, after 24 h the amount of carbon on the catalyst is about twice that required to produce $Fe_{2.2}C$ and the excess carbon forms a shell surrounding the iron carbide particle, as reported by Shroff et al. [91] (Figure 11.20). Our studies show that for large iron oxide particles the

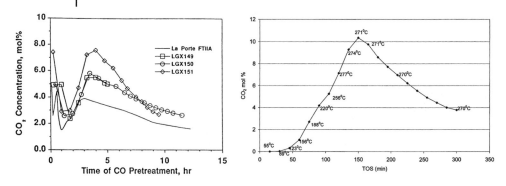

Figure 11.21 Left: Concentration of CO$_2$ in the reactor effluent versus time of Co pretreatment for the LaPorte FT IIA run, LGX149, LGX150 and LGX151. Right: CO$_2$ mol% versus TOS for 100Fe/5.1Si/2.0Cu/5K catalyst activated with CO.

reduction to Fe$_3$O$_4$ and the formation of the carbides are readily separated but that for the small nanoparticles the two steps merge and are accomplished more rapidly (Figure 11.21).

It appears that the iron catalyst during FTS in a slurry reactor attrites and attains a particle size that is about 1–3 µm in size. This is true when you begin with a large spray-dried particle, where the agitation in the solvent causes the exposed surface particles to attrite away [92] or, in contrast, when one starts with a small particle (3 nm) the size increases during the carbiding and then continues to grow during the synthesis to approach a size of 1 µm [93].

The iron catalyst precursor can be synthesized to have a long needle shape. This particle may be easier to effect wax separation without plugging the filter or blocking the filter pores. However, during activation the needle (>100 nm in length) does not carbide uniformly but instead breaks up, forming smaller spherical particles [94] (Figure 11.22).

During use the bulk phases change, with the carbides being converted to Fe$_3$O$_4$ until 40–60% of the iron carbide has been converted to the oxide within a few hundred hours. With a suitably promoted catalyst the carbide–oxide phases stabilize and remain essentially constant during a few thousand hours of operation. While these phase changes are occurring the activity undergoes a slow steady decline that does not follow the phase changes (Figure 11.23). Thus, it does not appear that some particles are completely oxidizing to Fe$_3$O$_4$ while others remain as the carbide form. If this was to happen then the conversion should decline in parallel with the phase changes, and this does not happen. Thus, we believe that the oxidation takes place at the core of the particles and this leaves a layer of the carbide which is exposed to the syngas (Figure 11.24). This model allows for the formation of the oxide phase without surface exposed to the reactant gases changing, hence the nearly constant conversion.

The phase changes that the nano-iron particles undergo depends upon the catalyst composition. For an unpromoted iron nanoparticle the iron carbide gradually

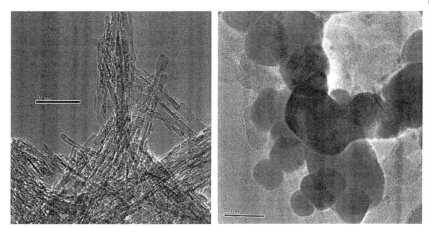

Figure 11.22 Left: HRTEM image of the synthesized iron oxyhydroxide particles. Needle-shaped morphology. Right: HRTEM images of catalyst particles collected after 24 h of FTS at 270 °C, 175 psig; ultrafine particles along with some larger-sized particles are evident [94].

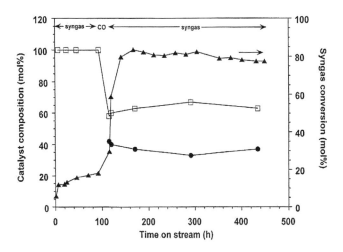

Figure 11.23 Fischer-Tropsch synthesis using iron catalyst initially pretreated with syngas, then CO only at about 120 h for 24 h, and then resuming normal synthesis (▲, CO conversion; □, Fe_3O_4, ●, iron carbides).

is converted to Fe_3O_4 during about 400 h of synthesis, and the CO conversion parallels this oxidation of the carbide phase (Figure 11.25). A potassium promoter was added to this iron catalyst and a different result was obtained. While the CO conversion declined even more rapidly with the potassium promoted sample than the one without potassium, the carbide phase was stable and may even have increased slightly during the run. Thus, the amount of carbide present in the sample is not a predictor of catalytic activity. A similar iron sample was prepared except that it

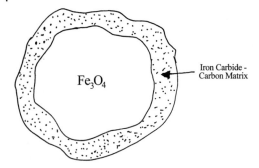

Figure 11.24 Schematic of the structure of an iron Fischer-Tropsch catalyst during use for FTS.

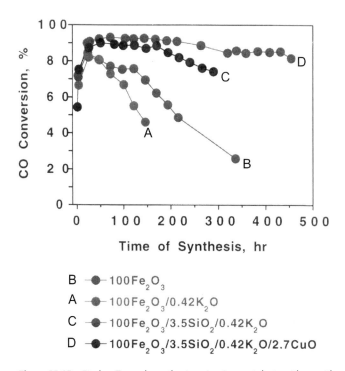

B —●— 100Fe$_2$O$_3$
A —●— 100Fe$_2$O$_3$/0.42K$_2$O
C —●— 100Fe$_2$O$_3$/3.5SiO$_2$/0.42K$_2$O
D —●— 100Fe$_2$O$_3$/3.5SiO$_2$/0.42K$_2$O/2.7CuO

Figure 11.25 Fischer-Tropsch synthesis using iron catalysts with or without promoters.

contained 4.4 atomic Si per 100 Fe. The Si-containing sample lost activity more slowly than the iron-only sample. The Si-containing sample with 1.4 K/Fe sample had a stable CO conversion and the iron carbide phases declined to a stable value of about 50%. These results clearly show that there is a synergism for the Si and K promoters since both are much better than either one alone.

For the K- and Si-promoted catalyst, the activation gas impacts the early FTS activity but all three activation procedures produce nearly the same CO conversion after about 300 h on-line [95]. This, together with the data described above, suggests that the surface layer of carbide is in a dynamic state, with carbon being lost and re-added as the reaction time increases.

11.11
Methanation

In practice, the catalytic activity is of less interest than the catalyst life. When methanation is accomplished at high temperatures, a relatively low surface area catalyst is utilized wherein the activity is sacrificed for long life. This generally puts the Ni-alumina high-temperature methanation catalyst outside the normal nano-particle size range. The high temperature methanation catalyst falls in the nano-range initially but undergoes growth during the methanation reaction. Thus, catalyst samples taken from the first, second and third reactors after 2200 h on-stream had mean nickel crystal sizes of 75–100 nm, 27–40 nm and 16–19 nm, respectively [96].

One of the classics of heterogeneous catalysis is the comparison of the methanation activity of single crystal and supported nickel catalysts [97]. Goodman indicated that: "The close correspondence between catalytic studies using single metal crystals and results for high-area, supported catalysts attests to the structure insensitivity of the methanation reaction." This correspondence is shown in the plot for the two types of catalysts (Figure 11.26). Another feature of the single crystal work was the poisoning of methanation single crystal catalysts. Phosphorous showed a decline in activity that was nearly linearly related to the surface concentration of the poison. However, sulfur showed a much stronger poisoning effect than phosphorous with less than 0.1 monolayer of S eliminating essentially all of the methanation activity (Figure 11.27). Goodman, in agreement with many others, considered methanation to be a surface-insensitive reaction.

van Meerten *et al.* [98] found that the methanation reaction with Ni-silica had an activity per unit area of nickel that strongly depended on the mean nickel crystal size (0.5–13 nm; Figure 11.28). The maximum activity was attained with about 4 nm Ni particles. Concurrent low-field magnetization measurements indicated that larger crystals grew at the expense of the smaller ones; these measurements were supported by hydrogen chemisorption data and from high-field magnetization measurements. A high partial pressure of CO and low temperature favored crystal growth, suggesting transport by nickel tetracarbonyl.

Recent studies [99] indicated that the methanation reaction is highly structure-sensitive and also is sensitive to hydrogen. Under ultrahigh vacuum conditions, with no hydrogen present, the dissociation proceeds through a direct route in which only under-coordinated sites (e.g., steps) are active. Under methanation conditions, the dissociation also proceeds most favorably over under-dissociated sites, but through a COH species. They compared their data to those reported earlier by Goodman *et al.* [100] (Figure 11.29). The Anderson *et al.* data do not exactly overlay those of Goodman *et al.*, especially in the low pressure region, because their data depend on hydrogen pressure. However, the qualitative data

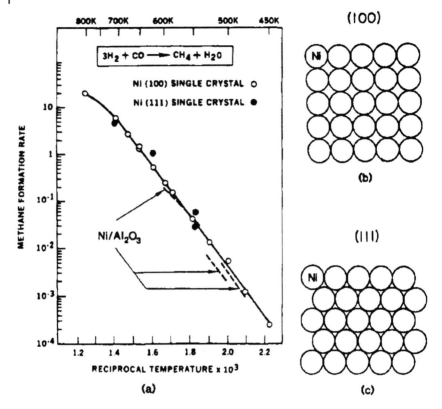

Figure 11.26 (a) A comparison of the rate of methane synthesis over single-crystal nickel catalysts and supported Ni/Al$_2$O$_3$ catalysts at 120 torr total reactant pressure. (b) Atomic conformation of a Ni(100) surface. (c) Atomic conformation of a Ni(111) surface [97].

are in excellent agreement. They conclude that the hydrogenation rate cannot keep up with the dissociation rate at high temperatures so there is a poisoning effect by carbon.

11.12
Nanocatalysis for Bioenergy

Most of the feedstocks that have the large volume needed for our energy use are solids. The most useful catalysts today for converting these biomaterials are solids. Solid–solid reactions are not easily accomplished. This can be illustrated for the direct liquefaction of coal where solid coal and solid heterogeneous catalysts are utilized. In the direct liquefaction of coal, Keogh and Davis [101] found that both the thermal and catalytic reactions followed a common pathway. Thus, solid coal was converted to preasphaltenes, then to asphaltenes and finally to oil and gas, the

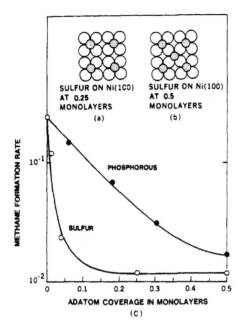

Figure 11.27 Methanation rate as a function of sulfur and phosphorus coverage on a Ni(100) catalyst. Pressure = 120 torr, H_2/CO = 4/1, reaction temperature = 600 K [97].

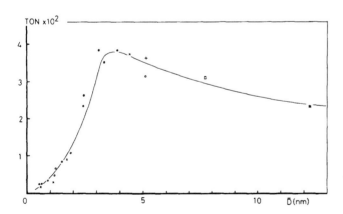

Figure 11.28 Methanation of CO/H_2. Initial activity (molecules CH_4 per second per surface Ni atom) of a series of nickel-silica catalysts as a function of the initial mean crystallite size (518 K, p_{CO} = 9.3 kPa, p_{H_2} = 93 kPa): (○) earlier measurements, reduction at 725 K; (×) reduction at 725 K (NZ 1, 5, 29, 54); (+) reduction at 875 K (NZ 54); (□) reduction a t975 K (NZ 54); (●) re-reduction at after reduction at 725 K, methanation at 519 K and oxidation at 725 K; (■) re-reduction at 975 K (NZ 54) after reduction at 725 K, methanation at 519 K and oxidation at 725 K [98].

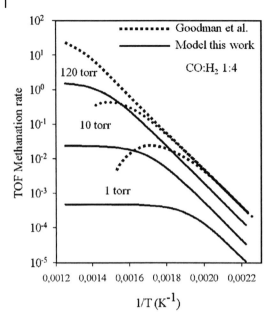

Figure 11.29 Comparison between the experimental data obtained on Ni(100) and the present model where carbon poisoning is taken into account. The absolute values are only quantitatively in agreement for the higher pressures while larger deviations are observed for lower pressures. Note, however, that the model captures qualitatively the onset of carbide poisoning of the steps although it is not taking the formation of islands into account [99].

desired products. In the case of bituminous coal, about 10% of the mass is present as oil and gas that is contained in the pore structure and is released rapidly as the material is heated to the reaction temperature of about 450 °C. The remaining solid is converted to very high molecular weight preasphaltenes that have low solubility. These preasphaltenes are converted to asphaltenes, a lower molecular material but still a solid at room temperature. Finally, the asphaltenes are converted to oils, as shown in Figure 11.30. Thus, the catalyst is effective only when the solid is first converted to an intermediate that is able to be dispersed in the solvent used for coal liquefaction.

In 2007, a group met to define basic research needs for catalysis for energy production with emphasis on biomass [102]. The conclusion of this group was that the challenge is to understand the chemistry by which cellulose- and lignin-derived molecules are converted to fuels and to use the knowledge to identify the needed catalysts. To obtain energy densities that are equivalent to those obtained from crude oil today, oxygen must be removed from the biomass. In general, each carbon atom in the biomass is bonded to an oxygen (–CHOH–) whereas each carbon in today's transportation fuel is bonded to one or two hydrogens (–CH– or –CH$_2$–); thus, one oxygen must be removed from each carbon in the biomass. The dominant pathways considered today require hydrogen. Thus, hydrogen must be

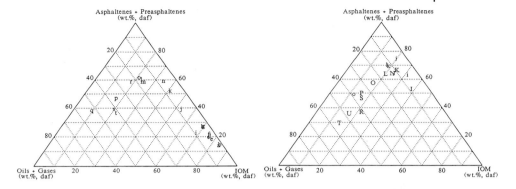

Figure 11.30 Left: Thermal liquefaction pathway of a bituminous Western Kentucky No. 6 coal (data obtained at residence times of 2.5 to 90 min and reaction temperatures 598, 658, 700 and 718 K). Right: Catalytic liquefaction pathway of a bituminous Western Kentucky No. 6 coal (data obtained with Shell 324 and molybdenum naphthenate catalysts, residence times 5 to 60 min and temperatures of 658, 700 and 718 K) [101].

derived from the biomass or from some other source (e.g., nuclear for electrolysis of water). Deriving hydrogen from the biomass makes the process inefficient with respect to carbon and energy utilization. Even so, the incentive to make use of biomass is great since the United States could harvest 1.3 billion tons of biomass annually (n.b., 1.0 ton = 1.02 t). Converting this resource to ethanol would produce more than 60 billion gallons/year (n.b., 1 gallon = 3.78 L) and this is enough to replace 30% of the United States use of gasoline [102]. In addition, the use of only biomass to generate transportation fuels results in no emission of CO_2, another attractive feature of fuels from biomass.

Most pathways today involve the conversion of the biomass to syngas. Once syngas is produced, the process from that point on is the same as the processes described above and would use the same catalysts.

The group [102] outlined steps for the complete and efficient destruction of targeted chemical linkages in biomass. These include: (i) selective thermal deconstruction of biomass, (ii) acid hydrolysis of biomass, (iii) or enzymatic hydrolysis of the biomass. There are problems to be overcome for each of these options. Slow thermal deconstruction of biomass leads primarily to gaseous products and charcoal. To obtain high yields of liquid products the pyrolysis must be rapid; flash pyrolysis with residence times of <2 s at 400–650 °C may produce up to 75 wt% moisture-free organic liquids on a dry mass basis [103]. These flash pyrolysis products are unstable and must be quickly upgraded, usually by adding hydrogen. Liquefaction produces a higher heating value product than flash pyrolysis does and a more stable product. However, liquefaction requires reasonably high temperatures and pressures, usually in the presence of a catalyst and hydrogen. New, more selective catalysts are needed.

Acid hydrolysis of the cellulosic components of biomass targets C–O–C bonds to produce glucose. This process has been investigated using mineral acid catalysts for sugar production to then make ethanol. This acid pretreatment is not as efficient as enzymatic hydrolysis. Normal heterogeneous catalysis may then follow the acid catalysis hydrolysis.

Today ethanol is produced using biological catalysts (yeasts). The space–time yield for bio-catalysts is only about 10^{-3}–10^{-5} that of catalysts used in today's petroleum refineries. Liquid phase catalysis offers the opportunity to approach the space–time yield of today's refinery but it greatly complicates the plant. The multi-step process requires acid catalysts for dehydration reactions, base catalysts for aldol condensation, metal catalysts for hydrogenation reactions and bifunctional metal/acid catalysts for dehydration/hydrogenation reactions [102]: "The figure illustrates how aqueous-phase processing can be used to produce selectively targeted alkanes from glucose. This is a multi-step process that requires acid catalysts for dehydration reactions, base catalysts for aldol condensations, metal catalysts for hydrogenation reactions, and bifunctional metal/acid catalysts for dehydration/hydrogenation reactions. Furthermore, the final step requires a four-phase reactor: (1) a gas phase containing hydrogen, (2) a solid phase composed of a bifunctional catalyst containing metal and acid sites, (3) an aqueous phase containing the sugar reactant, and a liquid alkane phase used to remove hydrophobic species from the catalyst surface before they react further to form carbonaceous deposits that deactivate the solid catalyst [102]." To accomplish this while approaching the cost of producing fuels from petroleum is a challenge.

While today's picture for biomass transformation to transportation fuels looks very complicated, one should remember that petroleum refining has become much more complicated during the past 100 years. Initially the process was limited to a distillation and that was followed by thermal cracking. Only after more than 30 years was catalytic cracking introduced, and another 20 years passed before the successful naphtha reforming process to produce high octane gasoline was accomplished. Heteroatom removal was first introduced to protect catalysts whose activity was poisoned by sulfur and has now progressed in 50 years to where it is used to produce fuels with only 10 ppm sulfur or less. The overall processing of crude is progressing to where each carbon number fraction may be handled by a separate process, at least for carbon number fractions through the gasoline range.

11.13
The Future

The United States DOE assembled a team of experts in 2004 to consider nanoscience research for energy needs. They outlined nine research targets in energy-related science and technology in which nanoscience is expected to have the greatest impact [104]:

1) Scalable methods to split water with sunlight for hydrogen production
2) Highly selective catalysts for clean and energy-efficient manufacturing
3) Harvesting solar energy with 20% power efficiency and 100 times lower costs

4) Solid-state lighting at 50% of the present power consumption
5) Super-strong, lightweight materials to improve efficiency of cars, airplanes and the like
6) Reversible hydrogen storage materials operating at ambient temperatures
7) Power transmission lines capable of one gigawatt transmission
8) Low-cost fuel cells, batteries, thermoelectrics, and ultra-capacitors built from nanostructured materials
9) Materials synthesis and energy harvesting based on the efficient and selective mechanisms of biology.

To accomplish these targets, the participants recognized six fundamental and vital crosscutting nanoscience research themes:

1. Catalysis by nanoscale materials
2. Using interfaces to manipulate energy carriers
3. Linking structure and function at the nanoscale
4. Assembly and architecture of nanoscale structures
5. Theory, modeling, and simulation for energy nanoscience
6. Scalable synthesis methods.

For catalysis by nanoscale materials they write [104]: "The research challenge in nanoscience for catalysis is learning to tune the energy landscape of the chemical reactants as they interact with the nanostructured catalyst materials. Drawing from the lessons of biology, nanostructured materials must be designed to match both the structural conformation of the reactants and to control the reaction pathway to the desired product. To accomplish this, new and efficient methods of *in situ* characterization and rapid throughput testing of catalytic properties will be required. The choice of materials, structural parameters, and the experimental design must be guided by a continually improving fundamental understanding of the structure–function relationships of the nanostructured catalysts." In other words, the new nanocatalysts must have a productivity that is 10^3–10^5 times that of today's enzymes.

There have been significant advances in introducing more complex separations into petroleum refining. For the lower boiling fractions of transportation fuels some refineries now accomplish what is close to separation by carbon number. In this sense, we now have a "nanoscale feedstock." While some progress has been made toward developing specific catalysts to meet the requirements of the very specific feedstock, much more progress is needed. There is reason to believe that the advances in nanocatalysis will lead the way to accomplish this goal. Today's refinery is operated very differently from 50 years ago. The changes are driven by the need for very different fuel properties to meet rapidly changing motor requirements and because of the much more stringent environmental requirements for refinery operations.

Change is evident, driven at least partly by the high cost of crude and partly by the limitations imposed by carbon emissions. These changes are in their infancy today. A significant fraction of the gasoline sold in the United States now contains oxygenates, and in a few areas fuel containing 85% ethanol is available. As the requirements for catalysis emphasize more and more selectivity, nanocatalysis should become even more important in the fuels industry.

References

1. Astruc, D. (2008) *Nanoparticles and Catalysis* (ed. D. Astruc), Wiley-VCH Verlag GmbH, Weinheim, pp. 1–48.
2. Mills, G.A., Heinemann, H., Milliken, T.H., and Oblad, A.G. (1953) *Ind. Eng. Chem.*, **45**, 134–137.
3. Prater, C.D., and Weisz, P.B. (1957) *Adv. Catal.*, **9**, 575–586.
4. Stevenson, S.A., Dumesic, J.A., Baker, R.T.K., and Ruckenstein, E. (1987) *Metal-Support Interactions in Catalysis, Sintering, and Redispersion*, van Nostrand Reinhold Co., New York.
5. Goeke, R.S., and Datye, A.K. (2007) *Top. Catal.*, **46**, 3–9.
6. Darling, T.A., and Moss, R.L. (1966) *J. Catal.*, **5**, 111–115.
7. van Hardeveld, R., and van Montfoort, A. (1966) *Surf. Sci.*, **4**, 396–430.
8. Clark, J.K.A., Farren, G., and Rubalcava, H.E. (1967) *J. Phys Chem.*, **71**, 2376–2377.
9. Carter, J.L., Cusumano, J.A., and Sinfeld, J.H. (1966) *J. Phys. Chem.*, **70**, 2257–2263.
10. Boudart, M., Aldag, A.W., Pak, C.D., and Benson, J.E. (1968) *J. Catal.*, **11**, 35–45.
11. Cormak, D., and Moss, R.L. (1969) *J. Catal.*, **13**, 1–11.
12. van Hardeveld, R., and Hartog, F. (1969) *Surf. Sci.*, **15**, 189–230.
13. Kluksdahl, H.E. (1968) U.S. Patent 3,415,737.
14. Dowden, D.A., and Reynolds, P.W. (1950) *Discuss. Faraday Soc.*, **8**, 184–189, disc., 189–190.
15. Sachtler, W.M.H., Dorgelo, G.H.H., and Jongepier, R. (1965) *J. Catal.*, **7**, 100–102.
16. Sachtler, W.M.H., and Dorgelo, G.J.H. (1965) *J. Catal.*, **4**, 654–664.
17. Davis, B.H., and Antos, G.J. (2004) Characterization of naphtha reforming catalysts, in *Catalytic Naphtha Reforming*, 2nd edn (eds G.J. Antos and A.M. Aitani), Marcel-Dekker, New York, pp. 199–274.
18. Meitzner, G., Via, G.H., Lytle, F.W., and Sinfelt, J.H. (1982) *J. Chem. Phys.*, **78**, 882–889.
19. Bazin, D., Dexpert, H., Lagarde, P., and Bournonville, J.P. (1986) *J. Phys.*, **47**, C8–C293.
20. Bensaddik, A., Caballero, A., Bazin, D., Dexpert, H., Didillon, B., and Lynch, J. (1997) *J. Appl. Catal. A: Gen.*, **162**, 171–180.
21. Biloen, P., Helle, J.N., Berbeck, H., Dautzenberg, F.M., and Sachtler, W.M.H. (1980) *J. Catal.*, **63**, 112–118.
22. Sachtler, W.M.H. (1984) *J. Mol. Catal.*, **25**, 1–12.
23. Davis, B.H. (1974) Bimetallic catalyst preparation, U. S. Patent 3,840,475, October 8.
24. Kawi, S., Chang, J.-R., and Gates, B.C. (1993) *J. Catal.*, **142**, 585–601.
25. Srinivasan, R., and Davis, B.H. (1992) *Platinum Metal. R.*, **36**, 151–163.
26. Somorjai, G.A., Tao, F., and Park, J.Y. (2008) *Top. Catal.*, **47**, 1–14.
27. Bernard, J.R., and Rees, L.W. (1980) *Proceedings of the Fifth International Conference on Zeolites*, Heyden, London, p. 686.
28. Jacobson, R.L., Kluksdahl, H.E., McCoy, C.S., and Davis, R.W. (1969) *Proc. Am. Petrol. Inst., Div. Refining, 34th Midyear Mtg., Chicago*, May 13.
29. Tauster, S.J., and Steger, J.J. (1990) *J. Catal.*, **125**, 387–389.
30. Besoukhanova, C., Guidot, J., Barthomeuf, D., Breysse, M., and Bernard, J.R. (1981) *J. Chem. Soc. Faraday Trans.*, **77**, 1595–1604.
31. Han, W., Kooh, A.B., and Hicks, R.F. (1993) *Catal. Lett.*, **18**, 219–225.
32. Menachery, P.V., and Haller, G.L. (1998) *J. Catal.*, **177**, 175–188.
33. McVicker, G.B., Kao, J.L., Ziemiak, J.J., Gates, W.E., Robbins, J.L., Treacy,

M.M.J., Rice, S.B., Vanderspurt, T.H., Cross, V.R., and Ghosh, A.K. (1993) *J. Catal.*, **139**, 48–61.
34 Iglesia, E., and Baumgartner, J.E. (1993) *Stud. Surf. Sci. Catal.*, **75**, 993–1006.
35 Jacobs, G., Padro, C.L., and Resasco, D. (1998) *J. Catal.*, **179**, 43–55.
36 Derouane, E.G., and Vanderveken, D.J. (1988) *Appl. Catal.*, **45**, L15–L22.
37 Triantafillou, N.D., Deutsch, S.E., Alexeev, O., Miller, J.T., and Gates, B.C. (1996) *J. Catal.*, **159**, 14–22.
38 Triantafillou, N.D., Miller, J.T., and Gates, B.C. (1995) *J. Catal.*, **155**, 131–140.
39 Grau, J.M., Daza, L., Seoane, X.L., and Arcoya, A. (1998) *Catal. Lett.*, **53**, 161–166.
40 Lang, W.H., Mikovsky, R.J., and Silvestri, A.J. (1971) *J. Catal.*, **20**, 293–298.
41 Mikovsky, R.J., Silvestri, A.J., Dempsey, E., and Olson, D.H. (1971) *J. Catal.*, **22**, 371–378.
42 Olson, D.H., Mikovsky, R.J., Shipman, G.F., and Dempsey, E. (1972) *J. Catal.*, **24**, 161–169.
43 Silvestri, A.J., and Smith, R.L. (1973) *J. Catal.*, **29**, 316–318.
44 Iglesia, E., Baumgartner, J., Price, G.L., Rose, K.D., and Robbins, J.L. (1990) *J. Catal.*, **125**, 95–111.
45 Davis, B.H. (1999) *Catal. Today*, **53**, 443–516.
46 Arcoya, A., Seoane, X.L., and Grau, J.M. (2005) *Appl. Catal. A: Gen.*, **284**, 85–95.
47 Shi, B., and Davis, B.H. (1995) *J. Catal.*, **157**, 626–630.
48 Shi, B., and Davis, B.H. (1996) *Studies in Surface Science and Catalysis*, vol. **101** (B), Elsevier, Amsterdam, pp. 1145–1154.
49 Leliveld, R.G., and Eijsbouts, S.E. (2008) *Catal. Today*, **130**, 183–189.
50 Lauritsen, J.V., and Besenbacher, F. (2006) *Adv. Catal.*, **50**, 97–147.
51 Heiz, U., and Landman, U. (eds) (2007) *Nanocatalysis*, Springer, Berlin.
52 Lauritsen, J.V., Kibsgaard, J., Helveg, S., Topsøe, H., Clausen, B.S., and Besenbacher, F. (2007) *Nat. Nanotechnol.*, **2**, 53–58.
53 Topsøe, H., Clausen, B.S., Candia, R., Wivel, C., and Mørup, S. (1981) *J. Catal.*, **68**, 433–452.
54 Zeuthen, P., and Skyum, L. (2004) *Hydrocarbon Eng.*, **11**, 65–68.
55 Topsøe, H., Hinnemann, B., Nørskov, J.K., Lauritsen, J.V., Besenbacher, F., Hansen, P.L., Hytoft, G., Egeberg, R.G., and Knudsen, K.G. (2005) *Catal. Today*, **107–108**, 12–22.
56 Besenbacher, F., Mrorson, M., Clausen, B.S., Helveg, S., Hinnemann, B., Kibsgaard, J., Lauritsen, J.V., Moses, P.G., Nørskov, J.K., and Topsøe, H. (2008) *Catal. Today*, **130**, 86–96.
57 de Jong, K.P., van den Oetelaar, L.C.A., Vogt, E.T.C., Eijsbouts, S., Koster, A.J., Friedrich, H., and De Jough, P.E. (2006) *J. Phys. Chem. B*, **110**, 10209–10212.
58 Breysee, M., Geantet, C., Afanasien, P., Blanchard, J., and Vrinat, M. (2008) *Catal. Today*, **130**, 3–13.
59 Cortés-Jácome, M.A., Escobar, J., Angeles Chavez, C., López-Salinas, E., Romero, E., Ferrat, G., and Toledo-Antonio, J.A. (2008) *Catal. Today*, **130**, 56–62.
60 Plank, C.J. (1983) *Heterogeneous Catalysis: Selected American Histories*, Series 222 (eds B.H. Davis and W.P. Hettinger Jr.), ACS Symp, pp. 253–273.
61 Venuto, P.B., and Habib, E.T., Jr. (1979) *Fluid Catalytic Cracking with Zeolite Catalysts*, Marcel Dekker, Inc, New York.
62 Davis, M.E., Saldarriaga, C., Montes, C., Garces, J., and Croweder, C. (1988) *Nature*, **331**, 698–699.
63 Kresge, C.T., Leonowicz, M.E., Roth, W.J., Vartuli, J.C., and Beck, J.S. (1992) *Nature*, **359**, 710–712.
64 Notari, B. (1993) *Catal. Today*, **18**, 163–172.
65 Ratnasamy, P., Srinivas, D., and Knözinger, H. (2004) *Adv. Catal.*, **48**, 1–169.
66 Lunsford, J.H. (1991) *Fluid Catalytic Cracking. III. Concepts in Catalyst Design*, Series 452 (ed. M.L. Occelli), ACS Symp, pp. 1–11.
67 Dessau, M., and Haag, W.O. (1985) *Int. Congr. Catal. Proc., 8th*, vol. 2, pp. 305–316.

68. Occelli, M.L. (ed.) (2007) Fluid catalytic cracking. VII: Materials, methods and process innovations, in *Studies in Surface Science and Catalysis*, vol. **166**, Elsevier B.V., Amsterdam, pp. 1–356, and earlier volumes.
69. Coonradt, H.L., and Garwood, W.E. (1964) *Ind. Eng. Chem., Process. Des. Dev.*, **3**, 38–45.
70. Sullivan, R.F., and Scott, J.W. (1983) *Heterogeneous Catalysis: Selected American Histories*, Series 222 (eds B.H. Davis and W.P. Hettinger Jr.), ACS Symp, pp. 293–316.
71. Jacobs, G., and Davis, B.H. (2007) *Catalysis*, vol. **20**, RSC Pub., London, pp. 122–285.
72. Jacobs, G., and Davis, B.H. (2007) *Appl. Catal. A: Gen.*, **333**, 192–201.
73. Haruta, M., Yamada, N., Kobayahsi, T., and Iijma, S., (1989) *J. Catal.*, **115**, 301–309.
74. Bond, G.C., and Thompson, D.T. (1999) *Catal. Rev. Sci. Eng.*, **41**, 319–388.
75. Andreeva, D., Idakiev, V., Tabakova, T., and Andrew, A. (1986) *J. Catal.*, **158**, 354–355.
76. Fu, Q., Weber, A., and Flytzani-Stephanopoulos, M. (2001) *Catal. Lett.*, **77**, 87–95.
77. Fu, Q., Saltsburg, H., and Flytzani-Stephanopoulos, M. (2003) *Science*, **301**, 935–938.
78. Fu, Q., Deng, W., Saltsburg, H., and Flytzani-Stephanopoulos, M. (2005) *Appl. Catal. B: Environ.*, **56**, 57–68.
79. Chinchen, G.C., Waugh, K.C., and Whan, D.A. (1986) *Appl. Catal.*, **25**, 101–107.
80. Kurr, P., Kasarkin, I., Girgsdies, F., Trunschke, A., Schlögl, R., and Ressler, T. (2008) *Appl. Catal. A: Gen.*, **348**, 153–164.
81. Hansen, P.L., Wagner, J.B., Helveg, S., Rostrup-Nielsen, J.R., Clausen, B.S., and Topsøe, H. (2002) *Science*, **295**, 2053–2055.
82. Fujitani, T., and Nakamura, J. (2000) *Appl. Catal. A: Gen.*, **191**, 111–129.
83. Choi, Y., Futagumi, K., Fujitani, T., and Nakamura, J. (2001) *Appl. Catal. A: Gen.*, **208**, 163–167.
84. Meitner, G., and Iglesia, E. (1999) *Catal.Today*, **53**, 433–441.
85. Jansen, W.P.A., Beckers, J., Heuvel, J.C.V.D., Denier, A.W., Gon, V.D., Blick, A., and Brongersma, H.H. (2002) *J. Catal.*, **210**, 229–236.
86. Poils, E.K., and Brands, D.S. (2000) *Appl. Catal. A: Gen.*, **191**, 83–96.
87. Frost, J.C. (1988) *Nature*, **334**, 577–580.
88. Kasatkin, I., Kurr, P., Kniep, B., Trunschke, A., and Schlögl, R. (2007) *Angew. Chem.*, **119**, 7465–7468.
89. Kurtz, M., Bauer, N., Wilmer, H., Hinrichsen, O., and Muhler, M. (2004) *Chem. Eng. Technol.*, **27**, 1146–1150.
90. Croy, J.R., Mostafa, S., Liu, J., Shon, Y., and Roddan Cuenya, B. (2007) *Catal. Lett.*, **118**, 1–7.
91. Shroff, M.D., Kalakkad, D.S., Coulter, K.E., Köhler, S.D., Harrington, M.S., Jackson, N.B., Sault, A.G., and Datye, A.K. (1995) *J. Catal.*, **156**, 185–207.
92. Srinivasan, R., Xu, L., Spicer, R.L., Tungate, F.L., and Davis, B.H. (1996) *Fuels Sci. Technol. Int.*, **114**, 1337–1359.
93. Sarkar, A., Seth, D., Dozier, A.K., Neathery, J.K., Hamdeh, H.H., and Davis, B.H. (2007) *Catal. Lett.*, **117**, 1–17.
94. Sarkar, A., Dozer, A.K., Graham, U.M., Thomas, G., O'Brien, R.J., and Davis, B.H. (2007) *Appl. Catal. A: Gen.*, **326**, 55–64.
95. Luo, M., and Davis, B.H. (2003) *Fuel Proc. Technol.*, **83**, 49–65.
96. Woodward, C. (1977) *Hydrocarbon Process.*, **130**, 136–137.
97. Goodman, D.W. (1984) *Acc. Chem. Res.*, **17**, 194–200.
98. van Meerten, R.Z.C., van Nessebrooij, A.H.G.M., Beaumont, P.F.M.T., and Coenen, J.W.E. (1983) *Surf. Sci.*, **135**, 565–579.
99. Anderson, M.P., Abild-Pedersen, F., Remediakis, I.N., Bligaard, K.T., Jones, G., Engbak, J., Lytken, O., Horch, S., Nielsen, J.H., Schested, J., Rostrup-Nielsen, J.R., Nørskov, J.K., and Chorkendorff, I. (2008) *J. Catal.*, **255**, 6–19.
100. Goodman, D.W., Kelley, R.D., Madey, T.E., and Yates, J.T., Jr. (1980) *J. Catal.*, **63**, 226–234.

101 Keogh, R.A., and Davis, B.H. (1994) *Energy Fuels*, **8**, 289–293.

102 Bell, A.T., Gates, B.C., and Ray, D. (2007) Basic Research Needs: Catalysis for Energy, Report from the U.S. Department of Energy, Basic Energy Sciences Workshop, August 6–8.

103 Hube, G.W., Cortright, R.D., and Dumesic, J.A. (2004) *Angew. Chem. Int. Ed.*, **43**, n1549–n1551.

104 Nanoscience Research for Energy Needs, Office of Basic Energy Sciences, Department of Energy, June 2005.

12
Surface-Functionalized Nanoporous Catalysts Towards Biofuel Applications
Hung-Ting Chen, Brian G. Trewyn, and Victor S.-Y. Lin

12.1
Introduction

The rapid consumption of petroleum and the soaring oil prices have increased the demand of finding an alternative energy source in the foreseeable future. To date several energy resources such as hydrogen, solar and biofuels are considered potential candidates for the replacement of fossil fuels. Although adsorbents for hydrogen storage have been extensively studied, the challenge of compressing such small and chemically inert gas molecules into a portable device for everyday use hinders its commercialization. Solar cells, converting sunlight into electricity by photovoltaic effect, initially were thought to provide a clean and renewable energy; however, low conversion efficiency and expensive manufacturing of solar plants reduce their usage. In contrast, biofuel, derived from biomass, is a more realistic renewable energy resource that can immediately be used in the engine of modern transportation vehicles. Biodiesel, a biodegradable, non-toxic alternative diesel fuel, is made from renewable biological sources such as vegetable oils, animal fats and microalgal oils. Biodiesel was recently accepted as a viable alternative to traditional petroleum-derived solvents, which are of environmental concern and are under legislative pressure to be replaced by biodegradable substitutes. Although interest in biodiesel is rapidly increasing, the process by which biodiesel is synthesized has not changed much in recent years. Technically, biodiesel is a methyl ester of fatty acid, produced from triglycerides through transesterification, as shown in Scheme 12.1. Currently, biodiesel is made commercially by an energy- and labor-intensive process wherein oil feedstocks are reacted with methanol at elevated temperature (about 60–65 °C), often under pressure, in the presence of sodium methoxide as the catalyst, to yield fatty acid methyl esters and glycerol. Isolation of the desired methyl soyate from the highly caustic (toxic) catalyst and other products, such as glycerol, involves a precise neutralization process with strong acids, such as hydrochloric acid (HCl), and extensive washes with water to remove the resulting sodium chloride (NaCl) salt. Also, the glycerol must be separated from the sodium chloride salt by vacuum distillation in an energy-intensive operation for this high-boiling product. Obviously, utilization of heterogeneous

Nanotechnology for the Energy Challenge. Edited by Javier Garcia-Martinez
© 2009 WILEY-VCH Verlag GmbH & Co. KGaA, Weinheim
ISBN: 978-3-527-32401-9

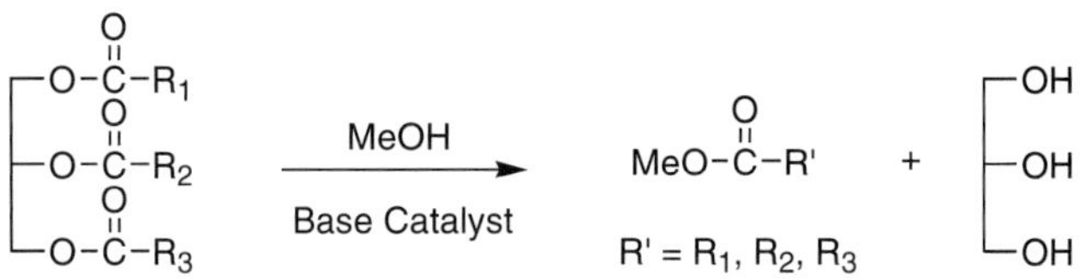

Scheme 12.1

$$\begin{bmatrix} O\text{-}\overset{O}{\overset{\|}{C}}\text{-}R_1 \\ O\text{-}\overset{O}{\overset{\|}{C}}\text{-}R_2 \\ O\text{-}\overset{O}{\overset{\|}{C}}\text{-}R_3 \end{bmatrix} \xrightarrow[\text{Base Catalyst}]{\text{MeOH}} \text{MeO-}\overset{O}{\overset{\|}{C}}\text{-}R' \quad + \quad \begin{bmatrix} \text{OH} \\ \text{OH} \\ \text{OH} \end{bmatrix}$$

$R' = R_1, R_2, R_3$

catalysts substantially simplifies the purification procedure, reduces the manufacturing costs and minimizes the production of chemical waste. Moreover, heterogeneous catalysts can be recycled in sequential reactions, making them economically competitive with homogeneous catalysts.

Conventional heterogeneous catalysis focuses mainly on reactions that take place at the *gas-solid interface* with purified reactant feedstocks under elevated pressure and temperature. In the past few decades, several important catalytic systems have been developed for converting *petroleum*-based feedstocks into energy and useful chemicals under these conditions. However, applying those metal- or alloy-based heterogeneous catalysts to fit into *solution-phase* reactions for biological feedstock has been difficult to demonstrate. As the public and legislative pressure gradually increases in environmental and energy consideration, the ability to design *catalysts that can operate efficiently under less energy-demanding reaction conditions at the liquid-solid or even solid-solid interfaces* become a major challenge for the 21st century catalysis science. As the structures and sources of feedstocks become more complex and diverse, the selectivity of the next generation of catalysts needs to be significantly improved from that of their predecessors to handle *reactants that are water-soluble and with a variety of heteroatoms and functionalities, such as biologically derived feedstocks*. In the case of enzymes, the desired reactivity and selectivity of the catalytically active site are achieved by *synergistic catalysis between different functional groups, such as general acids and bases, that are spatially well-positioned in the 3D-controlled cavity*. The question of how to engineer such a sophisticated catalytic system with superior mechanical and chemical stability than enzymes for industrial applications is a grand challenge. Herein, we discuss the development of heterogeneous catalysts, specifically mesoporous nanomaterials for controlled, selective catalysts for the ultimate goal of biofuel production. In discussing this ultimate goal, we describe in some detail the important steps and achievements that have been reached thus far in the journey towards a clean, renewable, sustainable fuel source.

12.1.1
"Single-Site" Heterogeneous Catalysis

Recently, a new direction in heterogeneous catalysis has been developing that bridges the gap between homogeneous and heterogeneous catalysis. This so-called "single-site heterogeneous catalysis" approach involves the immobilization of a well-defined spatially isolated, homogeneous catalyst on the solid support through various molecular interactions. As defined by John M. Thomas, the "single site" (or catalytically active center) is spatially isolated homogeneous molecular catalysts on the solid surface [1, 2]. Each site is isolated without any interaction among them and is a structurally well-characterized species, which possesses the same interaction energy between it and reactants. The catalytic performance of such heterogeneous catalysts can be carefully tailored since the structure-activity relationship of each catalytic site can be extracted by studying the homogeneous analog in the solution state. Single-site heterogeneous catalysis brings molecular insight to the design of new catalyst and even allows for the discovery of new reactions.

12.1.2
Techniques for the Characterization of Heterogeneous Catalysts

The first step to understanding the activity of heterogeneous catalysts is to fully characterize the structures of both supported material and catalytic sites, whose subtle deviation might deeply influence the performance. Thanks to the availability of advanced surface characterization tools, the images of surface structures can be explicitly probed and analyzed. Some useful spectroscopic techniques for surface characterization are summarized and listed in Table 12.1. The IR (usually *in situ*) and UV-Vis spectroscopy are used to monitor the modification of the support and catalyst sites after a series of chemical treatments. The extended X-ray adsorption fine structure (EXAFS) provides interatomic distances and average coordination number; whereas the X-ray adsorption near-edge structure (XANES) indicates the oxidation state and the geometry of the metal complex when applicable. Nitrogen sorption measurements, such as BET or BJH methods, provide information on

Table 12.1 Characterization techniques of heterogeneous catalysts.

Characterization methods	Information
IR spectroscopy	Functional group determination
UV-Vis spectroscopy	Adsorption spectrum of immobilized molecules
Extended X-ray adsorption fine structure (EXAFS)	Interatomic distances and average coordination number of surface metal complex
X-ray adsorption near-edge structure (XANES)	Oxidation state and the geometry of surface metal complex
N_2 sorption analysis	Surface area, porosity and pore size distribution of supported materials
Solid-state NMR spectroscopy	Detail chemical structure, molecular motion and surface interaction of immobilized molecules

the surface area, porosity and pore size distribution of the catalyst. Solid-state NMR spectroscopy is the most powerful tool for structure determination, by providing the detail chemical structure, molecular motion and interactions that exist on the surface.

12.2
Immobilization Strategies of Single-Site Heterogeneous Catalysts

12.2.1
Supported Materials

A judicious choice of support material contributes to the performance of heterogeneous catalysts. The material properties of support define the potential chemistry of immobilization strategies, allowing for control of the surface reactivity. Excellent catalyst support should have the following features:

1) Chemically, thermally and mechanically stable structure.
2) Large surface area and abundant surface functionalities.
3) Accessible catalytic sites and fast diffusion of substrate to the sites.
4) Selectively functionalizable interior and exterior surfaces.

The organic polymers, especially polystyrene-type resins, represent one of most popular commercialized support. This resin is commonly synthesized by copolymerization of active monomer, such as halogenated methyl styrene, in the presence of cross linkers. With a low content of cross linkers, the resulting polymer requires a compatible solvent that swells for reactants to access internal active sites. For instance, low reactivity is typically observed for such supports in protic, highly polar media, such as water or alcohol. Increasing the amount of cross linkers solidifies polymer resins, diminishing the swelling problems. However, it sometimes suffers from brittleness and a poor loading capacity of active sites due to ill-defined internal pore structures. Comparing to its organic counterpart, inorganic solid supports, such as silica, zeolite, alumina, zirconia, zinc oxide, clays, etc., generally show thermal and mechanical resistance. Moreover, the relatively high surface area and appropriate pore sizes maintain their competitive advantages over others.

Among the inorganic solid supports, silica is the most popular matrix due to its low cost, availability, mechanical robustness and feasible synthesis. The silicas consist of fully condensed silanoxy bridges ($\equiv$Si–O–Si$\equiv$) in the framework and silanol functional groups ($\equiv$Si–OH) on the surface. Three types of silanol groups are present on the silica surface: isolated, vicinal (hydrogen bonded) and geminal. The existence and relative amount of surface silanol groups is easily determined by IR spectroscopy. The density of silanol groups per gram of silica depends on surface area. Generally, a higher surface area results in a higher population of silanol groups. Silicas synthesized through the sol-gel route contain more silanol groups, due to incomplete condensation of molecule precursors. Thermal treatment of silica samples induces further condensation of silanol groups to form silanoxy bridges. This condensation process has been shown to be reversible, converting back to silonals by hydration. The scope of this review is mainly restricted in the silica supports, while at the end of chapter, we briefly discuss recent development of heterogeneous catalysts on other non-silica supports.

Notably, the structure and morphology of silicate support have a vital influence on the catalytic activity. Granular silica gel, widely used as desiccants, was frequently seen in the early literature serving as supports. Although several heterogeneous catalytic systems on silica gel were successfully demonstrated, the relative low surface area of such materials limited its usage. Later on, the zeolite replaced silica gel as a popular support. Zeolite is a crystalline alumisilicate with specific internal microporous (pore size <2 nm) structures. The strong Lewis acidic alumina center on the zeolite surface makes it widely used in industry to catalyze various important reactions. In addition, regular microporous channels of zeolite modulate the selectivity based on molecular sizes. Conversely, the size constraint of micropores considerably retards the diffusion of large molecules, which renders zeolite only reasonable for small substrates. A major breakthrough in the fabrication of large porous silica material was the development of the MCM family of mesoporous silicas (pore diameter = 2~50 nm) by the Mobil Corporation [3, 4]. By utilizing surfactants as structure-directing templates researchers at Mobil generated a range of MCM-type mesoporous silica structures with tunable pore size and pore morphology, such as MCM-41 and MCM-48 silicas consisting of hexagonal channels and cubic pores, respectively. This attractive material possessed some promising properties: such as high surface area, narrow pore size distribution and controllable pore size range from 2 to 100 nm, leading to many potential applications, such as catalysis, sensors, nanoelectronics, enzyme encapsulation and drug delivery. Over the past decade, several other mesoporous silica materials have also been developed with ordered porous structures, such as SBA- [5], MSU- [6] and FSM-type [7] of mesoporous silicas. The typical synthesis of these structurally well-defined mesoporous silicas is based on a surfactant micelle templating approach, as shown in the Figure 12.1. In an acidic or basic aqueous solution, organic surfactants, such as Pluronic P123 triblock copolymer or cetyltrimethylammonium bromide (CTAB), first form self-assembled rod-shaped micelles. These micelles serve as a structure-directing template that can interact with oligomeric silicate anions via hydrogen bonding or electrostatic interaction during the condensation reaction of tetraethoxysilane (TEOS). By either calcination or acid extraction, the organic surfactants are removed, leaving an inorganic mesoporous silica framework. Depending on the specific synthetic condition, a disordered, hexagonal, or cubic pore structure of mesoporous silica is obtained. Due to the virtue of high surface area and tunable pore size of mesoporous silicas, heterogeneous catalysts based on mesoporous silicas have flourished exponentially in the literature since 1992.

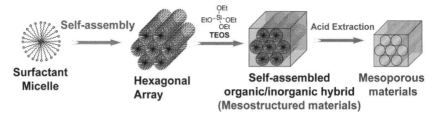

Figure 12.1 Formation mechanism of mesoprous silica materials.

12.2.2
Conventional Methods of Functionalization on Silica Surface

Two distinct methodologies have been developed for the immobilization of homogeneous catalysts on silica supports, based on types of intermolecular interaction between catalysts and supports. The following section contains several representative samples and a comparison of each method is also given.

12.2.2.1 Non-Covalent Binding of Homogeneous Catalysts

Homogeneous catalysts can be adsorbed on the silica surface through non-covalent interactions, such as van der Waals or electrostatic interactions, by which it provides a straightforward way to immobilize those catalysts. Even so, this weak interaction is so fragile that the catalyst readily leaches into the solution. Mazzei et al. first reported immobilization of a cationic rhodium complex onto a series of clays by adsorption [8]. The water or alcohol pre-treatment of mineral clays swelled the interlayer distance, allowing efficient intercalation of a large Rh complex (roughly 2 nm) into intercrystal space. The Rh intercalated clays catalyzed hydrogenation of (Z)-α-acetamidocinnamic acid to yield a product of enantiometric excess (e.e.) up to 72%. As expected, the activity and selectivity dropped dramatically upon recycling and reuse. Introducing capability of hydrogen bonding into the skeleton of catalysts improved the stability of supported catalysts. Bianchini et al. reported a chiral rhodium complex featuring a sulfonic functional group anchored on the silica surface through this approach, as shown in Figure 12.2 [9]. Catalytic asymmetric hydrogenation of dimethyl itaconate, ethyl trans-β-(methyl) cinnimate and α-(acetamido) acrylate was performed, which gave rise to products of low to moderate e.e. Multiple reactions were repeated to show no loss of catalytic activity when washing with non-polar solvent, whereas catalysts leaked considerably by washing with protic, polar solvent (methanol, ethanol). Later, it was reported that the specific triflate counter ion ($CF_3SO_3^-$) of metal complex facilitates the immobilization of catalysts. Broene et al. immobilized cationic rhodium complex of Me-DuPhos on the surface of mesoporous silica [10]. The activity and selectivity of the immobilized Rh catalyst was comparable with the homogeneous analog (Scheme 12.2). Flach et al. reported another strategy of entrapping a Rh complex into immobilized surfactants on the surface through hydrophobic interactions (Figure 12.3) [11]. No metal leaching was observed while repeating the

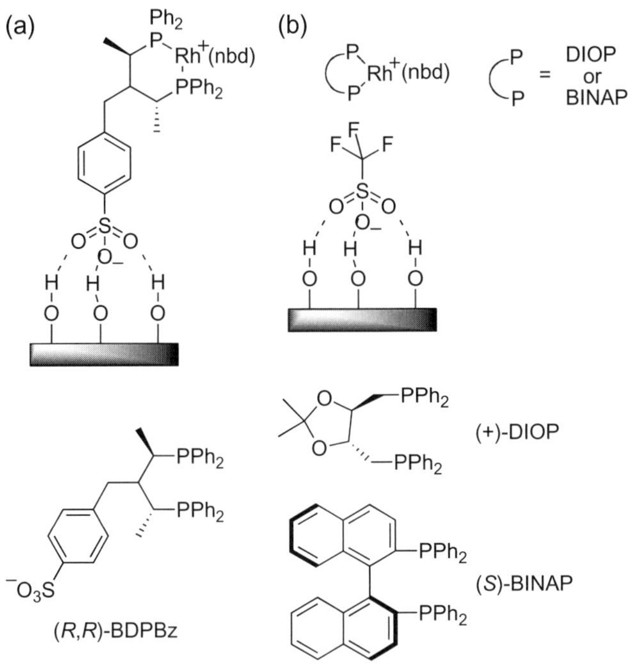

Figure 12.2 Immobilization of rhodium complex by adsorption through hydrogen bonding between surface silanol groups and pedant sulfonic acid (a) or triflate couterion of metal complex (b).

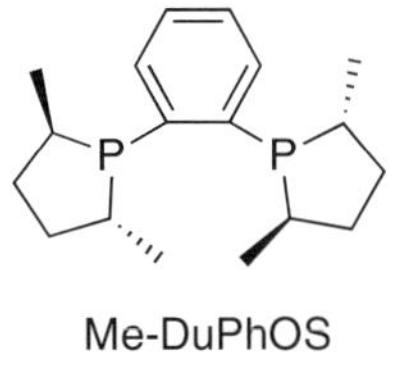

Me-DuPhOS

Scheme 12.2

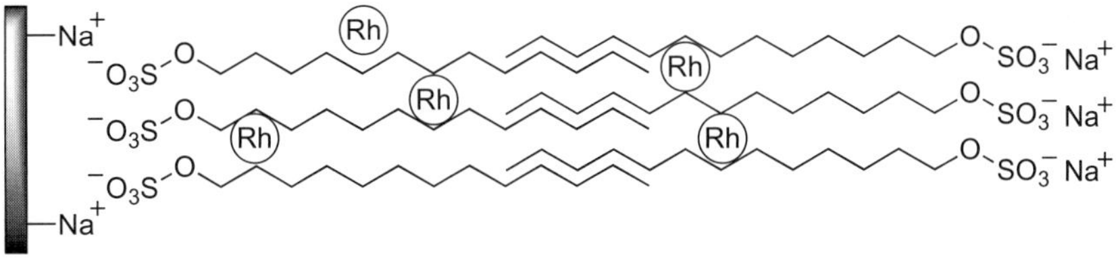

Figure 12.3 Entrapment of Rh complex through hydrophobic interaction in water.

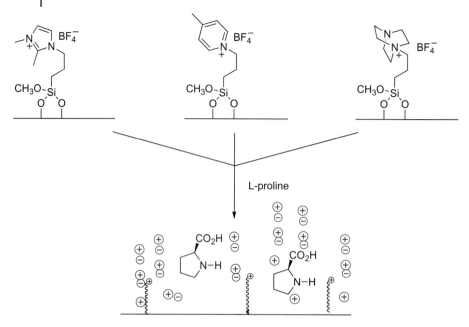

Figure 12.4 Heterogeneous proline catalyst anchored through surface adsorption by the assembly of amphiphilic ionic liquid film.

hydrogenation experiment ten times in water. Recently, a supported ionic liquid methodology was developed, by which method the catalysts attached on silica surface mediated by bipolar ionic liquids. Gruttadauria and co-workers reported a proline immobilized silica through this route (Figure 12.4) [12]. Notably, this surface-modified silica catalyst maintains high recyclability as well as enantioselectivity, much higher than that obtained by simple adsorption method. Moreover, this catalytic system avoids the viscosity problem when using ionic liquids as solvents on an industrial scale.

Encapsulation of catalysts by self-assembling inside the solid support is one method usually limited to immobilizing metal salen complex in the faujasite-type zeolite. The faujasite-type zeolite (zeolite X or Y) contains structurally well-defined inner cavities with 12 Å diameters, connected by smaller windows about 7.4 Å wide. The small precursors freely diffuse through the pores forming the desired metal salen and are later caged in the cavity (Figure 12.5). This so-called "ship-in-a-bottle" synthesis, initially reported by Herron and Stucky, was quickly adopted to prepare other metallosalens [13, 14]. Unfortunately, the encapsulated metallosalen gave lower *e.e.* compared to the homogeneous metallosalen. Both the reaction catalyzed by uncomplexed metal and the distorted geometry of the complex-substrate intermediate as the consequence of encapsulation were hypothesized as reasons for the deterioration of enantioselectivity.

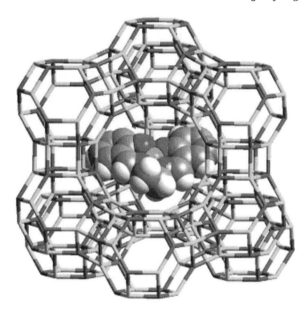

Figure 12.5 Representation of a *"ship in the bottle"* heterogeneous catalyst: a Co(Salen) complex was entrapped inside the cavity of zeolite Y. Light blue, Co; dark blue, N; red, O.

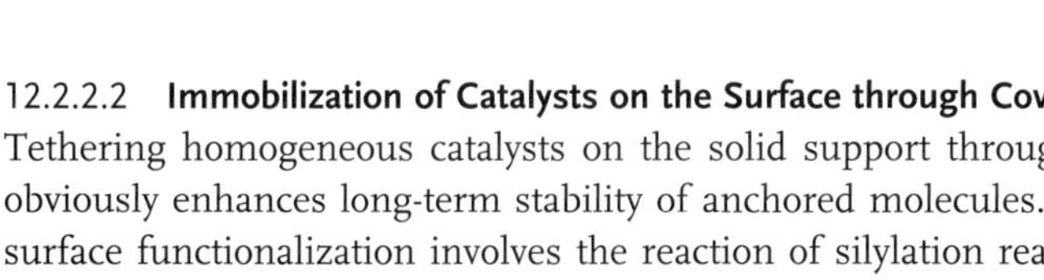

Trichlorosilane   Trialkoxysilane   disilazane

Scheme 12.3

12.2.2.2 Immobilization of Catalysts on the Surface through Covalent Bonds

Tethering homogeneous catalysts on the solid support through covalent bonds obviously enhances long-term stability of anchored molecules. The chemistry of surface functionalization involves the reaction of silylation reagents and surface silanol, which are germinal or isolated due to the reactivity. The chemical structures of the most common silylation reagents are listed in Scheme 12.3. Initially, cholorosilanes were used for surface immobilization; however, the discharge of hydrochloric acid caused damage to the silica structure. When the integrity of the porous silica was compromised, milder trialkoxysilanes were developed for the covalent attachment of homogeneous catalysts. The tripodal organosilanes, which are the most common due to synthetic reasons, usually do not condense completely, giving rise to mono- or dipodal anchoring of functional groups on the surface. This is attributed to the difficulty of geometrically locating three neighboring surface silanols close enough to bond with same silane. Even so, the anchored functional group is sufficiently bonded to the silica surface and avoids deterioration. Most recently, the silazane derivatives were found to be excellent silylation reagents. Silazane silylations are characterized by: (i) mild reaction condition, (ii) relative slow surface reaction and (iii) ease of releasing ammonium byproducts. Moreover, surface functionalization by silazanes expresses a dramatic steric influence. This enormous sterical dependence provides a way for multi-functionalization of silica surface. In terms of materials, the amount of available surface silanols relies on the synthetic condition of silicas. The silica obtained through a low-temperature process maintains more surface silanol than one through high-temperature calcination. Depending on the sequence of preparing single-site heterogeneous catalysts, the surface functionalization can be categorized into two methods: post-synthesis grafting and co-condensation.

12.2.2.3 Post-Grafting Silylation Method

Among various surface functionalization methods, the post-synthesis grafting method is the most popular approach for covalently incorporating functionalities on the silica surface. As illustrated in Figure 12.6, homogeneous catalysts can be immobilized on the surface of pre-synthesized silica supports through a silylation reaction in the moisture-free condition, by which self-condensation of organosilanes can be prevented. The structure of porous silicas remains intact after surface functionalization. However, it has been determined that most materials functionalized via the grafting method contain an inhomogeneous surface coverage of catalytic sites. This result has been attributed to the diffusion-dependent mass transport issue associated with these 3D porous materials. Given that the reactivity of surface silanol groups are indeed diffusion-driven, most catalytic sites introduced by grafting are located on the kinetically most accessible regions, such as the external surface and pore openings. In most extreme circumstances, congregated organic silanes might block the entrance of internal pores. The post-grafting of homogeneous catalysts is not only restricted to the silylation reaction but could

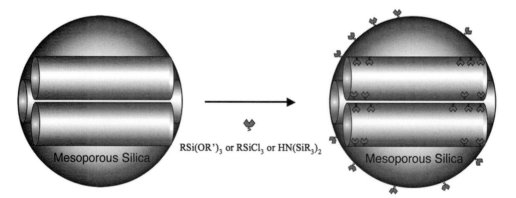

Figure 12.6 Schematic representation of surface functionalization through post-grafting method.

be achieved through surface organometallic chemistry. The preparation of these organometallic functionalized heterogeneous catalysts relies on the substitution of transition metal complex by silanol groups. Unlike grafting with organosilanes, the silanol groups directly attach to the metal center without any spacers. A variety of organometallic precursors have been discovered for direct surface substitution. Group IV and first-row metal complexes with alkyl ligands tend to graft on silica with the concomitant evolution of small alkanes (Figure 12.7). Since more than one alkyl ligand is present in the complex, multi-substitution could happen in principle. The control of the substitution and structure of the final surface-bound complex through this approach could be accomplished by carefully choosing appropriate metals and silicas. Thermal pre-treatment of silica samples determines the type of the majority of the surface silanol groups. For example, a mixture of mono- and bi-siloxy surface tantalum complexes [($\equiv$SiO)$_x$-Ta(=CHt-Bu)(CH$_2t$Bu)$_{3-x}$, n = 1 or 2] was found on a silica support thermal pre-aged at 500 °C upon grafting of Ta(= CHt-Bu)(CH$_2t$Bu)$_3$. In contrast, grafting the same precursors on silica, aged at 700 °C, gives exclusively a mono-siloxy complex [15]. This is simply due to the majority of isolated silanols that were formed upon thermal treatment at high temperature. Other transition metal precursors with labile ligands such as chloro- and alkoxyl groups are also good for surface immobilization. For instance, Maschmeyer *et al.* demonstrated the synthesis of a single-site titanium complex anchored MCM-41 catalyst [16]. The catalytic site was formed by reacting an internal silica surface with a particular titanocene dichloride bearing Ti–Cl bonds for the surface reaction and cyclopentadienyl groups as spatially blocking ligands for active site isolation (Figure 12.8). The experiments, including *in situ* X-ray diffraction, XANES and density functional theory calculation, demonstrated the conversion of the titanium precursor from an immobilized tripodal surface-bonded Ti species to an isolated anchored $\equiv$Ti–OH. Titanium oxide bonds (Ti–O–Ti) were absent in the EXAFS data, which again indicated the site isolation of catalytic centers. Furthermore, the anchored metal complex was able to transform into a more reactive catalyst through several surface organometallic chemistries, such as solvolysis, hydrogenolysis and pseudo-Witting reactions. The identification of the surface organometallic complex was spectroscopically

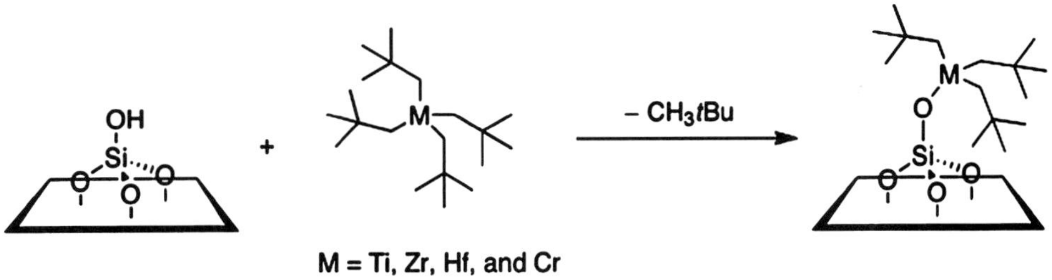

M = Ti, Zr, Hf, and Cr

Figure 12.7 Surface organometallic chemistry of group IV and first-row transition metal complex with tetraneopentyl ligands and silanol groups.

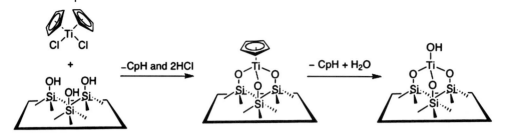

Figure 12.8 Surface immbilization of titanium complex by a sequential reaction of Cp_2TiCl_2 and silanol groups.

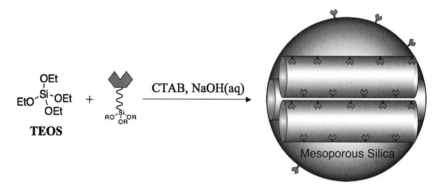

Figure 12.9 Schematic representation of incorporating organic functionality through co-condensation method.

characterized by the corresponding complex precursor reacted with polyhedral oligomeric silsesquioxanes (POSS), whose structure mimics vicinal, germinal and isolated surface silanol groups. Simulation through an appropriate POSS model established the detailed structure of immobilized surface species and provides information to understanding the reactivity and selectivity of the resulting heterogeneous catalysts at the molecular level.

12.2.2.4 Co-Condensation Method

Another immobilization approach, commonly used for preparing functionalized mesoporous silica supported catalysts, is the co-condensation method. This is a direct synthesis method where a given organoalkoxysilane is introduced to an aqueous solution of CTAB and TEOS during the condensation reaction (Figure 12.9). To efficiently incorporate organic functional groups on the silica surface, the organosilane precursors need to compete with silicate anions to interact favorably with the surfactant micelles by either electrostatic or other non-covalent interactions during the acid- or base-catalyzed condensation of silicate. Therefore, the choice of organosilane precursors for the co-condensation reaction is limited to those with organic functional groups that are soluble in water and can tolerate the extreme pH conditions that are required for the synthesis of mesoporous silicas and the subsequent removal of surfactants. In addition, attempting to incorporate bulky organosilanes usually interferes with silica condensation, which results in a less stable mesoporous silica structure. Also the amount of functional groups introduced by the co-condensation method often cannot exceed 25% surface coverage without destroying the structural integrity and the long-range periodicity of the synthesized materials. Despite these limitations, it has been investigated that the spatial distribution of the pore surface-immobilized organic groups in the mesoporous silica materials functionalized by the co-condensation method are more homogeneous than those of the post-synthesis grafting method, as recently reviewed by Stein and co-workers [17–19]. Ingenious utilization of interfacial force can help to design a heterogeneous catalyst with more control in terms of loading and spatial distribution of surface immobilized functional groups. Radu et al. reported an interfacial design of a modified co-condensation method to yield thiol functionalized mesoporous silicas [20]. The loading of surface thiol groups in the mesoporous silica prepared through this method was three times more than the conventional method. The increase of organic loading was associated with the enhanced interfacial electrostatic interaction between cationic cetyltrimethylammonium bromide (CTAB) and complementary charged organosilanes. To achieve this, the authors synthesized two different anionic organosilanes with disulfide linkages, 3-(3'-(trimethoxysilyl)-propyldisulfanyl)propionic acid (CDSP-TMS) and 2-(3-(trimethoxysilyl)-propyldisulfanyl)ethanesulfonic acid sodium salt (SDSP-TMS). The anionic functional groups of these organosilanes electrostatically matched with the cationic CTAB surfactant micelles in a NaOH-catalyzed condensation reaction of TEOS, as depicted in Figure 12.10. The affinity of CTAB surfactant micelles and anionic organosilanes should follow the sequence: citrate $< CO_3^{2-} < SO_4^{2-} < CH_3CO_2^- < F^- < OH^- < HCO_2^- < Cl^- < NO_3^- < Br^- < CH_3C_6H_4SO_3^-$, as discovered by Larsen and Magid [21, 22]. They reported that the anionic lyotropic series for interaction with the CTAB surfactant micelle is based on the enthalpy of transfer of the salt from water to a solution of 0.1 M CTAB. Thus, they concluded that anions less hydrated than bromide, such as sulfonate, would be able to replace bromide and bind tightly to the cetyltrimethylammonium head group of the CTAB molecule, thereby effectively mitigating the repulsion between these cationic head groups and stabilizing the micelle structure. Therefore, a higher loading of organic functional group was observed in the sulfonate functionalized mesoporous silica. Three organically functionalized mesoporous silica nanoparticles (MSN), MSN-COOH, MSN-SO$_3$H and MSN-SH, were synthesized by using CDSP-TMS, SDSP-TMS and mercaptopropyl trimethoxysilane (MP-TMS), respectively. Treating the resulting silica nanomaterials with a reducing agent, dithiothreitol (DTT), restored latent thiol functionality by disulfide exchange. The amount of thiol groups on the mesoporous silica surface was determined by both elemental analysis and UV-Vis spectroscopy. UV-Vis spectroscopy was utilized to calculate the byproduct (2-pyridothione) in the supernatant produced from the disulfide formation reaction of thiol functionalized MSN with aldrithiol-2 (Scheme 12.4). The amount of thiol groups measured from elemental analysis followed a trend in the order:

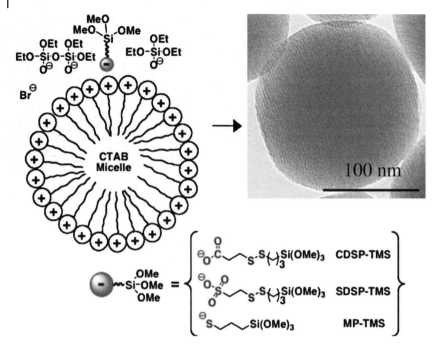

Figure 12.10 Schematic representation of using interfacial interaction between anionic organosilianes and cationic surfactants to control functionalization of the mesoporous silica nanoparticle (MSN) materials.

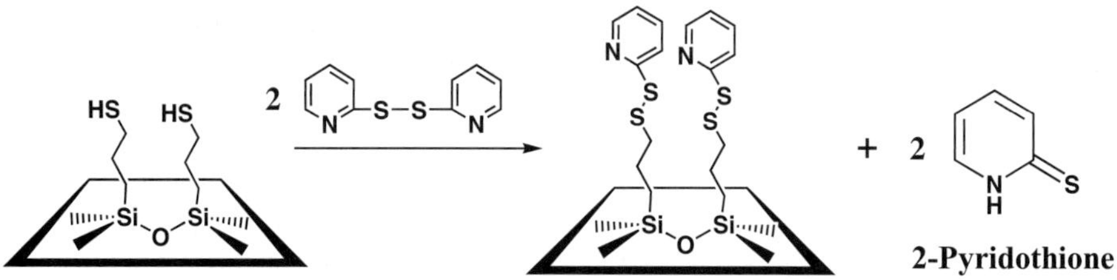

Scheme 12.4

MSN-COOH < MSN-SO$_3$H < MSN-SH (Table 12.2); whereas the one calculated from UV-Vis spectra exhibited an altogether different order: MSN-SH (0.56 ± 0.01 mmol g^{-1}) < MSN-COOH (0.97 ± 0.01 mmol g^{-1}) < MSN-SO$_3$H (1.56 ± 0.01 mmol g^{-1}). The discrepancy of quantification obtained from the two methods originated from a basic difference between the two measurements. Obviously, the elemental analysis data provided the total loading of thiol groups in the mesoporous silica catalysts no matter where the thiol functionality was located, either exposed on the surface or buried inside the framework. However,

Table 12.2 Elemental analysis of the organically functionalized MSNs.

Materials	C%	H%	S%
MSN-SH	13.23 ± 0.01	2.74 ± 0.01	10.09 ± 0.01
MSN-COOH	8.60 ± 0.01	2.33 ± 0.01	5.98 ± 0.01
MSN-SO$_3$H	9.79 ± 0.01	2.60 ± 0.01	7.95 ± 0.01

only chemically accessible thiol functional groups on the surface were counted in the UV-Vis measurement with aldrithiol-2 as a molecular probe. The higher thiol content in the MSN-SO$_3$H demonstrated that the spatial orientation and distribution of organic functional groups on the silica surface is possible by precise control of the interfacial interaction.

12.2.3
Alternative Synthesis of Immobilized Complex Catalysts on the Solid Support

The immobilization of homogeneous asymmetric catalysts for enantioselective organic reactions is an interesting direction in the formation of single-site heterogeneous catalysts due to the ability to recycle expensive chiral auxiliaries and valuable enantiopure compounds in the pharmaceutical industry. For most modern transition metal-mediated asymmetric reactions, a bulky and complicated chiral ligand with various functionalities is necessary. This is necessary to introduce chiral centers in the transition state by binding to a substrate through coordination, which increases electrondeficiency, and selectively blocking the trajectory of incoming nucleophile from a certain face. It is obviously synthetically challenging to introduce such sterically hindered chiral metal complexes on the surface; hence two distinct strategies of sequential and convergent approaches are widely used for this purpose. It is a challenge to both incorporate these sterically hindered metal complexes and preserve their reactivity on the surface. In the sequential approach, the organometallic was constructed stepwise through solid-phase synthesis. Starting from reactive functional groups already anchored, the metal complex was prepared in the solution phase with a pendent functional group for immobilization on the surface in the convergent route. For comparing both advantages and drawbacks, an example of immobilized metallosalen is described below. The salen manganese complex, initially reported by Jacobsen and co-workers, serves as an effective catalyst for asymmetric epoxidation of simple *cis*-alkenes, which represents a milestone in the asymmetric synthesis after the discovery of famous titanium-based Sharpless epoxidation [23]. Although the reactivity of a chiral salen complex is superior, the formation of inactive μ-oxo salen dimers turns out to be a common reason for deactivation. One possibility to avoid this deactivation pathway is to isolate each metallosalen by surface immobilization. Baleizáo *et al.* reported a synthesis of heterogenized salen chromium silica supported catalyst from sequentially imine formation to synthesize the salen

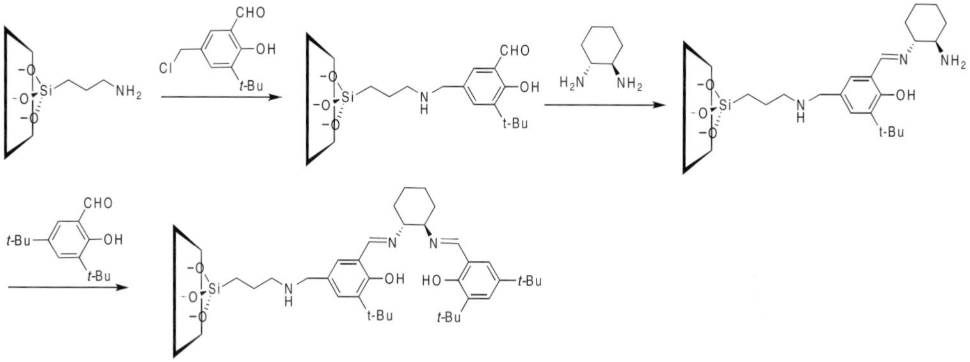

Scheme 12.5

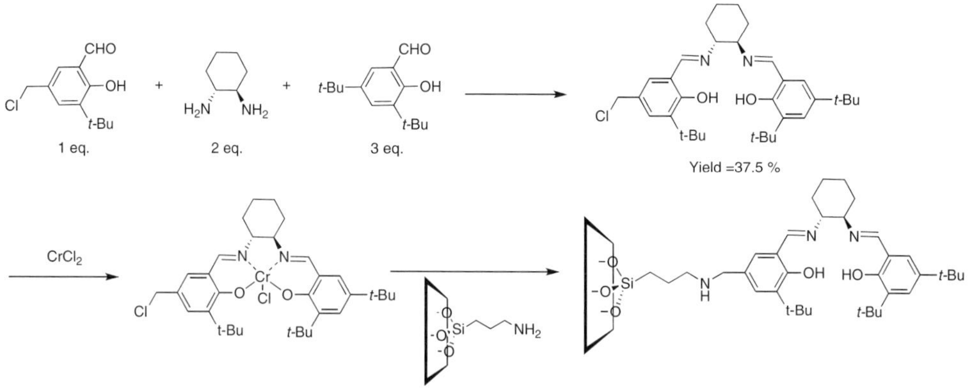

Scheme 12.6

ligand, followed by metallization with Mn, Cr, Co and Cu (Scheme 12.5) [24]. Apparently, simple separation of the product from the reaction mixture is expected to be an advantage of solid-phase synthesis. However, the major disadvantage of such an approach is from the uncertainty of reaction yield and characterization of surface species. Alternatively, heterogenization of the same salen chrominum complex reported by Baleizáo et al. was followed by the convergent route, in which the metal complex pre-synthesized in solution was tethered through substitution of amine functionalized mesoporous silica and asymmetric metal salen (Scheme 12.6). The chemical structure of metal complex precursors was ambiguously identified through well-developed characterization technique in solution, which gave more control over the surface-immobilized catalysts. Adversely, this approach required an enormous synthetic effort for the preparation of precursors, perhaps multiple steps were involved due to the sophisticated structure of the chiral ligand.

12.3
Design of more Efficient Heterogeneous Catalysts with Enhanced Reactivity and Selectivity

12.3.1
Surface Interaction of Silica and Immobilized Homogeneous Catalysts

Maintaining the reactivity and selectivity of heterogenized homogeneous catalysts is extremely important, since the catalytic activity of the heterogeneous system tends to decrease compared to homogeneous analogs. This reduced reactivity is understood as the restricted diffusion of substrates to reach the catalytic sites, which is often observed in the porous silica support. However, the retarded diffusion cannot be attributed to considerable alteration of selectivity, in particular enantioselectivity when asymmetric reactions are of interest. Even though the silica surface is usually considered to be inert, there is plenty of evidence supporting that some weak physical interactions exist between surface silanol groups and anchored organic functionality. For instance, the 3-aminopropyl functional group is one of the most popular surface-modifying silylation reagents useful in biological applications. It has been known for a long time that the amine functional group decreases the surface nucleophilicity. This observation was later investigated by solid-state NMR spectroscopy, where experiments indicated the presence of surface hydrogen bonding between immobilized amine functional groups and silanols, which explained the reduction of reactivity. The similar double hydrogen bonding phenomena of urea and silanols were also documented in the literature; Defreese *et al.* exploited this interaction to synthesize a mesoporous silica material with micropores formed by imprinting carbamate protected amino groups [25]. This negligible surface interaction slightly intervened and altered the conformation of immobilized catalysts on the surface compared to that in solution. The subtle alteration of the anchored complexes might change the optimized transition state in the reaction, giving an unpredictable enantioselectivity. When investigating a homogeneous catalyst for immobilization and conversion to heterogeous, one needs to examine the catalyst comprehensively due to all possible surface interactions.

One simple way to evade those surface interactions is to extend the spacer of catalytic moiety. The chiral copper bis(oxazoline) complex (BOX) is of tremendous interest for immobilization on the surface due to its high enantioselectivity in a variety of chemical reactions, such as cyclopropanation, Friedel–Crafts hydroxy-alkylation and the Diels–Alder reaction. Burguete *et al.* introduced two vinyl groups or vinylbenzene groups into methylene bridges of a chiral Cu(II)-BOX for immobilization by reacting with a thiol functionalized silica surface [26]. This intelligent coupling step involved a radical chain reaction that tolerated most functionalities and efficiently incorporated Cu(II)-BOX on the silica surface. The reaction studied was cyclopropanation of styrene (Scheme 12.7). The regioselectivity (*cis/trans ratio of substrates*) remained intact, while the enantioselectivity of heterogenous catalysts significantly dropped from the range of 29–80% obtained for the modified catalyst

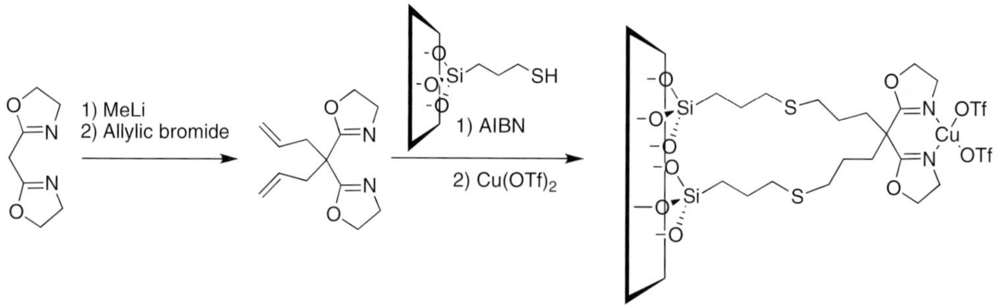

Scheme 12.7

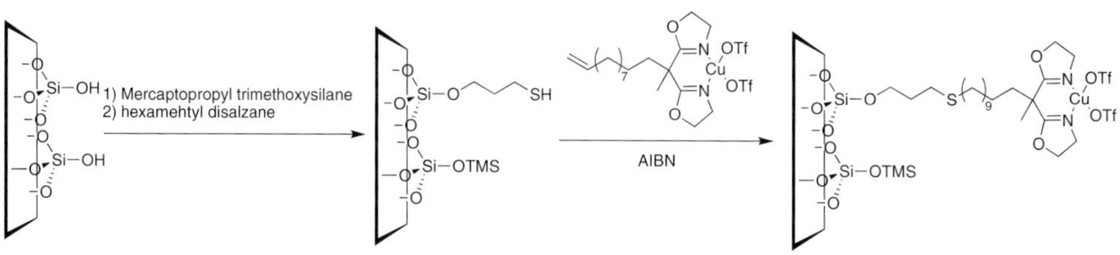

Scheme 12.8

performed in the homogeneous solution to 9–29% for both *cis-* and *trans-* isomeric products. A similar functionalization approach was adapted by Corma *et al.* to immobilize a Cu(II)-BOX complex containing a longer tether on the mesoporous silica surface (Scheme 12.8) [27]. The alkylation of 1,3-dimethoxybenzene with 3,3,3-trifluoropyruvate was catalyzed by Cu(II)-BOX functionalized mesoporous silica to investigate the reactivity gave an *e.e.* of 82% at 77% conversion, which was slightly lower than the homogeneous case. The resulting heterogeneous copper catalyst was recyclable and no metal content in the solution was detected. This study implied several important considerations regarding the design of heterogeneous catalysts. First, the elongated spacer minimized the support–complex interaction and second, the isolation of catalytic sites was preferred in order to prevent complex–complex interaction. Nevertheless, not all surface interactions existing on the interface is lethal to the catalytic activity; there are some cases showing promotion effect of surface interaction. Goettmann *et al.* recently designed a tethered rhodium complex system displaying a synergistic effect by metal complex–support interaction [28, 29]. The immobilization strategy relied on grafting a phophorous ligand on the mesoprous silica via a triethoxysilyl substituent, followed by the complexation with rhodium. The homogeneous analog of Rh complex showed both phosphine and oxygen in one of three ethoxy groups coordinated to a metal center by the evidence from ^{31}P NMR and low-temperature ^{1}H NMR measurement. The chemical shift of the heterogeneous catalyst in solid-state ^{31}P

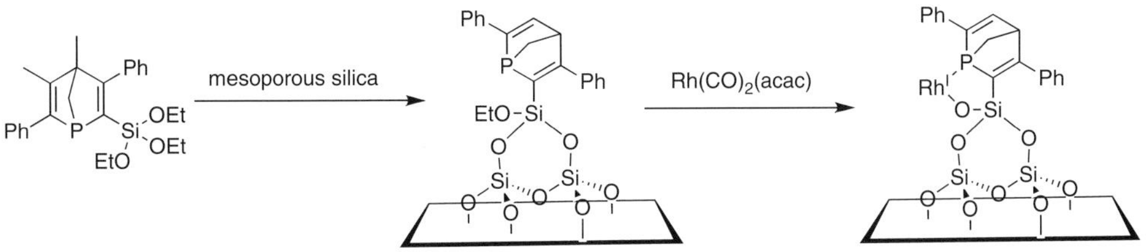

12.3 Design of more Efficient Heterogeneous Catalysts with Enhanced Reactivity and Selectivity

Figure 12.11 Immobilized Rh(I) phosphine complex for hydrogenation. Carbonyl ligand on Rh are omitted for clarity.

NMR resonance agreed with that of the homogeneous precatalyst, which implied a similar bidentate coordination environment. The counterion of the immobilized catalysts dissociated upon the complexation, based on IR and NMR studies. All structural analysis suggested that the rhodium complex anchored on the surface in a dipodal fashion with the third arm of negative charged oxy group, after hydrolyzation, binds to the rhodium metal as denoted in Figure 12.11. A hydrogenation of 1-hexene catalyzed by both catalysts gave an unprecedented result. The homogeneous metal complex showed no catalytic activity of hydrogenation, while the same reaction catalyzed by the heterogeneous catalyst gave a turn-over number (TON) in the range of 15–72. Since the steric effect offered by the silica surface cannot be responsible for this reactivity enhancement, the authors concluded that by covalently immobilizing the Rh center close to the surface, the electronic state of the organic ligand was manipulated in a way that promoted catalytic activity at the metal center.

12.3.2
Reactivity Enhancement of Heterogeneous Catalytic System Induced by Site Isolation

As mentioned in Section 12.2.2.2, the post-grafting approach is a straightforward functionalization method, but usually results in multiple catalytic sites due to undesired intermolecular interactions. For instance, grafting simple aminopropyl organosilanes on the silica surface creates ill-defined amine sites with potentially different chemical reactivities, due to the hydrogen bonds between basic amine sites and acidic surface silanols, as well as self-condensation. While attempting to further functionalize through this surface amine functional group, the steric hinderance or hydrogen bonding from proximal amine sites can prevent high conversion and also give different catalytic sites. Jones and co-workers reported a clever design, which utilized a bulky 3,3,3-triphenylpropanal to protect the aminopropyl trimethoxysilane by forming an imine linkage (Figure 12.12). [30] The steric constraint of the resulting tritylimine organosilanes forced space between individual amino functional groups when grafting on SBA-15, even at high loadings. Furthermore, the residual silanols were capped with trimethylsilyl (TMS) groups to passivate the surface. This passivation step served two reasons: first, it diminished

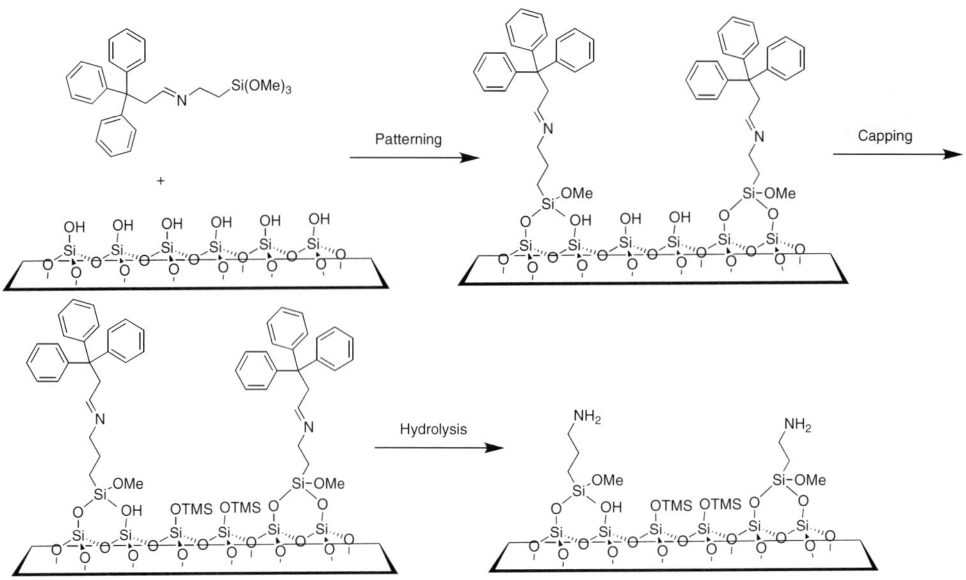

Figure 12.12 Single-site immobilized amine functionalized catalyst isolated through steric hindered protecting groups.

possible problematic hydrogen bonds between isolated amino groups and the silica surface and second, capping silanol groups with TMS provided a more hydrophobic surface which hindered the water adsorption on the surface, improving the recyclability of the heterogeneous catalysts. The author demonstrated that isolated single amine sites could be attained by examining the chemical reactivity of the amine group with chlorodimethyl(2,3,4,5-tetramethyl-2,4-cyclopentadiene-1yl)silane (Cp-silane). The resulting site-isolated Cp-amino functionalized SBA-15 material was subjected to coordination with Ti metal to yield a heterogeneous Ti functionalized mesoporous silica catalyst for ethylene polymerization. This catalyst showed higher reactivity than similar material prepared by the traditional grafting method.

12.3.3
Introduction of Functionalities and Control of Silica Support Morphology

As the homogeneous catalyst is immobilized on the surface, the catalytic activity of this heterogeneous system is no longer only governed by the anchored catalyst itself. The mass transport of substrates into the catalytic sites, controlled by the physical structure of the solid support, also determines the catalytic performance. For this reason, mesoporous silica nanoparticles (MSN) are of particular interest because of the enormous internal surface area, tunable large pore sizes and short diffusion length. Therefore, only by controlling both surface functionalization and particle/pore morphology of this solid support can we deconvolute and study all

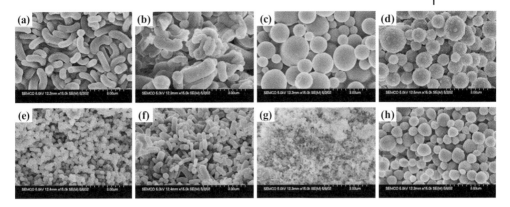

Figure 12.13 FE-SEM micrographs of (a) AP-MSN, (b) AAP-MSN, (c) AEP-MSN, (d) UDP-MSN, (e) ICP-MSN, (f) CP-MSN, (g) AL-MSN, (h) pure MCM-41 silica. The image magnification is same for all of the images (scale bar = 3 μm).

interconnected properties of the heterogeneous catalyst system. Huh et al. demonstrated such control using the aforementioned interfacial co-condensation method to synthesize a series of MSNs functionalized with 3-aminopropyl trimethoxysilane (APTMS), N-(2-aminoethyl)-3-aminopropyl trimethoxysilane (AAPTMS), 3-[2-(2-aminoethylamino)ethylamino]propyl trimethoxysilane (AEPTMS), ureidopropyl trimethoxysilane (UDPTMS), 3-isocyanatopropyl triethoxysilane (ICPTES), 3-cyanopropyl triethoxysilane (CPTES) and allyl trimethoxysilane (ALTMS) [31]. Here, these mono-functionalized mesoporous silicas are referred as X-MSN, where X represents the corresponding organotrialkoxysilane. The particle morphology of the organically functionalized MSN were found to strongly depend on the organic precursors, as showed in FE-SEM micrographs (Figure 12.13). Generally, the MSNs functionalized with hydrophilic organic precursors resulted in larger particle sizes, whereas the MSNs functionalized with hydrophobic organosilanes formed smaller particles. The ^{13}C and ^{29}Si solid-state NMR spectra of these MSN materials confirmed the covalently bonded organic functional groups on the mesoporous silica surface. The textual properties of these MSNs with different organic functional groups are summarized in Table 12.3. The powder XRD spectra of these materials indicated different pore structures were formed from regular hexagonal mesopores in the case of hydrophobic silanes to disordered worm-like pore structure by hydrophilic organosilanes, as depicted in Figure 12.14. The formation mechanism of functionalized silicas is shown in Scheme 12.9, which explains the aforementioned organosilane dependence by interfacial interactions between surfactants, oligomeric silicate and organosilanes. For non-polar organosilanes (R_2), intercalation of their hydrophobic groups into the micelles stabilizes the formation of long individual cylindrical micelles. This uniform assembly of organosilanes at the Gouy–Chapman region of micelles would facilitate the rapid cross-linking/condensation between the "micelle-oriented" trialkoxysilyl groups in the basic aqueous solution. The resulting "side-

Table 12.3 Structural properties of the organically functionalized mesoporous silica materials.

Sample	d_{100} (Å)[a]	a_0 (Å)[a]	S_{BET} (m²g⁻¹)[a]	V_p (cm³g⁻¹)[a]	W_{BJH} (Å)[a]	$d_{pore\ wall}$ (Å)[a]	Amount of organic group (%)[b]
AP-MSN	39.8	46.0	721.7	0.45	23.7	22.3	12
AAP-MSN	41.3	47.7	664.6	0.48	25.9	21.8	5
AEP-MSN	38.4	44.4	805.8	0.57	26.0	18.4	7
UDP-MSN	43.7	50.5	1022.4	0.78	28.6	21.9	6
ICP-MSN	39.8	46.0	840.1	0.66	25.8	20.2	14
CP-MSN	39.4	45.5	1012.5	0.68	23.5	22.0	10
AL-MSN	33.7	38.9	1080.5	0.65	19.7	19.2	11
MCM-41[c]	38.1	44.0	767.1	0.55	25.5	18.5	–

a) The BET surface area (S_{BET}), the mesopore volume (V_p), and the mean mesopore width (W_{BJH}) were obtained from the nitrogen adsorption/desorption data. The d_{100} numbers represent the d-spacing corresponding to the main (100) XRD peak. The unit cell size (a_0) is calculated from the d_{100} data using $a_0 = 2d_{100}/3^{1/2}$. The pore wall thickness ($d_{pore\ wall} = a_0 - W_{BJH}$).
b) The amounts of organic functional groups incorporated to the silica materials were estimated from the ^{29}Si DPMAS.
c) Pure MCM-41 silica synthesized under the same reaction condition without any addition of organoalkoxysilane.

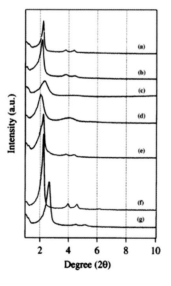

Figure 12.14 XRD spectra of the (a) AP-MSN, (b) AAP-MSN, (c) AEP-MSN, (d) UDP-MSN, (e) AL-MSN, (f) CP-MSN, (g) ICP-MSN, (h) MCM-41 silica without organic functional groups.

12.3 Design of more Efficient Heterogeneous Catalysts with Enhanced Reactivity and Selectivity

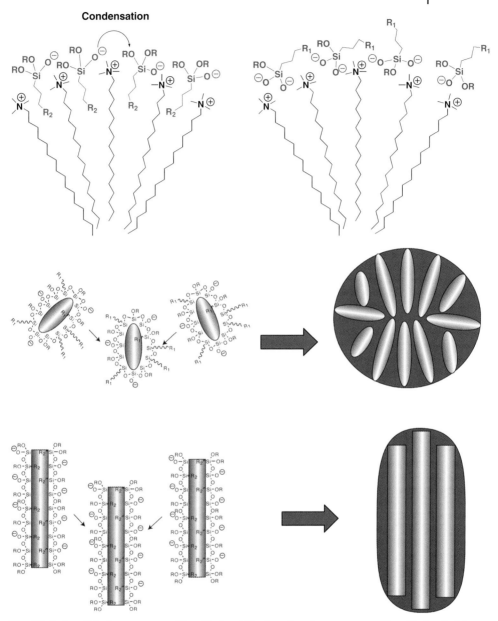

R = Methyl or Ethyl groups; functional groups

R_1 = Hydrophilic functional groups;

R_2 = Hydrophobic

Scheme 12.9

on" packing of the silicate-coated cylindrical micelles would give rise to small rod-like nanoparticles. A similar phenomenon was also observed by Cai and coworkers, while utilizing NaOH and NH_4OH as catalysts to manipulate the particle morphology of the MCM-41 silicas without organic functional groups [32]. In contrast, hydrophilic polar organoalkoxysilane precursors (R_1), would inhibit the formation of long micelles and reduce their tendency toward side-on condensation. The difference in the rate of condensation between organosilicate-coated micelles versus that of the free silicate (TEOS) molecules would likely be small. Because of the lack of thermodynamic incentives for the silicate-coated micelles to pack in an ordered fashion, such co-condensation reactions should yield particles with randomly oriented pore structures. A similar phenomenon was recently reported by Mann and co-workers, who observed the growth of the mesoporous silica particle in the direction that is perpendicular to the pore-alignment upon the introduction of amine-containing organoalkoxysilanes [33].

12.3.4
Selective Surface Functionalization of Solid Support for Utilization of Nanospace Inside the Porous Structure

The nanovoid inside the porous silicas indicates a 3D well-defined cavity where the chemical reaction might be manipulated through this spatially confined area and then induce a different control in reactivity and selectivity. However, a selective surface functionalization method differentially modifying the external and internal surface is a prerequisite for realizing the aforementioned application. Shephard *et al.* demonstrated that multiple grafting steps of mesoporous silica with different organic functionalities are possible by a kinetic control strategy [34]. By carefully controlling the reaction temperature and time of grafting diphenyl dichlorosilane (Ph_2SiCl_2) with a calcined mesoporous silica support, the silylation (or passivation) of diphenyl silyl groups preferential occurred on the kinetically most accessible external silica surface. Sequentially the internal channels were decorated with aminopropyl trimethoxysilane. To visualize distribution of organic functional groups, tethered aminopropyl group was utilized as an anchor point for a ruthenium cluster, which served as a stain in high-resolution transmission electron microscopy (HRTEM). Based on the HRTEM results, the author concluded that the aminopropyl groups are almost entirely located on the internal surface through this kinetic control approach. De Juan *et al.* utilized an alternative approach to address the selectivity of surface functionalization [35]. The first grafting step is carried out on as-synthesized MCM-41 with surfactant template-filled mesopores. Due to the steric restrictions in the surfactant-filled channels, the grafted organosilane is mainly deposited on the exterior surface. After surfactant removal by extraction, the interior mesochannel surface is restored and ready for sequential immobilizations of a second functionality. However, caution needs to be heeded in the assumption that silylation reagents are excluded from the channels of as-made MCM-41, particularly when high concentrations of less-hindered silylation reagents are involved. Johnson *et al.* applied a similar sequential grafting

12.3 Design of more Efficient Heterogeneous Catalysts with Enhanced Reactivity and Selectivity

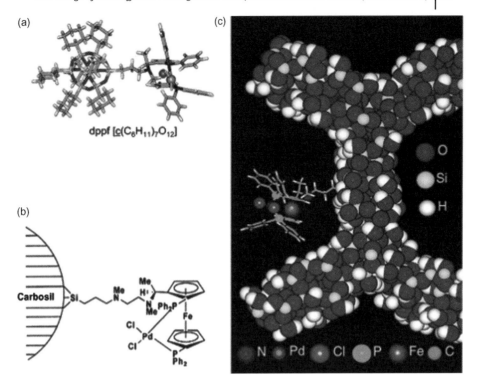

Figure 12.15 Schematic representation of dffp-diamine palladium dichloride catalysts attached to (a) silsesquioxane structure (homogeneous form), (b) non-porous silica (Carbosil), (c) confined MCM-41 solid support.

method to immobilize a palladium-1,1'-bis(diphenylphosphino)ferrocene (Pd-dppf) complex on the inner mesochannels of MCM-41 (Pd-MCM41). [36] The catalytic activity of the resulting Pd-MCM41 is investigated by the allylic amination (the "Trost–Tsuji" reaction) of cinnamyl acetate and benzylamine, as shown in Figure 12.15. For comparison, an unconfined Pd immobilized heterogeneous catalyst, consisting of the same complex immobilized on the concave surface of the non-porous silica support, and a homogeneous analog, prepared by linking a Pd-dppf complex with the cyclohexyl silsesquioxane, were employed as controls. Unlike the homogenous analog or Pd-dppf immobilized silica particle, by which the reaction yielded a majority of the linear product, a branched amination product with 51% selectivity is produced by Pd-MCM41, as indicated in Table 12.4. Meanwhile, the enantioselectivity of the branched amination product is also dramatically changed from 43% by solid silica particle support to 95% by porous MCM-41 support. The increase of regio- and enantioselectivity in the mesoporous silica supported catalyst is rationalized by the authors to the confinement effect of supported silica structure. Even though the regioselectivity of the reaction catalyzed by Pd-MCM41 is not high, this study explicitly demonstrated the improvement of

Table 12.4 Allylic amination of cinnamyl acetate and benzylamine using (S)-1-[(R)-1,2′-bis(diphenylphosphino) ferrocenyl]ethyl-N,N′-dimethylethylenediamine palladium dichloride in homogenous form, tethered to non-porous silica or confined inside the MCM-41.

Ph⌒⌒OAc + Ph⌒NH₂ →[Pd catalyst] Ph⌒⌒NHCH₂Ph (a) + Ph⌒⌒(NHCH₂Ph) (b)

Pd catalysts	Conversion (%)	a (%)	b (%)	e.e. (%)
Homogeneous	76	99	–	–
Non-porous silica	98	98	2	43
MCM-41 confined	99	50	50	95

heterogeneous catalyst by clever utilization of confined nanospace inside the porous silica support.

Given that the MCM-41-type mesoporous silica only contains 1D parallel channels, the diffusion of substrates into the internal mesopores is controlled by the pore entrance. The physical constraint of pore size can be used to manipulate the substrate diffusion depending on the relative size of substates to pore diameters. The larger the molecules are, the slower diffusion will be. Considering this size-dependent diffusion, one can achieve unique selectivity of heterogeneous catalyst if most catalytic sites are located inside the porous channels. Chen et al. reported a 4-(dimethylamino)pyridine functionalized mesoporous silica nanoparticle (DMAP-MSN) material, which exhibited a superior reactivity and product selectivity for several industrially important reactions, such as Baylis–Hillman, acylation and silylation [37]. In contrast to post-grafting method, the selective functionalization was accomplished through the aformentioned co-condensation condition, which is a one-pot synthesis to obtain a heterogeneous catalyst with spatially well-organized catalytic sites situated in the 3D cavity, by adding 4-(N-[3-(triethoxysilyl) propyl]-N-methyl-amino)pyridine (DMAP-TES) and tetraethoxysilane to an aqueous solution of sodium hydroxide with low concentration of CTAB to yield the DMAP-MSN catalyst, as depicted in Scheme 12.10. The resulted DMAP-MSN catalyst showed a spherical particle shape with an average diameter of 400 nm. The chemical shifts obtained from the ^{13}C solid-state NMR spectra matched well with the solution data of the organic precursors, as shown in Figure 12.16, which confirmed the presence of the DMAP functionality on the mesoporous silica surface. It was interesting to note that the heterogeneous MSN catalyst synthesized this way only contains the free base form without any indication of protonated or hydrogen-bonded species present on the surface, which explained the remarkable reactivity of DMAP-MSN (Table 12.5). The catalytic performance of the DMAP-MSN catalyst was investigated by three different nucleophilic reactions, Baylis–Hillman, acylation and silylation reactions, and compared with the homogeneous catalyst. For the Baylis–Hillman reaction, the reactivity of α,β-unsaturated ketone

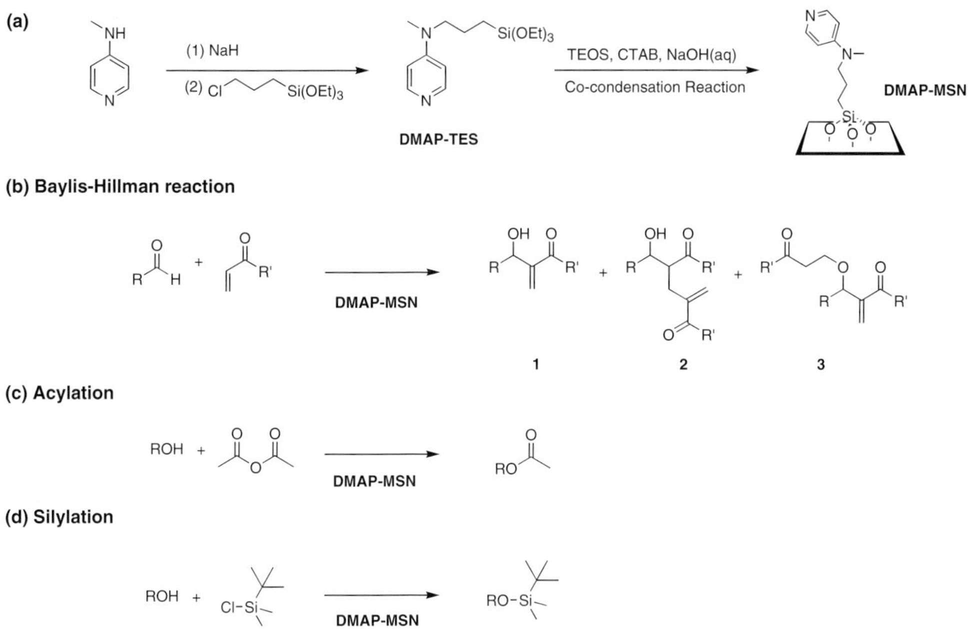

12.3 Design of more Efficient Heterogeneous Catalysts with Enhanced Reactivity and Selectivity

Scheme 12.10

(a) Synthesis of DMAP-MSN via DMAP-TES intermediate (NaH; Cl-propyl-Si(OEt)₃; then TEOS, CTAB, NaOH(aq) Co-condensation Reaction)

(b) Baylis-Hillman reaction — producing compounds 1, 2, 3 via DMAP-MSN

(c) Acylation: ROH + acetic anhydride → RO-acetate, DMAP-MSN

(d) Silylation: ROH + Cl-Si(tBu)(Me)₂ → RO-Si, DMAP-MSN

Table 12.5 ^{13}C chemical shifts observed in DMAP-MSN and in the reference compounds (DMAP-TES, 4NMe₂-Py).

Sample	Solvent	C1	C2	C3	C5	C6	C7,C11	C8,C10
DMAP-TES	CDCl₃	7.8	20.1	54.0	37.5	153.5	106.6	149.8
DMAP-MSN	Solid-state	8.8	19.3	53.0	35.8	153.8	106.1	147.8
4NMe₂-Py[a]	CDCl₃				38.4	153.6	106.0	149.1
	CF₃COOH in CDCl₃				40.1	153.6	105.6	139.9
	CF₃SO₃H in CD₃NO₂				47.3	157.8	120.6	145.6

a) Data taken from ref [28].

substrates follow the order: methyl vinyl ketone > cyclopentenone > cyclohexenone. In the case of the activated aldehyde, 4-nitrobenzaldehyde, only the desired product was obtained within high yield in the experiment. In contrast, the same reaction catalyzed by the homogeneous DMAP molecules, resulted in a mixtures of products, including the diadduct (compound **2**), Michael addition product

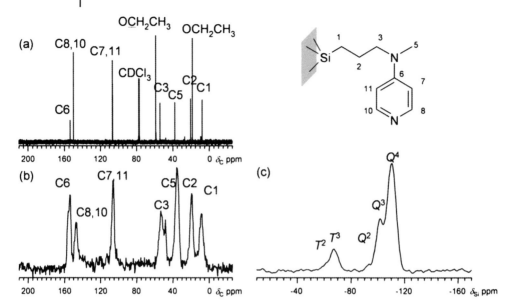

Figure 12.16 (a) ^{13}C NMR spectrum of DMAP-TES in CDCl$_3$ solution. (b) ^{13}C CPMAS spectrum of DMAP-MSN resulting from 12 000 scans acquired with a delay of 1 s in a 5 mm probe ($v_R = 10$ kHz). During each CP period of 1.5 ms, v_{RF}^H was ramped between 16 and 40 kHz (in 11 steps), while v_{RF}^H was set to 36 kHz. The v_{RF}^H fields of 83 and 65 kHz were applied to protons during initial excitation and high power decoupling, respectively. (c) ^{29}Si DPMAS spectrum of DMAP-MSN obtained with the same probe using CPMG acquisition (ten echoes). A total of 600 scans were collected with a delay of 300 s to allow the complete relaxation of ^{29}Si nuclei.

(compound **3**), and some oligomerized products (Scheme 12.11). Given that the diadduct **2** and the Michael addition side product **3** could only be generated from the Baylis–Hillman product **1**, the excellent product selectivity of DMAP-MSN could be attributed to the difference in the rate of diffusion to the "active sites." These active sites are located inside the pores between the aldehyde reactant and compound **1**, which would serve as the reactant for the undesired side reactions. The DMAP-MSN has been recycled more than ten times without losing any catalytic reactivity. Moreover, the TON of the DMAP-MSN catalyst was as high as 3340 for 24 days.

12.3.5
Cooperative Catalysis by Multi-Functionalized Heterogeneous Catalyst System

Unlike artificial catalysts, natural enzymes consist of multi-functionalities inside the pocket employed as catalytic sites to synergistically activate the substrate through general acid and base moieties or recognition sites providing specificity through hydrogen bonding. We are hoping to incorporate multifunctionalities into the heterogeneous catalyst to mimic the biological system and efficiently catalyze

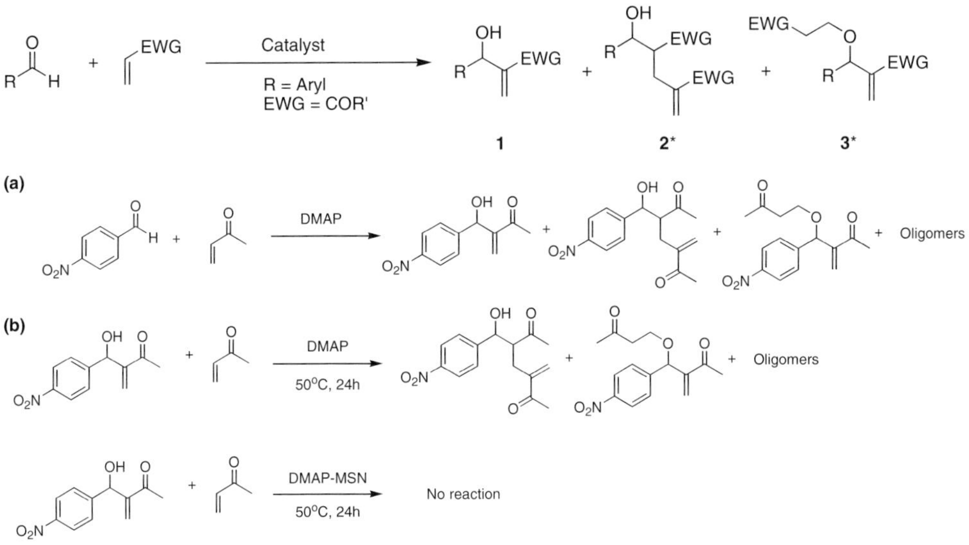

12.3 Design of more Efficient Heterogeneous Catalysts with Enhanced Reactivity and Selectivity

Scheme 12.11

chemical reactions with enhanced reactivity and selectivity. As mentioned previously, a comprehensive comparison of heterogeneous catalyst system can be concluded if both the factors of the immobilized catalytic sites and the structure of the catalyst support could be separated. By introducing two organosilanes with different structure-directing abilities as precursors of the co-condensation reaction, Lin and co-workers utilized one precursor with *stronger structure-directing ability* to create the desired pore and particle morphology and employ the *other* for selective immobilization of catalysts [38]. Through this strategy a series of multi-functionalized mesoporous silica catalysts with *control of both morphology and functionalization* are achieved. As a proof of principle, a series of bi-functionalized MSN materials with 3-[2-(2-aminoethylamino)ethylamino] propyl (AEP) and 3-cyanopropyl (CP) functional groups were prepared by varying the molar ratio of organosilane precursors, AEPTMS and CPTMS, while keeping the total amount of organosilanes the same (12.8% molar ratio to TEOS). The SEM micrographs of all AEP/CP-MSNs showed exclusively spherical particles (Figure 12.17). Given the fact that the mono-functionalized AEP-MSN and CP-MSN are spherical and rod-shaped particles, respectively, the shape of bifunctional AEP/CP-MSNs is governed by the structure-directing ability of the AEPTMS precursor in the co-condensation reaction. The powder XRD spectra and TEM micrographs of AEP/CP-MSNs materials indicated that all bifunctionalized MSNs share the same wormhole-like porous structures as observed in the case of the mono-functionalized AEP-MSN. The relative concentrations between the two organic functional groups measured by solid-state NMR could be manipulated by initial molar ratios of organosilanes during the synthesis.

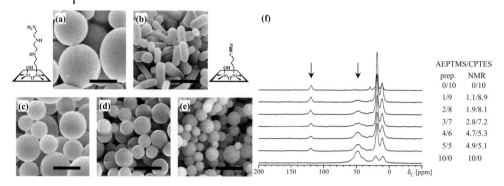

Figure 12.17 SEM images of (a) AEP-MSN, (b) CP-MSN, (c) 5/5 AEP/CP-MSN (d) 3/7 AEP/CP-MSN, (e) 1/9 AEP/CP-MSN. Scale bar = 1 μm. (f) ^{13}C CPMAS spectra of monofunctionalized (top and bottom traces) and bifunctionalized (middle traces) AEP/CP-MPs. Arrows highlight the resonances that are unique for each species and thus were used for quantitative analysis. The numbers represent the molar ratio between two components used for preparation (left column) and obtained form analysis of NMR spectra (right column).

12.3.6
Tuning the Selectivity of Multi-Functionalized Hetergeneous Catalysts by Gatekeeping Effect

The ability to anchor two types of functionalities on mesopore walls without changing the particle and pore morphologies allows researchers to further use multifunctionalized mesoporous silicas as new heterogeneous catalyst support. Herein, one functionality can be a dedicated catalytic functional group through the immobilization of homogeneous catalyst of interest and the other(s) will serve as "cofactors" to manipulate the reactivity and/or selectivity of the resulting catalytic system. If these auxiliary groups are chosen properly, they can regulate the diffusion of substrates in and out of the pores. In such a multi-functional catalyst, the selectivity of the reaction depends on the affinity of substrates and the gatekeeper. To investigate the influence of gatekeepers, three bifunctionalized MSN catalysts were synthesized under similar condition to give the heterogeneous catalyst system functionalized with a common 3-[2-(2-aminoethylamino)ethylamino]propyl (AEP) group and three different gatekeeper groups, ureidopropyl (UDP)-, mercaptopropyl (MP)- or allyl (AL)-functionalities [39]. As depicted in Figure 12.18, the particle morphology and mesoporous structure of all three MSN are controlled to be spherical in shape and disordered wormhole mesopores. Moreover, three bifunctional MSNs exhibit similar BET surface area measured by N_2 sorption analysis range from 805 to ~703 m^{2}g^{-1}, While the BJH pore distribution measurement shows larger pores as 26.0 and 22.9 Å in the AEP/UDP- and AEP/MP-MSN and smaller pores ca. 15 Å in the AEP/AL-MSN. The total amount of all organic groups in each MSN catalyst, quantified through solid-state ^{29}Si NMR measurement, is

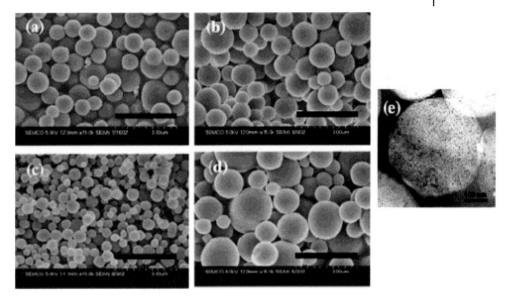

Figure 12.18 SEM images of (a) AEP/UDP-MSN, (b) AEP/MP-MSN, (c) AEP/AL-MSN, (d) AEP-MSN. TEM micrograph of the ultramicrotomed sample of AEP/AL-MSN (e). Scale bar is 3 μm for (a–d) and 100 nm for (e).

approximately the same (1.0 mmol g^{-1} in AEP/UDP-MSN, 1.4 mmol g^{-1} in AEP/MP-MSN and 1.3 mmol g^{-1} in AEP/AL-MSN). The relative molar concentration between AEP and the secondary groups were also very similar (1.17, 1.04 and 1.13 for AEP/UDP-MSN, AEP/MP-MSN and AEP/AL-MSN).

A competitive nitroaldol (Henry) reaction was performed on these bifunctional MSNs in the presence of equal concentrations of 4-hydroxybenzaldehyde and one of the three 4-alkoxybenzaldehydes (**5, 6, 7**) as shown in Figure 12.19. The reaction selectivity catalyzed by each MSN was given by the ratios of the nitroalkene products (**13/12, 14/12, 15/12**). As depicted in Figure 12.20, no selectivity for any combination of reactants was observed when the reaction was catalyzed by the mono-functionalized AEP-MSN or the hydrophilic AEP/UDP-MSN catalysts. An increase of reaction selectivity towards the hydrophobic products of alkoxybenzaldehydes (**5, 6, 7**) was clearly obtained in the cases of AEP/MP-MSN and AEP/AL-MSN catalysts, where the catalytic AEP groups are situated in the mesopores decorated with hydrophobic groups. The results suggested that the secondary hydrophobic group (MP, AL) plays a significant role in preferentially allowing more hydrophobic reactants to penetrate into the mesopores and react with the AEP catalytic group. To verify this solvation hypothesis, a pair of benzaldehydes (**4, 7**) with different polarities were tested for their relative solubility in 1-propanethiol, which simulated the microenviroment of the mercaptopropyl-functionalized mesopores. The different solubility of hydrophilic and hydrophobic benzaldehydes in the 1-propanethiol implied that the selectivity of our catalysts

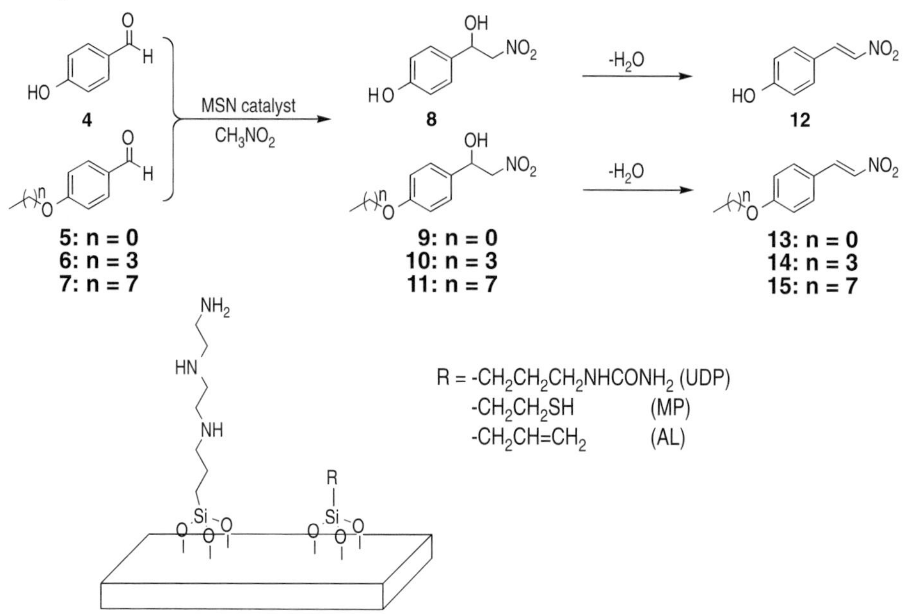

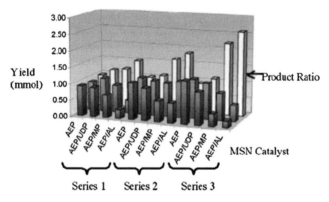

Figure 12.19 Competitive nitroaldol (Henry) reaction of equal amount of hydrophilic and hydrophobic benzaldehydes catalyzed by AEP/UDP- and AEP/MP-, AEP-AL-MSNs, respectively.

Figure 12.20 Histogram of the competitive nitroaldol reactions. Blue bars: Yield of **12** (mmol). Red bars: Yield of **13**, **14**, or **15** (mmol). White bars: Molar ratio of products (**13/12**, **14/12**, or **15/12**). Series 1, 2, 3 are experiments conducted with reactants **4** and **5**, **4** and **6**, **4** and **7**, respectively.

12.3 Design of more Efficient Heterogeneous Catalysts with Enhanced Reactivity and Selectivity

Table 12.6 Textual properties and organic loading of bifunctional AEP/UDP-MSNs with various initial concentrations of organosilane precursors.

Initial ratio of AEP to UDP organosilane	S_{BET} (m^2 g^{-1})$^{a)}$	W_{BJH} (Å)$^{a)}$	Total amount of organic functionalites (mmol g^{-1})$^{b)}$	Relative ratio of incorporated AEP- to UDP groups$^{c)}$
2/8	938.7	27.8	1.3	2.5/7.5
5/5	759.6	22.9	1.0	5.4/4.0/6.0
8/2	830.4	25.9	1.5	6.7/3.0/3.0

a) The BET surface area (S_{BET}) and the mean mesopore width (W_{BJH}) were obtained from the nitrogen sorption analysis.
b) The total amount of organic functional group was measured from solid-state ^{28}Si NMR spectrum.
c) The relative ratio of individual AEP and USP functionalities was calculated from solid-state ^{13}C NMR measurement.

most likely originated from the variation of physicochemical properties of the bifunctionalized mesopores, such as polarities and hydrophobicity.

12.3.7
Synergistic Catalysis by General Acid and Base Bifunctionalized MSN Catalysts

Simultaneous activation of the nucleophile and electrophile by both general acid and base groups located in the active site is a common phenomenon seen in enzymatic catalyzed reactions but rarely found in the artificial catalytic system; the formation of a synergistic heterogeneous catalyst system requires precise control of the relative amounts and spatial distribution of multi-functionalities on the surface.

To achieve this goal, Lin and co-workers developed a novel bifunctional heterogeneous catalyst, whose system consisted of relative amount of a general acid, ureidopropyl (UDP) group, and a base, 3-[2-(2-aminoethylamino)ethylamino]propyl (AEP) group, functioned cooperatively to catalyze several chemical reactions [40]. These bifunctionalized MSN catalysts with an initial molar ratio of organosilane precursors (AEP/UDP = 2/8, 5/5, 8/2) had several structural characteristics, such as spherical particle shape and disordered mesopores, in accordance with aforementioned AEP/CP-MSNs. All three MSN catalysts with relative ratios of AEP and UDP groups exhibited similar textual properties obtained from adsorption analysis, and the concentration of individual and total organic groups measured by ^{13}C and ^{28}Si solid-state NMR spectra are summarized in Table 12.6.

Three chemical reactions, aldol, Henry and cyanosilylation, were chosen to examine the catalytic activity of these bifunctionalized silicas as depicted in Figure 12.21. A common electrophile, 4-nitrobenzaldehyde, and three different nucleophiles (acetone, nitromethane, trimethylsilyl cyanide) were used as reactants. The UDP group in the bifunctional MSNs served as a general acid activating carbony

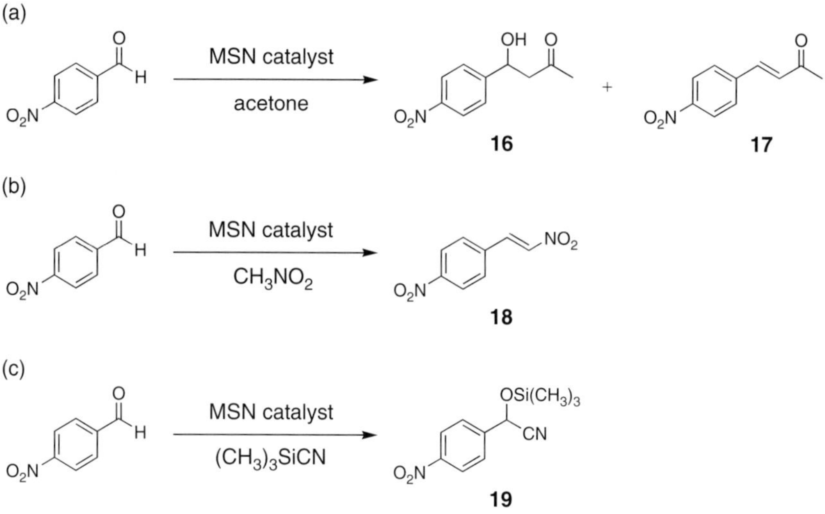

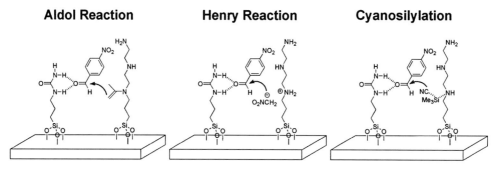

Figure 12.21 Three model reactions catalyzed by the MSN catalysts: aldol reaction (a), Henry reaction (b) and cyanosilylation (c).

Scheme 12.12

groups of the substrates, through double hydrogen bonding, and the AEP group functioned as a general base. The AEP group is capable of: (i) generating the enamine with acetone in the aldol condensation, (ii) deprotonating nitromethane in the Henry reaction and (iii) facilitating the formation a hypervalent silicate nucleophile with trimethylsilyl cyanide in cyanosilylation, could cooperatively catalyze these reactions as depicted in Scheme 12.12. Indeed, the TON of all AEP/UDP-MSNs was higher than that of mono-functionalized AEP-MSN.

The reaction rates of all three reactions were significantly accelerated (up to four times) by the AEP/UDP-MSN catalysts. Among different ratios of the AEP and UDP groups, 2/8 AEP/UDP-MSNs was the most reactive bifunctional catalyst in all three reactions. This result suggests that the activation of carbonyl groups could

12.3 Design of more Efficient Heterogeneous Catalysts with Enhanced Reactivity and Selectivity

be the rate-determining step in these heterogeneous reactions. In contrast, the reactions catalyzed by a physical mixture of AEP-MSN and UDP-MSN, showed significantly smaller TONs in comparison with those of the 5/5 AEP/UDP-MSN catalyst confirming the synergistic effect between the AEP and UDP groups. To verify that the activity enhancement observed here was not caused by the "site isolation effect" of the AEP group, the authors studied the catalytic performance of two other bifunctionalized MSN catalysts (2/8 and 5/5 AEP/CP-MSNs). These two MSN catalysts are intentionally functionalized with catalytic AEP groups along with a "chemically inert" cyanopropyl (CP) group. Assuming the rate enhancement was resulted from the "site isolation effect", the TONs of reactions catalyzed by AEP/CP-MSNs should give the similar result as AEP/UDP-MSNs. In contrast, the TONs of AEP/CP-MSNs were significantly lower than those of the AEP/UDP-MSNs indicated that the surface dilution effect could not account for the observed rate enhancements by the AEP/UDP-MSNs. These results support the hypothesis that the superior rate enhancements in these reactions catalyzed by the bifunctional acid-base MSN catalysts are most likely originated from a cooperative effect between the base (AEP) and the general acid (UDP) that are anchored on the mesopore surface.

In addition to utilizing mesoporous materials for selective carbonyl activation for chemical synthesis, Lin and co-workers developed a new cooperative catalytic system comprising a series of bifunctional mesoporous mixed oxide materials for biodiesel production (unpublished data). These mesoporous calcia silicate (MCS) materials contain both Lewis acidic and basic sites for the synthesis of biodiesel from various free fatty acid (FFA)-containing oil feedstocks, such as animal fats and restaurant waste oils. By converting both soybean oil and animal fats (high FFA) this catalyst demonstrated that the acid and base functionalities could cooperatively catalyze both the esterification of FFAs and the transesterification of oils with short-chain alcohols (e.g., methanol, ethanol) to form alkyl esters (biodiesel). The reactivity and recyclability of these heterogeneous solid catalysts have been investigated. In the case of soybean oil, the catalyst can be recycled 20 times without any decrease in reactivity. These nanoporous mixed oxides could serve as new selective catalysts for many other important reactions involving carbonyl activation, in addition to biodiesel production.

Utilization of a heterogeneous solid acid–base catalyst for the synthesis of biodiesel could circumvent the problem of catalyst separation and convert the free fatty acids in the crude FFA-containing feedstocks to biodiesel, so that the saponification during the transesterification reaction could be prevented. The MCS catalysts proved to be effective in the esterification of biomass feedstocks to biodiesel methyl esters. Current technologies require the use of pretreatment methods to remove FFAs from feedstocks prior to catalysis. The cooperative acid-base characteristics of the mixed oxide MCS catalysts however eliminate the need for pretreatment measures – both soybean oil and poultry fat sources can be utilized in esterification reactions without additional processes to remove FFAs. An important advantage of utilizing heterogeneous materials as catalysts is the possibility of recycling these solids. In this work, catalyst recycling was achieved by simple filtration of the mixture at the end of the reaction. The recovered catalyst was used again under the same reaction conditions. The MCS catalyst was reused up to 20 times for the soybean oil transesterification reaction and up to eight times for the poultry fat transesterification reaction. Remarkably, there is no significant loss of activity in each case. The excellent recyclability of these catalysts indicates that they are stable and there is no leaching of calcium. This is most likely due to the unique structure obtained via a co-condensation reaction, which yields a very stable and structurally homogenous calcium silicate mixed oxide material.

12.4
Other Heterogeneous Catalyst System on Non-Silica Support

The non-silica inorganic support recently induced high attitude of interest due to the intrinsic properties of these materials, such as fluorescence and magnetism. The magnetic nanoparticles like iron oxide (Fe_2O_3) has been extensively studied as the catalyst support due to its nano-size range, facilitating suspension in the solution and magnetic property, allowing easy separation from reaction media when applying an external magnetic field. Dálaigh et al. synthesized a robust DMAP immobilized iron oxide catalyst with superior reactivity, recyclability and stability [41]. The immobilization of DMAP-TES was performed on the surface of the iron oxide–silica (core-shell) magnetic nanoparticle composite, leading to a spherical heterogeneous catalyst of 60 nm in diameter. An acylation of secondary alcohols catalyzed with extremely low loading (as low as 1 mol%) of DMAP functionalized magnetic particles expressed the high reactivity of such a system. More than 80% yield was obtained repeatedly from the sequential recycling experiment, showing the robustness of the DMAP functionalized iron oxide catalyst. The recovery of the heterogeneous catalyst by magnetic decantation is both convenient and efficient. Zheng et al. prepared a diamino acid (His, Asp) functionalized iron oxide catalyst mimic of RNAase proteins for the hydrolysis of Paraoxon, a biocide widely used for crop protection and also a structural analog of chemical warfare agents such as sarin, soman and VX (Figure 12.22) [42]. The amino acid segments were incorporated directly on the iron oxide through amidation of pre-coated dopamine molecules. The surface-attached acidic and basic amino acids on the surface worked cooperatively to efficiently hydrolyze the phosphoester bond of Paraoxon and ester linkage of p-nitrophenyl acetate in the neutral pH aqueous solution at 37 °C without a heavy metal. The magnetic core of the heterogeneous catalyst again offers a facile recovery by magnetic decantation.

12.5
Conclusion

In this review, we have outlined the strategies of surface functionalization and summarized recent development of heterogeneous catalysts on the inorganic

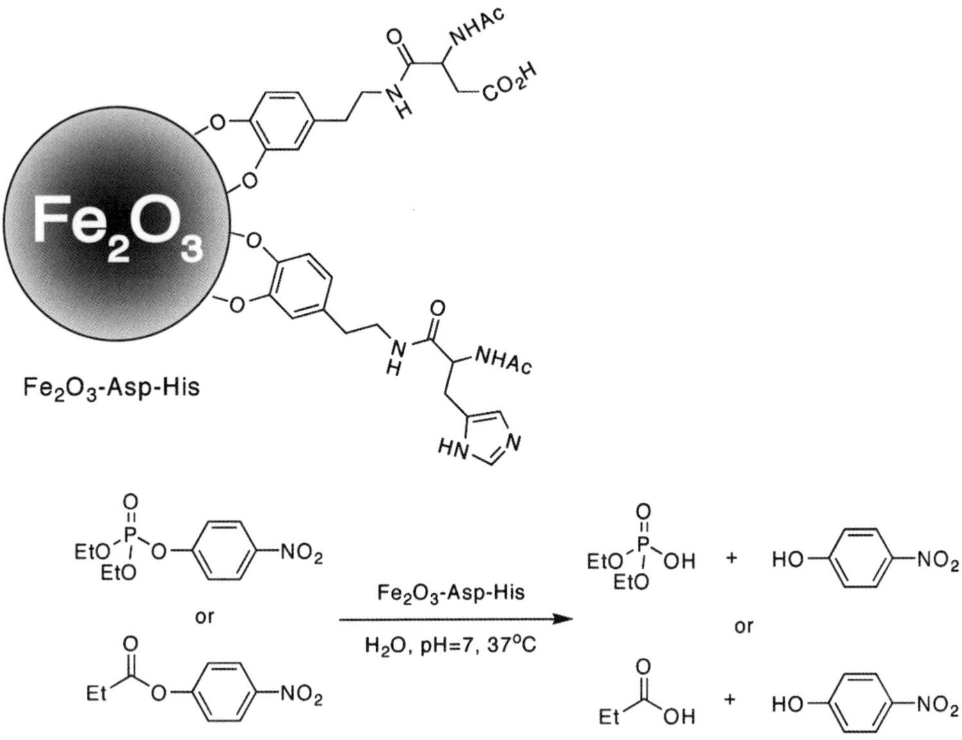

Figure 12.22 Hydrolysis of phosphoester and ester bond catalyzed by cooperative Asp-His bifunctional magnetic nanoparticles

support mainly on silica surface, with some examples of non-silica supports. Both the immobilized homogenous molecular catalyst and the structural properties of supported materials significantly govern the activity of heterogeneous catalyst system. Precise control of the spatial distribution of catalytic sites and the particle/pore morphologies of supported materials are important criteria to achieve a superior heterogeneous catalyst system. By mimicking natural enzymes, multifunctionalized heterogeneous catalysts have been synthesized; and several important catalytic principles, such as gatekeeping and cooperative catalysis, can be realized by using these novel systems as heterogeneous catalysts with high selectivity and efficiency. We envision that further development of this new synthetic method will lead to the control of the spatial location and distribution of organic functional groups in a variety of structurally well-defined mesoporous metal oxide materials that are important for many catalytic applications.

References

1 Thomas, J.M., and Raja, R. (2008) *Acc. Chem. Res.*, **41**, 708–720.
2 Thomas, J.M. (2008) *J. Chem. Phys.*, **128**, 182502/1–182502/19.
3 Beck, J.S., Vartuli, J.C., Roth, W.J., Leonowicz, M.E., Kresge, C.T., Schmitt, K.D., Chu, C.T.W., Olson, D.H., Sheppard, E.W., et al. (1992) *J. Am. Chem. Soc.*, **114**, 10834–10843.
4 Kresge, C.T., Leonowicz, M.E., Roth, W.J., Vartuli, J.C., and Beck, J.S. (1992) *Nature*, **359**, 710–712.
5 Zhao, D., Feng, J., Huo, Q., Melosh, N., Frederickson, G.H., Chmelka, B.F., and Stucky, G.D. (1998) *Science*, **279**, 548–552.
6 Bagshaw, S.A., Prouzet, E., and Pinnavaia, T.J. (1995) *Science*, **269**, 1242–1244.
7 Inagaki, S., Koiwai, A., Suzuki, N., Fukushima, Y., and Kuroda, K. (1996) *Bull. Chem. Soc. Jpn*, **69**, 1449–1457.
8 Mazzei, M., Marconi, W., and Riocci, M. (1980) *J. Mol. Catal. A: Chem.*, **9**, 381–387.
9 Bianchini, C., Burnaby, D.G., Evans, J., Frediani, P., Meli, A., Oberhauser, W., Psaro, R., Sordelli, L., and Vizza, F. (1999) *J. Am. Chem. Soc.*, **121**, 5961–5971.
10 de Rege, F.M., Morita, D.K., Ott, K.C., Tumas, W., and Broene, R.D. (2000) *Chem. Commun.*, 1797–1798.
11 Flach, H.N., Grassert, I., and Oehme, G. (1994) *Macromol. Chem. Phys.*, **195**, 3287–3301.
12 Gruttadauria, M., Riela, S., Lo Meo, P., D'Anna, F., and Noto, R. (2004) *Tetrahedron Lett.*, **45**, 6113–6116.
13 Herron, N., Wang, Y., Eddy, M.M., Stucky, G.D., Cox, D.E., Moller, K., and Bein, T. (1989) *J. Am. Chem. Soc.*, **111**, 530–540.
14 Herron, N., Stucky, G.D., and Tolman, C.A. (1986) *J. Chem. Soc. Chem. Commun.*, 1521–1522.
15 Vidal, V., Theolier, A., Thivolle-Cazat, J., and Basset, J.-M. (1995) *J. Chem. Soc. Chem. Commun.*, 991–992.
16 Gianotti, E., Frache, A., Coluccia, S., Thomas, J.M., Maschmeyer, T., and Marchese, L. (2003) *J. Mol. Catal. A: Chem.*, **204–205**, 483–489.
17 Lim, M.H., and Stein, A. (1999) *Chem. Mater.*, **11**, 3285–3295.
18 Stein, A. (2003) *Adv. Mater.*, **15**, 763–775.
19 Stein, A., Melde, B.J., and Schroden, R.C. (2000) *Adv. Mater.*, **12**, 1403–1419.
20 Radu, D.R., Lai, C.-Y., Huang, J., Shu, X., and Lin, V.S.Y. (2005) *Chem. Commun.*, 1264–1266.
21 Larsen, J.W., and Magid, L.J. (1974) *J. Am. Chem. Soc.*, **96**, 5774–5782.
22 Larsen, J.W., and Magid, L.J. (1974) *J. Phys. Chem.*, **78**, 834–839.
23 Chang, S., Galvin, J.M., and Jacobsen, E.N. (1994) *J. Am. Chem. Soc.*, **116**, 6937–6938.
24 Baleizao, C., and Garcia, H. (2006) *Chem. Rev.*, **106**, 3987–4043.
25 Defreese, J.L., and Katz, A. (2006) *Micropor. Mesopor. Mater.*, **89**, 25–32.
26 Burguete, M.I., Fraile, J.M., Garcia, J.I., Garcia-Verdugo, E., Herrerias, C.I., Luis, S.V., and Mayoral, J.A. (2001) *J. Org. Chem.*, **66**, 8893–8901.
27 Corma, A. (1997) *Chem. Rev.*, **97**, 2373–2419.
28 Goettmann, F., Boissiere, C., Grosso, D., Mercier, F., Le Floch, P., and Sanchez, C. (2005) *Chem. Eur. J.*, **11**, 7416–7426.
29 Goettmann, F., Grosso, D., Mercier, F., Mathey, F., and Sanchez, C. (2004) *Chem. Commun.*, 1240–1241.
30 McKittrick, M.W., Jones, C.W. (2004) *J. Am. Chem. Soc.*, **126**, 3052–3053.
31 Huh, S., Wiench, J.W., Yoo, J.-C., Pruski, M., and Lin, V.S.Y. (2003) *Chem. Mater.*, **15**, 4247–4256.
32 Cai, Q., Luo, Z.-S., Pang, W.-Q., Fan, Y.-W., Chen, X.-H., and Cui, F.-Z. (2001) *Chem. Mater.*, **13**, 258–263.
33 Sadasivan, S., Khushalani, D., and Mann, S. (2003) *J. Mater. Chem.*, **13**, 1023–1029.
34 Shephard, D.S., Zhou, W., Maschmeyer, T., Matters, J.M., Roper, C.L., Parsons, S., Johnson, B.F.G., Duer, M.J. (1998) *Angew. Chem. Int. Ed.*, **37**, 2719–2723.
35 De Juan, F., and Ruiz-Hitzky, E. (2000) *Adv. Mater.*, **12**, 430–432.

36 Johnson, B.F.G., Raynor, S.A., Shephard, D.S., Mashmeyer, T., Mashmeyer, T., Thomas, J.M., Sankar, G., Bromley, S., Oldroyd, R., Gladden, L., Mantle, M.D. (1999) *Chem. Commun.*, **13**, 1167–1168.

37 Chen, H.-T., Huh, S., Wiench, J.W., Pruski, M., and Lin, V.S.Y. (2005) *J. Am. Chem. Soc.*, **127**, 13305–13311.

38 Huh, S., Wiench, J.W., Trewyn, B.G., Song, S., Pruski, M., and Lin, V.S.Y. (2003) *Chem. Commun.*, 2364–2365.

39 Huh, S., Chen, H.-T., Wiench, J.W., Pruski, M., and Lin, V.S.Y. (2004) *J. Am. Chem. Soc.*, **126**, 1010–1011.

40 Huh, S., Chen, H.-T., Wiench, J.W., Pruski, M., and Lin, V.S.Y. (2005) *Angew. Chem. Int. Ed.*, **44**, 1826–1830.

41 Dalaigh, C.O., Corr, S.A., Gun'ko, Y., and Connon, S.J. (2007) *Angew. Chem. Int. Ed.*, **46**, 4329–4332.

42 Zheng, Y., Duanmu, C., and Gao, Y. (2006) *Org. Lett.*, **8**, 3215–3217.

13
Nanotechnology for Carbon Dioxide Capture[1]

Richard R. Willis, Annabelle Benin, Randall Q. Snurr, and Özgür Yazaydın

13.1
Introduction

The Earth is warming. Data, such as the average near-surface temperatures over the past 150 years shown in Figure 13.1, are quite clear on this point [1]. It is also clear this is not a new phenomenon. That is, the earth has warmed and cooled significantly many times throughout history. For example, as recently as the Middle Ages (Figure 13.2), the Earth was as warm as it is today [2]. In fact, if we go back even further in time, we see an almost regular oscillation in average earth temperature (Figure 13.3) over somewhat longer time periods [3]. Thus one might conclude, having only these data available, that we are simply experiencing a natural climate "upturn". However, the *rate* of temperature increase is much higher than at any other time in history, and additional data suggest that there is an anthropogenic contribution to this change in climate. That contribution manifests itself as greenhouse gas (GHG), such as carbon dioxide (CO_2), emissions.

GHGs cause atmospheric warming by making it more difficult for heat to dissipate to space easily. The mechanism of heat retention is based upon the absorption of infrared wavelength radiation that is emitted by the warmed Earth, which is of course different from the warming mechanism experienced in a real greenhouse. Nevertheless, the *greenhouse effect* moniker for atmospheric warming has stuck. GHGs in the atmosphere absorb the longer wavelength radiation from Earth and in turn warm the atmosphere. Meanwhile, GHGs also emit radiation both to space and to the Earth. This emission to the Earth, coupled with the absorption of the longer wavelength radiation described above, are the dual causes of the greenhouse effect. The largest contributors to atmospheric warming are water (over half of the contribution), carbon dioxide (CO_2; up to one-quarter of the contribution), methane (~10%) and ozone (~5%). Nitrogen and oxygen, which make up about 99% of the atmosphere, are not GHGs.

---

1) The views and opinions expressed in this chapter are those of the authors and are not necessarily those of UOP LLC or Honeywell International Inc.

Nanotechnology for the Energy Challenge. Edited by Javier Garcia-Martinez
© 2009 WILEY-VCH Verlag GmbH & Co. KGaA, Weinheim
ISBN: 978-3-527-32401-9

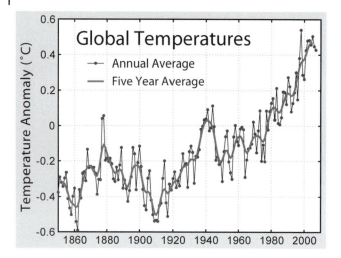

Figure 13.1 Global mean surface temperature anomaly relative to the period 1961 to 1990. Reprinted with permission from Robert Rohde.

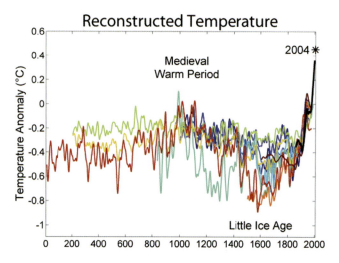

Figure 13.2 Two millennia of mean surface temperatures according to different reconstructions, each smoothed on a decadal scale. The unsmoothed, annual value for 2004 is also plotted for reference. Reprinted with permission from Robert Rohde.

Atmospheric carbon dioxide and other GHG levels are increasing at a rate that parallels global temperature increases (Figure 13.4) [4]. The correlation is striking. It is also important to point out that current levels are higher than have been observed over the past 450 000 years and that many projections show levels reaching 500 ppm or higher by 2050. Changes resulting from such an increase in atmospheric CO_2 concentration include a projected boost in average temperature

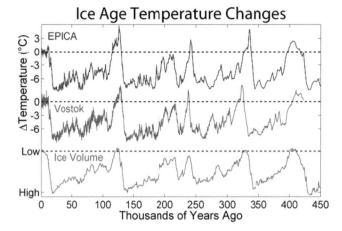

Figure 13.3 Curves of reconstructed temperature at two locations in Antarctica and a global record of variations in glacial ice volume. Today is defined as the left side of the graph. Reprinted with permission from Robert Rohde.

by as much as 6 °C (Figure 13.5) [5]. Such a large increase in global temperature could result in many devastating outcomes, such as severe coastal flooding and the literal loss of major cities such as London and New York to rising ocean levels. Changes in weather patterns could also severely affect food production and availability of fresh drinking water.

The source of the increased CO_2 in the atmosphere can be traced to human activity over the past 150 years or so. Since the start of the Industrial Revolution, atmospheric CO_2 has increased from about 280 to nearly 390 ppm today. The contributions to anthropogenic carbon dioxide emissions by sector[6] are shown in Figure 13.6. As can be seen in Figure 13.6, burning fossil fuels for energy accounts for over 61% of GHG emissions. Thus, using renewable fuel sources, and/or wind, solar and nuclear for power generation are means of achieving a reduction in CO_2 emissions. Another way would be to capture CO_2 at the emission source. The easiest way to make a significant impact in CO_2 capture is to go after the largest and most stationary emission sources. Electricity and heat generation, mostly from burning coal, is the largest subset at nearly 25% of the total GHG emissions (Figure 13.6). Thus our target CO_2 capture application is coal-fired power plants. Finally, one aspect of CO_2 capture that cannot be overlooked is the increasing demand expected for power as large countries such as India and China continue to develop at rapid rates. In China in particular, coal-burning power plants are being constructed at a rapid rate. Each new power plant represents an opportunity to incorporate CO_2 capture technology from the design phase. At a minimum, plans for CO_2 capture technology to be added in the future could be incorporated into new plant designs. The alternative for older power plants throughout the world is adding CO_2 capture technology later in a retrofit mode.

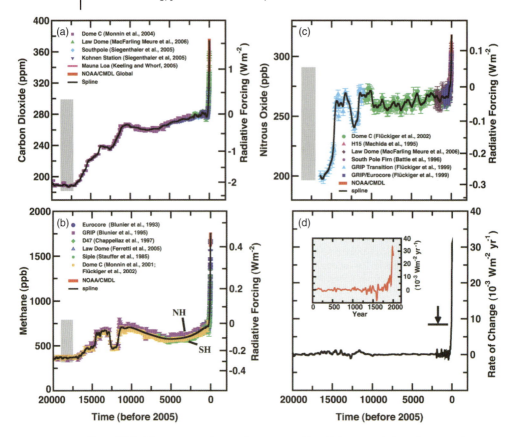

Figure 13.4 The concentrations and radiative forcing by (a) carbon dioxide (CO_2), (b) methane (CH_4), (c) nitrous oxide (N_2O) and (d) the rate of change in their combined radiative forcing over the last 20 000 years reconstructed from Antarctic and Greenland ice and fern data (symbols) and direct atmospheric measurements (panels a, b, c, red lines). The gray bars show the reconstructed ranges of natural variability for the past 650 000 years. The rate of change in radiative forcing (panel d, black line) has been computed from spline fits to the concentration data. The width of the age spread in the ice data varies from about 20 years for sites with a high accumulation of snow such as Law Dome, Antarctica, to about 200 years for low-accumulation sites such as Dome C, Antarctica. The arrow shows the peak in the rate of change in radiative forcing that would result if the anthropogenic signals of CO_2, CH_4, and N_2O had been smoothed corresponding to conditions at the low-accumulation Dome C site. The negative rate of change in forcing around 1600 shown in the higher-resolution inset in panel d results from a CO_2 decrease of about 10 ppm in the Law Dome record. Reprinted from Fig. 6.4 in ref. [4] with permission from Cambridge University Press.

13.1 Introduction | 363

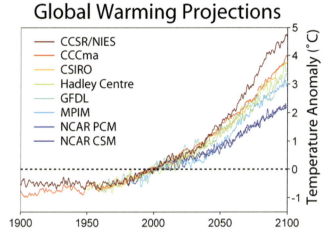

Figure 13.5 Calculations of global warming prepared in or before 2001 from a range of climate models under the SRES A2 emissions scenario, which assumes no action is taken to reduce emissions. Reprinted with permission from Robert Rohde.

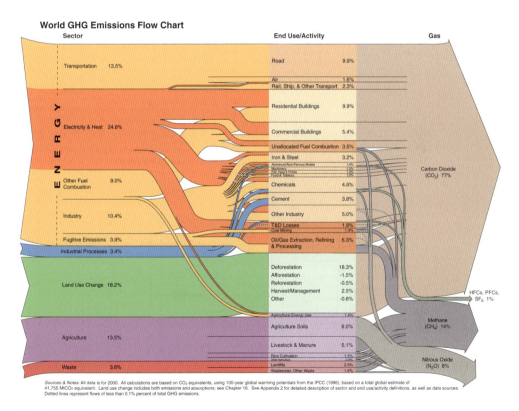

Figure 13.6 World greenhouse gas (GHG) emissions flow chart. Reprinted with permission from Ref. [6]. Copyright 2008 World Resources Institute.

13.2
CO₂ Capture Processes

The three major CO_2 capture technologies are pre-combustion, oxy-combustion and post-combustion. Simplified block-flow diagrams for the technologies are provided in Figures 13.7–13.9. For pre-combustion, the most typical application would be in an *integrated gas combined cycle* coal-fired power plant. The term "pre-combustion" indicates that the CO_2 is removed from the gas stream before it is combusted to generate power. An added benefit to this technology is hydrogen

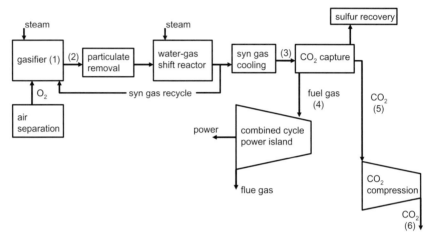

Figure 13.7 A proposed pre-combustion carbon dioxide capture flow scheme. Conditions: (1) ~800 psia, (2) raw syngas is about 30% H_2, 40% CO, 10% CO_2, (3) shifted syn gas is about 55% H_2, ~1% CO, 40% CO_2, (4) 25 °C and ~700 psia, (5) 25 °C and 25 psia, (6) 35 °C and 1500 psia.

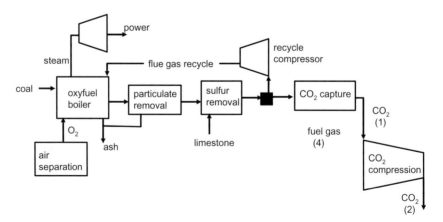

Figure 13.8 A proposed oxycombustion carbon dioxide capture flow scheme. Conditions: (1) 25 °C and 15 psia, (2) 35 °C and 2200 psia.

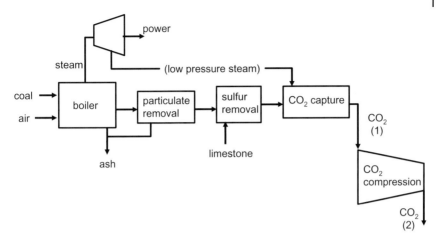

Figure 13.9 A proposed post-combustion carbon dioxide capture flow scheme. Conditions: (1) 25 °C and 15 psia (2) 35 °C and 1500 psia.

co-generation. As can be seen in Figure 13.7, coal is gasified with steam in the presence of oxygen to generate a raw synthesis (syn) gas composed of about 10% CO_2. Note that the large volume of oxygen required for this process comes from an air separation unit. Particulates are then removed, and the syn gas is shifted (via the water-gas shift catalytic reaction) to a hydrogen and CO_2-rich "shifted" stream. It is at this point where CO_2 would be removed from the stream before combustion and power generation. The stream will be at ~700 psi (n.b., 700 psi = 4830 kPa) and depending upon how much the stream is cooled, could be from ambient temperature to ~350 °C. Those CO_2 capture technologies that work best at high pressure, or at high pressure and relatively high temperature would be best for pre-combustion. If the syn gas must be cooled significantly, that would add to the overall process cost. Other challenges – and hence, expenses – include the construction and use of water-gas shift catalysis process in order to increase hydrogen in the syn gas, and the large pressure loss across the CO_2 capture part of the process. However, because current well-established CO_2 capture liquid-phase absorption technology, such as the UOP Selexol™ process, could be utilized in early generation installations, pre-combustion remains a key technology option when new power plants that must capture CO_2 are planned.

The relative amount of installed pre-combustion IGCC technology strictly for power generation is small, as only five demonstration plants at a 250–350 MW scale exist in the world today [7]. However, the technology is proven – three of the plants have been in operation more than ten years – and it is expected that additional plants will be designed and built when carbon capture at power plants is regulated into existence. Pre-combustion gasifers can also be designed to handle feedstocks other than coal, such as petroleum residues and biomass. In fact, as the desire to process heavier and heavier feedstocks increase, several refineries are planning to add IGCC technology to future refining capacity expansions.

Another emerging CO_2 capture technology for coal-fired power plants is oxy-combustion. In the oxy-combustion case (Figure 13.8), coal is combusted in pure oxygen or in oxygen diluted in recycled CO_2 to generate essentially steam and CO_2. The steam is utilized to make power, and the CO_2-rich portion of the stream is treated to remove particulates and sulfur contaminants. It is conceivable that if the CO_2 stream is pure enough, it could be compressed and sent to sequestration without being "captured". The downsides to this technology are boiler materials of construction issues, large flue gas recycle rates and the fact that cryogenic oxygen production at the front end is expensive and energy-intensive.

CO_2 capture for pre-combustion and oxycombustion technologies could be designed into future power plant configurations. Post-combustion technology, however, could be added to hundreds of existing units worldwide (Figure 13.9). In this case, coal is burned in air to produce steam to drive turbines and generate power. Because the coal is burned in air (rather than pure oxygen, or in oxygen diluted with recycle CO_2, as in the oxy-combustion case, above), the flue gas is rich in nitrogen and contains about 10–15% CO_2. After particulate matter, sulfur (as SO_x) and perhaps some moisture, the CO_2 could be captured. The challenges for this technology are that the CO_2 to be captured is diluted in nitrogen and is at low pressure, and that most CO_2 capture technologies have a large parasitic load on the power plant in order to regenerate the capture technology. One additional challenge is that some power plants may not have enough available real estate to add CO_2 capture technology. Despite these challenges, the largest and most immediate impact on global carbon mitigation from coal-fired power plants will come from post-combustion CO_2 capture.

Several technologies have been proposed for the CO_2 capture technologies described above, including amines (e.g., methanolamine; MEA) absorption, chilled ammonia (ammonium carbonate solution), dry sorbents (e.g., sodium, potassium carbonate), ionic liquids, various membranes (including facilitated membranes, carbonic anhydrase enzymatic membranes, membrane-supported ionic liquids) and many other solid adsorbents (e.g., zeolites, lithium zirconates, metal organic frameworks; MOFs) [8]. Since this chapter is devoted to the nano aspects of CO_2 capture, our focus is on solid sorbents which are either nano in size or possess at least one nano feature, such as nanoporosity. Owing to their extremely high surface areas and propensity to adsorb large volumes of gaseous molecules, porous coordination polymers (PCPs) or metal organic frameworks (MOFs) are featured in detail. Post-combustion CO_2 capture is emphasized because this nanotechnology application is where the largest impact on CO_2 capture can be made in the near future.

13.3
Nanotechnology for CO_2 Capture

The nano size regime, between a few nanometers and up to 100 nm or so, has attracted a great deal of attention in recent years owing to the unique and useful properties achieved by various materials in a diverse and growing number of fields and applications. Traditionally, this size range is larger than typically studied by most chemists, and smaller than where many materials scientists or engineers focus their work. It was when scientists working in such diverse fields as catalysis, polymers and colloidal science began to demand better characterization tools and techniques – and of course that these tools and techniques became available – that the nano field really began to expand significantly. Other chapters in this book offer a diverse summary of most of these technologies and the impacts they have and will have on society. In this chapter, our focus is on materials which possess nano features that can be exploited to enable facile CO_2 capture.

The United States Department of energy (DOE) has more specifically identified several technologies with medium to long term potential for CO_2 capture. The DOE target is 90% CO_2 capture at no more than 35% increase in energy cost by the year 2012 [9]. Most of these technologies do not possess a nano aspect. For example, a project utilizing "dry, regenerable" sorbents, such as sodium or potassium carbonate, is focused on an overall engineering solution rather than on any particular nano or nanomaterial characteristic [9].

Membrane Approaches to CO_2 Capture In contrast, polymeric and other gas-permeable membranes often possess features which are nano. As mentioned above, there are several approaches to using membranes for CO_2 capture, among them facilitated[10] and other dual-functional membranes [11], supported enzyme membranes [12] and supported ionic liquids on polymeric supports [13]. In general, membrane approaches are less than ideal for classic coal-fired power plant CO_2 capture, however, because the flue gas is a low pressure stream and hence the driving force for an efficient membrane separation is very low. Nevertheless, a brief summary of facilitated membrane approaches for CO_2 capture is provided here [14].

Facilitated membranes utilize a chemical reaction between some part of the membrane and CO_2 in order to enhance permeability through the membrane. Concurrently, the active phase of the membrane is selected such that it reacts with CO_2 and does not react with any other gases present on the feed side, such as nitrogen or oxygen in the case of flue gas. This adds significantly to the selectivity for CO_2 over any other gases present in the feed stream. Any gas other than CO_2 can only pass through the membrane via diffusion. A recently described facilitated membrane schematic is provided as Figure 13.10 [15]. In this example, an amine functionality is added to the polymeric backbone. The amine reacts with CO_2 and water to form a bicarbonate anion. The bicarbonate anion then releases the CO_2 via the reverse reaction, and thus it effectively transports the CO_2 to the permeate side of the membrane.

One limitation to this sort of technology for CO_2 capture from flue gas is the necessity to keep the membrane "wet". Loss of water to either side of the membrane essentially shuts down the bicarbonate shuttle across the membrane. Since the feed gas is likely saturated or nearly so with respect to water vapor, the feed side of the membrane likely remains wet. However, it may be difficult to keep the

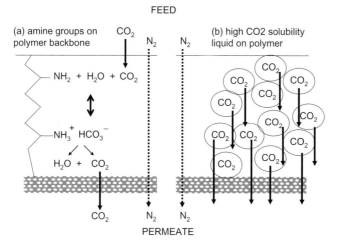

Figure 13.10 Membrane examples facilitated by: (a) an amine functionality attached to polymer backbone and (b) a high CO_2 solubility liquid trapped inside pores of polymeric membrane.

permeate side wet. This is owing to another limitation for membranes in the flue gas application – a low driving force across the membrane. One way to increase driving force is to pull a vacuum on the permeate side of the membrane. This tends to dry out the membrane, however. Another proposed path forward is to use a steam sweep gas on the permeate side in addition to slight vacuum [16].

Ionic liquids (ILs) supported on polymeric membranes can also serve as the active phase for selective CO_2 capture technology. CO_2 has been shown to be very soluble in several ILs [17], and thus the mechanism of CO_2 capture combines the high capacity of physical sorbents with the ease of use available via membrane technology. Current limitations to this emerging technology are the same as those mentioned above for facilitated membranes, plus an issue with absorption kinetics. That is, absorption rates are rather slow owing to the relatively high viscosity of ILs.

High-Temperature CO_2 Adsorbents Two adsorbent systems with nano characteristics have appeared in the literature recently. One is a perovskite-like $BaCe_{0.9}Y_{0.1}O_{2.95}$ (BCY) compound whose structure changes upon CO_2 adsorption at 700–1000 °C (Figure 13.11) [18]. Desorption occurs reversibly at 1400 °C in air. Unfortunately, these temperatures are extreme even for pre-combustion CO_2 capture. The theoretical maximum of 14 wt% CO_2/g sorbent is at the lower end of the likely capacity requirement for flue gas CO_2 capture.

Likewise, nanocrystalline lithium- [19], sodium- [20] and potassium-doped[21] lithium zirconates from the Chen group adsorb CO_2 at 575 °C or so and require at least 600 °C for desorption. Measured capacities are about 25 wt% CO_2/g adsorbent, which makes this technology a possibility for the pre-combustion CO_2 capture

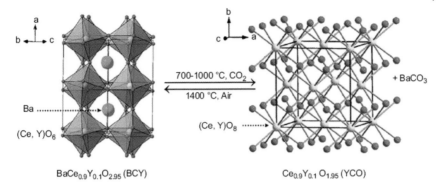

Figure 13.11 Idealized crystal structure showing the transformation of the perovskite-like structure $BaCe_{0.9}Y_{0.1}O_{2.95}$ (BCY) into a fluorite-like Y_2O_3-doped CeO_2 (YCO) and $BaCO_3$ in CO_2 at 700 °C. Reprinted from Ref. [18]. Copyright 2007, with permission from Elsevier.

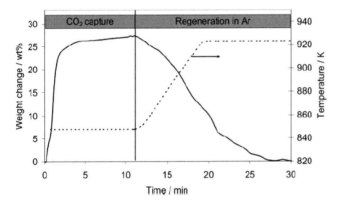

Figure 13.12 CO_2 capture at a CO_2 pressure of 1 atm (101 kPa) and 848 K on Li_2ZrO_3 and temperature-programmed regeneration in argon. Reprinted with permission from ref. [19b]. Copyright 2006 American Chemical Society.

application. Representative CO_2 adsorption data on Li_2ZrO_3 are provided as Figure 13.12 [19b]. As mentioned above, however, the extremely high adsorption (575 °C) and regeneration temperatures (650 °C) mean that technology has to be utilized right after water-gas shift with some additional heat required. It is also unclear how many times the zirconates can be regenerated without significant deactivation by sintering or other mechanisms at these high temperatures.

Zeolites and Molecular Sieves Carbon dioxide adsorption on zeolites, molecular sieves and related nanoporous materials has been measured [22]. A summary of representative adsorption isotherms at ambient temperature for example materials is provided in Figure 13.13, and a summary of properties for these materials is provided as Table 13.1 [23]. Nano-sized zeolites have not been utilized for CO_2

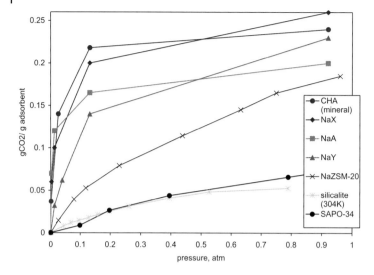

Figure 13.13 Representative CO_2 adsorption isotherms on several zeolites and molecular sieves at 298 K.

Table 13.1 Textural properties and CO_2 adsorption data for selected zeolites and molecular sieves.

Material	Surface area, sq m/g	Pore size(s), Å	Pore volume, cc/g	q, kJ/mol[a]	CO_2 capacity @ 0.1 atm, wt%	CO_2 capacity @ 1 atm, wt%	Reference
CHA (mineral)	510	4.3	0.31	34	20	24.5	[23a]
NaX	700	7.8	0.3	46	18	27	[23b]
NaA	600	4	0.28	46	16	20.5	[23c]
NaY	700	7.8	0.3	31	11.5	24	[23d]
NaZSM-20	700	7.8	0.28	31	5	19	[23e]
Silicalite	500	5.3	0.11[b]	25	1.5	5.3	[23f]
SAPO-34	440	3.8	0.23	24	0.9	7	[23g]

a) q = heat of CO_2 adsorption.
b) Micropore volume only.

capture, as it is expected that the results would not be significantly better than for more typical micron-sized crystals. This is likely because most measurements are equilibrium measurements where difference in crystal size would not affect overall performance. Adsorption rates might be more favorable on nanozeolites, however, but none of these types of measurements have appeared in the open literature. Since adsorption rates on typical micron-sized zeolites tend to be fast, the difference might be negligible and/or difficult to measure.

In general, alkali and alkaline earth modified zeolites adsorb significant quantities of CO_2 (approaching 20 wt% CO_2/g adsorbent) at up to 1 atmosphere pressure (see 13X results in Figure 13.13). The rapid rise in CO_2 adsorption at low pressure indicates a strong interaction between the adsorbent and adsorbate. For 13X zeolite, CO_2 heat of adsorption values as high as 63 kJ/mol have been reported [24]. A value this large suggests that it may be difficult to remove the adsorbed CO_2 during regeneration unless the pressure is lowered to near perfect vacuum. This would be very difficult and expensive to undertake at an industrial scale. Formation of carbonates on the zeolites, which could occur in the presence of water vapor, would also limit the lifetime of a 13X CO_2 adsorbent. It should also be emphasized that most low to medium Si/Al zeolites will adsorb water selectively over CO_2 at low pressure and ambient temperature operation. Since untreated flue gas is likely to be saturated with respect to water vapor, these zeolites are not practical for this application. One report[25] describes a greater than 70% reduction in CO_2 capacity as water vapor in the feed is increased from zero to about 1 mol% water vapor.

In contrast, high Si/Al or pure silica zeolites have less steep isotherms (see silicalite results in Figure 13.13) and lower heats of adsorption than the lower Si/Al zeolites described above [26]. Unfortunately, these materials do not adsorb enough CO_2 at low pressure to be useful for post-combustion flue gas CO_2 capture. Indeed, an ideal situation would occur when 15–20 wt% CO_2 is adsorbed per gram of adsorbent at low pressure. Coupled with this type of steeply rising, linear isotherm at low pressure, a smaller heat of adsorption, ~40–45 kJ/mol, has been postulated to be ideal for post-combustion flue gas operation. These parameters are important to keep in mind as we continue with the discussion below.

13.4
Porous Coordination Polymers for CO_2 Capture

Porous coordination polymers (PCPs)[27], metal organic frameworks (MOFs)[28] and porous hybrid solids (PHSs)[29] are three of the most common ways of referring to metal and/or metal clusters linked with polydentate organic molecules to form zero-, one-, two- and three-dimensional (3D) extended structures [30]. These hybrid inorganic–organic materials[31] possess high surface area and pore volume and are crystalline compounds with relatively high thermal and contaminant stability. As such, these materials are among the highest-capacity gaseous molecule adsorbents known [32]. Owing to their inherent low density and high pore volumes [28a], these compounds have been shown to be excellent adsorbents for such gases as hydrogen[33], methane[34] and carbon dioxide. Facile adsorption of the latter is of particular interest here. Because currently proposed CO_2 capture technology imposes a significant cost burden on delivering electricity or fuels, a PCP-based technology has the potential to change CO_2 capture economics, enabling practical CO_2 sequestration and accelerating the widespread use of CO_2 capture in the power generation industry.

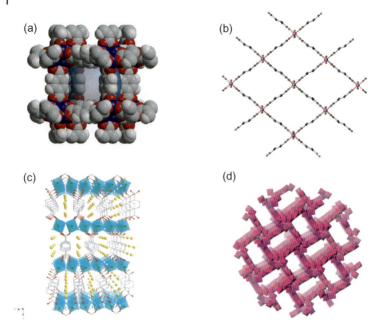

Figure 13.14 Examples of PCPs with different dimensionalities. (a) 0D = MOF-5, a.k.a., IRMOF-1; Zn = blue, oxygen = red, carbon = gray, hydrogen = white. (b) 1D = MIL-53; Al = pink, oxygen = red, carbon = black. (c) 2D = MIL-71 (courtesy Gerard Ferey); V = green spheres inside blue tetrahedra, oxygen = red, carbon = gray, intercalated water oxygen atoms = yellow. (d) 3D = MIL-77; all inorganic framework shown as Ni-containing octahedra and tetrahedra.

We use PCP to refer to the materials in this chapter because it is the most concise, yet complete description of these compounds. Also, for our purposes here, the "dimensionality" is that of the *inorganic portion* of the extended structure. In other words, the prototypical PCP, MOF-5[35] (a.k.a., IRMOF-1, Figure 13.14a), where each inorganic cluster is connected to six others via the benzenedicarboxylate (BDC) linker, is considered "0D". That is, we call this 0D because the inorganic clusters are not directly linked to the other inorganic clusters within the structure. By contrast, an example of a 1D extended structure, MIL-53 [36], is shown in Figure 13.14b. The MIL-53 structure is composed of infinite chains (in 1D) of corner-sharing octahedral aluminum atoms. Unlike IRMOF-1, where the linkers hold the structure together, for MIL-53 the bridging hydroxyls are the primary connectivity between the aluminum atoms. The BDC linkers in MIL-53 are used to connect the chains to other chains to form a 3D network. An example 2D structure is MIL-71 shown in Figure 13.14c [37]. This type of structure is characterized by somewhat dense, non-porous layers "pillared" to one another via the linkers. The pillars open up the structure in two dimensions. Porous 3D structures are rare. An example is MIL-77[38] shown in Figure 13.14d, where the infinite chains of OH and F atom bridged vanadium atoms are connected in three dimensions by the BDC linkers. The chains are not arranged in a "linear" fashion as in MIL-53 described above, but in a rather undulating manner by corner-sharing of the V-containing octahedra.

In order to operate effectively in this environment, a PCP adsorbent must possess high hydrothermal stability, high stability to contaminants, high selectivity for CO_2 over nitrogen and other gases and moderate CO_2 heat of adsorption. If the CO_2 heat of adsorption on a particular PCP is low, its CO_2 capacity is low. If its heat of adsorption is too high, or if an energetically favorable chemical reaction between part of the PCP linker or coordinatively unsaturated, or an "open" metal site and CO_2 occurs, the CO_2 is difficult if not impossible to remove and the PCP is difficult or impossible to regenerate. Of course, easy regeneration is a key to commercial reality. Also on the process side, high CO_2 capacity, coupled with "tunable" isotherms for a given set of PCP materials, provides multiple compositions for use in commercial CO_2 capture technology. Because each MOF material has a unique and distinct isotherm shape, a wide variety of process options is available to the end user.

Preliminary Screening of PCPs for CO_2 Capture Typically PCPs consist of transition metal vertices attached multi-dimensionally over coordination space to other metal vertices by organic linker molecules. After removal of reaction solvent(s), a vast and well-developed pore structure is generated. The resulting porosity is adjustable by simply changing the length or composition of the molecules used to link the metal vertices. Well-ordered openings, channels and pockets in the structures are from a few angstroms to tens of angstroms. These characteristics make MOF materials very much like zeolites, other molecular sieves and some well-ordered mesoporous materials and transition metal oxides. Unlike zeolites, however, where typically only oxygen atoms link Si and Al vertices together, the chemical functionality of the MOF framework is easily tailored by specific modification of the organic linker molecules.

PCP materials have been utilized in CO_2 capture research [39]. When compared to other materials, such as zeolites and activated carbons (e.g., MAXSORB™), the PCPs, such as MOF-177, have significantly higher gravimetric CO_2 capacity at room temperature and at pressure up to about 35 bar (n.b., 35 bar = 3500 kPa; Figure 13.15a). Expressed volumetrically, however, the capacity advantage, while still observed, is much less dramatic (Figure 13.15b) for MOF-177 over 13X zeolite. But there is much more to the CO_2 capture story than CO_2 capacity alone. For example, as mentioned above, the shape of the CO_2 isotherm on 13X suggests a very strong and high heat of adsorption. This means that the interaction between CO_2 and the zeolite is so strong that complete desorption of CO_2 by simply lowering pressure is very difficult. In contrast, the more linear isotherm shape of CO_2 on MOF-177 suggests a lower heat of adsorption and hence more complete desorption once the pressure is lowered. In fact, it is this gradual slope or modest rise in equilibrium CO_2 capacity as pressure is increased that is the key adsorption feature possessed by many PCP materials.

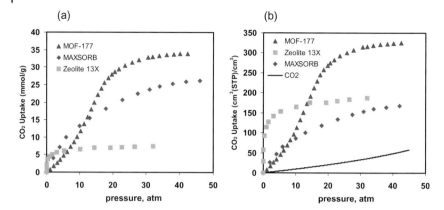

Figure 13.15 Gravimetric and volumetric comparisons of CO_2 uptake in MOF-177, MAXSORB™ activated carbon, and 13X zeolite.

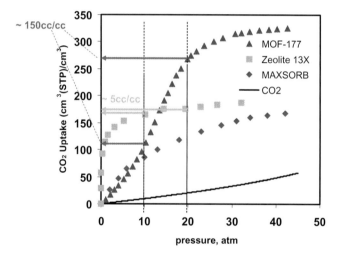

Figure 13.16 A detailed look at Fig. 15b showing significantly more "active" CO_2 capacity for MOF-177 versus 13X zeolite between 10 and 20 atm pressure.

In addition to the fact that typical zeolite CO_2 adsorption isotherms tend to have a very steep rise in CO_2 capacity versus pressure, zeolites tend to hit their somewhat lower CO_2 capacities at relatively low pressure. Also, the extremely steep slope severely limits the potential commercial pressure swing adsorption (PSA) operation mode to relatively low pressure and at small pressure change (delta). This limitation for zeolite 13X versus MOF-177 can be seen in Figure 13.16. In this example, we see much more CO_2 (150 vs 5 cm^3/cm^3) can be adsorbed at 20 atm and released at 10 atm for MOF-177 than for zeolite 13X. Therefore, in general, PCP materials, with a more gradual isotherm slope, allow for more flexibility in

Table 13.2 Typical coal-fired power plant flue gas composition.

Component	Mol%	ppm
Carbon dioxide	7 to 15	–
Nitrogen	65 to 75	–
Oxygen	2 to 12	–
Water	5 to 15[a]	–
Sulfur dioxide	–	2 to 400
Sulfur trioxide	–	1 to 10
NO_x	–	1 to 400
Particulates	–	0.1 to 0.5 grains/scf[b]

a) Saturated.
b) scf = standard cubic feet.

operating pressure and in pressure delta loading for pressure swing adsorption (PSA) operation. Also, as mentioned above, 13X zeolite preferentially reacts with water vapor present in flue gas. This means hydrophobic PCPs that adsorb CO_2 and not water are desired for this application.

Typical coal-fired power plant flue gas contains about 10–15% CO_2 in mostly nitrogen at about atmospheric pressure and temperatures from ambient to 50 °C (see Table 13.2). This means that an effective CO_2 adsorbent captures as much CO_2 as possible at 298 K and 0.1 atm, followed by easy desorption (regeneration) upon significantly lowering the pressure. At these very low adsorption and desorption pressures, the most practical mode of operation is VSA, or vacuum swing adsorption. The "ideal" isotherm for such reversible CO_2 capture at low pressure and ambient temperatures is a linear rise with a steep slope to at least a few tenths of an atmosphere of CO_2 pressure. The slope ideally approaches vertical, but does not become vertical, with a real value of ~1.0 to 1.5 g CO_2 adsorbed per gram adsorbent/atm CO_2. This slope establishes a goal of about 15 wt% CO_2 captured at 298 K and 0.1 atm CO_2 pressure. The "ideal" isotherm is discussed in more detail below.

General Survey of PCPs for CO_2 Capture It is the experience of the authors that more than 200 coordination polymer-related scientific publications and patents are issued each month. Many of the scientific papers report sorption properties, and along with nitrogen and hydrogen, CO_2 is a common adsorbate. Typical equilibrium isotherm measurements are carried out gravimetrically or volumetrically at dry ice (195 K), water ice (273 K), or near ambient (295–305 K) temperatures. Pressures range from sub-atmospheric to ≥30 atm CO_2. The higher pressures are utilized only at the higher temperatures, but high-quality low-pressure adsorption isotherms are collected at all temperatures. A survey of CO_2 adsorption isotherm data for many PCPs is provided below. The conditions are

195 K at up to 1 atm pressure, 273 K at up to 1 atm pressure, ~298 K at up to 35 atm pressure and ~298 K at up to 1 atm pressure. A summary of reported physical and chemical characteristics and of CO_2 saturation capacities for each PCP at the various conditions is provided in Tables 13.3 (195 K) [40], 13.4 (273 K) [41], and 13.5 (≥298 K) [42].

On the adsorption kinetics side, most authors report relatively fast approach to equilibrium during adsorption studies. For some PCPs where pore size is close to that of CO_2, slow kinetics are expected. For example, ErPDA possesses elliptical pores which are only slightly wider at the narrowest point (3.40 Å, see Figure 13.17) than the kinetic diameter of CO_2 (3.30 Å) [41h].

CO_2 Adsorption at 195 K As mentioned in the Introduction, CO_2 must be removed from flue gas at ambient temperature and atmospheric pressure. It follows that these would be the best conditions to evaluate CO_2 adsorption on PCPs in the laboratory. However, for many laboratories, it is more convenient to perform CO_2 adsorption experiments at dry-ice temperature 195 K or water-ice temperature 273 K. As we see below, however, results from such experiments may not directly read on CO_2 capture at ambient temperature. At 195 K, the PCPs that adsorb the most CO_2 are typically those with the largest surface area and pore volume (isotherms in Figure 13.18). These are the so-called supernanoporous PCPs. The best PCP at 195 K and approaching 1 atm pressure is thus UMCM-1 (Figure 13.19a), which saturates at nearly 2.5 times more CO_2 than its own weight! The next best PCPs at 195 K at saturation conditions are MOF-177 (Figure 13.19b) and the IRMOF samples.

Interestingly, the PCP ranking is different at lower pressure. For example, the samples that look more interesting at 0.1 atm and 195 K (Figure 13.18b) are PCN-17 (Figure 13.19c), MOF-508b (Figure 13.19d) and $Cu_2(CNC)_2dpt$ (Figure 13.19e). These PCPs display a type I isotherm at 195 K, which suggests a stronger interaction between these PCPs and CO_2 than for the supernanoporous PCPs. In fact, all PCPs except for the "super-nanoporous" UMCM-1, the IRMOFs and MOF-177 display type I isotherms at 195 K. This means that while the surface area, and hence saturation CO_2 loading, might be lower for a given PCP, its adsorption characteristics might be more desirable in a given lower pressure regime. Figure 13.18c shows a plot of CO_2 at saturation pressure versus surface area. The roughly linear correlation is as expected under these experimental conditions [43]. The slope suggests that for every 100 sq m/g surface area increase, a given PCP should pick up about 2.4 wt% CO_2. This compares to about 0.25 wt% per 100 sq m/g reported for hydrogen at 30 atm [43a]. Results reported in Figure 13.18b and in Table 13.3 show that the CO_2 capacity at 0.1 atm tracks fairly well with measured surface area and especially with pore volume values for samples PCN-17 through Ni-bpe. It is unlikely that open metal sites play a significant role in the CO_2 adsorption behavior at 195 K because only a few of the PCPs discussed here possess or have the potential to possess open metal sites after activation. One exception is Zn_3NTB (Figure 13.19f), which has good CO_2 adsorption despite modest surface area/pore volume (see Table 13.3).

Table 13.3 Structural description, textural properties and CO_2 adsorption data at 195 K for selected PCPs.

PCP	Inter-penetrated?	Exposed metal sites?	Very high CO_2 over N_2 selectivity reported?	Surface area, sq m/g	Pore size(s), Å	Pore volume, cc/g	q, kJ/mol[a]	CO_2 capacity @ 0.1 atm, wt%	CO_2 capacity @ 1 atm, wt%	Reference
MOF-177	No	No	No	4500	11	1.59	NA	10.7	190	[39a, 40a]
IRMOF-1	No	No	No	2900	12	1.04	17	11.8	150	[39a]
IRMOF-6	No	No	No	2500	7.5	0.6 cc/cc	NA	19.4	116	[39a]
PCN-17	Yes	No	Yes	820	3.5	0.34	NA	29	44	[40b]
MOF-508b	Yes	No	No	946	4×4	NA	14.9	27.5	32.7	[40c]
Zn2(CNC)2dpt	Yes	No	Yes	342	3.7	0.19	NA	21.6	29	[40d]
Zn3NTB	No	Yes	No	419	4.6	0.15	NA	20.6	28	[40e]
ZnDPT	No	No	Yes	313	4.1	0.16	NA	15.7	19	[40f]
Mn-formate	No	No	Yes	297	4.5	NA	NA	15.5	20	[40g]
CuFMAbpe	Yes	No	Yes	100	2.0×3.2	NA	NA	15.3	19.2	[40h]
ZnADCbpe	Yes	No	Yes	100	3.4	NA	NA	14.1	25.5	[40i]
PCN-5	Yes	No	No	225	NA	0.13	NA	14	21	[40j]
MOF-2	No	No	No	270	NA	0.086	NA	11.5	13.5	[40k]
MAMS-1	No	No	No	250	3.0 to 3.4	NA	NA	9.5	17.9	[40l]
CID-1	interdigitated	No	Yes	300	5×6	NA	NA	9.4	10.8	[40m]
Cd-aptz	interdigitated	Yes	Yes	210	5.5	0.12	NA	7.8	13.3	[40n]
MnNDC	No	Yes	No	227	4	0.11	NA	7.8	12.9	[40o]
PCN-13	No	No	Yes	150	3.5	0.1	NA	6.3	10.4	[40p]
Ni-bpe	Yes	No	Yes	243	4.8×6.5	NA	NA	5.3	7	[40q]
TbBDC	No	No	No	NA	5	0.032	NA	3	3.6	[40r]

a) q = heat of CO_2 adsorption.

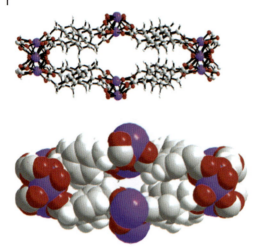

Figure 13.17 Cross-section of the open channel in [Er(PDA)$_{1.5}$]: ball-and-stick model (top), space-filling model (bottom). Color scheme: Er = blue, oxygen = red, carbon = gray, hydrogen = white. Reprinted with permission from Ref. [41h]. Copyright 2003 American Chemical Society.

Another important point to keep in mind is that most gas streams such as coal-fired power plant flue gas contain a mixture of at least two and sometimes many components. This means adsorption selectivity for one gas over another is as critical as adsorption loading at a given set of conditions for most relevant commercial processes. For flue gas, where over 85% of stream is nitrogen (Table 13.2), selectivity for CO_2 over nitrogen is required to make a viable commercial process. Several of the MOFs listed in Tables 13.3–13.5 (specially noted) adsorb CO_2 but do not adsorb nitrogen owing to small pore apertures. PCN-17 and Cu_2CNC_2dpt are two examples from Table 13.3 that adsorb very little nitrogen, yet still adsorb a reasonable amount of CO_2 at 195 K. Unfortunately, most of the other "CO_2-over-nitrogen selective" PCPs do not possess large surface area/pore volume and hence do not adsorb large quantities of CO_2, either. Also, high CO_2 over nitrogen selectivity at 195 K may not hold as adsorption temperature is increased. Additional examples of CO_2 over nitrogen selectivity at higher temperatures are discussed below.

CO_2 Adsorption at 273 K The CO_2 adsorption results recorded at 273 K are interesting (Figure 13.20). It would be interesting to compare 273 K ranking with those at 195 K. Unfortunately, data are available at 195 K and 273 K for only two PCPs – IRMOF-1 and MnNDC (Figure 13.21a). In both cases, CO_2 adsorption is relatively poor relative to the other samples analyzed (Figures 13.18 and 13.20, Tables 13.3 and 13.4). In other words, the samples rank in the bottom half at both temperatures. Therefore it is unfortunately not possible to draw additional general conclusions regarding PCP performance at these two temperatures.

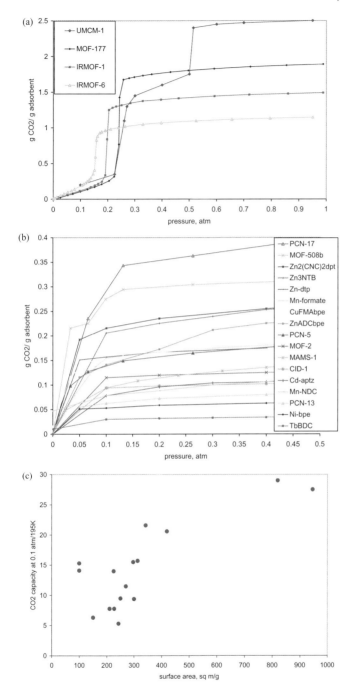

Figure 13.18 Plot of CO_2 adsorption data for several PCPs at 195 K. (a) supernanoporous PCPs. (b) CO_2 adsorption for other PCPs at 195 K. (c) A plot of CO_2 capacity at 0.1 atm CO_2 pressure versus measured surface area for the PCPs in Fig. 18(b). See Table 13.3 for additional information.

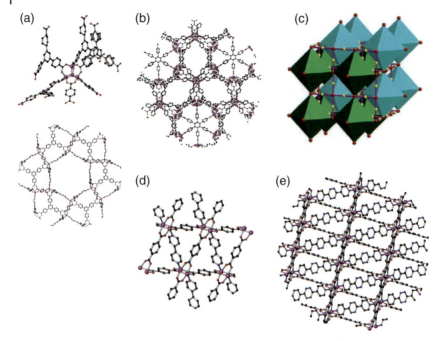

Figure 13.19 Structures of selected PCPs reported to adsorb CO_2 at 195 K. (a) UMCM-1; (top) SBU showing Zn_4O cluster connected to four BTB and two BDC linkers and (bottom) mesopore generated in UMCM-1; Zn = pink, oxygen = red, carbon = black. (b) MOF-177; Zn = pink, oxygen = red, carbon = black. (c) PCN-17; through sulfate bridges coordinatively linked interpenetrated framework (yellow spheres represent sulfur and red spheres represent the square-planar SBU). (d) MOF-508b; Zn = pink, oxygen = red, carbon = black, nitrogen = blue. (E) $Zn_2(CNC)_2dpt$; Zn = pink, oxygen = red, carbon = black, nitrogen = blue. PCN-17 structure from Ref. [40]. Copyright Wiley-VCH Verlag GmbH & Co. KGaA. Reproduced with permission.

The "best" sample at atmospheric pressure and 273 K is DyBTC. The isotherm for this PCP is unique, as it is essentially linear over the pressure range measured. DyBTC is an open, 1D channel structure with pore apertures of about 5.5 Å (see Figure 13.21b). Removal of water via activation results in the formation of open metal sites that were shown to be Lewis acidic by FTIR experiments with pyridine as basic probe molecule [41c]. Presumably these open metal sites are utilized to adsorb significant quantities of CO_2 at low pressure. The authors do not speculate as to why the CO_2 adsorption isotherm does not suggest a stronger interaction with CO_2 compared to other PCPs, however. A strong adsorbent–adsorbate interaction would be expected to result in the more common type I isotherm. The more linear isotherm for DyBTC shown in Figure 13.20 suggests a strong adsorbate–adsorbate interaction[44] not found with any of the other PCP adsorbents.

In fact, several of the next best CO_2 adsorbents, particularly those that adsorb a reasonable amount of CO_2 at lower pressure, provide type I isotherms. For example, Pd-pymo-F and EuBa PCP (Figure 13.21 c and d, respectively) reach CO_2

13.4 Porous Coordination Polymers for CO$_2$ Capture | 381

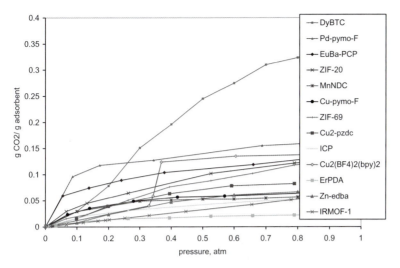

Figure 13.20 Plot of CO$_2$ adsorption data for several PCPs at 273 K. See Table 13.4 for additional information.

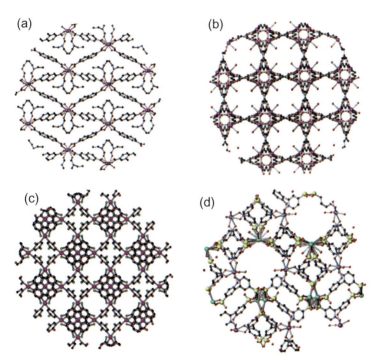

Figure 13.21 Structures of selected PCPs reported to adsorb CO$_2$ at 273 K. (a) MnNDC; Mn = pink, oxygen = red, carbon = black, nitrogen = blue. (b) DyBTC; Dy = pink, oxygen = red, carbon = black. (c) Pd-pymo-F; Pd = pink, oxygen = red, carbon = black, fluorine = green, nitrogen = blue. (d) EuBaPCP; Eu = pink, Ba = green, S = yellow, oxygen = red, carbon = black, nitrogen = blue.

Table 13.4 Structural description, textural properties and CO_2 adsorption data at 273 K for selected PCPs.

PCP	Inter-penetrated?	Open metal sites?	Very high CO_2 over N_2 selectivity reported?	Surface area, sq m/g	Pore size(s), Å	Pore volume, cc/g	q, kJ/mol[a]	CO_2 capacity @ 0.1 atm, wt%	CO_2 capacity @ 1 atm, wt%	Reference
Pd-pymo-F	No	Yes	No	600	4.8 + 8.8	NA	NA	10	16.4	[41a]
EuBa PCP	No	Yes (Eu)	No	718	6.2	0.25	21	6.5	15	[41b]
DyBTC	No	Yes	No	655	5.5	NA	NA	4	37.3	[41c]
ZIF-20	No	No	No[b]	800	2.8	0.27	NA	3.7	13.1	[41d]
MnNDC	No	Yes	No	191	4	0.068	NA	2.96	6.1	[40]
Cu-pymo-F	No	Yes	No	NA	2.9	NA	49–55	2.95	7.2	[41e]
ZIF-69	No	No	No	1070	4.4	NA	NA	1.9	13.3	[39c]
Cu2-pzdc	No	No (but has unique uncoordinated oxygen sites)	No	NA	4x6	NA	31.9	1.6	8.5	[41f]
ICP	No	No	Yes	225	NA	NA	NA	1.55	5.9	[41g]
Cu2(BF4)2(bpy)2	No	No	No	NA	4.6[c]	NA	NA	1.3	14	[37b]
ErPDA	No	Yes	Yes	"Low"	3.4	0.027	30.1	0.8	2.4	[41h]
Zn-edba	Yes	No (but Td Zn)	No	650	12	NA	NA	0.75	7.3	[41i]
IRMOF-1	No	No (but Td Zn)	No	2900	12	1.04	17	0.7	6.9	[39a]

a) q = heat of CO_2 adsorption.
b) Very slow diffusion for N_2 on ZIF-20 make it potentially kinetically selective for CO_2 over nitrogen!
c) This is interlayer spacing that changes to 6.8 Å upon CO_2 adsorption; also, 7.7 Å pores open up after CO_2 exposure.

saturation fairly early in their respective adsorption isotherms (Figure 13.20). Each of these PCPs, like DyBTC and most of the samples tested at 273 K (see Table 13.4), possess open metal sites upon activation (EuBa PCP) or simply as a result of being four-coordinate, square-planar geometry (Pd-pymo-F) when a six-coordinate octahedral arrangement is possible. Evidence that the open metal sites are Lewis acidic in nature in either of these PCPs is not provided [41b,c]. For EuBa PCP, the role of "extra-framework" chloride ion on gas adsorption is unclear. Likewise, it is not clear how stable the square planar palladium atoms would be to coordination by CO_2, water and other gases, particularly at increased pressure.

Carbon dioxide adsorption at 273 K was also measured for two other unique materials. One of these materials, $Cu(BF_4)_2(bpy)_2$, was generated by dehydration of an essentially non-porous 3D coordination polymer $\{[Cu(BF_4)_2(bpy)(H_2O)_2]*bpy\}$ to a 2D square-grid type PCP [37]. An interesting feature noted by the authors of [37] is that the $Cu(BF_4)_2(bpy)_2$ expands considerably upon CO_2 adsorption. The measured volume expansion was 6.6 vol% at 1 atm CO_2 pressure. It is postulated that CO_2 adsorption occurs via CO_2 clathrate formation between the square grid layers. As can be seen in Figure 13.20 and Table 13.4, its CO_2 adsorption is not as good as most other PCPs, however.

The other interesting material, prepared by Mirkin and co-workers, is not a crystalline PCP, but rather an ICP – an amorphous infinite coordination polymer [41g]. ICPs are prepared from organometallic ligands, such as metalated salen ligands, and metal ion connecting nodes, such as those provided by zinc acetate. The reaction scheme to form the zinc-containing ICP is provided as Figure 13.22. The resultant amorphous material was shown to adsorb a small amount of CO_2 at 258 K (Figure 13.20, Table 13.4), but essentially no nitrogen at 77 K. This suggests a molecular sieving mechanism of adsorption, probably via a disordered pore structure with an effective pore size centered around 3.4–3.6 Å. Like several other PCPs discussed in this chapter, this ICP material is mentioned here *not* because it has a large CO_2 capacity, but rather because it is a unique material that adsorbs at least some CO_2 and essentially no nitrogen. Since this is one of the first compounds of this type described in the literature, it suggests that better adsorbents may be prepared from ICPs in the future by carefully tailoring the chemistry of the organometallic ligands and metals used to connect them and other reaction parameters.

A couple of other points bear mentioning at this point. First, unlike the PCPs examined at 195 K, most of the PCPs evaluated at 273 K are not interpenetrated framework structures. It is unclear why there are fewer data for interpenetrated structures at 273 K than at 195 K. By contrast, most of the PCPs evaluated at 273 K possess open metals sites after activation, while those tested at 195 K do not. In this case, individual authors might have expected an enhanced degree of interaction between the open metal sites on the respective PCPs and CO_2 than for other PCPs without open metal sites. Expressed experimentally as heat of adsorption, this "enhanced interaction" is somewhat evident for those PCPs where heats of adsorption have been determined. Comparing heats of adsorption data in Tables

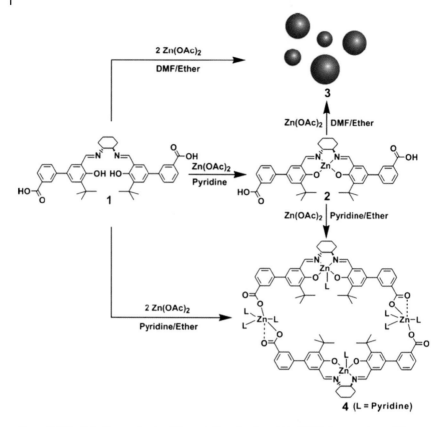

Figure 13.22 Selective synthesis of the metallo-salen ligand Zn(AFSL) 2, amorphous ICP particle 3, and crystalline metallomacrocycle 4. DMF:N,N-dimethyl formamide. Ref. [41g]. Reproduced with permission.

13.3 and 13.4, one can see in general that values are higher for the PCPs evaluated at 273 K (17–55 kJ/mol) versus 195 K (15–17 kJ/mol). However, one sample which stands out is the EuBa MOF which adsorbs a relatively large amount of CO_2 at 273 K despite a fairly low (21 kJ/mol) CO_2 heat of adsorption value. The combination of relatively high surface area, pore volume, moderate pore size and open metal sites likely contribute to the favorable CO_2 adsorption behavior for this PCP.

CO_2 Adsorption at 298 K Ambient temperature CO_2 adsorption at elevated pressure is dominated by the supernanoporous PCPs mentioned above (e.g., MOF-177, MIL-101) and related materials (Figure 13.23). One of the related materials, MIL-53, is discussed in detail in the PCP "breathing" section below. For now it is interesting to point out how the CO_2 adsorption for MIL-53 is different when it is fully activated (free from any solvent or guest molecules in its pores) to a "hydrated" state, where water occupies space within the pores. As can be seen in Figure 13.23,

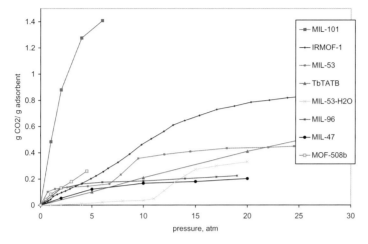

Figure 13.23 Plot of CO_2 adsorption data for several PCPs at 298 K and higher. See Table 13.5 for additional information.

the isotherm shape at low pressure changes from Type I for the fully activated MIL-53, to a "low adsorbate–adsorbent interaction" dominated shape. Only after about 10 atm CO_2 pressure have been applied do we observe significant CO_2 uptake for the hydrated MIL-53. Obviously, water adsorption has altered the CO_2 adsorption characteristics for MIL-53.

Favorable CO_2 adsorption at more realistic flue gas conditions (low pressure, ambient to slightly above ambient temperature, see Figure 13.24) is dominated by one particular PCP structure type. These PCPs are sometimes referred to as m-MOF-74, m-CPO-27, or m/DOBDC, where m designates Zn, Co, Ni, or Mg and DOBDC refers to the linker, dioxybenzenedicarboxylate. For our purposes, we stick with the most generic nomenclature, m/DOBDC. Note that the linker is added to the synthesis mixture as the dihydoxydicarboxylic acid and that the acid and hydroxyl functional groups are deprotonated by the basic reaction mixture in order to form the resultant PCP. The structure of m/DOBDC (Figure 13.25a) consists of infinite-rod secondary building units bound by DOBDC resulting in 1D hexagonal pores about 11 Å in diameter [42a]. The chains of metal atoms are held together by the linker coordinated through the carboxylic acid and deprotonated hydroxyl groups. Surface areas range between 1495 and 816 m^2/g for the different metals, which is consistent with 35 to 38 adsorbed nitrogen molecules per unit cell for each PCP. In other words, the lighter Mg/DOBDC has a higher surface area because it is expressed on a per gram basis.

In contrast, the enhanced CO_2 adsorption observed for the Mg/DOBDC (Figure 13.24 and Table 13.5) versus the other DOBDC PCPs is not related to the smaller mass of Mg. That is, Mg/DODBC saturates with about 12 CO_2 molecules per unit cell, while Co and Ni versions take up about seven CO_2 molecules and the Zn version about four CO_2 molecules per unit cell[42a]. It is postulated that CO_2

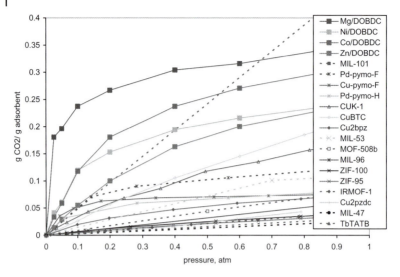

Figure 13.24 A closer look at the lower pressure region of the plot in Figure 13.23, plus CO_2 adsorption data for several additional PCPs. See Table 13.5 for additional information.

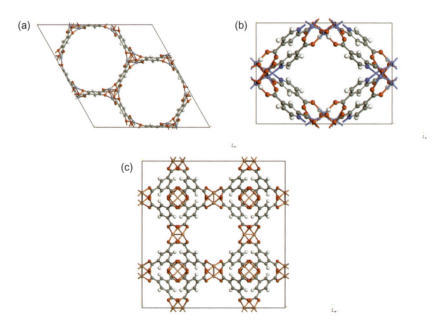

Figure 13.25 Structures of selected PCPs reported to adsorb CO_2 at 298 K. (a) M/DOBDC, where M = Mg, Zn, Ni or Co; M = blue, oxygen = red, carbon = gray, selected H = white. (b) CUK-1; Co = light blue, oxygen = red, carbon = gray, nitrogen = dark blue, selected H = white. (c) CuBTC; Cu = copper, oxygen = red, carbon = gray, selected H = white.

Table 13.5 Structural description, textural properties and CO_2 adsorption data at 298 K for selected PCPs.

PCP	Inter-penetrated?	Open metal sites?	Very high CO_2 over N_2 selectivity reported?	Surface area, sq m/g	Pore size(s), Å	Pore volume, cc/g	q, kJ/mol[a]	CO_2 capacity @ 0.1 atm, wt%	CO_2 capacity @ 1 atm, wt%	Reference
Mg-MOF-74	No	No	No	1495	11	NA	47	23.7	35	[42a]
Ni-MOF-74	No	No	No	1070	11	NA	41	12	25	[42a,b]
Co-MOF-74	No	No	No	1080	11	NA	37	11.8	32	[42a,c]
Zn-MOF-74	No	No	No	816	11	NA	NA	5.5	25	[42d]
MIL-101	No	No	No	2800 to 4200	12 and 16	1.37 to 2.15	44	5	48	[42e]
Pd-pymo-F	No	Yes	No	600	4.8 + 8.8	NA	NA	5	22.6	[41a]
Cu-pymo-F	No	Yes	No	NA	2.9	NA	49	5	7.5	[41e]
Pd-pymo-H	No	Yes	No	600	4.8 + 8.8	NA	NA	3.2	8	[42f]
CUK-1	No	No	Yes	630	11.1	0.26	NA	3	7.7	[42g]
CuBTC	No	Yes	No	1300	7 to 8	0.66	30	2.6	22.5	[42h]
Cu2-lpbz	Yes	Yes	No	660	3.4	0.25	27.5	2	7.7	[42i]
MIL-53	No	No	No	1200	4 to 8.5	NA	35 (Al)	0.1	0.22	[36]
MOF-508b	Yes	No	No	946	4 × 4	NA	14.9	0.9	8.4	[40c]
MIL-96	No	No	Yes	NA	2.5 to 3.5	NA	33	0.7	6.5	[42j]
ZIF-100	No	No	Yes	595	3.35	NA	NA	0.6	4.1	[42e]
ZIF-95	No	No	Yes	1050	3.65	NA	NA	0.5	3.6	[39d]
IRMOF-1	No	No	No	2900	12	1.04	17	0.4	4.3	[39a]
Cu2-pzdc	No	No	No	NA	4 × 6	NA	31.9	0.35	5.5	[41g]
MIL-47	No	No	No	930	10.5 × 11	NA	22	0.05	5.3	[42k-l]
TbTATB	No	No	No	1783	13 and 17	NA	NA	0.05	2	[42m]

a) q = heat of CO_2 adsorption.
b) But this PCP has unique uncoordinated oxygen sites.

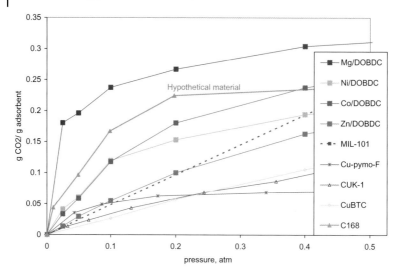

Figure 13.26 A plot featuring data from Figure 13.24 plotted with the "ideal" 298 K adsorbent, hypothetical C_{168}.

adsorption is enhanced in Mg/DOBDC PCPs owing to the increased ionic character of the Mg-O bond relative to transition metal oxygen bonds in the other DOBDC PCPs. The strength of interaction between CO_2 and the Mg–O bond is reflected in the relatively high heat of CO_2 adsorption of 47 kJ/mol. In fact, only one other PCP (Cu-pymo-F at 49 kJ/mol) has been reported to have a CO_2 heat of adsorption close to that for Mg/DOBDC or any of the other DOBDC PCPs (see Table 13.5). The next closest CO_2 heat of adsorption reported to date is for MIL-96 at 33 kJ/mol. Interestingly, MIL-96 does not adsorb much CO_2 at ambient temperature.

The other interesting PCPs at 298 K are the pymo-based ones (Pd-pymo-F in Figure 13.21c) CUK-1 (Figure 13.25b) and Cu-BTC (Figure 13.25c). The pymo-based PCPs have a strong interaction with CO_2 as evidenced by the type I isotherm shapes (Figure 13.24). However, saturation capacity is fairly low for these materials. CUK-1 also provides a type I isotherm, but with a significantly higher saturation loading and pressure. Cu-BTC has an even higher saturation pressure and loading, as evidenced in Figure 13.24, which shows CO_2 adsorption on Cu-BTC measured to several atmospheres. Most of the other PCPs in Table 13.5 and Figure 13.24 are somewhat to significantly inferior to those mentioned above.

A plot displaying only the best performers at 298 K and a theoretical material, C168 Schwarzite[45] is provided as Figure 13.26. The C168 material possesses an ideal isotherm for CO_2 adsorption at 298 K. It is expected that a material with such an isotherm would have a CO_2 heat of adsorption around 45 kJ/mol, and its working capacity of about 15 wt% CO_2 between 0 and 0.1 atm is considered acceptable performance in the flue gas CO_2 capture application. As can be seen in Figure 13.26, several PCP materials either meet or exceed the performance of the theoretical C168 material. This suggests great promise for PCPs in this application. The next hurdles will be maintenance of this performance in more realistic flue gas conditions, where nitrogen, water vapor, oxygen and other contaminants may affect PCP CO_2 adsorption.

Molecular Modeling of CO_2 Adsorption on PCPs As in other areas of nanotechnology, molecular modeling has proven to be a powerful tool for understanding adsorption in nanoporous materials, including CO_2 capture. Here, we provide a short snapshot of properties that can be calculated and the insights that can emerge from molecular simulation, focusing on adsorption in PCPs. More extensive reviews of molecular modeling in PCPs can be found elsewhere [46]. Prediction of adsorption isotherms, mixture selectivities, and heats of adsorption can be useful in the screening of existing or hypothetical PCPs for CO_2 capture. In addition, molecular-level information on how molecules interact with one another and with the PCP surface can aid in the design and tailoring of new materials.

Molecular modeling of small molecules such as CO_2 in nanoporous materials is generally based on classical mechanics, where the energetic interactions among all atoms in the system are described by a "force field." The PCP structure is typically assumed to be rigid, with the atoms placed at the positions reported from X-ray crystallography. The two most important contributions to the energy are the long-range electrostatic interactions and the short-range van der Waals interactions. These can be modeled by placing partial charges and Lennard–Jones sites on all atoms in the system and assuming that they interact in a pair-wise fashion [47]. The partial charges on CO_2 reflect its quadrupole moment. Given a model such as this, the properties of interest are calculated using large-scale computer simulations based on the principles of statistical mechanics. The most common method is grand canonical Monte Carlo (GCMC). GCMC simulations are based on the equilibrium conditions that the temperature and chemical potential of the gas inside and outside the adsorbent must be equal. Only the adsorbed phase is simulated, and the temperature, volume and chemical potential are kept fixed while the number of sorbate molecules fluctuates. The chemical potential can be related to the temperature and pressure of the external gas phase, for example, with an equation of state. During the course of the simulation, randomly generated configurations satisfying the thermodynamic conditions imposed by the chemical potential and temperature are sampled by performing random moves of molecules, as well as insertions of new molecules and deletions of existing molecules. One calculates the number of molecules in the adsorbed phase by averaging over the sampled configurations.

One of the primary outputs of a GCMC simulation is the adsorption isotherm. Figure 13.27 shows a comparison of simulated and experimental adsorption isotherms for CO_2 in IRMOF-1 over a wide range of temperatures [44]. The simulated isotherms are in very good agreement with the experimental data and are able to capture the complex shapes of these isotherms. It should be noted that the results in Figure 13.27 are pure predictions, with no fitting of any parameters to the experimental data. Walton *et al.* also reported excellent agreement between

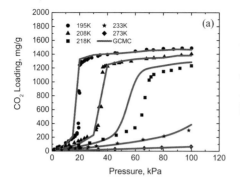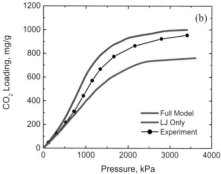

Figure 13.27 (a) Comparison of GCMC simulations and experimental adsorption isotherms for CO_2 in IRMOF-1 at various temperatures. (b) Comparison of GCMC simulations and experimental adsorption isotherms for CO_2 in IRMOF-1 at 298 K. The top curve was calculated from a model which included electrostatic effects. The bottom curve was calculated considering only the Lennard-Jones interactions. Reprinted with permission from Ref. [44]. Copyright 2008 by the American Chemical Society.

simulated and experimental CO_2 isotherms in IRMOF-3 and MOF-177 [44]. With increasing temperature, the sharp steps in the IRMOF-1 isotherms gradually smooth out, but a distinct inflection is still observed at room temperature [Figure 13.27b]. Walton et al. found that the electrostatic interactions between CO_2 molecules were crucial to reproduce the inflection behavior of the isotherm. To show this, they did something that would not be possible in a real experiment; they turned off the electrostatic interactions between CO_2 molecules. The results in Figure 13.27b show that the full model (with electrostatic interactions) captures the inflection, but when the electrostatic interactions are turned off the isotherm shows no inflection, thus revealing the importance of electrostatic interactions for this system.

Another interesting study involving isotherms with steps was reported by Ramsahye et al. [48] for CO_2 adsorption in MIL-53 (Al). In the work of Walton et al. on IRMOF-1, the framework did not undergo any structural change, but MIL-53(Al) was observed to have a structural interchange between a narrow-pore form (MIL-53np [Al]) and a large-pore form (MIL-53lp [Al]) during CO_2 adsorption, according to microcalorimetry and adsorption experiments [49]. Ramsahye and co-workers shed light on how this breathing effect is connected to CO_2 adsorption by performing GCMC simulations and calculating the adsorption isotherms of CO_2 for both versions of MIL-53(Al) up to 30 bar. As expected, the isotherms showed completely different behavior, as the narrow-pore MIL-53np (Al) adsorbed only three CO_2 molecules per unit cell, while the large-pore MIL-53lp (Al) adsorbed up to nine CO_2 molecules. See Figure 13.28a. The experimental isotherm (triangles in Figure 13.28b) shows a sharp increase around 6 bar. When Ramsahye et al. combined the simulated adsorption data below 6 bar from MIL-53np (Al) with the data above 6 bar from MIL-53lp (Al), a very nice match with the experimental

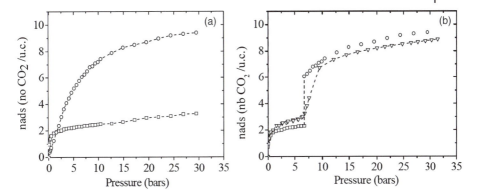

Figure 13.28 (a) Simulated absolute adsorption isotherms for CO_2 in two different forms of MIL-53, MIL-53np (Al) (squares) and MIL-53lp (Al) (circles). (b) The simulations (circles) are compared with the experimental data (triangles) [48b]. Reproduced by permission of The Royal Society of Chemistry.

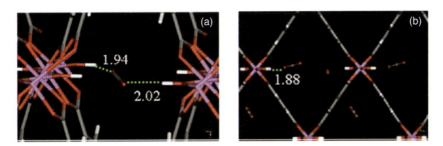

Figure 13.29 (a) The double interaction of a CO_2 molecule with the MIL-53np (Al) structure and (b) the interaction of a CO_2 molecule with a single μ_2-OH group in MIL-53lp (Al) [48b]. Reproduced by permission of The Royal Society of Chemistry.

isotherm was obtained (Figure 13.28b). With a similar approach they combined the heats of adsorption from the two structures, which again yielded good agreement with experimental heats of adsorption.

CO_2 adsorption in MIL-47 (V) [50], the vanadium analog of MIL-53lp (Al), does not show any breathing behavior. Besides the difference of the metal used, MIL-47 (V) lacks the μ_2-OH groups present in MIL-53 (Al). Snapshots from the simulations revealed that these μ_2-OH sites are responsible for triggering the structural shift observed in MIL-53 (Al). CO_2 molecules were observed to bridge the μ_2-OH groups of MIL-53np (Al) which are preferential adsorption sites and present on both sides of the pore, as shown in Figure 13.29a. Note that such a stable configuration was not observed for the adsorbed CO_2 molecules in the larger pores of MIL-53lp (Al) (Figure 13.29b). This stable bridging configuration for CO_2 molecules was only possible since the pore dimension is only 8.30 Å in MIL-53np (Al). As the pressure increases all of the μ_2-OH sites are occupied by the CO_2 molecules

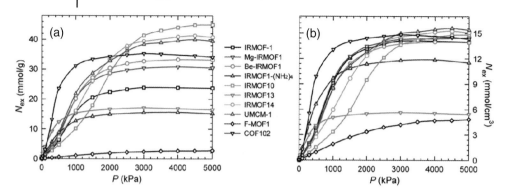

Figure 13.30 (a) Gravimetric and (b) volumetric isotherms of CO_2 adsorption in IRMOF-1, Mg-IRMOF-1, Be-IRMOF-1, IRMOF-1-$(NH_2)_4$, IRMOF-10, IRMOF-13, IRMOF-14, UMCM-1, F-MOF-1, and COF-102 [51a]. Reproduced by permission of The Royal Society of Chemistry.

and CO_2/CO_2 interactions start to dominate the energy of the system. As the saturation capacity of MIL-53np (Al) is reached, interactions between the CO_2 molecules tend to break the interactions between the CO_2 molecules and the μ_2-OH groups leading to configurations with weaker sorbate–sorbent interactions, eventually giving way to the transition from the narrow-pore structure to the large-pore structure.

As noted above, if modeling is sufficiently predictive, it can also be used to screen existing materials for new applications or even to screen hypothetical structures created on the computer which have not been synthesized yet. A nice example of this was reported by Babarao et al. [51]. This group screened different types of existing PCP materials for CO_2 adsorption. Moreover, they created hypothetical structures in two ways. First, they changed the metal oxide corners in IRMOF-1 from Zn_4O to Mg_4O or Be_4O. Second, they added amine groups to the linkers of IRMOF-1 to create a PCP with a new functional group. The simulated isotherms are presented in Figure 13.30. In addition, they investigated correlations of CO_2 adsorption with properties such as the framework density, free volume, porosity, accessible surface area and heat of adsorption for these materials. This type of structure–function correlation may be useful for screening of PCPs for CO_2 adsorption and other applications [43a].

GCMC simulations can also be used to predict mixture adsorption and thus selectivities. Measuring mixture adsorption in nanoporous materials is tedious, so researchers have long sought simple theoretical models for predicting mixture adsorption from single-component isotherms. The most common method is the ideal adsorbed solution theory (IAST), developed by Myers and Prausnitz [52]. IAST can be used with either experimental or simulated single-component isotherms as inputs. It often works well, although deviations are frequently seen at high loadings [53]. Several groups have used GCMC simulations to test the applicability of IAST for predicting mixture adsorption in PCPs [45a, 54]. For example, Bae et al. [55] simulated mixture and single-component adsorption isotherms for

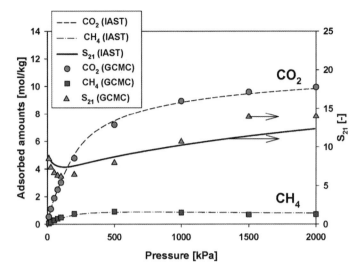

Figure 13.31 Verification of the IAST calculations by GCMC simulations for equimolar mixtures of CO_2 and CH_4 at 296 K. The IAST calculation was based on single-component GCMC isotherms. Reprinted with permission from Bae et al. [55]. Copyright 2008 by the American Chemical Society.

CO_2 and CH_4 in mixed-ligand MOFs. They compared the simulated mixture results with those calculated by using IAST with the single-component simulated isotherms as inputs. As shown in Figure 13.31, the agreement is quite good, indicating that IAST works well for this system. Keskin et al. [46b] reviewed similar work and concluded that in almost all cases investigated to date, IAST works well for predicting mixture adsorption in PCPs from single-component data. This is a useful conclusion, given the difficulty in measuring mixed-gas adsorption experimentally.

Molecular simulations also provide a wealth of molecular-level information. Siting locations of molecules within a framework can easily be extracted from the simulation output to reveal preferred adsorption sites. For example, Babarao et al. [51b] investigated the CO_2 siting in an covalent–organic frameworks (COFs) [56]. Like PCPs, COFs are nanoporous materials synthesized in a building-block approach but without the use of metal corners. In their study, Barbarao et al. report CO_2 density distribution maps calculated by accumulating the locations visited during the GCMC simulation by the CO_2 centers of mass (Figure 13.32). Similar results have been reported by other researchers for other PCPs [52b, 57].

Future Prospects: Flue Gas Contaminant Stability We have already emphasized the importance of high CO_2/N_2 selectivity for a given CO_2 capture adsorbent. However, as can be seen in Table 13.2, typical flue gas also contains a small amount of oxygen, SO_2, SO_3, NO_x compounds and particulates. How readily a PCP adsorbs such contaminants, and just as importantly how readily it releases

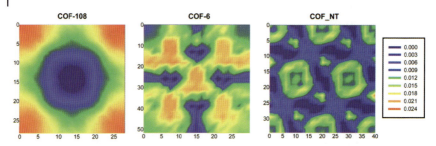

Figure 13.32 Density distribution maps for the center-of-mass of CO_2 molecules in COF-108, COF-6 and COF_NT at 1000 kPa [51b]. Reproduced by permission of The Royal Society of Chemistry.

the contaminants during the regeneration cycle, is of critical importance. If a PCP irreversibly adsorbs or reacts with a contaminant such that it cannot be removed under normal regeneration process conditions, it would not be a suitable candidate for CO_2 capture. This is because the number of useful active sites would not be available for CO_2 capture the longer the PCP was utilized in the CO_2 capture process.

Most significantly, the flue gas is saturated with respect to water vapor (Table 13.2). Even if a bed of low-cost adsorbent is placed ahead of the PCP adsorbent to remove as much water vapor as possible, it is still critical that the PCP is able to effectively adsorb CO_2 in the presence of water vapor. However, the addition of pre-drying would increase the cost of an already expensive process. Therefore, it would be best if the PCP could capture CO_2 in water-saturated flue gas streams. Of course the PCP must be stable to moisture in the first place. It has been reported that many PCPs, including IRMOF-1[58] are not moisture-stable. Further, the long-term stability of PCPs to even low levels of oxygen, such as in typical flue gas (Table 13.2), has not been reported to date. Because oxygen has a larger kinetic diameter than CO_2, it is of course possible that PCPs that are selective for CO_2 over nitrogen could have similar selectivity for CO_2 over oxygen.

It is expected that any PCP based adsorption system with capacity for CO_2 would also have capacity for SO_x compounds, and to a lesser extent NO_x compounds. Therefore, if the PCP CO_2 capture technology did not follow a limestone bed to remove SO_x, the captured CO_2 stream would require secondary treatment to remove the SO_x and other contaminants from the CO_2-rich effluent stream produced during regeneration of the adsorbent bed. Many power plants currently have technology installed to remove SO_x, NO_x and mercury from flue gas. This means that the MOF-based CO_2 adsorption system would best be situated after particulate removal, but before the final "scrubbing" equipment. If the PCP removes most or all of the trace contaminants, it is possible that the rest of the flue gas, which is mostly nitrogen, could be vented to the atmosphere without having to pass through the current scrubbers. However, having the limestone bed before the PCP technology would likely enable the PCP to work better at capturing CO_2.

References

1. (a) Brohan, P., Kennedy, J.J., Harris, I., Tett, S.F.B., and Jones, P.D. (2006) Uncertainty estimates in regional and global observed temperature changes: a new dataset from 1850. *J. Geophys. Res.*, **111**, D12106.
(b) Jones, P.D., New, M., Parker, D.E., Martin, S., and Rigor, I.G. (1999) Surface air temperature and its variations over the last 150 years. *Rev. Geophys.*, **37**, 173–199.
(c) Rayner, N.A., Brohan, P., Parker, D.E., Folland, C.K., Kennedy, J.J., Vanicek, M., Ansell, T., and Tett, S.F.B. (2006) Improved analyses of changes and uncertainties in marine temperature measured in situ since the mid-nineteenth century: the HadSST2 dataset. *J. Clim.*, **19**, 446–469.
(d) Rayner, N.A., Parker, D.E., Horton, E.B., Folland, C.K., Alexander, L.V., Rowell, D.P., Kent, E.C., and Kaplan, A. (2003) Globally complete analyses of sea surface temperature, sea ice and night marine air temperature, 1871–2000. *J. Geophys. Res.*, **108**, 4407.

2. The reconstructions used, in order from oldest to most recent publication are:
(a) (dark blue 1000-1991): Jones, P.D., Briffa, K.R., Barnett, T.P., and Tett, S.F.B. (1998) *Holocene*, **8**, 455–471.
(b) (blue 1000-1980): Mann, M.E., Bradley, R.S., and Hughes, M.K. (1999) *Geophys. Res. Lett.*, **26** (6), 759–762.
(c) (light blue 1000-1965): Crowley, T.J., and Lowery, T.S. (2000) *Ambio*, **29**, 51–54. Modified as published in Crowley, T.J. (2000) *Science*, **289**, 270–277.
(d) (lightest blue 1402-1960): Briffa, K.R., Osborn, T.J., Schweingruber, F.H., Harris, I.C., Jones, P.D., Shiyatov, S.G., and Vaganov, E.A. (2001) *J. Geophys. Res.*, **106**, 2929–2941.
(e) (light green 831-1992): Esper, J., Cook, E.R., and Schweingruber, F.H. (2002) *Science*, **295** (5563), 2250–2253.
(f) (yellow 200-1980): Mann, M.E., and Jones, P.D. (2003) *Geophy. Res. Lett.*, **30** (15), 1820.
(g) (orange 200-1995): Jones, P.D., and Mann, M.E. (2004) *Rev. Geophys.*, **42**, RG2002.
(h) (red-orange 1500-1980): Huang, S. (2004) *Geophys. Res. Lett.*, **31**, L13205.
(i) (red 1-1979): Moberg, A., Sonechkin, D.M., Holmgren, K., Datsenko, N.M., and Karlén, W. (2005) *Nature*, **443**, 613–617.
(j) (dark red 1600-1990): Oerlemans, J.H. (2005) *Science*, **308**, 675–677.
(k) (black 1856-2004): Instrumental data was jointly compiled by the Climatic Research Unit and the UK Meteorological Office Hadley Centre. Global Annual Average data set TaveGL2v was used.
(l) Documentation for the most recent update of the CRU/Hadley instrumental data set appears in: Jones, P.D., and Moberg, A. (2003) *J. Clim.*, **16**, 206–223.

3. (a) Petit, J.R., Jouzel, J., Raynaud, D., Barkov, N.I., Barnola, J.M., Basile, I., Bender, M., Chappellaz, J., Davis, J., Delaygue, G., Delmotte, M., Kotlyakov, V.M., Legrand, M., Lipenkov, V., Lorius, C., Pépin, L., Ritz, C., Saltzman, E., and Stievenard, M. (1999) Climate and atmospheric history of the past 420 000 years from the Vostok Ice Core, Antarctica. *Nature*, **399**, 429–436.
(b) EPICA Community Members (2004) Eight glacial cycles from an Antarctic ice core. *Nature*, **429** (6992), 623–628. doi: 10.1038/nature02599
(c) Lisiecki, L.E., and Raymo, M.E. (2005) A pliocene-pleistocene stack of 57 globally distributed benthic d18O records. *Paleoceanography*, **20**, PA1003.
(d) Hearty, P.J., and Kaufman, D.S. (2000) Whole-rock aminostratigraphy and Quaternary sea-level history of the Bahamas. *Quater. Res.*, **54**, 163–173.
(e) Karner, D.B., Levine, J., Medeiros, B.P., and Muller, R.A. (2002) Constructing a stacked benthic $\delta^{18}O$ record. *Paleoceanography*, **17**, 1030–1035.

4. Solomon, S., Qin, D., Manning, M., Alley, R.B., Berntsen, T., Bindoff, N.L., Chen, Z., Chidthaisong, A., Gregory, J.M., Hegerl, G.C., Heimann, M., Hewitson, B., Hoskins, B.J., Joos, F., Jouzel, J., Kattsov, V., Lohmann, U., Matsuno, T., Molina, M., Nicholls, N., Overpeck, J., Raga, G., Ramaswamy, V., Ren, J., Rusticucci, M., Somerville, R.,

Stocker, T.F., Whetton, P., Wood, R.A., and Wratt, D. (2007) Technical summary, in *Climate Change 2007: The Physical Science Basis. Contribution of Working Group I to the Fourth Assessment Report of the Intergovernmental Panel on Climate Change* (eds S. Solomon, D. Qin, M. Manning, Z. Chen, M. Marquis, K.B. Averyt, M. Tignor, and H.L. Miller). Cambridge University Press, Cambridge and New York, p. 25.

5 Houghton, J.T., Ding, Y., Griggs, D.J., Noguer, M., van der Linden, P.J., Dai, X., Maskell, K., and Johnson, C.A. (2001) *Climate Change 2001: The Scientific Basis. Contribution of Working Group I to the Third Assessment Report of the Intergovernmental Panel on Climate Change*, Cambridge University Press, Cambridge.

6 Baumert, K.A., Herzog, T., and Pershing, J. (2005) *Navigating the Numbers: Greenhouse Gas Data and International Climate Policy*, World Resources Institute.

7 Santos, S. (2008) From Current Status and Development in Carbon Capture Technologies for Power Generation, presented at the *29th Annual European AIChE Colloquium*, 12 June.

8 (a) Plasynski, S. (2007) From the Carbon Sequestration Technology Roadmap and Program Plan, US Department of Energy/National Energy Technology Laboratory.
Other good background: (b) Aaron, D., and Tsouris, C. (2005) *Sep. Sci. Tech.*, **40**, 321.

9 (a) Hoffman, J.S., and Pennline, H.W. (2000) Proceed. Annual Inter. Pittsburgh Coal Conf., 17th, 12-1.
(b) Green, D.A., Turk, B.S., Gupta, R.P., Portzer, J.W., McMichael, W.J., and Harrison, D.P. (2004) *Int. J. Environ. Tech. Manag.*, **4** (1–2), 53.

10 Selected examples of facilitated membrane references: (a) Cussler, E.L., Aris, R., and Bhown, A. (1989) *J. Membr. Sci.*, **43**, 149.
(b) Matsuyama, H., Teramoto, M., and Sakakura, H. (1996) *J. Membr. Sci.*, **114**, 193.
(c) Huang, J., Zou, J., and Ho, W.S.W. (2008) *Ind. Eng. Chem. Res.*, **47**, 1261.

11 (a) Xomeritakis, G., Liu, N.G., Chen, Z., Jiang, Y.-B., Koehn, R., Johnson, P.E., Tsai, C.-Y., Shah, P.B., Khalil, S., Singh, S., and Brinker, C.J. (2007) *J. Membr. Sci.*, **287** (2), 157.
(b) Xomeritakis, G., Tsai, C.-Y., and Brinker, C. (2005) *J. Sep. Purif. Tech.*, **42** (3), 249–257.
(c) Xomeritakis, G., Braunbarth, C.M., Smarsly, B., Liu, N., Kohn, R., Klipowicz, Z., and Brinker, C.J. (2003) *Micropor. Mesopor. Mater.*, **66** (1), 91.

12 (a) Cowan, R.M., Ge, J.-J., Qin, Y.-J., McGregor, M.L., and Trachtenberg, M.C. (2003) *Ann. N.Y. Acad. Sci.*, **984**, 453.
(b) Chen, H., Cowan, R.M., Smith, D.A., Trachtenberg, M.C. (2008) Abstracts of Papers, 236th ACS National Meeting, Philadelphia, PA, August 17–21, FUEL-102.

13 (a) Ilonich, J.B., Luebke, D.R., Myers, C., and Pennline, H.W. (2006) Proceed. Annual Inter. Pittsburgh Coal Conf., 23rd, 24.4/1.
(b) Ilonich, J.B., Myers, C., Pennline, H.W., and Luebke, D.R. (2007) *J. Membr. Sci.*, **298**, 41.

14 Scholes, C.A., Kentish, S.E., and Stevens, G.W. (2008) *Recent Patents Chem. Eng.*, **1**, 52.

15 Haag, M.-B., Kim, T.-J., and Li, B. WO 05089907.

16 (a) Kim, T.-J., Li, B., and Hagg, M.-B. (2004) *J. Polym. Sci. B: Polym. Phys.*, **42** (23), 4326.
(b) Hagg, M.-B., and Knoph, I. (2005) World Congr. Chem. Engin., 7th, Glasgow, United Kingdom, July 10–14, pp. 84610/1–84610/10.

17 (a) Anderson, J.L., Dixon, J.K., and Brennecke, J.F. (2007) *Acct. Chem. Res.*, **40**, 1208.
(b) Soutullo, M.D., Odom, C.I., Wicker, B.F., Henderson, C.N., Stenson, A.C., and Davis, J.H., Jr. (2007) *Chem. Mater.*, **19** (15), 3581.

18 Sneda, B.R., and Thangadurai, V.J. (2007) *J. Solid State Chem.*, **180**, 2661.

19 (a) Fernandez-Ochoa, E., Ronning, M., Grande, T., and Chen, D. (2006) *Chem. Mater.*, **18**, 6037.
(b) Fernandez-Ochoa, E., Ronning, M., Grande, T., and Chen, D. (2006) *Chem. Mater.*, **18**, 1383.

20 Zhao, T., Fernandez-Ochoa, E., Ronning, M., and Chen, D. (2007) *Chem. Mater.*, **19**, 3294.

21 Fernandez-Ochoa, E., Ronning, M., Yu, X., Grande, T., and Chen, D. (2008) *Ind. Eng. Chem. Res.*, **47**, 434.

22 CO_2 on zeolites and molecular sieves examples: (a) Merel, J., Clausse, M., and Meunier, F. (2008) *Ind. Eng. Chem. Res.*, **47**, 209.
(b) Pillai, R.S., Peter, S.A., and Jasra, R.V. (2008) *Micropor. Mesopor. Mater.*, **113**, 268.
(c) Ghezini, R., Sassi, M., and Bengueddach, A. (2008) *Micropor. Mesopor. Mater.*, **113**, 370.
(d) Li, P., and Tezel, H. (2007) *Micropor. Mesopor. Mater.*, **98**, 94.
(e) Jadhav, P.D., Rayalu, S.S., Biniwale, R.B., and Devotta, S. (2007) *Curr. Sci.*, **92** (6), 724.
(f) Siriwardane, R.V., Shen, M.-S., Fisher, E.P., and Losch, J. (2005) *Energy Fuels*, **19**, 1153.
(g) Harlick, P.J.E., and Tezel, H. (2004) *Micropor. Mesopor. Mater.*, **76**, 71.
(h) Chue, K.T., Kim, J.N., Yoo, Y.J., Cho, S.H., and Yang, R.T. (1995) *Ind. Eng. Chem. Res.*, **34**, 591.

23 (a) Breck, D.W. (1974) *Zeolite Molecular Sieves: Structure, Chemistry and Use*, John Wiley & Sons, Inc., New York, p. 657.
(b) Breck, D.W. (1974) *Zeolite Molecular Sieves: Structure, Chemistry and Use*, John Wiley & Sons, Inc., New York, p. 611.
(c) Breck, D.W. (1974) *Zeolite Molecular Sieves: Structure, Chemistry and Use*, John Wiley & Sons, Inc., New York, p. 607.
(d) Breck, D.W. (1974) *Zeolite Molecular Sieves: Structure, Chemistry and Use*, John Wiley & Sons, Inc., New York, p. 614.
(e) Pires, J., Brotas de Carvalho, M., Ramoa-Ribeiro, F., Derouane, E.G. (1993) *J. Mol. Catal.*, **85** (3), 295.
(f) Dunne, J.A., Mariwala, R., Roa, M., Sircar, S., Gorte, R.J., and Myers, A.L. (1996) *Langmuir*, **12**, 5888.
(g) Li, S., Falconer, J.L., and Noble, R.D. (2004) *J. Membr. Sci.*, **241**, 121.

24 Siriwardane, R.V., Shen, M.-S., Fisher, E.P., and Losch, J. (2005) *Energy Fuels*, **19**, 1153.

25 Majchrzak-Kuceba, I., and Nowak, W. (2005) *Proc. Internat. Conf. Circulating Fluidized Beds, 8th, Hangzhou, China*, p. 748.

26 Silicalite examples: (a) Golden, T.C., and Sircar, S. (1994) *J. Colloid. Interface Sci.*, **162**, 182.
(b) Goj, A., Sholl, D.S., Akten, E.D., and Kohen, D. (2002) *J. Phys. Chem. B*, **106**, 8367.
(c) Harlick, P.J.E., and Tezel, F.H. (2003) *Sep. Purif. Tech.*, **33**, 199.
(d) Delgado, J.A., Uguina, M.A., Sotelo, J.L., Ruiz, B., and Gomez, J.M. (2006) *Adsorption*, **12**, 5.

27 (a) Kitagawa, S., Kitaura, R., and Noro, S. (2004) *Angew. Chem. Int. Ed.*, **43**, 2334–2375.
(b) Kitaura, R., Kitagawa, S., Kubota, Y., Kobayashi, T.C., Kindo, K., Mita, Y., Matsuo, A., Kobayashi, M., Chang, H.C., Ozawa, T.C., Suzuki, M., Sakata, M., and Takata, M. (2002) *Science*, **298**, 2358.
(c) Uemura, T., Hiramatsu, D., Kubota, Y., Takata, M., and Kitagawa, S. (2007) *Angew. Chem.*, **46**, 4987.
(d) Kaneko, W., Ohba, M., and Kitagawa, S. (2007) *J. Am. Chem. Soc.*, **129**, 13706.
(e) Tanaka, D., Horike, S., Kitagawa, S., Ohba, M., Hasegawa, M., Ozawa, Y., and Toriumi, K. (2007) *Chem. Comm.*, 3142.
(f) Tanaka, D., and Kitagawa, S. (2008) *Chem. Mater.*, **20**, 922.

28 (a) Yaghi, O.M., Li, H., Davis, C., Richardson, D., and Groy, T.L. (1998) *Acc. Chem. Res.*, **31**, 474.
(b) Eddaoudi, M., Moler, D.B., Li, H., Chen, B., Reineke, T.M., O'Keefe, M., and Yaghi, O.M. (2001) *Accts. Chem. Res.*, **34**, 319.
(c) Yaghi, O.M., O'Keeffe, M., Ockwig, N.W., Chae, H.K., Eddaoudi, M., and Kim, J. (2003) *Nature*, **423**, 705.
(d) Rowsell, J.L.C., and Yaghi, O.M. (2004) *Micropor. Mesopor. Mater.*, **73**, 3.
(e) Ockwig, N.W., Delgado-Friedrichs, O., O'Keeffee, M., and Yaghi, O.M. (2005) *Accts. Chem. Res.*, **38**, 176.
(f) Rowsell, J.L.C., Spencer, E.C., Eckert, J., Howard, J.A.K., and Yaghi, O.M. (2005) *Science*, **309**, 1350.

29 (a) Ferey, G., Serre, C., Mellot-Draznieks, C., Millange, F., Surble, S., Dutour, J., and Margiolaki, I. (2004) *Angew. Chem. Int. Ed.*, **43**, 6296.

(b) Ferey, G., Mellot-Draznieks, C., Serre, C., Millange, F., Dutour, J., Surble, S., and Margiolaki, I. (2005) *Science*, **309**, 2040.
(c) Ferey, G., Mellot-Draznieks, C., Serre, C., and Millange, F. (2005) *Accts. Chem. Res.*, **38**, 217.
(d) Volkringer, C., Popov, D., Loiseau, T., Guillou, N., Ferey, G., Haouas, M., Taulelle, F., Mellot-Draznieks, C., Burghammer, M., and Riekel, C. (2007) *Nat. Mater.*, **6**, 760.
(e) Millange, F., Serre, C., Guillou, N., Ferey, G., and Walton, R.I. (2008) *Angew. Chem. Int. Ed.*, **47**, 1.
(f) Horcajada, P., Serre, C., Maurin, G., Ramsahye, N.A., Balas, F., Vallet-Rigi, M., Sebban, M., Taulelle, F., and Ferey, G. (2008) *J. Am. Chem. Soc.*, **130**, 6774.
(g) Ferey, G. (2007) *Stud. Surf. Sci. Catal.*, **168** (Introduction to Zeolite Molecular Science and Practice), 327.

30 Ferey, G. (2008) *Chem. Soc. Rev.*, **37** (1), 191.

31 (a) Fletcher, A.J., Thomas, K.M., and Rosseinsky, M.J. (2005) *J. Solid State Chem.*, **8**, 2491–2510.
(b) Kepert, C.J., Prior, T.J., and Rosseinsky, M.J. (2000) *J. Am. Chem. Soc.*, **122**, 5158–5168.
(c) Côté, A.P., and Shimizu, G.K.H. (2003) *Coord. Chem. Rev.*, **245**, 49–64.
(d) Duren, T., Sarkisov, L., Yaghi, O.M., and Snurr, R.Q. (2004) *Langmuir*, **20**, 2683–2689.
(e) Schlichte, K., Kratzke, T., and Kaskel, S. (2004) *Micropor. Mesopor. Mater.*, **73**, 81–88.
(f) Rowsell, J.L.C., and Yaghi, O.M. (2004) *Micropor. Mesopor. Mater.*, **73**, 3.

32 Some adsorption examples: (a) Snurr, R.Q., Hupp, J.T., and Nguyen, S.T. (2004) *AIChE J.*, **50**, 1090.
(b) Garberoglio, G., Skoulidas, A.I., and Johnson, J.K. (2005) *J. Phys. Chem. B*, **109**, 13094–13103.
(c) Panella, B., Hircher, M., Putter, H., and Muller, U. (2006) *Adv. Funct. Mater.*, **16**, 520.
(d) Lee, J.Y., Pan, L., Kelly, S.P., Jagiello, J., Emge, T.J., and Li, J. (2005) *Adv. Mater.*, **17**, 2703.
(e) Chen, B., Ockwig, N.W., Millward, A.R., Contreras, D.S., and Yaghi, O.M. (2005) *Angew. Chem. Int. Ed.*, **44**, 4745.
(f) Kitagawa, S., and Uemura, K. (2005) *Chem. Soc. Rev.*, **35**, 109.
(g) Li, H., Eddaoudi, M., Groy, T.L., and Yaghi, O.M. (1998) *J. Am. Chme. Soc.*, **120**, 8571.
(h) Eddaoudi, M., Li, H., and Yaghi, O.M. (2000) *J. Am. Chem. Soc.*, **122**, 1391.
(i) Foy, A.G., Matzger, A.J., and Yaghi, O.M. (2006) *J. Am. Chem.Soc.*, **128**, 3494.
(j) Sun, D., Ma, S., Ke, Y., Collins, D.J., and Zhou, H.-C. (2006) *J. Am. Chem. Soc.*, **128**, 3896.
(k) Lee, J.Y., Li, J., and Jagiello, J. (2005) *J. Solid State Chem.*, **178**, 2527.
(l) Wang, Q.M., Shen, D., Bulow, M., Lau, M.L., Fitch, F.R., Lemcoff, N.O., and Semanscin, J. (2002) *Micropor. Mesopor. Mater.*, **55**, 217.

33 See Zhao, D., Yuan, D., and Zhou, H.-C. (2008) *Energy Environ. Sci.*, **1**, 22 and references therein.

34 Eddaoudi, M., Kim, J., Rosi, N., Vodak, D., Wachter, J., O'Keefe, M., and Yaghi, O.M. (2002) *Science*, **295**, 469.

35 Li, H., Eddaoudi, M., O'Keefe, M., and Yaghi, O.M. (1999) *Nature*, **402**, 276.

36 (a) Serre, C., Millange, F., Thouvenot, C., Nogues, M., Marsolier, G., Louer, D., and Ferey, G. (2002) *J. Am. Chem. Soc.*, **124**, 13519.
(b) Millange, F., Serre, C., and Ferey, G. (2002) *Chem Comm.*, 822.

37 (a) Barthelet, K., Adil, K., Millange, F., Serre, C., Riou, D., and Ferey, G. (2003) *J. Mater. Chem.*, **13**, 2208.
(b) Kondo, A., Noguchi, H., Ohnishi, S., Kajiro, H., Tohdo, A., Hattori, Y., Xu, W.-C., Tanaka, H., Kanoh, H., and Kaneko, K. (2006) *Nano Lett.*, **6** (11), 2581.

38 Guillou, N., Livage, C., Drillon, M., and Ferey, G. (2003) *Angew. Chem. Int. Ed.*, **42** (43), 5314.

39 Examples of CO_2 adsorption studies: (a) Millward, A.R., and Yaghi, O.M. (2005) *J. Am Chem Soc.*, **127**, 17998.
(b) Matzger, A.J., Walton, K.S., Dubbeldam, D., Snurr, R.Q. (2006) Abst. Papers American Filtration Society, Pittsburgh, PA.

(c) Banerjee, R., Phan, A., Wang, B., Knobler, C., Furukawa, H., O'Keefe, M., and Yaghi, O.M. (2008) *Science*, **319**, 939.
(d) Wang, B., Cote, A.P., Furukawa, H., O'Keefe, M., and Yaghi, O.M. (2008) *Nature*, **453**, 207.

40 (a) Chae, H.K., Siberio-Perez, D.Y., Kim, J., Go, Y., Eddaoudi, M., Matzger, A.J., O'Keefe, M., and Yaghi, O.M. (2004) *Nature*, **427**, 523.
(b) Ma, S., Wang, X.S., Yuan, D., and Zhou, H.-C. (2008) *Angew. Chem. Int. Ed.*, **47**, 4130.
(c) Chen, B., Liang, C., Yang, J., Contreras, D.S., Clancy, Y.L., Lobkovsky, E.B., Yaghi, O.M., and Dai, S. (2006) *Angew. Chem. Int. Ed.*, **45**, 1390.
(d) Xue, M., Ma, S., Jin, Z., Schaffino, R.M., Zhu, G.-S., Lobkovsky, E.B., Qiu, S.-L., and Chen, B. (2008) *Inorg. Chem.*, **47**, 6825.
(e) Suh, M.P., Cheon, Y.E., and Lee, E.Y. (2007) *Chem. Eur. J.*, **13**, 4208.
(f) Li, J.-R., Tao, Y., Yu, Q., Bu, X.-H., Sakamoto, H., and Kitagawa, S. (2008) *Chem. Eur. J.*, **14**, 2771.
(g) Dybtsev, D.N., Chun, H., Yoon, S.H., Kim, D., and Kim, K. (2004) *J. Am. Chem. Soc.*, **126**, 32.
(h) Chen, B., Ma, S., Zapata, F., Fronczek, F.R., Lobkovsky, E.B., and Zhou, H.-C. (2007) *Inorg. Chem.*, **46**, 1233.
(i) Chen, B., Ma, S., Hurtado, E.J., Lobkovsky, E.B., and Zhou, H.-C. (2007) *Inorg. Chem.*, **46**, 8490.
(j) Ma, S., Wang, X.-S., Manis, E.S., Collier, C.D., and Zhou, H.-C. (2007) *Inorg. Chem.*, **46**, 3432.
(k) Eddaoudi, M., Li, H., and Yaghi, O.M. (2000) *J. Am. Chem. Soc.*, **122**, 1391.
(l) Ma, S., Sun, D., Wang, X.-S., and Zhou, H.-C. (2007) *Angew. Chem. Int. Ed.*, **46**, 2458.
(m) Horike, S., Tanaka, D., Nakagawa, K., and Kitagawa, S. (2007) *Chem. Commun.*, 3395.
(n) Zou, Y., Hong, S., Park, M., Chun, H., and Lah, M.S. (2007) *Chem. Commun.*, 5182.
(o) Moon, H.R., Kobayashi, N., and Suh, M.P. (2006) *Inorg. Chem.*, **45**, 8672.
(p) Ma, S., Wang, X.S., Collier, C.D., Manis, E.S., and Zhou, H.-C. (2007) *Inorg. Chem.*, **46**, 8499.
(q) Maji, T.K., Matsuda, R., and Kitagawa, S. (2007) *Nat. Mater.*, **6**, 142.
(r) Reineke, T.M., Eddaoudi, M., O'Keefe, M., and Yaghi, O.M. (1999) *Angew. Chem. Int. Ed.*, **38**, 2590.

41 (a) Navarro, J.A.R., Barea, E., Salas, J.M., Masciocchi, N., Galli, S., Sironi, A., Ania, C.O., and Parra, J.B. (2007) *J. Mater. Chem.*, **17**, 1939.
(b) Chandler, B.D., Cramb, D.T., and Shimizu, G.H.K. (2006) *J. Am. Chem. Soc.*, **128**, 10403.
(c) Guo, X., Zhu, G., Li, Z., Sun, F., Yang, Z., and Qiu, S. (2006) *Chem. Commun.*, 3172.
(d) Hayishi, H., Cote, A.P., Furukawa, H., O'Keefe, M., and Yaghi, O.M. (2007) *Nat. Mater.*, **6**, 501.
(e) Navarro, J.A.R., Barea, E., Rodriguez-Dieguez, A., Salas, J.M., Ania, C.O., Parra, J.B., Masciocchi, N., Galli, S., and Sironi, A. (2008) *J. Am. Chem. Soc.*, **130**, 3978.
(f) Matsuda, R., Kitaura, R., Kitagawa, S., Belosludov, R.V., Kobayashi, T.C., Sakamoto, H., Chiba, T., Takata, M., Kawazoe, Y., and Mita, Y. (2005) *Nature*, **436**, 238.
(g) Jeon, Y.-M., Armatas, G.S., Heo, J., Kanatzidis, M.G., and Mirkin, C.A. (2008) *Adv. Mater.*, **20**, 2105.
(h) Pan, L., Adams, K.M., Hernandez, H.E., Wang, X., Zheng, C., Hattori, Y., and Kaneko, K. (2003) *J. Am. Chem. Soc.*, **125**, 3062.
(i) Gadizkwa, T., Zeng, B.-S., Hupp, J.T., and Nguyen, S.T. (2008) *Chem. Commun.*, 3672.

42 (a) Caskey, S.R., Wong-Foy, A.G., and Matzger, A.J. (2008) *J. Am. Chem. Soc.*, **130**, 10870.
(b) Dietzel, P.D.C., Panella, B., Hirscher, M., Blom, R., and Fjellvag, H. (2006) *Chem. Commun.*, 959.
(c) Dietzel, P.D.C., Morita, Y., Blom, R., and Fjellvag, H. (2005) *Angew. Chem. Int. Ed.*, **44**, 6354.
(d) Rosi, N., Kim, J., Eddaoudi, M., Chen, B., O'Keefe, M., and Yaghi, O.M. (2005) *J. Am. Chem. Soc.*, **127**, 1504.

(e) Llewellyn, P.L., Bourrelly, S., Serre, C., Vimont, A., Daturi, M., Hamon, L., De Weireld, G., Chang, J.-S., Hong, D.-Y., Hwang, Y.K., Jhung, S.H., and Ferey, G. (2008) *Langmuir*, **24**, 7245.
(f) Navarro, J.A.R., Barea, E., Salas, J.M., Masciocchi, N., Galli, S., Sironi, A., Ania, C.O., and Parra, J.B. (2006) *Inorg. Chem.*, **128**, 2397.
(g) Yoon, J.W., Jhung, S.H., Hwang, Y.K., Humphrey, S.M., Wood, P.T., and Chang, J.-S. (2007) *Adv. Mater.*, **19**, 1830.
(h) Wang, Q.M., Shen, D., Bulow, M., Lau, M.L., Deng, S., Fitch, F.R., Lemcoff, N.O., and Semanscin, J. (2002) *Micropor. Mesopor. Mater.*, **55**, 217.
(i) Zhang, J.-P., and Kitagawa, S. (2008) *J. Am. Chem. Soc.*, **130**, 907.
(j) Loiseau, T., Lecroq, L., Volkringer, C., Marrot, J., Ferey, G., Haouas, M., Taulelle, F., Bourrelly, S., Llewellyn, P.L., and Latroche, M. (2006) *J. Am. Chem. Soc.*, **128**, 10223.
(k) Barthelet, K., Marrot, J., Riou, D., and Ferey, G. (2002) *Angew. Chem. Int. Ed.*, **41**, 281.
(l) Ramsahye, N.A., Maurin, G., Bourrelly, S., Llewellyn, P.L., Serre, C., Loiseau, T., Devic, T., and Ferey, G. (2008) *J. Phys. Chem. C*, **112**, 514.
(m) Park, Y.K., Choi, S.B., Kim, H., Kim, K., Won, B.-Y., Choi, K., Choi, J.-S., Ahn, W.-S., Won, N., Kim, S., Jung, D.H., Choi, S.-H., Kim, G.-H., Cha, S.-S., Jhon, Y.H., Yang, J.K., and Kim, J. (2007) *Angew. Chem. Int. Ed.*, **46**, 8230.

43 (a) Frost, H., Duren, T., and Snurr, R.Q. (2006) *J. Phys. Chem. B*, **110**, 9565.
(b) Duren, T., Millange, F., Ferey, G., Walton, K.S., and Snurr, R.Q. (2007) *J. Phys. Chem. C*, **111** (42), 15350.
(c) Walton, K.S., and Snurr, R.Q. (2007) *J. Am. Chem. Soc.*, **129** (27), 11989.

44 Walton, K.S., Millward, A.R., Dubbeldam, D., Frost, H., Low, J.J., Yaghi, O.M., and Snurr, R.Q. (2008) *J. Am. Chem. Soc.*, **130**, 406.

45 (a) Barbaro, R., Hu, Z., Jiang, J., Chempath, S., and Sandler, S. (2007) *Langmuir*, **23**, 659.
(b) Jiang, J., and Sandler, S. (2005) *J. Am. Chem. Soc.*, **127**, 8552.

46 (a) Snurr, R.Q., Yazaydin, A.O., Dubbeldam, D., and Frost, H. Molecular modeling of adsorption and diffusion in metal-organic frameworks, in *Metal-Organic Frameworks: Design and Application* (ed. L.R. MacGillivray), Wiley-VCH Verlag GmbH, in press.
(b) Keskin, S., Liu, J., Rankin, R.B., Johnson, J.K., and Sholl, D.S. (2009) Progress, opportunities, and challenges for applying atomically detailed modeling to molecular adsorption and transport in metal-organic framework materials. *Ind. Eng. Chem. Res.*, **48**, 2355–2371.

47 Fuchs, A.H., and Cheetham, A.K. (2001) *J. Phys. Chem. B*, **105**, 7375–7383.

48 (a) Ramsahye, N.A., Maurin, G., Bourrelly, S., Llewellyn, P.L., Devic, T., Serre, C., Loiseau, T., and Ferey, G. (2007) *Adsorption*, **13**, 461–467.
(b) Ramsahye, N.A., Maurin, G., Bourrelly, S., Llewellyn, P.L., Devic, T., Serre, C., Loiseau, T., and Ferey, G. (2007) *Chem. Commun.*, 3261–3263.

49 Bourrelly, S., Llewellyn, P.L., Serre, C., Millange, F., Loiseau, T., and Ferey, G. (2005) *J. Am. Chem. Soc.*, **127**, 13519–13521.

50 Barthelet, K., Marrot, J., Riou, D., and Ferey, G. (2002) *Angew. Chem. Int. Ed.*, **41**, 281–282.

51 (a) Babarao, R., and Jiang, J.W. (2008) *Langmuir*, **24**, 6270–6278.
(b) Babarao, R., and Jiang, J.W. (2008) *Energy Environ. Sci.*, **1**, 139–143.

52 Myers, A.L., and Prausnitz, J.M. (1965) *AIChE J.*, **11**, 121–127.

53 Murthi, M., and Snurr, R.Q. (2004) *Langmuir*, **20**, 2489–2497.

54 (a) Yang, Q.Y., Xue, C.Y., Zhong, C.L., and Chen, J.F. (2007) *AIChE J.*, **53**, 2832–2840.
(b) Yang, Q.Y., and Zhong, C.L. (2006) *ChemPhysChem*, **7**, 1417–1421.
(c) Yang, Q.Y., and Zhong, C.L. (2006) *J. Phys. Chem. B*, **110**, 17776–17783.

55 Bae, Y.-S., Mulfort, K.L., Frost, H., Ryan, P., Punnathanam, S., Broadbelt, L.J., Hupp, J.T., and Snurr, R.Q. (2008) *Langmuir*, **24**, 8592–8598.

56 (a) Cote, A.P., Benin, A.I., Ockwig, N.W., O'Keeffe, M., Matzger, A.J., and

Yaghi, O.M. (2005) *Science*, **310**, 1166–1170.
(b) Cote, A.P., El-Kaderi, H.M., Furukawa, H., Hunt, J.R., and Yaghi, O.M. (2007) *J. Am. Chem. Soc.*, **129**, 12914–12915.
(c) El-Kaderi, H.M., Hunt, J.R., Mendoza-Cortes, J.L., Cote, A.P., Taylor, R.E., O'Keeffe, M., and Yaghi, O.M. (2007) *Science*, **316**, 268–272.

57 Dubbeldam, D., Frost, H., Walton, K.S., and Snurr, R.Q. (2007) *Fluid Phase Equil.*, **261**, 152–161.

58 Kaye, S.S., Dailley, A., Yaghi, O.M., and Long, J.R. (2007) *J. Am. Chem. Soc.*, **129**, 14176.

14
Nanostructured Organic Light-Emitting Devices
Juo-Hao Li, Jinsong Huang, and Yang Yang

14.1
Introduction

Lighting is essential in human civilization. Lighting technology extends human activity to the night, bringing more conveniences to human life. However in doing so more energy is consumed. Today the energy issue attracts a lot of attention because of the limitation of fossil fuel resources and the endless expansion of human demand and consumption.

The appearance of light-emitting diodes (LEDs) provided a way to save energy without compromising the quality of life and civilization. Because of the high energy conversion from electricity to light, LEDs can provide the same illumination with less energy, compared with light bulbs and fluorescence tubes. Organic light-emitting diodes (OLEDs), including small-molecule and polymer-based light-emitting diodes (PLEDs), also aim for the same target using organic semiconductor materials.

OLEDs have a simple structure consisting of an emissive layer sandwiched by the metal electrodes. Given a forward bias, the excited organic molecules within the emissive layer release the energy and generate a light output. Figure 14.1 shows the basic structure of a OLED. Moreover, some unique characteristics such as flexibility and processibility for large area fabrication allow OLEDs to have potential applications in the fields of solid state lighting and display [1–6].

Typically, we use external quantum efficiency (EQE; the ratio of the number of photons emitted from the surface to the number of charge carriers injected) to evaluate the energy conversion of light-emitting devices [7, 8]. Higher quantum efficiency represents more efficient energy conversion from electricity to the light output. Several factors are related to the EQE of OLEDs, and a simple expression can be realized as:

$$\eta_{EQE} = \eta_{IQE} \cdot \chi = \gamma \cdot \eta_{ex} \cdot \phi_p \cdot \chi \tag{14.1}$$

where χ is the light out-coupling efficiency, η_{ex} is the fraction of total excitons formed which result in radiative transitions ($\eta_{ex} \sim 0.25$ for fluorescent molecular

---

Nanotechnology for the Energy Challenge. Edited by Javier Garcia-Martinez
© 2009 WILEY-VCH Verlag GmbH & Co. KGaA, Weinheim
ISBN: 978-3-527-32401-9

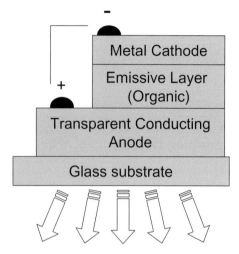

Figure 14.1 Basic structure of an OLED.

materials, 1.0 for phosphorescent dyes), γ is the electron-hole charge-balance factor ($\gamma \leq 1$) and φ_p is the intrinsic quantum efficiency for radiative decay. To improve the internal quantum efficiency (η_{IQE}), one can modify the device structures or utilize new materials to fulfill requirements for more photon generation (light). Due to the optical effects, most of the light is trapped in the device structure ($\chi \sim 20\%$ for a glass substrate with index of refraction $n = 1.5$). The modification of substrate surface or the adding of optical thin films can help extract light.

In this chapter, the factors mentioned above are discussed in separate sections to show the way to achieve efficient OLEDs and PLEDs. The concepts of quantum confinement are proposed in Section 14.2 for improving the internal quantum efficiency, and are carried out through multilayer structures. Meanwhile, the adding of nanostructured interfacial layer between the emissive layer and the metal electrode dramatically increases the charge injection and charge balance. In Section 14.3, phosphorescent materials are discussed for the improvement of photon generation. The difference between fluorescent and phosphorescent emitters is explained and some high-efficiency devices based on phosphorescent emitters are demonstrated. Also, solution-processed phosphorescent emitters are presented in this section. In Section 14.4, the tandem structure of OLEDs is discussed for improving the quantum efficiency through the series connecting of multiple emitting units. The enhancement of light out-coupling by modifying the substrate surface or the adding of optical thin films is presented in Section 14.5. Finally, a discussion is provided about the future possibility of nanostructured OLEDs and PLEDs, followed by a conclusion.

14.2
Quantum Confinement and Charge Balance for OLEDs and PLEDs

14.2.1
Multilayer Structured OLEDs and PLEDs

The basic structure of OLEDs and PLEDs is an organic emitter sandwiched between electrodes. Although this kind of structure provides advantages of simplicity and easy processing, the electronic structure is not favorable for the exciton formation and photon generation. The main drawbacks are the inefficiency of charge injection, transport and recombination. The energy level alignment in the sandwich structure shows the significant energy barrier between the emitter and the electrodes, which hinders the charge injection. Also, the organic emitters are usually not bipolar materials, implying the different ability of charge transport, resulting in broaden charge recombination and low recombination probability. However, excited molecules (excitons) close to the anode or the cathode increase the possibility of quenching effects at the interface of emitters and metal electrodes. Studies have shown that organic multilayer structures can significantly improve the OLED device performance due to the confinement of charge carriers inside an emissive layer by balanced charge injection and transport [9, 10]. Figure 14.2 compares the conventional OLEDs structure and multilayer structure. By introducing the unipolar materials of electron transport layer (ETL) and hole transport layer (HTL), the charge carrier can be confined within a narrow region, resulting in a high recombination probability. Another advantage of utilizing a multilayer structure is keeping the energy of the excited molecules from dissipating into the metal electrodes.

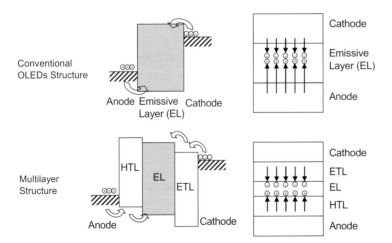

Figure 14.2 Schematic diagrams show the conventional OLED structure and multilayer structure. By introducing the unipolar materials of ETL and HTL, the charge carrier can be confined within a narrow region, resulting in a high recombination probability.

The function of a multilayer is clear, and the requirements for these nanostructured layers of ETL/HTL are their electronic characteristics, including transport properties and energy level. The HTL/ETL should have a small barrier for hole/electron injection from the electrodes. To achieve this purpose, the lowest unoccupied molecular orbital (LUMO) of the ETL and highest occupied molecular orbital (HOMO) of the HTL should match or have a similar value with the work function of cathode and anode, respectively. Also, the HTL should play the role of electron-blocking layer to prevent the electron passing through the emissive layer, and vice versa for the ETL. High hole/electron mobility is favorable for the charge transport in the OLED structure. In addition, the exciplex formation between HTL/ETL should be avoided. A high glass transition temperature and melting point would aid the thermal stability.

Scheme 14.1 shows several typical examples of HTL and ETL materials. Each compound has different characteristics and device performance when employed in OLEDs. For example, the phthalocyanine, CuPc, with an inonization potential (I_p) of around 5.0 eV (ITO anode, $I_p \approx 5.0$ eV), can serve as a good hole transport material; however, it has large absorption in the visible range [11]. The compound of m-MTDATA later proposed by Shirota et al. [12] has less absorption in the visible range and also can form a good contact for hole injection with the ITO anode. Therefore, it has been used widely as the HTL in OLEDs. However, it has been reported that there is around 36% of voltage drop through the electron transport layer, implying a higher operating voltage and power consumption. Thus, the development of ETL is also an urgent need and attracts a lot of attention. Several types of ETL have been developed, such as Alq$_3$, PBD, TAZ and the widely used material, Bphen. Alq$_3$ is a standard ETL because of its low cost and relatively good device lifetime. However, the mobility ($\approx 10^{-6}$ cm^2/Vs) is relatively low for an ETL. Oxadiazoles such as PBD have a higher mobility than Alq$_3$ but a poor lifetime. The derivatives of phenanthroline, such as Bphen, provide higher mobility ($\approx 10^{-5}$ cm^2/Vs) but also have a poor lifetime. Later Kathirgamanathan et al. [14] reported several derivatives of phenanthroline with very high mobility ($\approx 10^{-2}$ cm^2/Vs) and a lifetime comparable with Alq$_3$. Because of the continuing demands for lowering the power consumption, operating voltage and extending the lifetime of OLEDs, further studies and developments of the HTL/ETL are still needed.

14.2.2
Charge Balance in a Polymer Blended System

The design of a multilayer structure is desirable to enhance the internal quantum efficiency by raising the recombination probability and reducing the exciton-quenching phenomenon. However, for solution-processed OLEDs, such as PLEDs, the multilayer structure is quite difficult to achieve because the bottom layer washes away while the upper layer is being deposited. Therefore, a different strategy has been developed for improving the internal quantum efficiency with PLEDs. Here we demonstrate a polymer blended system that can be used for the

(a) HTL

Phthalocyanine, Copper complex
CuPc

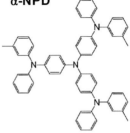
N, N'-Bis(naphthalen-1-yl)-N,N'-bis(phenyl)-9,9-diphenyl-fluorene
α-NPD

N, N'-Bis(3-methylphenyl)-N,N'-bis(phenyl)-benzidine
TPD

4,4',4"-Tris(N-3-methylphenyl-N-phenyl-amino)triphenylamine
m-MTDATA

(b) ETL

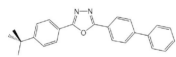

2-(4-Biphenylyl)-5-(4-tert-butylphenyl)-1,3,4-oxadiazole
PBD

4,7-Diphenyl-1,10-phenanthroline
Bphen

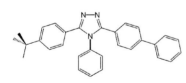

3-(4-Biphenylyl)-4-phenyl-5-tert-butylphenyl-1,2,4-triazole
TAZ

Tris(8-hydroxy-quinolinato)aluminium
Alq$_3$

Scheme 14.1 Typical (a) HTL and (b) ETL materials used in multilayer structure.

enhancement of device performance due to the characteristic of a better charge balance.

As mentioned in the introduction, most conjugated polymers are unipolar and have unbalanced charge transport properties, such as a hole mobility that is much larger than the electron mobility. Here, a general method is introduced to significantly increase the efficiency of PLEDs by controlling the charge injection and distribution through material processing and interface engineering in the device. By blending high band gap and low band gap polymers in proper ratios, we are able to introduce charge traps in the light-emitting polymer (LEP) layer. Similarly, by introducing an electron injection/hole-blocking layer, we are able to enhance the minority carrier (electron) injection and confine holes to the emissive layer. Efficient and balanced charge injections, as well as charge confinement, are attained simultaneously, and as a result high efficiency devices can be achieved. From 0.25% to 2.0% of poly[2-methoxy-5-(2´-ethyl-hexyloxy)-1,4-phenylene vinylene] (MEH-PPV) is blended with poly(9,9-dioctylfluorene) (PFO) as the active polymer layer for PLEDs. A Cs_2CO_3 electron injection (and hole-blocking) layer is used at the cathode interface. The emission from the device covers colors from white to yellow, depending on the blending ratio.

There are several benefits to using a polymer blend: (i) low bandgap LEP behaves as a dopant for energy transfer from the higher bandgap LEP, (ii) low bandgap LEP behaves as a charge trapping site to trap (and confine) the injected charges, which is particularly important in the low voltage regime where only one type of charge is often present and (iii) the trapped electrons in the low bandgap LEP eventually help the injection of holes and lead to self-balanced charge injection. When this LEP blend system is coupled with an electron injection (and hole-blocking) layer of $Ca(acac)_2$ [15] or Cs_2CO_3 [16] at the cathode interface, holes are blocked within the LEP layer as well. As a result, both electrons and holes are effectively confined in LEP layer rather than being extracted directly at the electrodes. Hence, efficient recombination occurs due to the overlapping distribution of electrons and holes (through the formation of excitons). All of these factors can help to increase the efficiency of PLED devices. The schematic profile for the energy structure is shown in Figure 14.3.

Four types of devices with the structure of ITO/poly(ethylene dioxy thiophene):polystyrene sulfonate (PEDOT:PSS)/PFO:MEH-PPV/Cs_2CO_3/Al were fabricated. In order to obtain the electron injection and hole-blocking layer through solution processing, Cs_2CO_3 was dissolved in 2-ethoxyethonal to form a dilute solution. The three layers of PEDOT:PSS, PFO:MEH-PPV and Cs_2CO_3 were formed sequentially by spin-coating one layer after another. The thickness of the polymer blend layer was between 80 and 100 nm. The color of EL can be modulated from yellow to white by changing the concentration of MEH-PPV from 2.0 wt% to 0.25 wt%.

The devices show very good performance. The current–voltage–luminance (I-V-L) plot is shown in Figure 14.4. The leakage current before light turn-on was low ($\sim 10^{-5}$ mA/cm^2), which is ideal for large area illumination application. Light emission was observed at a low applied external voltage of 2.3 V. The single

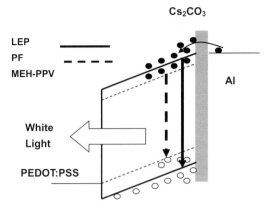

Figure 14.3 Schematic electronic energy profile for the proposed device architecture.

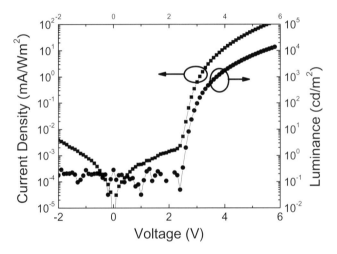

Figure 14.4 I–V–L plot of device with 0.5% MEH-PPV in PFO.

emission layer structure assures a low operating voltage; the emitting intensity for the 0.5 wt% device reaches 3000 and 10 000 cd/m^2 at voltages of 4.3 and 5.4 V, respectively.

The high performance of the device is attributed to the excellent balance of electrons and holes, and the charge confinement; in addition, the polymer system has a high PL efficiency. The power efficiency versus current density of the four devices is shown in Figure 14.5. The forward external quantum efficiency is calculated according to the luminous efficiency and EL spectra at 25 mA/m^2, which is also shown in Figure 14.5. The maxima are 6.0% for the white device (device C) at 110 cd/m^2 and 4.3% for the yellow device (device D) at 300 cd/m^2. The peak power efficiencies for the white and yellow devices are 16 lm/W (devices B, C) and

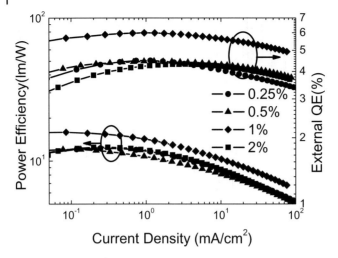

Figure 14.5 Characteristics of external efficiency and power efficiency as functions of current density for the four PLEDs.

12.5 lm/W (device D), respectively, at low current density. The power efficiency at 100 cd/m² is still as high as 15.3 lm/W (device B) and 12.1 lm/W (device D) for the white and yellow devices, respectively.

According to the assumption, the improvement of the devices' performance results from the combination of two factors: self-balanced efficient charge injection and charge confinement. A minor increase in luminous efficiency can be obtained when only one condition is satisfied. This is illustrated by the comparison of luminous efficiencies for three groups of devices, as shown in Figure 14.6. For those devices with the same device structure, the luminous efficiency reflects the degree of charge balance giving the same EL spectra. The three groups of devices are: MEH-PPV, PFO and 2 wt% MEH-PPV:PFO devices, using Ca or Cs_2CO_3 as the cathode for each. The reason why we chose the high percentage MEH-PPV sample for comparison is because the EL emission of this sample was contributed mainly from MEH-PPV, and the contribution for the improved efficiency from PFO emission is excluded. For the Ca cathode devices, the doped MEH-PPV:PFO device has an efficiency 3.5 times greater than the MEH-PPV device, which should be partially attributed to the increased PL efficiency of polymer film. The increase of MEH-PPV PL efficiency in the blend can be explained by the suppression of interchain species [17, 18]. In the diluted MEH-PPV:PFO solid solution, the MEH-PPV chains are effectively isolated by the PFO molecules, and interchain interactions are significantly reduced, which otherwise decrease the PL efficiency of MEH-PPV film by aggregation. For the MEH-PPV and PFO devices with two different cathodes, the luminous efficiencies of devices with a Cs_2CO_3 cathode are increased to 1.4 and 1.3 times over the devices with a Ca cathode; when both dopant and hole-blocking layer are used, the efficiency is improved to 11.2 cd/A, which is more than 3.0 times that of the MEH-PPV:PFO/Ca device. We note that

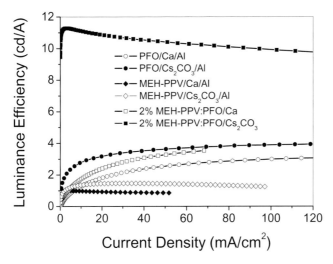

Figure 14.6 Luminance efficiencies of three groups of devices.

the efficiency improvement ratios are different after cathode modification for the single component and blend material (3.0 times improvement in the blend system compared to 1.3 times in the single-component polymer system). This apparent difference could be explained by the following scenario: because the Cs_2CO_3 hole-blocking layer is only a few nanometers thick, it can only partially block the holes and thus hole accumulation is limited. For devices using the polymer blend, however, the charge-trapping effect occurring on MEH-PPV molecules enhances the hole accumulation more than just the interface-blocking effect. The combined result is a better balance of electrons and holes confined inside the LEP layer and subsequently achieving much higher EL efficiency. This argument agrees with our statement: the considerable improvement of efficiency for PLEDs can be achieved when two conditions are simultaneously satisfied: (i) the use of a polymer blend system and (ii) balanced electrons and holes by the interface modification layer (Cs_2CO_3).

14.2.3
Interfacial Layer and Charge Injection

Charge injection occurs at the interface between the organic semiconductor and metal electrode. A Schottky barrier forms due to the charge accumulation between two materials of different electronic structure, resulting in the difficulty of charge injection from the metal to the organic semiconductor. Several strategies to solve this problem have been reported in the literature, including using a low work function metal, such as K, Na, Li, Mg and Ca [19–21]. Models have also been proposed to explain the origin of charge injection and relate it with the device performance [22–24]. An alternative way is inserting a ultra-thin layer between the organic semiconductor and metal electrode, such as LiF and CsF [25–27]. These nanostructured interface layers play a critical role in controlling the performance of organic light-emitting devices. The most famous example would be the utilizing of LiF between the emission layer, tris(8-hydroxyquinolate)aluminum (Alq$_3$), and the metal cathode, Al.

It is desirable to realize the function of the inserted interfacial layer because of its important role in dramatic improvement of device performance. Here the Cs$_2$CO$_3$ is used as an example to study the correlations between interfacial layer and charge injection. This section presents a detailed mechanism study of Cs$_2$CO$_3$ layer formed by either solution process or thermal deposition process on PLEDs using several different metal cathodes. First, the devices are characterized by current–voltage (I–V) curves to demonstrate the increased electron injection by Cs$_2$CO$_3$. Second, the increase of electron injection is explored by measuring the change of cathode work function in real devices. This is done by measuring the device built-in potential of the devices using photovoltaic measurement. Third, the interface electronic structures and the mechanism of work function variation are further studied by XPS/UPS. Fourth, analysis of high-performance devices with a Cs/Al cathode points to the same origin of the obtained results.

14.2.3.1 I–V Characteristics

In order to illustrate how the electron injection layer forms and functions, batches of PLED devices with various Cs$_2$CO$_3$ thicknesses were fabricated by thermal evaporation. For comparison, PLEDs using solution process Cs$_2$CO$_3$ were also fabricated. All the devices have the structure ITO/PEDOT/LEP/Cs$_2$CO$_3$/Al (or Ag), where the LEP is polymer blend of polyfluorene (PF) and poly[2-methoxy-5-(2′-ethyl- hexyloxy)-1,4-phenylene vinylene] (MEH-PPV). In order to test whether Al is obligatory for the metal cathode for high performance device, silver (Ag) was used as a cathode. Ag was chosen because Ag has a similar work function as Al, but is much less chemically reactive. The thickness of thermally evaporated Cs$_2$CO$_3$ was changed from 0.3 to 30.0 Å. The thickness of solution-processed Cs$_2$CO$_3$ was estimated by XPS absorption to be about 20 Å. In the manuscript, the terminologies of Cs$_2$CO$_3$ (sol)/Al, Cs$_2$CO$_3$ (sol)/Ag, Cs$_2$CO$_3$ (evp)/Al and Cs$_2$CO$_3$ (evp)/Ag stand for devices with solution-processed Cs$_2$CO$_3$/Al, solution-processed Cs$_2$CO$_3$/Ag, thermally evaporated Cs$_2$CO$_3$/Al cathode and thermally evaporated Cs$_2$CO$_3$/Ag cathode, respectively. The terminology of Cs$_2$CO$_3$ (evp) and Cs$_2$CO$_3$ (sol) are used to represent thermally evaporated Cs$_2$CO$_3$ and solution-processed Cs$_2$CO$_3$, respectively, because the thermally evaporated Cs$_2$CO$_3$ is indeed no longer Cs$_2$CO$_3$, as shown in the following sections.

Thermally Evaporated Cs$_2$CO$_3$ Figure 14.7 shows the current–voltage (I–V) plots for the Cs$_2$CO$_3$ (evp)/Al and Cs$_2$CO$_3$ (evp)/Ag devices with Cs$_2$CO$_3$ thickness varying from 0.3 to 30.0 Å. As one can see from the figure, the current increases with increasing Cs$_2$CO$_3$ thickness for both Al and Ag as the cathode metal. Within the thickness range of this testing, no decrease of current was observed at high Cs$_2$CO$_3$ thickness. In contrast to what Hasegawa et al. [28] observed, we did find some small but distinct difference between Cs$_2$CO$_3$ (evp)/Al devices and Cs$_2$CO$_3$

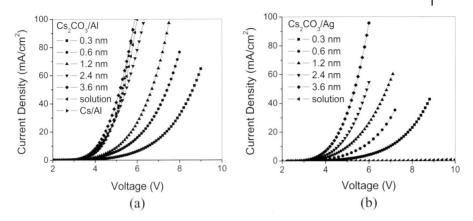

Figure 14.7 I–V curves of devices with different Cs$_2$CO$_3$ (thermally deposited) thicknesses using Al (a) and Ag (b) as cathode metals; the I–V curves of devices with solution-processed Cs$_2$CO$_3$ are also shown in each figure; I–V curve of Cs/Al device is shown in (a).

(evp)/Ag devices in terms of magnitude of injected current. For the devices with same parameters, Cs$_2$CO$_3$ (evp)/Al devices have larger current than Cs$_2$CO$_3$ (evp)/Ag devices. Since the work functions of Al and Ag are very close, the difference in device current is ascribed to their different chemical reactivity. Interface layer formed by reaction of Al with thermally evaporated Cs$_2$CO$_3$ can further increase the electron injection, which was proven by XPS/UPS measurement. One problem with Cs$_2$CO$_3$ (evp)/Ag devices is that Cs$_2$CO$_3$ (evp)/Ag devices are easier to short electrically, possibly due to the diffusion of Ag into the LEP layer.

Solution-Processed Cs$_2$CO$_3$ It is also shown in Figure 14.7 the I-V curve of Cs$_2$CO$_3$ (sol)/Al and Cs$_2$CO$_3$ (sol)/Ag devices. For the Cs$_2$CO$_3$ (sol)/Al device, the current is comparable to that of the device with 24 Å of Cs$_2$CO$_3$ (evp). This result is reasonable because the thickness of solution-processed Cs$_2$CO$_3$ is around 20 Å. But for the Cs$_2$CO$_3$ (sol)/Ag device, the current is as small as that of the device with only Al cathode. This result suggests that the solution-processed Cs$_2$CO$_3$ does not play any role in increasing the injection of electrons from Ag. Again this huge difference is ascribed to the different chemical reactivity of these two metals. We conclude that the reaction of Al with spin-coated Cs$_2$CO$_3$ is important to obtain the electron injection interface. Another conclusion is that the idea of doping of the polymer by Cs$_2$CO$_3$ can be rejected as the main reason for increased electron injection by Cs$_2$CO$_3$, in contrast to the situation of Cs$_2$CO$_3$ doped into Alq$_3$ [29]. When the polymer layer is doped by Cs$_2$CO$_3$ (sol) and forms a thin n-type region at the LEP/cathode interface, there should not be such a huge difference in current injection between Cs$_2$CO$_3$ (sol)/Ag and Cs$_2$CO$_3$ (sol)/Al devices.

14.2.3.2 Built-in Potential From Photovoltaic Measurement

To understand the mechanism responsible for the increased electron injection by Cs$_2$CO$_3$ (evp), the work function of Cs$_2$CO$_3$ (evp)/Al cathode is evaluated by

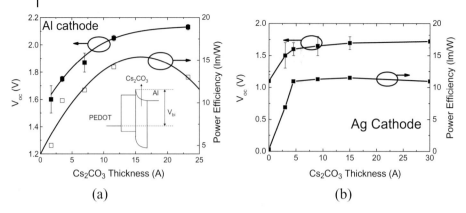

Figure 14.8 Open circuit voltage of devices with different Cs_2CO_3 (thermally deposited) thickness using Al (a) and Ag (b) as cathode metals.

photovoltaic measurement. Similar to electro-absorption measurement [30], photovoltaic measurement provides information about work function shift of electrode when there is no internal charge transfer [31, 32]. In this measurement, the photo-induced current is measured when the devices are subjected to 1.5 M global simulated sun illumination. In order to exclude the effect of leakage current, the dark current is subtracted from photocurrent to get the modified I-V curve. The open circuit voltage (V_{oc}) is obtained from a modified I-V curve when the current is equal to zero. The V_{oc} from photovoltaic measurement reflects the built-in potential in PLED devices. Generally the built-in potential is the difference of work function of anode and cathode in PLED devices in the absence of an interfacial dpole. Since the parameters of all devices are the same except the thickness of Cs_2CO_3 (evp), the dipole configuration at the interface is also the same for all devices, if there are any of them. So the shift of V_{oc} must be in scale with the change of cathode work function. Figure 14.8 shows the relationship of measured V_{oc} with the thickness of Cs_2CO_3 (evp) for both Al and Ag cathode. As one can see from the figure, V_{oc} increases with Cs_2CO_3 (evp) thickness for devices with Al or Ag cathode when Cs_2CO_3 is thin, and saturates when bulk Cs_2CO_3 (evp) forms. As we mentioned above, the higher V_{oc} indicates essentially the lower work function of Cs_2CO_3 (evp)/Al cathode. The PV measurement provides direct evidence that evaporated Cs_2CO_3 can reduce the work function of Al and Ag; and therefore increase electron injection. It has been reported that the low work function of thermally evaporated Cs_2CO_3 is its intrinsic property [33, 34]. Therefore, the observed evolution of work function with different Cs_2CO_3 thicknesses demonstrates how the evaporated Cs_2CO_3 covers the surface.

It is noted that the saturated V_{oc} of Cs_2CO_3 (evp)/Ag devices is smaller than that of Cs_2CO_3 (evp)/Al devices, which means that even lower work function can be realized in Cs_2CO_3 (evp)/Al cathode. This also agrees with I-V results in Figure 14.7. The power efficiencies of these devices at the luminance of 100 cd/cm^2 are also shown in Figure 14.8. The variation of power efficiency follows that of V_{oc}

when thickness of Cs_2CO_3 (evp) is less than 15 Å. It reaches a maximum at the Cs_2CO_3 (evp) thickness of 15 Å and drops again when the Cs_2CO_3 (evp) thickness is further increased. It is easy to understand that power efficiency increases with increasing Cs_2CO_3 (evp) thickness because electron injection becomes stronger. However when electron injection is so strong that the electron current in LEP layer overrides hole current, efficiency decreases with increasing Cs_2CO_3 (evp) thickness due to the imbalanced charges.

The V_{oc} of the Cs_2CO_3 (sol)/Al device coincides with the saturated V_{oc} of the Cs_2CO_3 (evp)/Al device. Once more, we find that the maximum power efficiency of the Cs_2CO_3 (sol)/Al device is the same as that of the Cs_2CO_3 (evp)/Al device, when their device parameters are optimized. This is not surprising because a similar product, Al–O–Cs, forms at the Cs_2CO_3 (sol and evp)/Al interface, which is responsible for the device performance enhancement.

14.2.3.3 XPS/UPS Study of the Interface

To further understand the mechanism on an atomic scale, the interface of Al and Cs_2CO_3 (sol and evp) is studied by XPS/UPS. The thermal-evaporated Cs_2CO_3 tends to decompose into cesium oxide and carbon dioxide; however, the solution-processed Cs_2CO_3 is unlikely to decompose. Therefore, an understanding of the fundamental mechanism is critical for achieving a reproducible device performance.

For the XPS/UPS measurement, the experiments were carried out in an Omicron Nanotechnology system with a base pressure of 2×10^{-10} torr. The deposition chamber and characterization chamber are interconnected. UPS spectra were taken using a He I line ($h\nu = 21.2$ eV) and an Mg Kα radiation source ($h\nu = 1253.6$ eV) was used in XPS measurements. Samples were biased at -5 V during UPS measurement to see the secondary electron edge. The Fermi energy of the system was measured by Ag and Au substrate before each measurement.

Thermally evaporated Cs_2CO_3 Thermally evaporated Cs_2CO_3 has been used to produce low work function surfaces [33, 34]. A low work function of around 1 eV was achieved by Cs_2CO_3 decomposition followed by delicate high-temperature heat treatment (550–600 °C). In contrast to the application of thermionic emission of hot electrons, we studied cathode Cs_2CO_3 decomposition for electron injection into organic material. It was suggested that Cs_2CO_3 decomposed into stoichiometric cesium oxide (CsO_2) doped by cesium peroxide (Cs_2O_2) during thermal evaporation, which is a kind of n-type semiconductor with an estimated band gap of 1.9 eV [34, 35]. Nevertheless this hypothesis has not been proven by experimental data, due to a lack of evidence to identify Cs_2O_2. We did observe carbon dioxide by mass spectrograph during the evaporation of Cs_2CO_3 in high vacuum, which suggests Cs_2CO_3 decomposed in this experimental condition. However the Cs/O atom ratio in these thermally evaporated films, as derived from XPS signal intensity, is 2.096. Hence the thermally deposited film contains cesium suboxide (e.g., $Cs_7O, Cs_4O, Cs_3O, Cs_7O_2$)-doped cesium oxide, or cesium-doped cesium oxide rather than cesium peroxide in this experimental condition. It is supposed to be

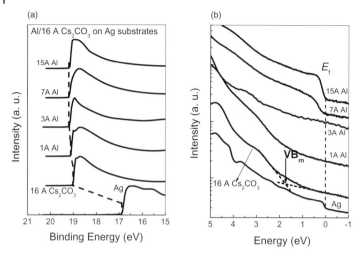

Figure 14.9 UPS spectra around (a) secondary electron cut-off, and (b) Fermi energy for the thermally deposited 16 Å of Cs_2CO_3 and subsequently deposited Al. with thickness ranging from 1 to 15 Å.

a kind of heavily doped n-type semiconductor [34, 35]. It is noted that Wu et al. [29] still observed a signal from C in the thermally deposited film, which might come from incomplete decomposition of Cs_2CO_3 during thermal evaporation under such a vacuum condition. In addition, the possibility of organic contaminants desorbed from the inner wall of the vacuum chamber should also be considered.

A silver substrate was prepared by depositing a thin layer of Ag metal on the silicon substrate. The Ag substrate was subjected to Ar sputtering to remove the absorbed molecules in air before the Cs_2CO_3 deposition. After depositing 16 Å of Cs_2CO_3 (evp) onto the substrate, 16 Å aluminum was deposited on cesium oxide. XPS/UPS spectra were measured after the deposition of each layer. Figure 14.9 shows the UPS spectra at secondary electron cut-off (a) and near the Fermi edge (b). There is a large shift (2.1 eV) of secondary electron cut-off towards a lower binding energy after Cs_2CO_3 deposition, which suggests the work function of the Ag substrate surface is lowered by 2.1 eV. Here, we feel it is better to adopt the term "effective work function" to describe the property of this surface, because thermally evaporated Cs_2CO_3 (mainly cesium oxide) is not a metal. Although there were some arguments that the Cs_2O-covered surface exhibits metallic behavior [36], we have not observed a clear Fermi edge in this experiments. In the past, the term "effective work function" was used to describe the thermionic emission cathode [34]. Effective work function of deposited cesium oxide surface is calculated to be 2.2 eV. In order to test whether this work function is its intrinsic (bulk material) property, we deposited 80 Å Cs_2CO_3 on Au. The same work function was obtained for the 16 Å deposited Cs_2CO_3, and the work function decreased slowly to 2.0 eV when the thickness of deposited Cs_2CO_3 increases to 80 Å. All these facts show that the low work function of thermally evaporated Cs_2CO_3 (cesium suboxide) is an intrinsic property, rather than its surface property. Therefore this low work function film is not subjected to substrates, which explains why metals such as Al, Au and Ag can be used as cathodes of OLED to obtain similar device performance [16]. We also noticed that the effective work function measured here is about 1 eV higher than that reported by A. H. Sommer [34]. One possible reason is that the thermally evaporated Cs_2CO_3 film has not been heat-treated at very high temperature. Fortunately this work function is low enough for the electrons to inject into most of organic materials, when there is no additional dipole formation at the interface to increase electron injection barrier.

The work function further decreases by a small value (0.16 eV) after 3 Å Al is deposited on the resulting cesium oxide layer. The work function does not decrease any further when the Al thickness is increased beyond 3 Å. Considering the lattice constant of Al, 3 Å Al is just a monolayer of Al atoms. The XPS results reveal that in fact this monolayer of Al exists in the form of oxide. Therefore, the result suggests that the first monolayer of Al reacts with cesium oxide, which decreases the work function further. Since oxygen atoms are bonded to cesium atoms and aluminum atoms simultaneously, metal oxide of Al–O–Cs is supposed to be formed in this monolayer, which has even lower work function than decomposed Cs_2CO_3. In fact, it has been realized that the formation of an element–O–Cs structure is a general method to obtain low (or negative) work function surface, such as C–O–Cs [37], W–O–Cs [38], Si–O–Cs [39], Ga–O–Cs [40] and so on. It seems that there is low possibility of chemical reaction between cesium oxide and Ag due to its low chemical reactivity, which might explain the fact that the efficiency of Ag cathode device is a little lower than Al cathode device. Nakamura *et al.* also found that the oxidation of cesium at Cs/Al interface was responsible for the enhanced electron injection, and oxided Cs/Ag cathode was less effective than oxided Cs/Al cathode in improving device efficiency [41].

Figure 14.9b shows the UPS spectra around the Fermi edge. The spectra is vertically shifted and plotted on a logarithm scale in order to make the Fermi edge clear. No Fermi edge is observed for the thermally deposited Cs_2CO_3, which means the thermally deposited Cs_2CO_3 is a kind of semiconductor rather than metal. Fermi edge becomes obvious after Cs_2CO_3 is covered by 7 Å Al, which again proves that the first monolayer of Al reacts with Cs_2CO_3 (evp). The valence band of deposited Cs_2CO_3 can be discerned and the valence band maximum (VB_m) is extracted, as shown in the figure. VB_m is about 1.8 eV below the Fermi energy, which is very close to the band gap of cesium oxide (1.9 eV). Hence, the thermally evaporated Cs_2CO_3 is indeed a kind of heavy n-doped semiconductor with high conductivity. This explains the higher power efficiency of devices with Cs_2CO_3 (evp) as electron injection layer compared to devices with other electron injection layers, because Cs_2CO_3 (evp) does not cause a voltage drop on this layer due to its high conductivity. Based on these data, the energy diagram between Al and deposited Cs_2CO_3 is plotted as Figure 14.10. As one can see from the diagram, the work function of Al is 4.3 eV, however the work function of deposited Cs_2CO_3 covered surface has a low work function of 2.2 eV due to the large displacement of vacuum level of 2.1 eV.

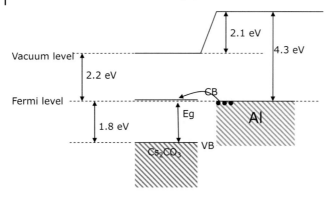

Figure 14.10 ENERGY structure at thermally deposited Cs$_2$CO$_3$/Al interface.

More importantly, there is no barrier for the injection of electrons from Al to the conduction band of the highly conductive n-type Cs$_2$CO$_3$ (evp).

The formation of an Al–O–Cs structure at the contact between Al and Cs$_2$CO$_3$ (evp) slightly lowers the bulk work function of Cs$_2$CO$_3$, the Al–O–Cs leads to small increase of PLEDs devices' efficiency with a thermally evaporated Cs$_2$CO$_3$ interfacial layer. However, this mechanism is crucially important in the solution-processed Cs$_2$CO$_3$ devices.

Solution-Processed Cs$_2$CO$_3$ The device with spin-coated Cs$_2$CO$_3$ has performance comparable to the best (or optimized) device with thermally evaporated Cs$_2$CO$_3$. At the first glance, it is surprising that the device with spin-coated Cs$_2$CO$_3$ can work so well, because spin-coated Cs$_2$CO$_3$ should not decompose into low work function cesium oxide. The doping of polymer by Cs$_2$CO$_3$ (sol) in this system is unlikely, because the device performance is poor when Ag is used as cathode metal. Hence, it is inferred that the reaction between Cs$_2$CO$_3$ (sol) and evaporated metal plays a crucial role for the reduced work function.

The reaction of Cs$_2$CO$_3$ (sol) with evaporated Al is proved by XPS. Figure 14.11 shows the emission of O 1s and Al 2s core levels. The first 2 Å of evaporated Al only shows a peak from oxide. As we mentioned before, this thickness is close to mono layer of Al atoms. The first layer of Al atoms forms strong chemical bonds with underneath Cs$_2$CO$_3$ (sol). Metallic Al begins to appear upon the second layer of Al atoms. For the solution process Cs$_2$CO$_3$, the emission from O 1s has one very broad peak. Another peak arises upon Al deposition onto Cs$_2$CO$_3$ (sol), which is assigned to the formation of O–Al bond. Undoubtedly there are plenty of O atoms which are bonded to both Al and Cs with the structure of Al–O–Cs. It plays a very crucial role of reducing cathode work function.

The UPS results of Al thin film deposited on solution-processed Cs$_2$CO$_3$ is shown in Figure 14.12. A layer of Ag was first coated on the silicon substrate to provide the reference; subsequently a thin layer of Cs$_2$CO$_3$ was spin-coated on the Ag layer. As one can see from Figure 14.12a, the work function of Ag surface

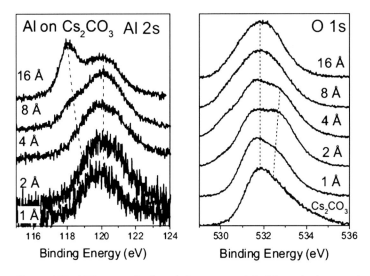

Figure 14.11 XPS spectra for the solution-processed Cs_2CO_3, and subsequently deposited Al with thickness ranging from 0.5 to 16.0 Å.

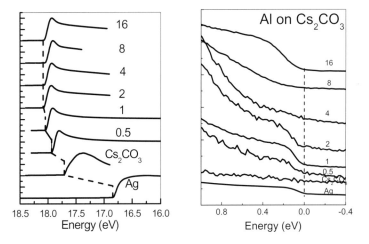

Figure 14.12 UPS spectra of the solution-processed Cs_2CO_3 and subsequently deposited Al. with thickness ranging from 0.5 to 16.0 Å. Si coated with Ag was used as substrate.

decreases by 0.8 eV after it is covered by spin-coated Cs_2CO_3. The work function of spin-coated Cs_2CO_3 surface is calculated to be 3.5 eV. It explains why high work function ITO could be used as cathode after covering a thin layer of Cs_2CO_3 (sol) on it [42]. The work function of Cs_2CO_3 (sol) covered substrate is reduced dramatically again by depositing a thin layer of Al on the spin-coated Cs_2CO_3 layer and it saturates after coverage of Al reaches 4 Å (one mono-layer of Al atoms). This saturation is reasonable because only the reaction of Al with Cs_2CO_3 (sol) results in reduced work function by formation of Al–O–Cs. There is nearly no reaction of second layer Al with Cs_2CO_3 (sol), as deduced from former XPS results. The lowest work function obtained (saturated work function) is 2.8 eV, which is higher than that of thermally evaporated Cs_2CO_3 (2.2 eV). However it is found that the real devices made by the two different processes have very similar characteristics: the same order of magnitude of current for same Cs_2CO_3 thickness, the same maximum efficiencies and the same saturated built-in potential shown in Figure 14.8. It was expected that the same work function should be obtained for these two different processes, because both processes get the same the final product of Al–O–Cs, which is the origin of lowest work function. One possible reason to explain this discrepancy is that the spin-coated Cs_2CO_3 film on Ag may *not* be continuous. Unlike real device fabrication, there is no surfactant mixed with Cs_2CO_3 (sol) in order to eliminate the photo emission from elements in surfactant such as O and C, which otherwise would make spectroscopy unclear. Nevertheless the work function of Al–O–Cs structure is believed to be around 2 eV.

Figure 14.12b shows the UPS spectra around the Fermi level. The valence band of Cs_2CO_3 (sol) is 2.3 eV below Fermi energy. It becomes quiet featureless after Al deposition, as a result of chemical reaction. One interesting phenomenon is observed according to the UPS spectra near the Fermi edge: the Fermi edge becomes very clear after 1 Å Al deposition. This means the surface is in fact metallic in nature. Similar metallic surfaces were found for cesium on diamond surface (C–O–Cs) [37] and Cs_2O/Cs_2O_2/GaAs surface (Ga–O–Cs) [41]. However the surface loses this metallic feature after the thickness of Al reaches 4 Å (one monolayer). The surface experiences a transition from metal (Ag)–non-metal (Cs_2CO_3)–metal–non-metal–metal (Al).

14.2.3.4 Comparison with Cs/Al Cathode

Cs has a particularly low work function of 2.1 eV, which is ideal for electron injection from cathode in OLEDs. The Cs/Al cathode has been demonstrated to posses an efficient electron-injection function [41, 43]. However there is some phenomenon in the Cs cathode device which cannot be simply explained by the low work function of Cs. Adachi *et al.* found that Cs loses its strong electron injection ability when its thickness exceeds 5 nm [43]. It was explained that the cathode had a low work function only in the presence of mixture of Cs and Al, or CsAl alloy, rather than Cs alone. The possibility of oxidation of Cs upon Cs deposition was also considered. Since Cs is extremely chemically reactive at room temperature, it is very easy to be oxidized by the residual oxygen in the chamber, absorbed oxygen by polymer film, or oxygen anion in organic molecule during the thermal deposition. The first monolayer of thermally deposited Cs exists very likely in the form of its oxide rather than pure metal. In fact XPS depth profile by A. Nakamura *et al.* showed that 10 atm% of oxygen coexisted at the same depth where Cs was observed [41]. In contrast, the device with a Cs/Ag cathode, but containing less oxygen in the interface layer, is much less efficient [41]. We also fabricated some devices using one monolayer of Cs between LEP and Al cathode. The performance

14.3
Phosphorescent Materials for OLEDs and PLEDs

14.3.1
Fluorescence and Phosphorescent Materials

As mentioned in the introduction, the carrier recombination is a crucial factor to the internal quantum efficiency. Theoretically, carrier recombination creates both singlet and triplet excitons at a ratio of one to three. Therefore, from Equation 14.1, even though the charge-balance factor and fluorescence efficiency is equal to 1 or 100%, the maximum internal quantum efficiency is 25%. The energy relaxation of triplet excitons is usually through the non-radiative pathway; however, in some metal–organic complexes can make use of both singlet and triplet excited states. Thus, the internal quantum efficiency can exceed the limitation of 25% and in principle may approach 100%.

Scheme 14.2 shows some typical metal–organic complexes used in phosphorescent OLEDs. The strategy of utilizing phosphorescent materials indeed provides

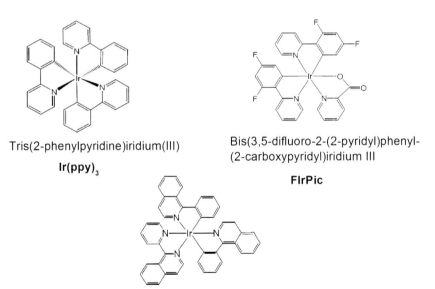

Tris(2-phenylpyridine)iridium(III)

Ir(ppy)$_3$

Bis(3,5-difluoro-2-(2-pyridyl)phenyl-(2-carboxypyridyl)iridium III

FIrPic

Tris(1-phenylisoquinoline)iridium(III)

Ir(piq)$_3$

Scheme 14.2 Several phosphorescent materials: green dopant Ir(ppy)$_3$, blue dopant FIrPic and red dopant Ir(piq)$_3$.

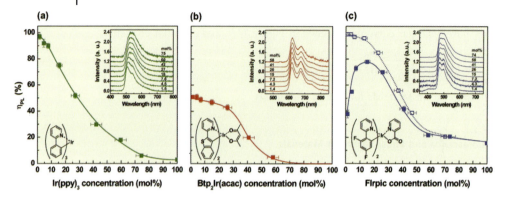

Figure 14.13 PL quantum efficiency versus dopant concentration in: (a) Ir(ppy)$_3$:CBP, (b) Btp$_2$Ir(acac):CBP and (c) FIrpic:CBP (■) and FIrpic:mCP (□). Insets show PL spectra of Ir(III) complex:CBP measured at each dopant concentration [48]. [Reprinted with permission from Appl. Phys. Lett. 2005, 86, 071104. Copyright 2005. American Institute of Physics.]

an approach to boost the efficiency of OLEDs and PLEDs. Baldo et al. first achieved internal and external efficiencies of 23 and 4%, respectively, with PtOEP-based devices [44], and Tsutsui et al. reported an external η of 13.7% and a power efficiency of 38.3 lumens/W with Ir(ppy)$_3$-based OLEDs [45]. More and more phosphorescent materials and OLEDs have been developed and exhibit dramatically breakthrough in device efficiency and operating lifetime [46, 47]. Adachi et al. reported 87 ± 7% phosphorescence efficiency with bis(2-phenylpyridine) iridium(III) acetylacetonate [(ppy)$_2$Ir(acac)] doped into 3-phenyl-4(1′-naphthyl)-5-phenyl-1,2,4-triazole, a maximum external quantum efficiency of 19.0 ± 1.0% and luminous power efficiency of 60 ± 5 lm/W were achieved. Later Kawamura et al. [48]. reported the maximum photoluminescence (PL) quantum efficiency (η_{PL}) values for Ir(ppy)$_3$, Btp$_2$Ir(acac) and FIrpic were, respectively, 97 ± 2, 51 ± 1 and 99 ± 1%, as shown in Figure 14.13. Nowadays the continuing development of phosphorescent materials has pushed the internal photoluminescence quantum efficiency to nearly 100%, and reach η_{EQE} of 26 [49], 27 [50] and 18% [51] for blue, green and red emissions.

Although the theoretical PL quantum efficiency is nearly 100%, from the aspect of device physics, other electrical characteristics such as charge balance still have to be considered to ultimately achieve higher η_{EQE} devices.

14.3.2
Solution-Processed Phosphorescent Materials

Considering the guest/host system with phosphorescent materials as dopants, solution-processed devices have the advantage of convenience for device fabrication. Moreover, it is also reported for highly efficient polymer phosphorescent light-emitting diodes (P-PhoLEDs) compared with small molecules OLEDs using vacuum deposition. The primary requirement of phosphorescent materials for

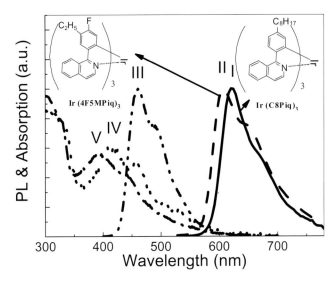

Figure 14.14 PL spectra of Ir(C8Piq)$_3$ (I), Ir(4F5MPiq)$_3$ (II), PF (III), and absorption spectra Ir(4F5MPiq)$_3$ (IV), Ir(C8Piq)$_3$ (V); Chemical structure of Ir(C8Piq)$_3$ and Ir(4F5MPiq)$_3$ are all shown in the figure, where "M" stands for methyl.

P-PhoLEDs is the solubility in comment solvents. However, as mentioned in Sections 14.2.2 and 14.2.3, the charge injection and charge balance are also necessary factors to achieve high quantum efficiency. Therefore, the choice of the host and guest material, ETL/HTL, and the interface engineering become of equal importance in P-PhoLEDs.

Intensive work has been devoted to high-efficiency green-emitting P-PhoLEDs [52–55], in which poly(N-vinylcarbazole) (PVK) is chosen as the host material for its high triplet energy level, and Ir(ppy)$_3$ or tris(2-4-tolyl phenylpyridine) iridium [Ir(mppy)$_3$] is the dopant. By improving its electrical properties and device energy structure, the highest power efficiency of 67 cd/A or a η_{EQE} of 18.8% has been realized based on Ir(mppy)$_3$ by Yang et al. [55]. They also reported highly efficient red [dye: Ir(piq)$_2$(acac)] and blue (dye: FIrpic) P-PhoLEDs with a η_{EQE} of 13.0 and 5.7%, respectively.

The strategy used to achieve high efficiency in PLEDs is different from that in small molecule OLEDs due to the difficulty of depositing multilayer, resulting in unbalanced electron and hole current. Therefore, a detail study for red emission P-PhoLEDs is demonstrated here to realize the mechanism and criteria for improving the η_{EQE}. Two red iridium based phosphorescence dyes, tris(1-phenylisoquinolinato-C2,N) iridium(III) derivatives, were used as dopants, and polyfluorene-derivative (PF) is used as host material.

Figure 14.14 shows the chemical structure of both compounds [56]. The highest occupied molecular orbital (HOMO) of these materials is measured by UPS and the lowest unoccupied molecular orbital (LUMO) is calculated based on the UV-vis absorption spectrum edge and HOMO. HOMO/LUMO levels of Ir(C8piq)$_3$ are

Table 14.1 HOMO/LUMO levels of the dopants used and device performance using these dopants.

	HOMO (eV)	LUMO (eV)	CIE	Luminance Efficiency (cd/A)[a]	Power efficiency (lm/W)[a]
Ir(piq)3	5.03	3.10	(0.68, 0.32)	NA	10/4.5[b]
Ir(4F5mpiq)3	5.24	2.97	(0.65, 0.35)	16.9/16	17.6/13.4
Ir(C8piq)3	4.9	2.8	(0.67, 0.32)	11.2/8.6	11.2/5.4
PF	5.8	2.9			

a) The former number is the maximum efficiency and the latter number is the efficiency measured at 1000 cd/m^2.
b) Data are obtained from Ref. [57].

derived from other alkyl-substituted Ir complexes (Ir(R-piq)$_3$). All these data are outlined in Table 14.1.

The devices have the following structure: indium tin oxide (ITO)/ poly(ethylene dioxy thiophene):polystyrene sulfonate (PEDOT:PSS)/LEP/Cs$_2$CO$_3$/Al. In the LEP layer, polyfluorene (PF) is chosen as host material, because PF is not only a good light emission material, but also a good host material for its large triplet-ground band gap, compatible with these two dopants. Furthermore, it shows good thin-film-forming ability and good electronic properties [2, 17, 18, 58, 59]. Cs$_2$CO$_3$ is a good candidate as interfacial layer between LEP and cathode, which helps the electron injection and blocks holes. Two procedures are done for each dopant: first, determining the dopant concentration, second, optimizing the thickness of LEP layer and Cs$_2$CO$_3$ electrons injection layer. These two procedures are discussed in detail below.

Optimizing the dopant concentration is an important step toward high device performance. When the dopant concentration is too low, the energy transfer from host to dopants is incomplete, and hence the color is not pure; in contrast, when the dopant concentration is too high, triplet–triplet annihilation (TTA) effect would reduce the quantum efficiency of devices. At the same time, high-density charge traps induced by high dopant concentration increase the device-driving voltage and hence reduce the power efficiency of devices. Energy transfer from host to dopants can occur via Förster or/and Dexter process. In these devices, both processes work simultaneously. As shown in Figure 14.14, the overlap between the absorption of dopants and emission spectrum of PF is significant, which allows efficient energy transfer from PF to dopants by Förster energy transfer. PL emission from PF can be eliminated at high dopant concentration of 8–10 wt%, which means complete energy transfer from PF to Ir(C8piq)$_3$. Since this concentration of 8–10% is relatively high, Dexter energy transfer is believed to also play important role. We note that the concentration of dopants to eliminate the emission from PF is different in PL from that in EL spectra. EL emission from PF (shown

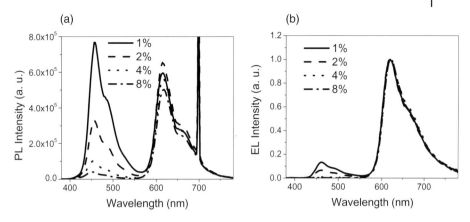

Figure 14.15 (a) PL and (b) EL spectra of PF:Ir(C8piq)$_3$ blend with different dopant concentration.

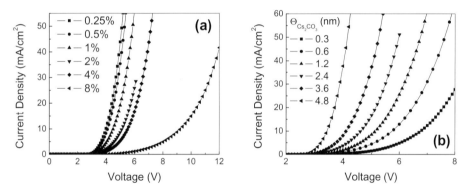

Figure 14.16 (a) I–V plot for the PF: Ir(4F5MPiq)$_3$ blend with different dopant concentration. (b) I–V plot of devices with different Cs$_2$CO$_3$ thickness while other conditions remain the same.

in Figure 14.15b) is eliminated at lower concentration of 4–6 wt%. Direct charge-trapping effects induced by dopants should be accounted as the reason for this phenomenon, which occurs only in devices when injected charges are transported through LEP. Dopants can directly act as trapping sites for electrons and holes and consequently form emission centers. To manifest the trapping effect of the dopants, we fabricated several devices which all have the same parameters, except for dopant concentration. A current–voltage (I–V) plot of devices with LEP of PF:Ir(4F5Mpiq)$_3$ blend is shown in Figure 14.16a. As one can see, the device current decreases with increasing dopant concentration. But it is not necessary that the device must have lower efficiency at higher dopant concentration. The trapping effect from the dopants increases the driving voltage of the devices, and at the same time, increases the quantum efficiency of the devices because of more efficient energy transfer and direct emission center formed by dopants. Fortunately, increased driving voltage caused by trapping effect can be partially compensated by supplying a sufficiently large electron current via efficient injection and transport, which is related to device performance optimization.

In order to obtain high device performance, electron and hole currents should be balanced. One main factor determining electron current in PLED devices is electron injection ability from cathode. In this work, electron injection was modulated to increase gradually with the thickness of Cs_2CO_3. The XPS/UPS study shows that the work function of Al cathode can be lowered gradually by increasing Cs_2CO_3 (thermally deposited) thickness. A very low work function of 2.2 eV is obtained when thickness of Cs_2CO_3 approaches 3–4 nm. A detailed study about it is discussed in the next chapter. Hole current in the devices should not change much with the thickness of Cs_2CO_3, because the hole injection barrier is unchanged. Therefore the perfect current balance point can be found by continuously increasing electron current. Figure 14.16b illustrates how electron currents in the devices can be modulated continuously by varying Cs_2CO_3 thicknesses. In these devices, all the devices parameters are kept the same, except for Cs_2CO_3 thickness. Currents in the devices increase monotonously with Cs_2CO_3 thickness. These increased currents in the devices are attributed to enhanced electron injection from the cathode by Cs_2CO_3.

As mentioned previously, direct trap of charge on dopants plays an important role in increasing the efficiency of devices. The probability of charge being trapped by the dopant in LEP is related to the LEP thickness. In this experiment, the thickness of LEP is varied from 90 to 150 nm by changing the spin rate of polymer film. The thickness of Cs_2CO_3 is varied gradually from 0.3 to 4.8 nm. These two parameters are changed independently to search for the optimized condition for each dopant and also to find out the variation rules. Here the devices using dopants Ir(4F5MPiq)$_3$ are adopted for demonstration. Figure 14.17 shows the maximum luminance efficiency and power efficiency of devices using dopant Ir(4F5MPiq)$_3$ with different LEP thicknesses and Cs_2CO_3 thicknesses. These data can be understood when we keep in mind that luminance efficiency is a characteristic for the extent of balance between electron and hole current. Since there is a large barrier for holes injection from PEDOT:PSS to PF (0.6–0.8 eV), the hole current in these devices is injection-limited. However, holes are still believed to be the majority carriers in devices with weak electrons injection because electrons sustain a severe trapping effect in PF [60, 61]. Cs_2CO_3 can facilitate the injection of electrons from cathode to the LEP layer, and the injection ability keeps increasing with increased Cs_2CO_3 thickness in the range of 0–4.8 nm (driving voltages keep dropping). As one can see from Figure 14.17, there are three stages for the change of luminance (power) efficiency with respect to the thickness of Cs_2CO_3. First, when the thickness of Cs_2CO_3 is low (0.3–0.6 nm), electron injection ability is weak. Excess hole current results in the low luminance efficiency at this stage. Power efficiencies are quite low because of the high driving voltages and low luminance efficiencies. Second, when the thickness of Cs_2CO_3 is higher (1.2–3.6 nm), electron-injection ability becomes much stronger. Best charge balance is achieved at this stage and

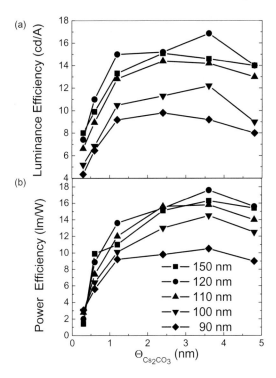

Figure 14.17 (a) Maximum luminance efficiency, (b) maximum power efficiency at different LEP thickness and Cs_2CO_3 thickness.

luminance efficiencies reach the maximum. The maximum power efficiency predominately occurs when the Cs_2CO_3 thickness is 3.6 nm for all LEP thickness because of the low driving voltage at higher Cs_2CO_3 thickness; when thickness of Cs_2CO_3 is increased (4.8 nm), both luminance efficiencies and power efficiencies drop, due to excess electrons injection. It should be mentioned that weak emission from PF appears when LEP is thin (80–100 nm), while it is completely eliminated from thick LEP devices. This can be explained by the trapping effect of dopants: thicker LEP can provide more opportunity for the trapping of charges when charges are transporting through it. Also the stronger trapping effect at thicker LEP can explain the higher efficiency at higher LEP thickness.

Finally, Table 14.1 summarizes the best device performance using the two dopants. The highest efficiencies of 16.9 cd/A and 17.6 lm/W at 220 cd/m^2 are obtained using dopant Ir(4F5Mpiq)$_3$ with red emission, and highest efficiencies of 11.2 cd/A and 11.2 lm/W are obtained for saturated red emission using dopant Ir(C8piq)$_3$. It should be mentioned that the efficiencies of these devices drop slowly with increasing brightness and still have high efficiencies of 16.0 cd/A and 13.4 lm/W at high brightness of 1000 cd/m^2, which is also shown in Table 14.1.

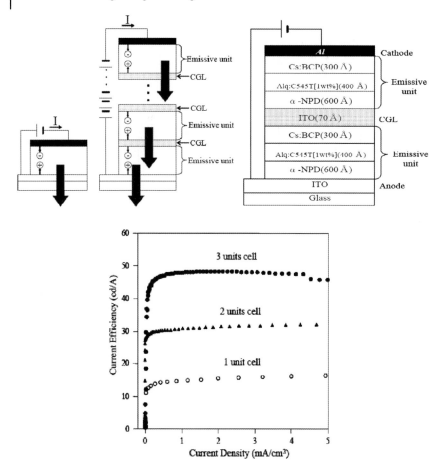

Figure 14.18 Structure of a conventional and a two-unit tandem cell, and the current efficiency of a tandem cell [62]. [Reprinted with permission from SID Digest, Copyright 2003, Society for Information Display.]

14.4
Multi-Photon Emission and Tandem Structure for OLEDs and PLEDs

Besides those factors discussed above, another strategy to further enhance the quantum efficiency is to integrate two emitting units in the same area, that is, a tandem structure. With the series connection, the current efficiency and quantum efficiency can be double with two connected units, triple with three connected units and so on because of more light output (multi-photon emission) generated from different emissive units in the same operating area. The schematic diagrams of structure and efficiency of tandem OLEDs are shown in Figure 14.18.

The structure of tandem OLEDs consists of at least two emitting units and the charge generation layer (CGL). Some previous studies revealed the potential of using vertically stacking multiple active regions in a device in series to achieve enhanced performance [63–69]. Matsumoto and Kido et al. demonstrated two and three unit tandem OLEDs using a conductive layer (ITO) or V_2O_5 as the CGL in 2003 [62]. Later Liao et al. published their tandem structure that uses an optically transparent doped organic "p-n" junction as the connecting unit between adjacent emitting units [70]. At the CGL, electrons and holes are generated upon voltage application, and injected to the adjacent organic emissive layers. Several kinds of CGL has been reported since then, such as MoO_3 [71, 72], LiF/Ca/Ag [73] and the p-n type of $Mg:Alq_3$/F4TCNQ:m-MTDATA [74]. The driving voltage is usually doubled or tripled when applying two or three units of emitting cell and so is the current efficiency. The major requirements to be a proper CGL can be summarized as:

1) The CGL has to act as the charge generation layer, as well as the role of conducting layer. A high resistance layer raises the driving voltage and reduces the device performance.

2) The CGL should be a transparent thin film to allow most of the visible light to pass through.

Tandem structure and multi-photon emission not only can significantly improve the device quantum efficiency, but also can be used to fabricate multi-color emission devices. For example, the combination of blue emission from unit 1 and orange emission from unit 2 can produce white emission OLEDs. Such an application allows the great potential of mixing colors in the field of display or lighting technology.

14.5
The Enhancement of Light Out-Coupling

To improve the external quantum efficiency of OLEDs and PLEDs, the light out-coupling efficiency is another important factor. From theoretical analysis on light-extraction efficiency of the organic light-emitting diodes, there are two major modes of light trapping: wave-guide mode and glass (substrate) mode [75, 76]. In the wave-guide mode, the light is confined in the organic active layer and indium tin oxide (ITO) due to differences in the refractive index ($n_{org} \approx 1.7$, $n_{ITO} \approx 2.0$, $n_{glass} \approx 1.5$). The emitted light passes through the organic and ITO layer but is trapped in the glass due to the total internal reflection at the interface of glass and air. The loss from both modes can range from 60 to ~80% of light generated, as shown in Figure 14.19.

Figure 14.20 demonstrates several ways to improve the light out-coupling efficiency. For the substrate mode, the mesa-shaped structure can extract the light trapped in a substrate by total internal reflection (TIR) [76], while the light can be out-coupled by a spherical structure laminated on the backside of the substrate by refraction [81]. Moller and Forrest [82] reported that adding ordered micro-lens

430 | 14 Nanostructured Organic Light-Emitting Devices

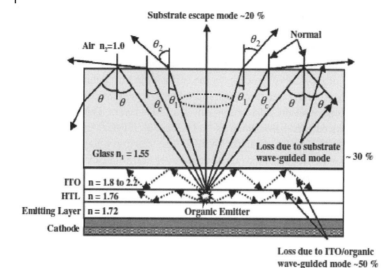

Figure 14.19 Schematic depicting the OLED structure and computational model [77]. [Reprinted from K. Saxena, D. S. Mehta, V. K. Rai, R. Srivastava, G. Chauhan, M. N. Kamalasanan, Journal of Luminescence 2008, 128, 525, Copyright 2008. with permission from Elsevier.]

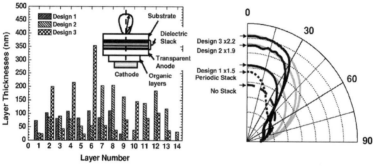

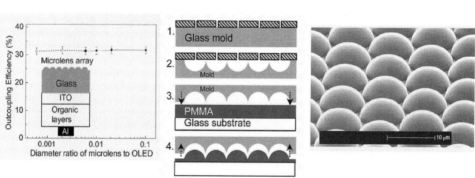

Figure 14.20 Two examples to improve the light out-coupling efficiency. The upper part employs the aperiodic dielectric mirrors in OLEDs, while the lower parts shows the way via microlenses fabricated by imprint lithography [78–80]. [Reprinted with permission from Journal of Applied Physics 2006, 100. 073106, and Applied Physics Letters 2007, 90. 241112 copyright 2006, 2007. American Institute of Physics.]

arrays (10 μm diameter poly-dimethyl-siloxane lenses attached to glass substrates) increases the light output of OLED by a factor of 1.5 over unlensed substrates.

In addition to reflection and refraction, efficiency enhancement by the use of optical films [83], diffraction as well as scattering was also studied. Fujita et al. [84] reported a factor of 1.5 increase from the use of a corrugated photonic crystal structure fabricated by etching the indium tin oxide anode layer of an OLED. Nakamura et al. [85] reported that the use of resin substrate dispersed with diffusive particles was effective. For the wave-guide mode between active layer and substrate, several ways are also proposed, including the use of subwavelength photonic crystals, [86] surface plasmons [87] and aperiodic dielectric mirrors [78]. Tetsuo et al. [ref] reported a two-fold increase in the out-coupling efficiency when applying a very-low-index aerogel layer [88]. In 2008, Sun et al. [79] employed both the microlenses and low-index grids in the structure of OLED to improve the light loss from substrate mode and wave-guide mode at the same time. The resulting out-coupling efficiency is 2.3 times that of a conventional OLED. Indeed, these methods provide another route to effectively improve the external quantum efficiency. However, some issues accompanying with the surface modification or the adding of optical thin films also need to be addressed, such as the distorted or shifted output spectra, enhancement at limited viewing angle and complicated or high-cost fabrication procedures.

14.6
Outlook for the Future of Nanostructured OLEDs and PLEDs

The development of nanostructured OLEDs has made tremendous progress for the past few decades, including efficiency and operation lifetime. After it was demonstrated that the power efficiency of OLEDs can achieve 100 lm/W [89] and the operation lifetime can also reach over 100 000 h, [90] the possibility of application of OLEDs and PLEDs in lighting and display technology has been proven. However, to compete with present technology and also to face challenges from the energy perspective, the external quantum efficiency of 25% represents that there is still much room for improvement. Newly designed and synthesized molecules with characteristics of high carrier mobility and proper band structure can further advance device performance. For example, the derivatives of phenanthroline and pyridine-containing triphenylbenzene have been proposed for the mobility range of to 10^{-3} to $\sim 10^{-2}\,cm^2/Vs$ and significant improvements have been confirmed. Further, some new discoveries, such as the discotic liquid crystals, particularly triphenylene derivatives, are confirmed to have very high hole mobility ($10\,cm^2/Vs$) and have been discovered to show an electroluminance property [91, 92]. Various new concepts or applications have also been proposed, such as OLED-based chemical and biological sensors [93], nanoscale OLEDs by spin-casting MEH-PPV into a cylindrical nanohole [94] and an organic laser diode [95, 96]. These concepts have just started and launched new directions separate from lighting and display.

14.7
Conclusion

In this chapter we demonstrate the nanostructured OLEDs and PLEDs and discuss most of the important factors related to the external quantum efficiency. Quantum confinement and charge balance can be carried out through the design of multilayer structures. The electronic characteristics of HTL/ETL have significant influences on the device performance. For solution-processed OLEDs, different strategies to achieve charge balance have been demonstrated by the polymer blended system. Studies also have revealed the importance of the surface of nanostructured materials and the interface between different thin films on the device physics. The research of phosphorescent materials and tandem structure truly accelerates the development of OLEDs and PLEDs toward practical application in the market. The enhancement of light out-coupling shows another important route to reach the goal of high quantum efficiency by optical engineering without altering the OLED structure.

With new materials and devices being developed, not only can the evolution of nanostructured devices be expected, but it also may create a new category of semiconductor materials heading toward different directions and possibilities in the future.

References

1 Gather, M.C., Kohnen, A., Falcou, A., Becker, H., and Meerholz, K. (2007) *Adv. Funct. Mater.*, **17**, 191.
2 Huang, J.S., Li, G., Wu, E., Xu, Q.F., and Yang, Y. (2006) *Adv. Mater.*, **18**, 114.
3 Burroughes, D.D.C.B.J.H., Brown, A.R., Marks, R.N., Mackay, R.H.F.K., Burns, P.L., and Holmes, A.B. (1990) *Nature*, **347**, 539.
4 Misra, A., Kumar, P., Kamalasanan, M.N., and Chandra, S. (2006) *Semicond. Sci. Technol.*, **21**, R35.
5 Sun, Y.R., Giebink, N.C., Kanno, H., Ma, B.W., Thompson, M.E., and Forrest, S.R. (2006) *Nature*, **440**, 908.
6 Li, J.H., Huang, J.S., and Yang, Y. (2007) *Appl. Phys. Lett.*, **90**, 173505-1–173505-3.
7 Moliton, A. (2005) *Optoelectronics of Molecules and Polymers*, Springer.
8 Shinar, J. (2003) *Organic Light-Emitting Devices*, Springer.
9 Adachi, T.T.C., and Saito, S. (1991) *Optoelectron: Devices Technol.*, **6**, 25–36.
10 Tang, C.W., Vanslyke, S.A., and Chen, C.H. (1989) *J. Appl. Phys.*, **65**, 3610.
11 VanSlyke, S.A., and Tang, C.W. (1995) U.S. Patent 4,720,432.
12 Shirota, Y., Kuwabara, Y., Inada, H., Wakimoto, T., Nakada, H., Yonemoto, Y., Kawami, S., and Imai, K. (1994) *Appl. Phys. Lett.*, **65**, 807.
13 Kathirgamanathan, V.A.P., Surendrakumar, S., Chan, Y.F., Antipan-Lara, J., Ganeshamurugan, S., Kumaraverl, M., Paramaswara, G., and Ravichandran, S. (2008) *SID Digest*, **8**, 47.1.
15 Xu, Q., Ouyang, J., and Yang, Y. (2003) *Appl. Phys. Lett.*, **83**, 4695.
16 Osato, Y., and Mizutani, H. (2004) *SID Digest*, **4**, 1863.
17 He, G.F., Li, Y.F., Liu, J., and Yang, Y. (2002) *Appl. Phys. Lett.*, **80**, 4247.
18 Liu, J., Shi, Y.J., and Yang, Y. (2001) *Appl. Phys. Lett.*, **79**, 578.
19 Choong, V.E., Mason, M.G., Tang, C.W., and Gao, Y.G. (1998) *Appl. Phys. Lett.*, **72**, 2689.
20 Gao, Y.L. (1999) *Acc. Chem. Res.*, **32**, 247.
21 Johansson, N., Osada, T., Stafstrom, S., Salaneck, W.R., Parente, V., dos Santos,

D.A., Crispin, X., and Bredas, J.L. (1999) *J. Chem. Phys.*, **111**, 2157.
22. Baldo, M.A., and Forrest, S.R. (2001) *Phys. Rev. B*, **64**, 085201-1–085201-17.
23. Campbell, A.J., Weaver, M.S., Lidzey, D.G., and Bradley, D.D.C. (1998) *J. Appl. Phys.*, **84**, 6737.
24. Parker, I.D. (1994) *J. Appl. Phys.*, **75**, 1656.
25. Brown, T.M., Friend, R.H., Millard, I.S., Lacey, D.J., Burroughes, J.H., and Cacialli, F. (2001) *Appl. Phys. Lett.*, **79**, 174.
26. Piromreun, P., Oh, H., Shen, Y.L., Malliaras, G.G., Scott, J.C., and Brock, P.J. (2000) *Appl. Phys. Lett.*, **77**, 2403.
27. Yang, X.H., Mo, Y.Q., Yang, W., Yu, G., and Cao, Y. (2001) *Appl. Phys. Lett.*, **79**, 563.
28. Hasegawa, S.M.T., Moriyama, T., Kimura, T., Takaya, I., Osata, Y., and Mizutani, H. (2004) SID Int. Symp. Digest Technol. Papers, 35, 154.
29. Wu, C.I., Lin, C.T., Chen, Y.H., Chen, M.H., Lu, Y.J., and Wu, C.C. (2006) *Appl. Phys. Lett.*, **88**, 152104.
30. Brown, T.M., Friend, R.H., Millard, I.S., Lacey, D.J., Burroughes, J.H., and Cacialli, F. (2000) *Appl. Phys. Lett.*, **77**, 3096.
31. Malliaras, G.G., Salem, J.R., Brock, P.J., and Scott, J.C. (1998) *J. Appl. Phys.*, **84**, 1583.
32. Choulis, S.A., Choong, V.E., Patwardhan, A., Mathai, M.K., and So, F. (2006) *Adv. Funct. Mater.*, **16**, 1075.
33. Briere, T.R., and Sommer, A.H. (1977) *J. Appl. Phys.*, **48**, 3547.
34. Sommer, A.H. (1979) *J. Appl. Phys.*, **51**, 1254.
35. Band, A., Albu-Yaron, A., Livneh, T., Cohen, H., Feldman, Y., Shimon, L., Popovitz-Biro, R., Lyahovitskaya, V., and Tenne, R. (2004) *J. Phys. Chem. B.*, **108**, 12360.
36. Wu, J.X., Li, F.Q., Zhu, J.S., Ji, M.R., and Ma, M.S. (2002) *J. Vac. Sci. Technol. A*, **20**, 1532.
37. Pickett, W.E. (1994) *Phys. Rev. Lett.*, **73**, 1664.
38. Desplat, J.L. (1973) *Surf. Sci.*, **34**, 588.
39. Martinel, R. (1974) *J. Appl. Phys.*, **45**, 1183.
40. Levine, J.D., and Gelhaus, F.E. (1967) *J. Appl. Phys.*, **38**, 892.
41. Nakamura, A., Tada, T., Mizukami, M., and Yagyu, S. (2004) *Appl. Phys. Lett.*, **84**, 130.
42. Li, G., Chu, C.-W., Shrotriya, V., Huang, J., and Yang, Y. (2006) *Appl. Phys. Lett.*, **88**, 253503.
43. Oyamada, T., Maeda, C., Sasabe, H., and Adachi, C. (2003) *Jpn J. Appl. Phys.*, **42**, 1535.
44. Baldo, M.A., O'Brien, D.F., You, Y., Shoustikov, A., Sibley, S., Thompson, M.E., and Forrest, S.R. (1998) *Nature*, **395**, 151.
45. Tsutsui, T., Yang, M.J., Yahiro, M., Nakamura, K., Watanabe, T., Tsuji, T., Fukuda, Y., Wakimoto, T., and Miyaguchi, S. (1999) *Jpn J. Appl. Phys.*, **38**, L1502.
46. Chi, Y., and Chou, P.T. (2007) *Chem. Soc. Rev.*, **36**, 1421.
47. Ikai Masamichi, H.T., Fujikawa, H., and Taga, Y. (2006) R&D Review of Toyota CRDL 41.
48. Kawamura, Y., Goushi, K., Brooks, J., Brown, J.J., Sasabe, H., and Adachi, C. (2005) *Appl. Phys. Lett.*, **86**, 071104-1–071104-3.
49. Su, E.G.S.J., Sasabe, H., and Kido, J. (2008) *Adv. Mater.*, **20**, 1.
50. Su, S.J., Chiba, T., Takeda, T., and Kido, J. (2008) *Adv. Mater.*, **20**, 2125.
51. Wu, C.H., Shih, P.I., Shu, C.F., and Chi, Y. (2008) *Appl. Phys. Lett.*, **92**, 233303-1–233303-3.
52. Lamansky, S., Djurovich, P.I., Abdel-Razzaq, F., Garon, S., Murphy, D.L., and Thompson, M.E. (2002) *J. Appl. Phys.*, **92**, 1570.
53. Vaeth, K.M., and Tang, C.W. (2002) *J. Appl. Phys.*, **92**, 3447.
54. Gong, X., Robinson, M.R., Ostrowski, J.C., Moses, D., Bazan, G.C., and Heeger, A.J. (2002) *Adv. Mater.*, **14**, 581.
55. Yang, X.H., Muller, D.C., Neher, D., and Meerholz, K. (2006) *Adv. Mater.*, **18**, 948.
56. Okada, S., Okinaka, K., Iwawaki, H., Furugori, M., Hashimoto, M., Mukaide, T., Kamatani, J., Igawa, S., Tsuboyama, A., Takiguchi, T., and Ueno, K. (2005) *Dalton Trans.*, **9**, 1583.
57. Tsuboyama, A., Iwawaki, H., Furugori, M., Mukaide, T., Kamatani, J., Igawa, S.,

Moriyama, T., Miura, S., Takiguchi, T., Okada, S., Hoshino, M., and Ueno, K. (2003) *J. Am. Chem. Soc.*, **125**, 12971.

58 Huang, J.S., Xu, Z., and Yang, Y. (2007) *Adv. Funct. Mater.*, **17**, 1966.

59 Xu, Q.F., Ouyang, J.Y., Yang, Y., Ito, T., and Kido, J. (2003) *Appl. Phys. Lett.*, **83**, 4695.

60 Huang, J.S., Hou, W.J., Li, J.H., Li, G., and Yang, Y. (2006) *Appl. Phys. Lett.*, **89**, 3.

61 Redecker, M., Bradley, D.D.C., Inbasekaran, M., and Woo, E.P. (1998) *Appl. Phys. Lett.*, **73**, 1565.

62 Matsumoto, T.N.T., Endo, J., Mori, K., Kawamura, N., Yokoi, A., and Kido, J. (2003) SID 03 Digest 27. 5L, 979.

63 Howard, G.W.J.A.W.E. (2002) U.S. Patent 6, 337, 492 B1.

64 Gu, G., Parthasarathy, G., Tian, P., Burrows, P.E., and Forrest, S.R. (1999) *J. Appl. Phys.*, **86**, 4076.

65 Shen, Z.L., Burrows, P.E., Bulovic, V., Forrest, S.R., and Thompson, M.E. (1997) *Science*, **276**, 2009.

66 Kim, J.K., Hall, E., Sjolund, O., and Coldren, L.A. (1999) *Appl. Phys. Lett.*, **74**, 3251.

67 Guo, X., Shen, G.D., Wang, G.H., Zhu, W.J., Du, J.Y., Gao, G., Zou, D.S., Chen, Y.H., Ma, X.Y., and Chen, L.H. (2001) *Appl. Phys. Lett.*, **79**, 2985.

68 Gu, G., Parthasarathy, G., Burrows, P.E., Tian, P., Hill, I.G., Kahn, A., and Forrest, S.R. (1999) *J. Appl. Phys.*, **86**, 4067.

69 Hosakawa, S.T.A.C. (2000) U.S. Patent 6, 107, 734.

70 Liao, L.S., Klubek, K.P., and Tang, C.W. (2004) *Appl. Phys. Lett.*, **84**, 167.

71 Chen, C.W., Lu, Y.J., Wu, C.C., Wu, E.H.E., Chu, C.W., and Yang, Y. (2005) *Appl. Phys. Lett.*, **87**, 241121.

72 Kanno, H., Holmes, R.J., Sun, Y., Kena-Cohen, S., and Forrest, S.R. (2006) *Adv. Mater.*, **18**, 339.

73 Sun, J.X., Zhu, X.L., Peng, H.J., Wong, M., and Kwok, H.S. (2005) *Appl. Phys. Lett.*, **87**, 093504.

74 Law, C.W., Lau, K.M., Fung, M.K., Chan, M.Y., Wong, F.L., Lee, C.S., and Lee, S.T. (2006) *Appl. Phys. Lett.*, **89**, 133511.

75 Chutinan, A., Ishihara, K., Asano, T., Fujita, M., and Noda, S. (2005) *Org. Electron.*, **6**, 3.

76 Gu, G., Garbuzov, D.Z., Burrows, P.E., Venkatesh, S., Forrest, S.R., and Thompson, M.E. (1997) *Opt. Lett.*, **22**, 396.

77 Saxena, K., Mehta, D.S., Rai, V.K., Srivastava, R., Chauhan, G., and Kamalasanan, M.N. (2008) *J. Lumin.*, **128**, 525.

78 Agrawal, M., Sun, Y., Forrest, S.R., and Peumans, P. (2007) *Appl. Phys. Lett.*, **90**, 241112-1–241112-3.

79 Sun, Y., and Forrest, S.R. (2008) *Nat. Photonics*, **2**, 483.

80 Sun, Y., and Forrest, S.R. (2006) *J. Appl. Phys.*, **100**, 073106-1–073106-3.

81 Madigan, C.F., Lu, M.H., and Sturm, J.C. (2000) *Appl. Phys. Lett.*, **76**, 1650.

82 Moller, S., and Forrest, S.R. (2002) *J. Appl. Phys.*, **91**, 3324.

83 Rahadian, F., Imai, K., and Fujieda, I. (2007) *Opt. Eng.*, **46**, 1536–1541.

84 Fujita, M., Ueno, T., Ishihara, K., Asano, T., Noda, S., Ohata, H., Tsuji, T., Nakada, H., and Shimoji, N. (2004) *Appl. Phys. Lett.*, **85**, 5769.

85 Nakamura, T., Fujii, H., Juni, N., and Tsutsumi, N. (2006) *Opt. Rev.*, **13**, 104.

86 Do, Y.R., Kim, Y.C., Song, Y.W., and Lee, Y.H. (2004) *J. Appl. Phys.*, **96**, 7629.

87 Feng, J., and Okamoto, T. (2005) *Opt. Lett.*, **30**, 2302.

88 Tsutsui, T., Yahiro, M., Yokogawa, H., and Kawano, K. (2001) *Adv. Mater.*, **13**, 1149.

89 Brian, J.E., and D'Andrade, W. (2008) *SPIE Optics and Photonics, San Diego*, Proceeding vol. 7051.

90 Kido, J. (2008) *SPIE Optics and Photonics, San Diego*, Proceeding vol. 7051.

91 Bacher, A., Bleyl, I., and Erdelen, C.H. (1997) *Adv. Mater.*, **9**, 1031.

92 Tanaka, S., Adachi, C., Koyama, T., and Taniguchi, Y. (1998) *Chem. Lett.*, **10**, 975–976.

93 Shinar, J., and Shinar, R. (2008) *J. Phys D Appl. Phys.*, **41**, 133001.

94 Yamamoto, H., Wilkinson, J., Long, J.P., Bussman, K., Christodoulides, J.A., and Kafafi, Z.H. (2005) *Nano Lett.*, **5**, 2485.

95 Dodabalapur, A., Rothberg, L.J., and Miller, T.M. (1994) *Appl. Phys. Lett.*, **65**, 2308.

96 Kozlov, V.G., Bulovic, V., Burrows, P.E., and Forrest, S.R. (1997) *Nature*, **389**, 362.

15
Electrochromic Materials and Devices for Energy Efficient Buildings
Claes-Göran Granqvist

15.1
Introduction

Global warming attracts worldwide attention, and ways to avoid its harmful consequences are the focus of scientists with various specializations, politicians and the general public [1]. Major changes in energy technology will become necessary, which will impact the global economy in ways that may be harmful but also open new business opportunities in the "clean tech" arena [2]. Furthermore, the changes related to energy technology must account for an increasing population, whose accumulation in mega-cities will lead to "heat islands", tending to aggravate global warming [3–5].

Reliance on fossil fuel leads to massive CO_2 emissions and must be curtailed, which will influence the use of energy in various sectors of society: in industry, for transport and in buildings. It is natural to put particular attention on the built environment since this sector uses as much as 30–40% of the primary energy worldwide [6]. This energy is used mainly for heating, cooling, lighting and ventilation. The energy demand for space cooling by air conditioning has grown exceptionally rapidly – by some 17% per year – in the European Union [7], and electrically powered air conditioning presently dominates the peak power during the summer in parts of Southern Europe as well as in the Southern United States. In a very hot climate, the electrical peak power may be completely dominated by air conditioning [8]. One should note that energy savings in the built environment often conserves energy *and* money [9], as elaborated in recent comprehensive studies on the marginal costs of CO_2 abatement [10, 11]. Another reason to emphasize the importance of the indoor environment is the fact that persons in industrialized countries tend to spend as much as 80–90% of his or her time there.

Why is there such a rapid growth in the energy spent on air conditioning? The driving force is in the increasing demands for indoor comfort, leading to each person's unwillingness to accept thermal discomfort due to temperatures perceived as too high or too low. Another reason for the boost in air conditioning lies in each person's wish to have unmitigated indoors–outdoors contact via large windows and glass façades. Large glazed areas have space cooling requirements,

Nanotechnology for the Energy Challenge. Edited by Javier Garcia-Martinez
© 2009 WILEY-VCH Verlag GmbH & Co. KGaA, Weinheim
ISBN: 978-3-527-32401-9

at least in commercial buildings in most parts of the world, but small windows are generally seen as leading to unsatisfactory indoor comfort and hence poor job satisfaction, with ensuing bad job performance. One way to improve the situation with regard to energy *and* indoor comfort is to have building envelopes with *variable throughput of visible light and solar energy*. The term "smart windows" was introduced for this kind of fenestration in 1984 [12].

"Smart windows" rely on one of the *chromogenic* technologies [13, 14], where the term "chromogenic"–introduced around 1990–is used to signal that the optical properties can be changed in response to some external stimulus. There are four major chromogenic technologies: *electrochromic* (EC; depending on electrical voltage or charge), *thermochromic* (depending on temperature), *photochromic* (depending on irradiation, mostly in the ultraviolet range) and *gasochromic* (depending on exposure to reducing/oxidizing gases). This chapter discusses EC materials and devices for buildings-related applications and puts particular emphasis on nanostructural features of the materials.

The energy savings in "smart windows" technologies have been investigated during the past several years. A simple "back-of-an-envelope" analysis can illustrate these savings by use of an analogy: consider a surface with arbitrary orientation but facing the Sun. Letting this surface be an EC-based "smart window" with a large range between the dark (colored) and transparent (bleached) states leads to a certain amount of saved energy; and letting the same surface be covered with today's best solar cells for terrestrial applications leads to energy production of a magnitude that is the same as the energy savings in the case of a "smart window" [15].

This analogy is very much simplified, though, and more realistic analyses must also consider the role of electric lighting and many other aspects, such as the specific uses of the buildings under consideration. Such an analysis was done in recent work for the California Energy Commission [16, 17]. The summary of that report pointed at two items in particular: that EC windows could yield an annual reduction of the peak cooling load by 19–26% when controlled for solar heat gain, and that the energy for lighting could decrease by 48–67% when the windows were controlled for visual comfort. The comparison was made with today's best static fenestration technology [18, 19] as a baseline. Furthermore, and very importantly, the users of the building strongly preferred the EC-based windows since they led to diminished glare, lower reflections in computer screens and less window luminance [17].

Another recent study from the Madison Gas and Electric Company gave a very schematic illustration of the energies for cooling and for electric lighting with a number of contemporary and future fenestration types [20]. Not surprisingly, clear glass gave a comparatively small need for artificial lighting but was disadvantageous with regard to cooling energy. Going to tinted and reflecting glass diminished the cooling energy but increased the demand for lighting. Chromogenic technologies–especially the one based on electrochromism–were found to have strong advantages for both cooling energy and electric lighting energy.

Generally speaking, the energy savings potential inherent in chromogenic technologies is still poorly understood and a number of options are ready for further study. For example, combinations of chromogenic fenestration and light-guiding seem to open avenues towards very energy efficient day-lighting via new concepts such as "light balancing" [21]. Also combinations of more than one chromogenic technology – such as electrochromics together with thermochromics – are virtually unexplored in the scientific literature.

15.2
Electrochromic Materials

Electrochromism was discovered in transition metal oxides in the late 1960s and early 1970s [22–25]. Applications were first sought in the field of information displays, but EC displays did not stand up to the then rapidly developing liquid crystal displays. The focus of the research on EC materials was then changed to windows during the middle of the 1980s when it became clear that the technology had "green" attributes and might be able to produce energy efficiency and comfort in buildings.

15.2.1
Functional Principles and Basic Materials

EC materials must be understood from the perspective of the devices wherein they are used. Figure 15.1 shows a standard design of an EC device [18, 22, 26–28]. There are five layers backed by one substrate or positioned between two substrates

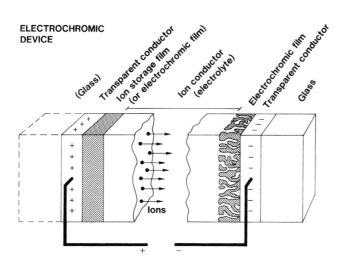

Figure 15.1 Basic design of an EC device. The transport of positive ions in an electrical field is indicated. From Granqvist [22].

by use of lamination. The substrates are normally of glass, but plastics work too, and flexible PET foil allows for interesting devices manufactured by potentially low-cost roll-to-roll techniques. The central part of the device conducts ions but not electrons. This can be an organic material, such as a transparent polymeric electrolyte or an ionic liquid preferably with adhesive properties; it can also be a thin film, for example, a porous oxide incorporating ions. The ions should be small in order to be easily mobile in an electric field, and protons (H^+) and Li^+ are of interest in most cases.

The ion conductor is in contact with an EC film which is able to conduct both ions and electrons (i.e., a mixed conductor). Tungsten oxide is a typical example. On the other side of the ion conductor is an ion storage film, which also is a mixed conductor for ions and electrons. Ideally this also has EC properties complementary to those of the first EC film. This central three-layer stack is positioned between two transparent electrically conducting films. Here In_2O_3:Sn (known as indium tin oxide; ITO) is often preferred since it has an unsurpassed combination of optical transparency and electrical conductivity [18, 29], but SnO_2-based films can be used and offer cost benefits especially if they are produced on glass by an inexpensive technique such as spray pyrolysis onto hot, solidifying glass in a float glass plant [18]. Metal-based transparent conductors have not been tried to any large extent but can be of considerable interest especially if the transmittance does not have to be extremely large [30]; we return to this in Section 15.3.2 below. Carbon nanotubes [31] and graphene [32] are other materials that may become of much interest in the future.

When a voltage is applied between the transparent conductors, ions can be transported between the EC film and the ion storage film. The charge of the ions is then balanced by electrons that are injected into or withdrawn from the EC film and ion storage film via the transparent conductors, and these electrons are the cause of the optical absorption, as discussed shortly. Reversal of the voltage or, with suitable materials, short-circuiting brings back the original properties. The coloration can be halted at any intermediate level, which means that the EC device has open-circuit memory. The fact that power is needed only to change the optical properties is important for window-type devices designed for energy savings. The memory effect hinges on the fact that the ion conductor in the middle of the EC device does not conduct electrons, which is easier to accomplish with a laminate layer with a thickness in the order of several microns than in a thin film. The voltage level needed to move the ions is of the order of 1 V dc, which can be easily obtained by solar cells, and a number of EC devices with integral solar-cell-based powering have been researched [33].

It should now be evident that the EC device resembles an electrical battery with a charging state corresponding to a degree of optical absorption. The analogy is a useful one, and the two types of devices share many pros and cons. For example both of them can easily degrade if they are mistreated by overcharging or overheating, but they also exhibit certain "self-healing" features which are poorly understood for EC devices as well as for batteries. Also both kinds of devices are unable to change their properties abruptly, and in the case of an EC device the time for

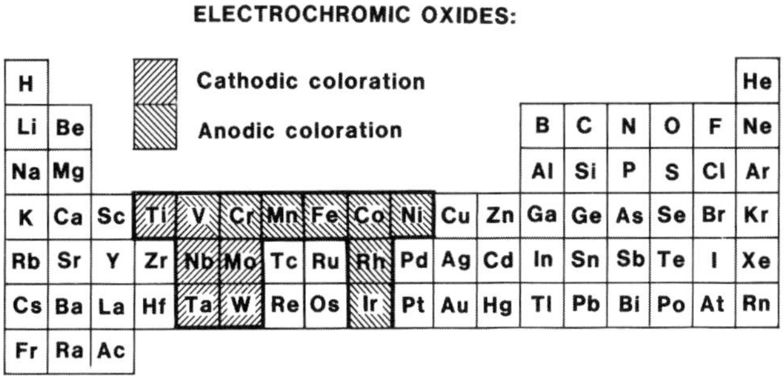

ELECTROCHROMIC OXIDES:

H																	He
Li	Be		Cathodic coloration									B	C	N	O	F	Ne
Na	Mg		Anodic coloration									Al	Si	P	S	Cl	Ar
K	Ca	Sc	Ti	V	Cr	Mn	Fe	Co	Ni	Cu	Zn	Ga	Ge	As	Se	Br	Kr
Rb	Sr	Y	Zr	Nb	Mo	Tc	Ru	Rh	Pd	Ag	Cd	In	Sn	Sb	Te	I	Xe
Cs	Ba	La	Hf	Ta	W	Re	Os	Ir	Pt	Au	Hg	Tl	Pb	Bi	Po	At	Rn
Fr	Ra	Ac															

Figure 15.2 Periodic table of the elements (apart from the lanthanides and actinides). The differently shaded boxes indicate transition metals whose oxides have clear cut cathodic and anodic electrochromism. From Granqvist [22].

going from a fully colored to a fully bleached state may vary from a few seconds (in a device with a size of some square centimeters) to minutes or even tens of minutes (for a window with a size of square meters).

It was pointed out above that it is desirable to have an EC film and an ion storage film with complementary properties. The underlying reason for this is that there are oxides with two types of EC properties: (i) those coloring under ion insertion and known as "cathodic" and (ii) those coloring under ion extraction and known as "anodic". The terminology clearly points at the kinship between EC technology and batteries. Figure 15.2 shows the metallic elements whose oxides exhibit the two kinds of coloration. Among the cathodic oxides, most attention has been on oxides of W, Mo and Nb; among the anodic oxides, those based on Ir and Ni stand out as most interesting. It should be noted, though, that Ir is very rare and precious and hence ill-suited for large-scale applications, although the properties can remain rather unchanged after dilution with cheaper Ta [34]. The only metal with qualitatively different properties is vanadium, for which V_2O_5 can exhibit cathodic and anodic features in different wavelength regions, but this is a special case to which we return briefly below. By combining, say, a cathodic EC film such as W oxide with an anodically coloring ion storage film such as Ni oxide, one can accomplish devices with both films becoming dark when a charge is moved from one to the other and both films bleaching when the original charging state is brought back [35]. This complementary feature can be used also to create color neutrality, provided that adequate combinations of cathodic and anodic films are used.

15.2.2
The Role of Nanostructure

Most of the EC oxides can be viewed as built from octagonal building blocks in different arrangements [22]. The spaces between these octahedra are large enough to allow at least some ion transport. Furthermore clusters of octahedra can aggregate into disordered and loosely packed clusters with significant intergranular

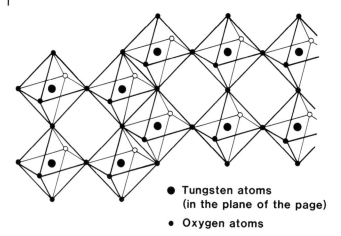

Figure 15.3 Schematic illustration of corner-sharing and edge-sharing octahedra in a W oxide crystal. From Granqvist [22].

spaces. The nanostructural features hence enter on two levels, as further elaborated below. Most of the discussion concerns W oxide, which is the most widely investigated EC material.

Figure 15.3 describes nanostructural features of W oxide and shows WO_6 octahedra with six oxygen atoms and a tungsten atom in the center. Stoichiometric WO_3 corresponds to a structure with each octahedron sharing corners with adjacent octahedra. W oxide has a tendency to form sub-stoichiometric phases in which some of the octahedra are not corner-sharing but edge-sharing, as is also illustrated in Figure 15.3. It is easily realized that a three-dimensional arrangement of octahedral "building blocks" leaves a three-dimensional network of "tunnels". They are wide enough to serve as conduits for small ions.

The schematic crystal structure in Figure 15.3 is in fact not entirely correct since it refers to a cubic structure of W oxide, which does not form except under high pressure [22]. The transition metal oxides are notorious for their large number of possible structures, and a tetragonal structure is appropriate for WO_3 crystals at normal temperature and pressure. It is shown in Figure 15.4 that the spaces between the octahedral "building blocks" are then larger than for the cubic structure, and an even more favorable structure with regard to the possibility of ionic movement is found in a hexagonal atomic arrangement. The latter structure seems to be common in thin films of W oxide [22].

Nanostructures formed in thin films of W oxides have been investigated many times. Figure 15.5 reports results for films made by evaporation onto substrates at different temperatures (τ_s) and represents modeling based on X-ray scattering [36]. Clearly the films exhibit cluster-type features with clusters growing in size as τ_s rises. The individual clusters are believed to be linked by hydrogen bonds via water molecules. The cluster size is ~3 nm for τ_s = 150 °C, and further cluster growth takes place at τ_s = 300 °C. The clusters start to interconnect at the latter

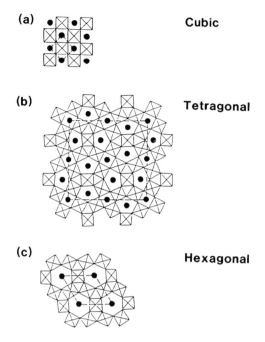

Figure 15.4 Atomic arrangements for W oxide with (a) cubic, (b) tetragonal and (c) hexagonal structure. Solid dots indicate sites available for ions in the open spaces between the WO_6 octahedra. Dashed lines show the extents of the unit cells. From Granqvist [22].

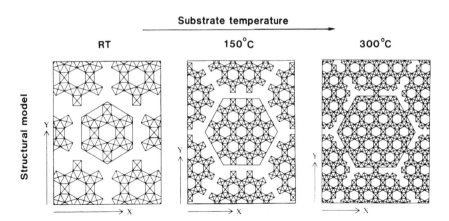

Figure 15.5 Structural models based on connected WO_6 octahedra for W oxide films made by evaporation onto substrates at room temperature (RT) and two elevated temperatures. Arrows in the x and y directions denote 2 nm. After Nanba and Yasui [36].

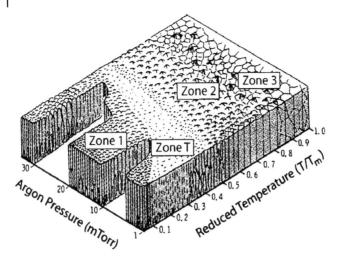

Figure 15.6 "Thornton diagram" illustrating nanostructures of thin films prepared by sputtering at different argon pressures and substrate temperatures. The melting point of the material is denoted T_m. From Thornton [38].

temperature so that a long-range ordered structure then starts to prevail. Two features in Figure 15.5 should be emphasized: (i) the existence of large open spaces between clusters, particularly in the absence of substrate heating, and (ii) the hexagonal nature of the individual clusters (as apparent from a comparison with Figure 15.4c). Both of there features are advantageous for ionic mobility in the oxides.

The structural models discussed above are believed to be essentially correct, at least for the cathodic EC oxides. For the anodic ones the situation is more complicated and the atomic distances are so small that any direct insertion on ions is unlikely; therefore it seems that film porosity and processes on the grains' surfaces are of overriding importance.

Nanofeatures are important for EC materials also as a consequence of the thin film deposition as such [37]. Virtually any thin film technology may be capable of achieving the desired properties, though with more or less difficulty. Regarding sputtering, the deposition parameters should be confined to those giving "zone 1" films in the well known "Thornton diagram" [38] in Figure 15.6 showing, schematically, what a sputter-deposited film looks like under the electron microscope. Thin films most often are prepared under conditions such as the ones of the "transition" zone denoted T. Those films are compact and it is possible to minimize grain-boundary scattering of the conduction electrons in a metallic film. The transparent conducting films in EC devices should be of this character [18, 29]. Nanocrystallinity and nanoporosity are found at higher pressure in the sputter plasma, such as in "zone 1". In this case it is possible to have ion conduction in the inter-columnar spaces. The specific nanotopography then accomplished is referred to as "parallel penniform", at least in the case of TiO_2 films [39]; a cross-

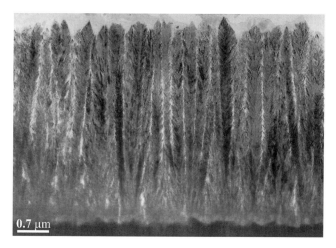

Figure 15.7 Cross-sectional transmission electron micrograph of a nanocrystalline and nanoporous TiO$_2$ film. From Rodrígues et al. [39].

sectional transmission electron image of a 4.2-µm film of this type is illustrated in Figure 15.7.

15.2.3
The Cause of Optical Absorption

The origins of optical absorption in EC oxides have been the subject of much research. It is a complicated subject for various reasons, one being that oxides are poorly defined with regard to crystallinity, as pointed out above, and can incorporate mobile ions and water molecules to varying degrees. Another difficulty ensues from the fact that even the intrinsic electronic structure for materials such as NiO has been debated for decades without any consensus having been reached. Here the optical absorption mechanisms for the EC materials is approached step by step, with the focus on a particularly interesting combination of cathodically and anodically coloring EC oxides: those based on W and Ni [35]. These particular oxides are discussed below with regard to devices.

Insertion and extraction of protons (hydrogen ions) and electrons in WO$_3$ can be represented by the simple electrochemical reaction [22]:

$$[WO_3 + H^+ + e^-]_{bleached} \leftrightarrow [HWO_3]_{colored}, \tag{15.1}$$

where it should be remembered that H$^+$ can be replaced by some other small ion (such as Li$^+$) and that the reaction should only be partial in order to be reversible [40, 41] so that the colored compound should be written H$_x$WO$_3$ with $x < 0.5$. For the Ni-based oxide, the corresponding reaction, which is expected to be confined to hydrous grain boundaries, is [42, 43]:

$$[Ni(OH)_2]_{bleached} \leftrightarrow [NiOOH + H^+ + e^-]_{colored}. \tag{15.2}$$

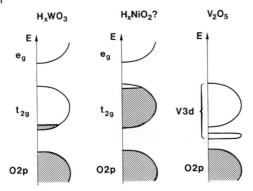

Figure 15.8 Schematic band structures for different classes of EC oxides. Shaded regions denote filled states and E signifies energy. From Granqvist [22, 45].

An understanding of the optical properties under ion and electron exchange and of the principle differences between the cathodic and anodic oxides can be developed by considering the electronic band structure appropriate for oxides comprised of octahedral "building blocks". The oxygen $2p$ bands are separated from the metal d levels, and octahedral symmetry leads to splitting of the d levels into bands with the conventional notation e_g and t_{2g}. We refer to the literature for a detailed discussion of these matters [44].

Figure 15.8 illustrates schematically the cases believed to be relevant for typical cathodic and anodic oxides [22, 45]. The left-hand panel—for H_xWO_3, say—shows that the $O2p$ band is separated from the split d band by an energy gap. Pure WO_3 has a full $O2p$ band and an empty d band and hence is transparent as any semiconductor characterized by a wide enough band gap. Insertion of small ions and accompanying electrons leads to a partial filling of the d band followed by optical absorption, as discussed below. The filling of the lowest states in the band gap permits optical transitions across the band gap only with a larger energy than in the case of undoped WO_3, which is the same mechanism as the one leading to band gap widening for example by Sn doping of In_2O_3 to make ITO [29, 46].

The middle panel in Figure 15.8 applies to the anodic oxides. Here the "pure" oxide has some unoccupied t_{2g} states, and insertion of ions and electrons may fill these states to the top of the band so that the material becomes characterized by a band gap between the e_g and t_{2g} states. Transparency then prevails provided that this gap is wide enough.

For V_2O_5, finally, the structure deviates sufficiently from the octahedra-based one in that the d band displays a narrow low-energy portion lying in the band gap, as illustrated in the right-hand part of Figure 15.8. Low-level doping of V_2O_5 leads to filling of this split-off band and subsequent doping makes the split-off band fully occupied so that the "effective" band gap is widened [47]. These features account for the fact that V_2O_5 is neither purely cathodic nor anodic.

The detailed mechanism for the optical absorption is considered next. It has been studied in greatest detail for W oxide. When ions and electrons are inserted, the electrons are localized on tungsten sites, so some of the W^{6+} sites are transformed to W^{5+} sites. By absorbing a photon, the inserted electrons can gain enough energy to be transferred to a neighboring site. Such transfer between sites i and j, say, can be written schematically as [22, 48, 49]:

$$W_i^{5+} + W_j^{6+} + \text{photon} \rightarrow W_i^{6+} + W_j^{5+}. \tag{15.3}$$

This mechanism is effective only as long as transfer can take place from a state occupied by an electron to an empty one available to receive the electron. If the ion and electron insertion are large enough this is no longer the case and, then, not only is transfer of the type $W^{5+} \leftrightarrow W^{6+}$ important but also $W^{4+} \leftrightarrow W^{6+}$ and $W^{4+} \leftrightarrow W^{5+}$ [40, 41]. In a practical situation, such "site saturation" effects may not be so important, though, since the possibilities to have a highly reversible ion exchange tends to limit the insertion levels to those for which $W^{5+} \leftrightarrow W^{6+}$ are dominant.

15.3
Electrochromic Devices

Many different types of EC devices have been studied over the years [22, 26, 27, 35]. Recently much interest has been devoted to combinations of cathodic W oxide and anodic Ni oxide joined via an electrolyte being an ion-containing inorganic thin film or an organic ion-conducting laminate; glass-based [50–64] as well as polyester-based [35, 65–71] devices have been investigated. A device of this type is discussed below. Some recent results on metal-based transparent conductors and on the possibilities of combining EC and thermochromic (TC) materials are treated as well.

15.3.1
Data on Foil-Based Devices with W Oxide and Ni Oxide

The manufacturing of EC devices, especially those based on polymer foils, is technically demanding, and five specific challenges have to be overcome [69, 70]:

1) The ITO must combine excellent electrical conductivity with very low optical absorption, which is challenging especially for films on temperature sensitive substrates such as polyester.

2) The EC and counter electrode films must exhibit well specified nanoporosities over large areas; this requires non-standard coating technologies, as touched upon above.

3) Viewing the EC device as a "thin film battery" makes it evident that charge insertion/extraction and charge balancing must be accomplished by properly controllable and industrially viable techniques, such as ozone gas treatment for the case of Ni oxide [72].

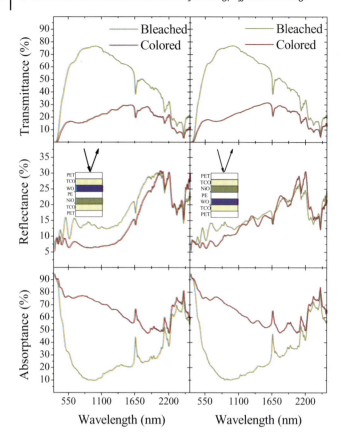

Figure 15.9 Spectral transmittance, reflectance and absorptance for the EC device shown in the inset. It incorporates films based on W oxide and Ni oxide, as well as transparent conducting oxides (TCOs), a polymer electrolyte (PE) and PET foils. From Avendaño and Granqvist [71].

4) The electrolyte must combine good ion conductivity with adhesiveness and high transparency for ultraviolet irradiation.

5) Long-term cycling durability demands adequate strategies for voltage and current control during coloration/bleaching – just as it does for charging/discharging of batteries.

All of these challenges can be successfully met, however, and EC technology may finally emerge as suitable for large-area, large-scale applications [73]. Specifically, Figure 15.9 shows optical properties for a laminated device with W oxide and Ni oxide [71]. The transmittance can be modulated between widely separated extrema both in the luminous and solar ranges, defined by $0.4 < \lambda < 0.7\,\mu m$ and $0.3 < \lambda < 3.0\,\mu m$, respectively, where λ denotes wavelength [35]. Specifically the luminous transmittance (averaged over the eye's sensitivity) is changed between

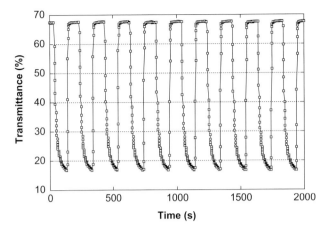

Figure 15.10 Transmittance versus time at $\lambda = 550$ nm for repeated coloring and bleaching of an EC foil device of the type shown in Figure 15.9.

57.5 and 17.0%, while the solar transmittance (averaged over a typical solar irradiance spectrum) is changed between 61 and 18%. Clearly the modulation is due to a change of the absorption rather than the reflection. The reflectance is not completely identical when the device is viewed from the two sides. The modulation ranges can be changed by having other film thicknesses or by use of suitable admixtures to the EC oxides. In particular, an addition of Al or Mg is able to enhance the transmittance of Ni-based films [74]. The transmittance modulation during several color/bleach cycles for a similar EC foil device, 240 cm² in size, is shown in Figure 15.10. It is evident that the optical changes are not abrupt but take place during a time span of tens of seconds for this size of device.

The transmittance through the foil can be increased via antireflection coatings. Figure 15.11 shows transmittance for a foil of the type depicted in Figure 15.9 after treatment on both sides with a sol-gel coating comprised of ~50-nm SiO_2 nanoparticles [70]. The transmittance at $\lambda = 550$ nm was enhanced from 73.1 to 79.1%, that is, by as much as six percentage units.

Some applications may require very low transmittance levels for EC devices in their states of maximum coloration. A contrast ratio, defined as $\zeta = T_{colored}/T_{bleached}$, between 3 and 6 is typical for the luminous or solar properties of EC devices [22, 26, 27], although $\zeta > 10$ has been stated in some earlier work [75–80]. Requiring $T_{bleached}$ to be high – which is often a strong demand for architectural uses [81] – tends to make $T_{colored}$ so large that the window is unable become dark enough. For other applications, such as automotive sun roofs, $T_{bleached} \approx 15\%$ may be desirable [82], in which case $T_{colored}$ becomes low, possibly even sufficient for privacy, but a low $T_{bleached}$ leads to severe limitations for most other applications. Putting EC devices in tandem is possible, however, and gives as a first approximation that $\zeta_{total} = \zeta_1 \times \zeta_2 \times \ldots \times \zeta_n$ for n superimposed EC devices. Practical demands based on weight and/or cost tend to restrict this option particularly for glass-based EC devices, but these restrictions are much less stringent for foil-based EC devices,

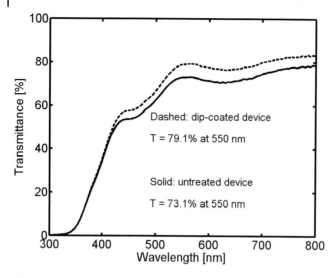

Figure 15.11 Spectral transmittance of an EC foil-type device with and without antireflection treatment. From Granqvist et al. [70].

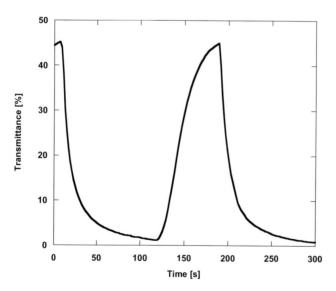

Figure 15.12 Optical modulation of an EC device with two superimposed foils. From Granqvist et al. [70].

such as those discussed above. Figure 15.12 shows transmittance at $\lambda = 550$ nm for color/bleach cycling of two superimposed EC foils of the type described earlier above. The transmittance is varied between 45.0 and 0.8%, implying $\zeta \approx 56$, which surpasses earlier results on ζ by a wide margin. The minimum $T_{colored}$ may be low enough to provide privacy. Irrespectively of whether privacy can be accomplished

Figure 15.13 "Smart window" prototype with four 30 × 30 cm² panels.

or not, the present results prove that very large modulation ranges, and very low values of $T_{colored}$ can be achieved with a tandem EC foil.

Figure 15.13 displays an EC window with four 30 × 30 cm² units with EC foil between glass panes. The upper two units are in their colored state, while the lower two are bleached. In particular, it should be noticed that an unmitigated view of the surroundings is possible, irrespective of the coloration state, which is very important for practical fenestration. Detailed measurements of spectrally resolved scattering showed that the diffuse transmittance was well below 1% [83], which is insignificant for practical purposes.

15.3.2
Au-Based Transparent Conductors

The data in Figure 15.9 show clearly that the bleached EC devices are transparent at $0.7 < \lambda < 3.0 \mu m$, that is, in the infrared part of the solar spectrum which carries ~50% of the solar energy. This property may be undesirable for windows to be used in a warm climate, and under such conditions it would be better to use transparent conductors that are infrared-reflecting though, of course, transparent for visible light. Such transparent conductors can be constructed from very thin metal films, as discussed next [30].

When a thin film grows on a dielectric substrate it undergoes a series of growth phases, as illustrated schematically in Figure 15.14 [84]. The initial formation of nuclei is followed by the growth of irregular metal "islands" which go through large-scale coalescence into a structure characterized by a meandering metallic network. The film then forms a "holey" structure which may transform into a fully continuous metal film. The thickness scale depends on the deposition technique and is more extended for conventional evaporation than for a film formed under ion bombardment (i.e., by ion assisted deposition) [84]. The latter conditions prevail also for sputter deposition [30].

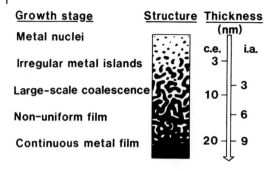

Figure 15.14 Growth stages and film structures for Au deposition onto glass by conventional evaporation (c.e.) and by ion-assisted (i.a.) growth. From Smith et al. [84].

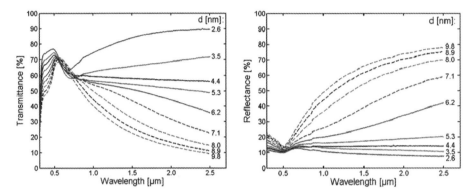

Figure 15.15 Spectral transmittance (left-hand panel) and reflectance (right-hand panel) for Au films made by sputtering onto glass to the shown thicknesses d. From Lansåker et al. [30].

The optical properties for films evolve in a way that is fully consistent with the growth stages, and Figure 15.15 displays spectral transmittance and reflectance for sputter-deposited Au films. The films with the smallest thickness d, expectedly, have high transmittance and low reflectance. After large-scale coalescence – occurring at $d \approx 6$ nm – the metallic properties are signaled by a large reflectance which then blocks infrared solar radiation. Such films have a resistivity that is about a factor ten less than the resistivity for good ITO films [30]. The optical and electrical data of the Au films can be reconciled with free-electron (Drude) theory by invoking a mean free path consistent with the film thickness.

The transmittance in the visible range is undesirably low, but this deficiency can be evaded if the Au film is embedded between dielectric films with thicknesses chosen so as to antireflect the metal. Figure 15.16 shows data for an 8-nm Au film between 55-nm films of sputter-deposited TiO_2, both as computed from the theory of thin film optics and as measured. Clearly the transmittance for visible

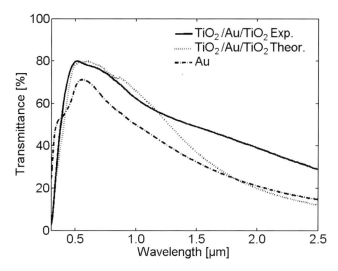

Figure 15.16 Computed and experimental spectral transmittance for a $TiO_2/Au/TiO_2$ stack with an 8-nm Au film and a 55-nm TiO_2 films. For comparison, data for the single Au film are reproduced from Figure 15.15. From Lansåker et al. [30].

light has been enhanced very significantly by the TiO_2 layers. Such three-layer films underwent electrochemical tests and were found to be durable under conditions simulating those characteristic for foil-type EC devices. Similar tests on Ag-based films, in contrast, led to rapid deterioration, as described in detail elsewhere [30].

15.3.3
Thermochromic VO_2-Based Films for Use with Electrochromic Devices

Thermochromic materials are able to change their optical properties reversibly when a "critical" temperature τ_c is exceeded. Combinations of TC coatings and EC devices can offer enhanced energy efficiency. Superimposed TC films on EC multilayer devices have been mentioned before, but the most interesting option with regard to windows in buildings seems partly to have escaped attention: this option combines an outer pane with an EC device and an inner pane with a TC coating having high luminous transparency and a critical temperature close to room temperature.

The most interesting TC materials for buildings-related applications are based on V dioxide [13, 14, 18]. VO_2 has a reversible structural transition at $\tau_c \approx 68\,°C$: below this temperature the material is monoclinic, semiconducting and relatively infrared-transparent; and above τ_c it is tetragonal, metallic and infrared-reflecting [85]. Having $\tau_c \approx 68\,°C$ clearly is too high for buildings-related applications, but τ_c can be decreased to a comfort temperature by doping, most commonly with tungsten [18, 86–88]. A more severe problem with thermochromic VO_2-based films is the difficulty to reach a sufficiently large luminous transmittance along with well

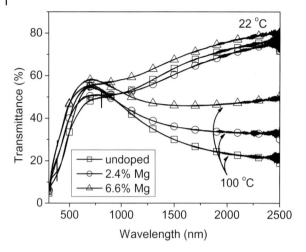

Figure 15.17 Spectral transmittance for 50-nm $Mg_xV_{1-x}O_2$ films with the shown compositions as measured at two temperatures. From Mlyuka et al. [93].

developed thermochromism [18]. Some improvement can be achieved by embedding the VO_2 film between antireflecting dielectric layers [89–92] in the same way as for the Au films discussed above, but the need for TC films with higher luminous transmittance has remained despite decades of research. However, as shown next, Mg-doping of VO_2 can enhance the transmittance and at the same time decrease τ_c to room temperature. These new TCs may bring superior fenestration – combining EC devices and TC films – closer to practical applications.

Figure 15.17 shows spectral transmittance at temperatures below and above τ_c for 50-nm $Mg_xV_{1-x}O_2$ films prepared by dc magnetron sputtering [93]. It is evident that the transmittance goes up as x is increased. Quantitative results for the luminous transmittance at room temperature are shown in Figure 15.18. Corresponding data taken above τ_c are very similar.

The value of τ_c was then inferred from changes of the infrared transmittance at $\lambda = 2300$ nm. Data shown in Figure 15.19 show that τ_c falls off approximately linearly with increasing doping level, which is consistent with data for W-doped VO_2 [86].

15.4
Conclusions and Remarks

Electrochromism was discovered and made publicly known many years ago [22, 23]. Possible applications of EC-based technology were realized already from the outset, but the progress of this technology has been slow. A number of reasons for this are given above. Buildings with "smart windows" based on electrochromism have been tried by the glass and coatings industry for decades, and

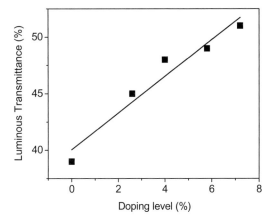

Figure 15.18 Luminous transmittance for 50-nm $Mg_xV_{1-x}O_2$ films with the shown compositions as measured 22 °C. From Mlyuka et al. [93].

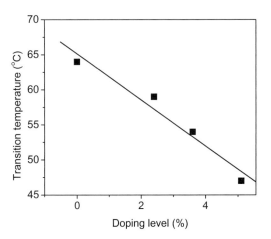

Figure 15.19 Critical metal-semiconductor transition temperature τ_c for 50-nm $Mg_xV_{1-x}O_2$ films with different compositions. From Mlyuka et al. [93].

more or less encouraging results have been reported. For one reason or another, these "smart windows" have not made it to the market except in a few niche products. Today, however, the scene seems to be changing and EC devices may well be on their way towards much larger applications [73]. Most EC devices that have been studied thus far have been based on glass, implying that large substrates and expensive coating units must be used. However, cheap production of EC devices is necessary if the technology will ever be used on a massive scale. An alternative route, using web coating by technologies allowing roll-to-roll manufacturing, is now employed for EC prototypes; this EC foil is discussed above.

What about the future of electrochromics? Windows with tunable optical properties have long been something of a Holy Grail in "high-tech" glass architecture [81]. Electrically regulated EC "smart windows" are particularly interesting, considering their potential to provide user-related operation. Windows of this type make it possible to combine increased indoor comfort for the occupant of the building (less glare, reduced thermal stress) with large energy efficiency (especially lowered air conditioning load in cooled spaces). Therefore there are strong incentives for their introduction. Durability, optical switching speed and size constraints constitute an interrelated problem complex which is not yet fully solved, but consensus seems to be growing that a satisfactory solution can indeed be achieved. Cost is another major concern, but recent progress in manufacturing technology appears to open new avenues towards inexpensive products. It should be realized that electrochromism is an enabling technology of relevance for a vast number of applications, so there may be multiple roads towards devices eventually used in buildings. Contemplating the combined blessings of comfort and energy savings, the recent advances in technology and the undeniable business opportunities, it may not be far-fetched to envisage a gradual introduction of EC "smart windows" on a large scale in the course of some decades, perhaps leading to such windows being the norm rather than the exception, in the same way that fenestration with "static" surface-coated glass, after many years of intense development [94], is today the norm for a number of markets.

The EC technology may be combined with facilities to direct daylight deeply into buildings by use of light-guiding devices. By equalizing the light level in a room, the eye – which tends to adjust to the brightest illumination – does not perceive deeper regions as disturbingly dark and hence there is less need for artificial lighting. Similar notions, though employing a Venetian blind system, have been discussed recently [95, 96]. Generally speaking, the EC technology leads to new vista in day-lighting, which is generally regarded as superior to artificial lighting by giving better task performance, improved visual comfort and positive mood effects, especially if glare problems are eliminated [97]. Particularly beneficial effects have been observed concerning student performance in day-lit schools [98] and increased sales in day-lit stores [99].

Looking into the future, membrane architecture [100, 101] may be merged with EC foil technology in order to allow lightweight buildings with little embodied energy. Thus one can envisage huge membranes allowing the flow of visible light and solar energy to be controlled and optimized. The principle possibilities offered by such membranes were in fact pointed out more than 50 years ago by visionaries such as Buckminster Fuller [102] and Frei Otto [103]. Advances in building materials – such as ethylene tetrafluoroethylene (ETFE) [104, 105] – make these visions much more realistic than in the past.

References

1. IPCC (2007) *Climate Change 2007: Mitigation. Contribution of Working Group III to the Fourth Assessment Report of the Intergovernmental Panel on Climate Change* (eds B. Metz, O.R. Davidson, P. Bosch, R. Dave, and L.A. Meyer), Cambridge University Press, Cambridge, UK & New York, NY, USA.
2. Stern, N. (2007) *The Economics of Climate Change: The Stern Report*, Cambridge University Press, Cambridge, UK.
3. Rosenfeld, A.H., Akbari, H., Romm, J.J., and Pommerantz, M. (1998) *Energy Build.*, **28**, 51–62.
4. Santamouris, M. (2001) *Energy and Climate in the Urban Built Environment*, James & James Sci. Publ., London, UK.
5. Kolokotroni, M., and Giridharan, R. (2008) *Solar Energy*, **82**, 986–998.
6. UNEP (2007) *Buildings and Climate Change: Status, Challenges and Opportunities*. United Nations Environment Programme, Paris, France.
7. Jäger-Waldau, A. (ed.) (2004) REF-SYST Status Report 2004, EUR 21297 EN; European, Commission Joint Research Centre (JRC), Ispra, Italy.
8. Darwich, M.A. (2005) *Kuwait J. Sci. Eng.*, **32**, 209–222.
9. Glicksman, L. (2008) *Phys. Today*, **61** (7), 35–40.
10. Vattenfall AB (2007) Moving Towards a Low Carbon Society, http://www.vattenfall.com/climate/www/ccc/ccc/index.jsp (accessed 10 January 2009).
11. Creyts, J., Derkach, A., Nyquist, S., Ostrowski, K., and Stephenson, J. (2007) Reducing U.S. Greenhouse Gas Emissions: How Much and at What Cost? U.S. Greenhouse Gas Abatement. Mapping Initiative Excecutive Report, http://mckinsey.com/clientservice/ccsi/pdf/US_ghg_final_report.pdf (accessed 10 January 2009).
12. Svensson, J.S.E.M., and Granqvist, C.G. (1984) *Proc. Soc. Photo-Opt. Instrum. Eng.*, **502**, 30–37; *Solar Energy Mater. Solar Cells* (1985), **12**, 391–402.
13. Lampert, C.M., and Granqvist, C.G. (eds.) (1990) *Large-Area Chromogenics: Materials and Devices for Transmittance Control*, SPIE–The International Society for Optical Engineering, Bellingham, WA, USA.
14. Granqvist, C.G. (1990) *Crit. Rev. Mod. Phys.*, **16**, 291–308.
15. Azens, A., and Granqvist, C.G. (2003) *J. Solid State Electrochem.*, **7**, 64–68.
16. Lee, E.S., Selkowitz, S.E., Clear, R.D., DiBartolomeo, D.L., Klems, J.H., Fernandes, L.L., Ward, G.J., Inkarojrit, V., and Yazdanian, M. (2006) *Advancement of Electrochromic Windows*, California Energy Commission, PIER, CEC-500-2006-052.
17. Clear, R.D., Inkarojrit, V., and Lee, E.S. (2006) *Energy Build.*, **38**, 758–779.
18. Granqvist, C.G. (2007) *Solar Energy Mater. Solar Cells*, **91**, 1529–1598.
19. Granqvist, C.G. (2009) *Int. J. Nanotechnol.*, **6**, 785–798.
20. Madison Gas and Electric (2007) http://www.mge.com/business/saving/madison/PA_6.html (accessed 10 January 2009).
21. Granqvist, C.G. (2007) *50th Annual Technical Conference Proceedings*, Society of Vacuum Coaters, Albuquerque, NM, USA, pp. 561–567.
22. Granqvist, C.G. (1995) *Handbook of Inorganic Electrochromic Materials*, Elsevier, Amsterdam, The Netherlands.
23. Deb, S.K. (1973) *Philos. Mag.*, **27**, 801–822.
24. Deb, S.K. (1995) *Solar Energy Mater. Solar Cells*, **39**, 191–201.
25. Deb, S. (2008) *Solar Energy Mater. Solar Cells*, **92**, 245–258.
26. Granqvist, C.G. (2000) *Solar Energy Mater. Solar Cells*, **60**, 201–262.
27. Granqvist, C.G., Avendaño, E., and Azens, A. (2003) *Thin Solid Films*, **442**, 101–211.
28. Monk, P.M.S., Mortimer, R.J., and Rosseinsky, D.R. (2007) *Electrochromism and Electrochromic Devices*, Cambridge University Press, Cambridge, UK.
29. Hamberg, I., and Granqvist, C.G. (1986) *J. Appl. Phys.*, **60**, R123–R159.

30 Lansåker, P.C., Backholm, J., Niklasson, G.A., and Granqvist, C.G. (2009) *Thin Solid Films* doi: 10.1016/j.tsf.2009.02.158.

31 Zhou, Y.-X., Hu, L.-B., and Grüner, G. (2006) *Appl. Phys. Lett.*, **88** (123109), 1–3.

32 Wang, X., Zhi, L., and Müllen, K. (2008) *Nano Lett.*, **8**, 323–327.

33 Lampert, C.M. (2003) *Solar Energy Mater. Solar Cells*, **76**, 489–499.

34 Backholm, J., and Niklasson, G.A. (2008) *Solar Energy Mater. Solar Cells*, **92**, 1388–1392.

35 Niklasson, G.A., and Granqvist, C.G. (2007) *J. Mater. Chem.*, **17**, 127–156.

36 Nanba, T., and Yasui, I. (1989) *J. Solid State Chem.*, **83**, 304–315.

37 Granqvist, C.G. (2008) *Reactive Sputter Deposition* (eds. D. Depla and S. Mahieu), Springer, Berlin, Germany, pp. 485–495.

38 Thornton, J.A. (1977) *Annu. Rev. Mater. Sci.*, **7**, 239–260.

39 Rodríguez, J., Gómez, M., Lu, J., Olsson, E., and Granqvist, C.G. (2000) *Adv. Mater.*, **12**, 341–343.

40 Berggren, L., and Niklasson, G.A. (2006) *Appl. Phys. Lett.*, **88** (081906), 1–3.

41 Berggren, L., Jonsson, J.C., and Niklasson, G.A. (2007) *J. Appl. Phys.*, **102** (083538), 1–7.

42 Avendaño, A., Azens, A., Niklasson, G.A., and Granqvist, C.G. (2005) *J. Electrochem. Soc.*, **152**, F203–F212.

43 Avendaño, E., Rensmo, H., Azens, A., Sandell, A., de Azevedo, G.M., Siegbahn, H., Niklasson, G.A., and Granqvist, C.G. (2009) *J. Electrochem. Soc.*, **156**, P132–P138.

44 Goodenough, J.B. (1971) *Progress in Solid State Chemistry*, vol. **5** (ed. H. Reiss), Pergamon, Oxford, UK, pp. 145–399.

45 Granqvist, C.G. (1994) *Solid State Ionics*, **70/71**, 678–685.

46 Hamberg, I., Granqvist, C.G., Berggren, K.-F., Sernelius, B.E., and Engström, L. (1984) *Phys. Rev. B*, **30**, 3240–3249.

47 Talledo, A., and Granqvist, C.G. (1995) *J. Appl. Phys.*, **77**, 4655–4666.

48 Schirmer, O.F., Wittwer, V., Baur, G., and Brandt, G. (1977) *J. Electrochem. Soc.*, **124**, 749–753.

49 Ederth, J., Hoel, A., Niklasson, G.A., and Granqvist, C.G. (2004) *J. Appl. Phys.*, **96**, 5722–5726.

50 Lee, S.-H., and Joo, S.-K. (1995) *Solar Energy Mater. Solar Cells*, **39**, 155–166.

51 Mathew, J.H.G., Sapers, S.P., Cumbo, M.J., O'Brien, N.A., Sargent, R.B., Raksha, V.P., Ladaherne, R.B., and Hichwa, B.P. (1997) *J. Non-Cryst. Solids*, **218**, 342–346.

52 Azens, A., Kullman, L., Vaivars, G., Nordborg, H., and Granqvist, C.G. (1998) *Solid State Ionics*, **113–115**, 449–456.

53 Lechner, R., and Thomas, L.K. (1998) *Solar Energy Mater. Solar Cells*, **54**, 139–146.

54 Nagai, J., McMeeking, G.D., and Saitoh, Y. (1999) *Solar Energy Mater. Solar Cells*, **56**, 309–319.

55 Karlsson, J., and Roos, A. (2000) *Solar Energy*, **68**, 493–497.

56 Azens, A., Vaivars, G., Veszelei, M., Kullman, L., and Granqvist, C.G. (2001) *J. Appl. Phys.*, **89**, 7885–7887.

57 Ahn, K.-S., Nah, Y.-C., Sung, Y.-E., Cho, K.-Y., Shin, S.-S., and Park, J.-K. (2002) *Appl. Phys. Lett.*, **81**, 3930–3932.

58 Ahn, K.-S., Nah, Y.-C., Park, J.-Y., Sung, Y.-E., Cho, K.-Y., Shin, S.-S., and Park, J.-K. (2003) *Appl. Phys. Lett.*, **82**, 3379–3381.

59 Person, C., Porqueras, I., Vives, M., Corbella, C., Pinyol, A., and Bertran, E. (2003) *Solid State Ionics*, **165**, 73–80.

60 Larsson, A.-L., and Niklasson, G.A. (2004) *Solar Energy Mater. Solar Cells*, **84**, 351–360.

61 Jonsson, A.K., Furlani, M., and Niklasson, G.A. (2004) *Solar Energy Mater. Solar Cells*, **84**, 361–367.

62 Larsson, A.L., and Niklasson, G.A. (2004) *Mater. Lett.*, **58**, 2517–2520.

63 Jonsson, A.K., Larsson, A.-L., Niklasson, G.A., and Strømme, M. (2005) *J. Electrochem. Soc.*, **152**, A377–A379.

64 Subrahmanyam, A., Kumar, C.S., and Karuppasamy, K.M. (2007) *Solar Energy Mater. Solar Cells*, **91**, 62–66.

65 Azens, A., Avendaño, E., and Granqvist, C.G. (2003) *Proc. Soc.*

Photo-Opt. Instrum. Engr., **5123**, 185–195.
66 Azens, A., Gustavsson, G., Karmhag, R., and Granqvist, C.G. (2003) *Solid State Ionics*, **165**, 1–5.
67 Azens, A., Avendaño, E., Backholm, J., Berggren, L., Gustavsson, G., Karmhag, R., Niklasson, G.A., Roos, A., and Granqvist, C.G. (2005) *Mater. Sci. Engr. B*, **119**, 214–223.
68 Granqvist, C.G., Niklasson, G.A., and Azens, A. (2007) *Appl. Phys. A*, **89**, 29–35.
69 Granqvist, C.G. (2008) *Solar Energy Mater. Solar Cells*, **92**, 203–208.
70 Granqvist, C.G., Green, S., Jonson, E.K., Marsal, R., Niklasson, G.A., Roos, A., Topalian, Z., Azens, A., Georén, P., Gustavsson, G., Karmhag, R., Smulko, J., and Kish, L.B. (2008) *Thin Solid Films*, **516**, 5921–5926.
71 Granqvist, C.G., Lansåker, P.C., Mlyuka, N.M., Niklasson, G.A., and Granqvist, C.G. (2009) *Solar Energy Mater. Solar Cells*, **93**, 2032–2039. doi: 10.1016/j.solmat.2009.02.026.
72 Azens, A., Kullman, L., and Granqvist, C.G. (2003) *Solar Energy Mater. Solar Cells*, **76**, 147–153.
73 Granqvist, C.G. (2006) *Nat. Mater.*, **5**, 89–90.
74 Avendaño, E., Azens, A., Niklasson, G.A., and Granqvist, C.G. (2004) *Solar Energy Mater. Solar Cells*, **84**, 337–350.
75 Ashrit, P.V., Benaissa, K., Bader, G., Girouard, F.E., and Truong, V.-V. (1993) *Solid State Ionics*, **59**, 47–57.
76 Schlotter, P., Baur, G., Schmidt, R., and Weinberg, U. (1994) *Proc. Soc. Photo-Opt. Instrum. Engr.*, **2255**, 351–355.
77 Michalak, F., von Rottkay, K., Richardson, T., Slack, J., and Rubin, M. (1999) *Electrochim. Acta*, **44**, 3085–3092.
78 Nishikitani, Y., Asano, T., Ushida, S., and Kubo, T. (1999) *Electrochim. Acta*, **44**, 3211–3217.
79 Chen, L.-C., and Ho, K.-C. (2001) *Electrochim. Acta*, **46**, 2151–2158.
80 Hauch, A., Georg, A., Baumgärtner, S., Opara Krašovec, U., and Orel, B. (2001) *Electrochim. Acta*, **46**, 2131–2136.
81 Wigginton, M. (1996) *Glass in Architecture*, Phaidon, London, UK.

82 Schütt, J., Giron, J.-C., Béteille, F., and Fanton, X. (2002) *Proceedings of the 4th International Conference on Coatings on Glass (4th ICCG)* (eds C.-P. Klages, H.J. Gläser, and M.A. Aegerter), Fraunhofer-Institut für Schicht- und Oberflächentechnik, Braunschweig, Germany, pp. 661–663.
83 Rönnow, D., Kullman, L., and Granqvist, C.G. (1996) *J. Appl. Phys.*, **80**, 423–430.
84 Smith, G.B., Niklasson, G.A., Svensson, J.S.E.M., and Granqvist, C.G. (1986) *J. Appl. Phys.*, **59**, 571–581.
85 Morin, F.J. (1959) *Phys. Rev. Lett.*, **3**, 34–35.
86 Sobhan, M.A., Kivaisi, R.T., Stjerna, B., and Granqvist, C.G. (1996) *Solar Energy Mater. Solar Cells*, **44**, 451–455.
87 Romanyuk, A., Steiner, R., Marot, L., and Oelhafen, P. (2007) *Solar Energy Mater. Solar Cells*, **91**, 1831–1835.
88 Picciarillo, C., Binions, R., and Parkin, I.P. (2008) *Thin Solid Films*, **516**, 1992–1997.
89 Jin, P., Xu, G., Tazawa, M., and Yoshimura, K. (2003) *Appl. Phys. A*, **77**, 455–459.
90 Kakiuchida, H., Jin, P., and Tazawa, M. (2008) *Solar Energy Mater. Solar Cells*, **92**, 1279–1284.
91 Mlyuka, N.R., Niklasson, G.A., and Granqvist, C.G. (2009) *Solar Energy Mater. Solar Cells*, **93**, 1685–1687; *Phys. Status Solidi A*.
92 Mlyyuka, N.R., Niklasson, G.A., and Granqvist, C.G. (2009) *Phys. Status Solidi A*, **206**, 2155–2160.
93 Mlyuka, N.R., Niklasson, G.A., and Granqvist, C.G. (2009) *Appl. Phys. Lett.*, to be published.
94 Gläser, H.J. (2008) *Appl. Opt.*, **47**, C193–C199.
95 Beltrán, L.O., Lee, E.S., and Selkowitz, S.E. (1997) *J. Illum. Eng. Soc.*, **26** (2), 91–106.
96 Martins-Mogo, B.G., and Beltrán, L.O. (2006) *Conference Proceedings: EuroSun 2006* (eds S. Burek, M.G. Hutchins, H. Lockhart-Ball, and S. Abrahamson), The Solar Energy Society, Abingdon, UK.
97 Gligor, V. (2004) The Luminous Environment and Office Productivity,

Tech. Lic. Thesis, Department of Electrical and Communication Engineering, Helsinki University of Technology, Espoo, Finland.
98 Heshong, L., Wright, R.L., and Okura, S. (2002) *J. Illum. Eng. Soc*, **31**, 101–114.
99 Heshong, L., Wright, R.L., and Okura, S. (2002) *J. Illum. Eng. Soc.*, **31**, 21–25.
100 Ishii, K. (ed.) (1999) *Membrane Designs and Structures in the World*, Shinkenchikusa-sha, Tokyo, Japan.
101 Koch, K.-M. (ed.) (2004) *Membrane Structures*, Prestel, Munich, Germany.
102 Gorman, M.J. (2005) *Buckminster Fuller: Designing for Mobility*, Skira, Milan, Italy.
103 Nerdinger, W. (ed.) (2005) *Frei Otto: Complete Works*, Birkhäuser, Basel, Switzerland.
104 Fernández, J.E. (2007) *Science*, **315**, 1807–1810.
105 LeCuyer, A. (2008) *ETFE: Technology and Design*, Birkhäuser, Basel, Switzerland.

Index

2
21st century, energy 3–6

a
AAO *see* anodic aluminum oxide
AAP-MSN, functionalization 340
AAPTMS (N-(2-aminoethyl)-3-aminopropyl trimethoxysilane) 339
absorption of light
– EC materials 443–445
– in nature 117
– layers 8–9
– shifting bandgaps 17–18
– tuning 8
– *see also* dye-sensitized solar cells; photovoltaics
acceptor-type, polymers 12
acid bifunctionalized MSN catalysts 351–354
acid hydrolysis 311
activated carbons 170, 182
– capacitance values 191
active film interface, nanostructuring 163–168
active materials
– electrochemical deposition 167
– nanostructuring 168–174
active surface area (ASA), irreversible capacity 199
acylation reaction 344–345
adsorption, chemisorption *see* chemisorption; physisorption
adsorption isotherms, for CO_2 389–392
AEP *see also* 3-[2-(2-aminoethylamino)ethylamino]propyl
AEP-MSNs
– functionalization 340, 347–350
– site isolation effect 353
– TON 352–353

AEP/UDP-MSNs 351–353
AEPTMS (3-[2-(2-aminoethylamino)ethylamino]propyl trimethoxysilane) 339, 347
aerogels 149
– electrochemical reactivity 168
AFC *see* alkaline fuel cell
AFM tips, parallel stamping 245
air conditioning, energy demand 435
AL-MSN, functionalization 340
Al–O–Cs structure, Cs_2CO_3 interfacial layer 418–420
aldol reaction, and bifunctionalized silicas 351–352
alkali and alkaline earth modified zeolites 371
alkaline fuel cell (AFC) 20
alloy, metal hydrides 144–147
allyl (AL) functionalities 348
allyl trimethoxysilane (ALTMS) 339
allylic amination, cinnamyl acetate and benzylamine 344
ALTMS (allyl trimethoxysilane) 339
amine functionality, membrane carbon dioxide capture 367–368
amine functionalized catalysts, immobilization 338
N-(2-aminoethyl)-3-aminopropyl trimethoxysilane (AAPTMS) 339
3-[2-(2-aminoethylamino)ethylamino]propyl (AEP), functional group 347
3-[2-(2-aminoethylamino)ethylamino]propyl trimethoxysilane (AEPTMS) 339
3-aminopropyl trimethoxysilane (APTMS) 339
angle of evaporation, shadow evaporation 261
anisotropic current properties 210–212

Nanotechnology for the Energy Challenge. Edited by Javier Garcia-Martinez
© 2009 WILEY-VCH Verlag GmbH & Co. KGaA, Weinheim
ISBN: 978-3-527-32401-9

Index

anodes
- based on nanostructured carbons 195–196
- based on Si/C composites 196–199
- irreversible capacity 199–201

anodic aluminum oxide (AAO) 241
anodic electrochromism 439
anodized alumina (AAO) membrane, shadow evaporation 259
Antarctica, variations of temperature and ice volume 361
anthracene, HRTM images 178
AP-MSN, functionalization 340
aprotic electrolyte 194
APTMS (3-aminopropyl trimethoxysilane) 339
arbitrary patterns, block copolymers 270
aromatization, naphtha reforming 288
arrays, linear of single-crystalline nanowires 249–252
ASA see active surface area
Asp-His bifunctional magnetic nanoparticles, hydrolysis of phosphoester and ester bond 355
asphaltenes 310
assembly, hierarchical of nanocrystals 268–269
atomic sulfur superstructure, epitaxy-enabling 221–223
attractive electrode materials 158–160
Au catalysts, water-gas shift (WGS) 297
Au nanostructures, shadow evaporation 262–263
Au nanowires
- production steps 250, 252
Au-based transparent conductors 449–451

b

ball models
- COMOS cluster 291
- hexagonal MoS_2 cluster 292
- MoS_2 291
band structures, EC oxides 444
bandgap
- dependence on nanorod diameter 62
- hydrogen production 114, 121, 127
- of nanosized particles 16
- photoelectrochemical cell 114
- semiconductors 34–36, 112
- thermoelectricity 51–52
barrier formation, carbon coating 200
base bifunctionalized MSN catalysts 351–354

batteries
- lithium-ion 160–161, 194–201
battery model, EC devices 438, 445
Baylis–Hillman reaction 344–345
BCY, perovskite-like structure 368–369
benzendicarboxylate (BDC) linkers 372
benzylamine, allylic amination 344
BET surface area 182
- carbon materials 191
- relation with active surface area 200
Bi, electrochemically deposited on Cu electrodes 167
Bi nanowire/nanorod 60–62
bifunctional catalysis
- hydrocracking 295
- naphtha reforming 283
bifunctional magnetic nanoparticles, Asp-His 355
bifunctional theory, fuell cells 83–84
bifunctionalized MSN catalysts, synergistic catalysis 351–354
Biloen's model 286
bimetallic catalysts, naphtha reforming 285
binding energy, physisorption 140–142
bioassays, patterning paper 272
biodiesel, transesterification 319–320
bioenergy, nanocatalysis 308–312
biofuels, applications of surface-functionalized nanoporous catalysts 319–355
biomass, for hydrogen production 111
biomass deconstruction 311
biopolymers, carbonization 193
biosensors, conjugated polymer nanowires 252
bis(oxazoline) complex (BOX), Cu 335–336
block copolymers 269–271
- lithography 267–268
blue dopant, FirPic 421
boundary problem 208–210
buildings, energy efficient 435–454
bulk
- ceramics 99
- in comparison to nanostructured systems 16, 101, 106, 114, 148, 151
- semiconductors 115
- thermoelectric materials see thermoelectric materials
bulk chemical synthesis 268
bulk heterojunctions, ordered 256
BZO nanocolumns, HRTEM-images 221
BZO nanodot, STEM-image 219

c

C ... see carbon ...
C168 Schwarzite 388
cables
– superconducting 205
cage structure
– hydrogen-bonded water framework 143
– metal-organic frameworks 143
capacitance
– electrochemical 181
– nitrogen-enriched composites 193
capacitance values 191
capacitors
– electric double layer 182–185
– electrochemical double-layer 169–174
– golden 180
capacity 171–172
– irreversible of carbon anodes 199–201
– reversible 194
– Si/C composites 198
capacity gains, conversion reactions 162
carbide-derived carbons (CDC) 171
carbon
– activated 170
– anodes 195–196, 199–201
– electric double layer capacitors 182–185
– for hydrogen storage 141–142
– in fuel cells 85–89, 97
– lithium insertion 196
– low-temperature 179
– nanoporous turbostractic 178
– nanotexture and surface functionality 177–180
– pseudocapacitance effects 185–188
– supercapacitors 170
carbon aerogels 149–150
carbon-based materials, pseudo-capacitors 185–194
carbon-based nanomaterials, electrochemical energy storage 177–204
carbon capacitance, increase with pore size 183
carbon-coated Si nanoparticles, voltage-composition curve 161
carbon coating, irreversible capacity 200
carbon cryogels 149–151
– differential scanning calorimetry 151
carbon dioxide adsorption
– at 195 K 376–378
– at 273 K 378–384
– at 298 K 384–389
– molecular modeling of 389–393
carbon dioxide capture 359–394
– high-temperature adsorbents 368–369
– membrane approaches 367–368
– oxycombustion flow scheme 364
– PCPs 371–394
– post-combustion flow scheme 365
– pre-combustion flow scheme 364
– processes 364–366
– zeolites and molecular sieves 369–371
carbon dioxide concentration, increase 360–362
carbon materials
– capacitance values 191
– nitrogen functionalities 180
carbon monolayer, graphene 271
carbon nanotubes, electrically conducting polymer (ECP) 188
carbon network, oxygen functionality 180
carbon network effects, heteroatoms pseudo-capacitive 190–194
carbonization, biopolymers 193
carbonization temperature, electric double layer capacitors (EDLC) 182
CAs see carbon aerogels
catalyst activity, methanation 309
catalyst regeneration 283
catalyst supports see support materials
catalysts
– bifunctional theory 83–84
– in fuel cells 20, 79–90
– heterogeneous see heterogeneous catalysts
– homogeneous see homogeneous catalysts
– methanol synthesis 300
– Ni-alumina 307
– Pt, dependencies on particle size 81
– X-ray diffraction patterns 85
catalytic activity
– gold, temperature dependence 299
– methanation 307
catalytically active center see single-site heterogeneous catalysis
cathodic electrochromism 439
cationic cetyltrimethylammonium bromide (CTAB) 331–332
CC see carbon cryogels; coated conductors
CDC see carbide-derived carbons
CDSPTMS (3-(3′-(trimethoxysilyl)-propyldisulfanyl) propionic acid) 331
ceramic electrocatalysts, high-temperature 101–103
ceramic films
– for photoelectrochemical devices 42
– nanoporous 41–42
CGL, charge generation layer (CGL) 428–429

Index

charge balance
– in a polymer blended system   406–411
– OLEDs and PLEDs   405–420
charge carrier recombination losses, inhibition   37
charge generation layer (CGL)   428–429
charge injection, and interfacial layer   411–420
charge storage capacity optimization   170–171
charge transfer   127
– in organic solar cells   12
– limitation   162
– pseudo-faradaic   191
charge transport, in EC devices   437–438
chemical nitrogen   179
chemical synthesis   268
chemisorption
– bond strenghts   140
– enthalpy diagram   146
– for hydrogen storage   144–151
chimie douce synthesis   158
chlorophyll   38
– light absorption   117
– natural solar energy conversion dye   38
chromium oxides, mesoporous   168–169
chromogenic technologies   436
cinnamyl acetate, allylic amination   344
clathrates, for hydrogen storage   143–144
cleanroom, energy usage   234
closed-loop structures   248–249
co-condensation   330–333
Co_2-concentration, Fischer-tropsch synthesis (FTS)   304
CO_2 ... see carbon dioxide ...
coal-fired power plants, flue gas composition   375
coated conductors, flux pinning   208
coated conductors (CC) wire technology   206
coated Si particles   160
coating, conducting carbon layer   158
coating strategies, new   160
collectors, current   163–168
color, as function of particle size of quantum dot   7
coloring, EC oxides   443–445
column orientation, film growth   221
competitive nitroaldol reaction   349–350
– see also Henry reaction
complex catalysts, synthesis on solid surface   333–334
complexity, electrochemical energy storage   155

composites, Si/C anodes   196–199
conducting polymers   188–190
conduction band   34, 112–113
– photoelectrochemical cell   114
– water splitting   123
conductivity, modification of inonic   158
conjugated polymer nanowires   252–253
conservative use of energy, unconventional approaches   231–280
controlled dimensions, nanowires   248–259
conversion of syngas, nanocatalysis   296
conversion reaction mechanism   162
conversion reactions   161–162
cooperative Asp-His bifunctional magnetic nanoparticles, hydrolysis of phosphoester and ester bond   355
cooperative catalysis, by multi-functionalized heterogeneous catalyst system   346–348
copper see Cu
copper bis(oxazoline) complex (BOX)   335–336
core/shell nanofibers   265–267
correlated pinning   215
cost effectiveness, nanocatalysis   281
covalent bonding, homogeneous catalysts   327–328
CP (3-cyanopropyl)   347
CP-MSN
– functionalization   340, 347–348
CPTES (3-cyanopropyl triethoxysilane)   339
CPTMS (3-cyanopropyl trimethylsilyl)   347
Cr_2O_3 electrodes, electrochemical performances   169
cracking
– nanocatalysis   293–295
– petroleum refining   282
cryogels   149
crystals
– nanosized   7–9, 73, 100, 113–115
Cs_2CO_3 408–411
– Al–O–Cs structure   418–420
– and Cs/Al cathode   420
– I–V characteristics   412
– interfacial layer and charge injection   411–420
– LEP layer   426–427
– open circuit voltage   414
– photovoltaic measurements   413–415
– solution-processed   413, 418–420
– thermally evaporated   412–413, 415–418
– UPS spectra   416, 419
– work function   413–415
– XPS spectra   419
– XPS/UPS study interface   415–420

Cs/Al cathode, and Cs_2CO_3 interfacial layer 420
Cu catalysts, methanol synthesis 299–300
Cu deposition, current collectors 164
Cu electrodes, nanostructured 167
Cu-nanostructured current collector
– cross section 165
– cross section with iron oxide deposits 166
$Cu(BF_4)_2$(bpy), carbon dioxide adsorption at 273 K 383
CuBTC, structure 386
CUK-1, structure 386
current, anisotropic 210–212
current collectors 164
– Cu-nanostructured 165–166
– nano-architectured 163–168
– nanostructuring 163–168
current density
– angular dependence 211
– dissipation-free 207
– magnetic field dependence 212
– YBCO thin films 209–210
current–voltage (I–V) plot, LEPs 425
cut-off potentials, dependence of capacities 198
3-cyanopropyl (CP), functional groups 347
3-cyanopropyl triethoxysilane (CPTES) 339
3-cyanopropyl trimethylsilyl (CPTMS) 347
cyanosilylation, and bifunctionalized silicas 351–352
cycleability, Si/C composites 197
cyclic voltammetry 186–187
cyclization, naphtha reforming 288
cyclopropanation, of styrene 335–336

d

defect types, current density 210
defects
– current density 211
– flux pinning 207–208
– nanosized on YBCO films 213
– naturally occurring 212–215
dehydrocyclization 289
dehydrogenation
– chemisorption 144–151
– physisorption 140–144
Dexter energy transfer 424
diffraction patterns, anthracene 178
dimensionality, of PCPs 372
dimensions, controlled of nanowires 248–259

4-(dimethylamino)pyridine functionalized mesoporous silica nanoparticle (DMAP-MSN) 344–346
dioxybenzenedicarboxylate *see* DOBDC
dip-pen nanolithography 245–246
direct methanol fuel cell (DMFC) 83, 91, 94–95
– filler 94
discharge/charge cycles, galvanostatic 197
DMAP immobilized iron oxide catalysts 354
DMAP-MSN
– ^{13}C chemical shifts 345
– *see also* 4-(dimethylamino)pyridine functionalized mesoporous silica nanoparticle
DMAP-MSN catalysts, TON 346
DMFC *see* direct methanol fuel cell
DOBDC (dioxybenzenedicarboxylate) PCPs 385, 388
m/DOBDC, structure 386
donor-type, polymers 12
doping
– absorption spectra 18
– anion 17
– cation 17
– photocatalysts 17–19
– of TiO_2 128
dotsize, wet chemical synthesis 268
double layer capacitance
– energy storage systems 170
– interface 168
double layer capacitors
– electrochemical 169–174
– supercapacitors 182–185
double-layer specific capacity, comparison with pseudo-capacity 185
DPN, production steps 246
dubinin-raduskevich micropore volume 183
DyBTC, carbon dioxide adsorption at 273 K 380–381
dye molecular engineering 38–40
– natural solar energy conversion dye 38
– photocurrent spectra 39
dye-sensitized solar cells 8–9, 33–45
– application of nanostructures in 37
– for artificial photosynthesis 16
– general device 36
– self-assembling dye 40–41

e

EC *see* electrochromic
ECP *see* electrically conducting polymer
EDCC *see* electric double cylinder capacitor

edge lithography, nanoskiving 246–259
edge sites, hydrotreatment 290
EDL *see* electric double layer
EDLC *see* electrochemical double-layer capacitor
EDX intensity of Sn, naphtha reforming 287
effective work function 416
EL *see* emissive layer
elastomer, soft mold production 240
electric double cylinder capacitor (EDCC) 184
electric double layer (EDL) 181
electric double layer (EDL) capacitors, supercapacitors 182–185
electric wire-in-cylinder capacitor (EWCC) 184
electrical conductivity, carbon electrode 184
electrically conducting polymer (ECP) 188
electrocatalysis of oxygen, fuell cells 80–83
electrochemical activity, capacity gains 162
electrochemical biosensors, conjugated polymer nanowires 252
electrochemical capacitors
– energy storage 177
– working principle 181
electrochemical cell, configuration 156
electrochemical double-layer capacitors 169–174
electrochemical energy storage 155–175
– carbon-based nanomaterials 177–204
– interfaces 163
electrochemical performances, Cr_2O_3 electrodes 169
electrochemical reactivity, vanadium oxide aerogels 168
electrochromic (EC) devices 445–452
– Au-based transparent conductors 449–451
– battery model 438, 445
– charge transport 437–438
– foil-based 445–449
– transmittance 446–448
– with thermochromic Vo_2-based films 451–452
electrochromic (EC) materials
– cause of optical absorption 443–445
– for energy efficient buildings 435–454
– functional principles and basic materials 437–439
– future 454
– role of nanostructure 439–443
electrochromic (EC) oxides
– band structures 444
– built from octagonal building blocks 439–440
– coloring 443–445
electrochromic (EC) technology 436
– membrane architecture 454
electrode geometry 164
electrode materials
– attractive 158–160
– electrochemical energy storage 155
electrodes
– nanostructured 163
– self-supported 163
– Si-based, rejected 160–161
– steel 163
electrolyte interfaces, nanostructuring 168–174
electrolyte interphase formation, solid 179
electrolytes 155
– aprotic 194
– capacitance 183
– in fuel cells *see* fuel cells
electron transport, in EC devices 437–438
electron transport layer (ETL) 405–407
electron-hole pairs 6
electron-sink, in photocatalytic degradation of organics 116
electrons, material dependent number 157
electrosorption
– hydrogen in carbon 185–188
– reversible 185
electrospinning 263–267
– energetic costs 267
– production steps 263
– scanned 264–265
– uniaxial 265
electrostatic attraction 185
embossing, green nanofabrication 238–244
emission spectra 7
emissions flow chart, greenhouse gases (GHGs) 363
emissive layer (EL) 405
– spectra 424–425
energetic costs 233–234
energy, conservative use 231–280
energy challenge 3–6
– potential solution 232
energy conversion, from electricity to light 403
energy conversion efficiency
– as function of potential and temperature 118

- in dye-sensitized solar cell 37
- organic solar cells 12
- photovoltaics standard 7–10, 36–37
- thermoelectricity 47

energy demand 435
energy density, storage cells 156
energy dependence, of electron density of states 56
energy efficient buildings, electrochromic materials and devices 435–454
energy harvesting, in photoelectrochemical devices 9
energy perspective, nanomaterials for superconductors 205–228
energy production 3–32
- nanotechnology 6

energy savings, in "smart windows" technologies 436
energy storage
- carbon-based nanomaterials 177–204
- electrochemical 155
- modes 185
- nanomaterials 158–162
- superconductors 205

energy storage systems (ESS), average performance 170
energy transport, superconductors 205
epitaxial self-assembly 270
epitaxy-enabling superstructure, control 221–223
epoxy, stabilization of nanostructures 258
epoxy slabs, ultramicrotome 249
EQE see external quantum efficiency
ErPDA
- pore size 376, 378

ESS see energy storage systems
ester bond, hydrolysis 355
ethanol production 312
ETL see electron transport layer
EuBaPCP, structure 381
evaporation, shadows 259–263
EWCC see electric wire-in-cylinder capacitor
excitons 6–8
extended X-ray adsorption fine structure (EXAFS), characterization of heterogeneous catalysts 321
external efficiency, PLEDs 410
external magnetic fields 207
external quantum efficiency (EQE) 403

f

fabrication, "zero-cost" 271–274
FCC see fluid bed catalytic cracking
Fe . . . see also iron

Fe catalysts
- Fischer-Tropsch synthesis 303, 306
- phase changes 304–305

Fe_3C, Fischer-Tropsch synthesis 304
Fe_3O_4 deposits, rate capability plots 166
fenestration technology 436
FFA see free fatty acid
fillers
- dependencies on pH 94
- in fuel cells 91–96
- inorganic 93

film interface, active 163–168
film-nanowire transition 255
film thickness 219
films 197
- current density 209–210
- graphene, preparation with scotch-tape method 271–272
- topographically patterned 248
- see also thin films

$FirPic_2$ (blue dopant) structure 421
Fischer-Tropsch catalyst, structure 306
Fischer–Tropsch synthesis (FTS), nanocatalysis 302–307
flue gas composition, coal-fired power plants 375
flue gas contaminants, stability of PCPs 393–394
fluid bed catalytic cracking (FCC) 282
fluorescence materials, PL quantum efficiency 421–422
flux lines, pinned and unpinned 208
flux pinning 207–208
- artificial introduction 215–216
- optimization 213
- vortex–vortex interactions 210

Förster energy transfer 424
foil-based EC devices, with W oxide and Ni oxide 445–449
force field, molecular modeling 389
formate mechanism, water-gas shift (WGS) 296–297
free fatty acid (FFA)-containing oil feedstocks 353
free path, length scales 54
free radicals 180
FTS see Fischer–Tropsch Synthesis (FTS)
fuel cells 20–26
- anodic reaction 83–85
- catalysts, ceramic electro- 101–103
- catalysts, non-precious 90
- catalysts, practical 85–89
- cathode reaction 80–83
- critical materials 24

– different types 20–22
– direct utilization of dry hydrocarbons in SOFCs 103–105
– electrochemical reactions 21
– electrolytes 90–91
– fillers 91–96
– general device 21
– high-temperature 20–21, 98–105
– low-temperature 20–21, 80–98
– membrane-electrode assembly 96–98
– membranes, high-temperature polymer electrolyte 91–96
– microbial 22–23
– nanomaterials for 79–109
– nanostructured electrodes 24–26
– particle size 83–84, 97
– probable applications 24
fuel production
– future of 312–314
– nanocatalysis 281–318
functionality, sp² carbons 177–180
functionalization, interface 168
functionalized silicas
– formation 339, 341

g
galvanostatic discharge/charge cycles 197
galvanostatic insertion/deinsertion 195
galvanostatic intermittent titration technique (GITT) 187
gasochromic technology 436
gatekeeping effect, multi-functionalized hetergeneous catalysts 348–351
GCMC *see* grand canonical Monte Carlo
GHGs *see* greenhouse gases
GITT *see* galvanostatic intermittent titration technique
glacial ice volume, variations of temperature and ice volume 361
glass (substrate) mode 429–431
glassy carbon, electrochemical capacitors 178
global mean surface temperature 360
global warming projections 363
gold *see* Au
golden capacitor 180
Grätzel cells *see* dye-sensitized solar cells
grafting, post-synthesis 328–330
grain boundaries, flux pinning 208
grain boundary problem 208–210
grand canonical Monte Carlo (GCMC) 389–390
graphene films, preparation with scotch-tape method 271–272

graphene layers, nanoporous carbon 179
graphite, lithium-ion batteries 194
gravimetric capacity 171
green approaches 238–271
green dopant, Ir(ppy)₃ 421
green nanofabrication 231–280
– design constraints 274
– future directions 271–274
greenhouse effect 359
greenhouse gases (GHGs) 359
– emissions flow chart 363
growth, temperature-driven of layers 163

h
hard mold, lifetime increase 239
hard mold production, green nanofabrication 238
hard pattern transfer elements 238–240
Henry reaction *see also* competitive nitroaldol reaction
– and bifunctionalized silicas 351–352
heteroatoms, pseudo-capacitive effects 190–194
heterogeneous catalysis
– single-site 320, 322–334
heterogeneous catalysts
– characterization of 321
– design of 335–354
– multi-functionalized 346–348, 348–351
– on non-silica support 354
– reactivity enhancement 337–338
– "ship in the bottle" 327
heterogeneous proline catalysts 326
heterojunction production, nanoskiving 255
heterojunctions
– bulk 256
– nanoskiving 256
– polymer, nanostructured 253–258
hexane, selective aromatization 288
hierarchical assembly, nanocrystals 268–269
high capacity anodes, search among metals 196
high CO₂ solubility liquid, membrane carbon dioxide capture 368
high-temperature catalyst, methanation 307
high-temperature fuel cells 98–105
– *see also* fuel cells
high-temperature superconductors (HTS)
– cables 205
– *see also* YBCO . . .

high-work-function electrode (HWFE), heterojunctions   253
high-temperature adsorbents, carbon dioxide capture   368–369
highest occupied molecular orbital   38
highest occupied molecular orbital (HOMO), phosphorescent materials   423–424
hole transport layer (HTL)   405–407
hollow inorganic tubes   259–261
hollow nanofibers   265–267
HOMO see highest occupied molecular orbital
homogeneous catalysts
– covalent bonding   327–328
– immobilization   322–334
– non-covalent binding   324–327
honeycomb structure, in photoelectrochemical devices   12
host structures, li-ions   158
HTL see hole transport layer
HTS see high-temperature superconductors
HWFE see high-work-function electrode
hybrid cells, thermal   7
hybrid configuration, electrodes   190
hybrid power generator, solar-thermoelectric   48
hydrides, covalent   145
hydrides, chemical, for hydrogen storage   147–148
hydrides, complex, for hydrogen storage   144–147
hydrides, metal, for hydrogen storage   144–147
hydrocracking
– nanocatalysis   295–296
– petroleum refining   282
hydrogels, as source material for nanoporous carbons   149
hydrogen
– electrosorption   186
– energy system   15
– partial pressure   289
– pseudocapacitance effects   185–188
– reversible electrosorption   185
hydrogen production   14–20, 111–136
– by semiconductor nanomaterials   113–131
– generation and application   112
– photoelectrochemical cells   115–117
– renewable cycle   19
– solar-to-hydrogen efficiency   16–19
– water splitting   16
hydrogen storage
– as solid   139
– by chemisorption   144–151

– by physisorption   140–144
– in carbon network   186–187
– desorption   145–147
– nanostructured materials for   139–153
– storage density   145
hydrogen-bonded water framework, clathrates   143
hydrogenolysis, naphtha reforming   288
hydrothermal techniques, film surface   44
hydrotreatment
– nanocatalysis   289–293
– petroleum refining   282

i

I–V plot see current–voltage plot
IAST see ideal adsorbed solution theory
IBAD see ion beam assisted deposition
ice volume, glacial   361
ICP (infinite coordination polymer), amorphous   383–384
ICP-MSN, functionalization   340
ICPTES (3-isocyanatopropyl triethoxysilane)   339
ideal adsorbed solution theory (IAST)   392–393
IGCC see integrated gas combined cycle coal
immobilization
– amine functionalized catalysts   338
– of rhodium complex   325, 337
immobilization strategies, single-site heterogeneous catalysis   322–334
immobilized homogeneous catalysts, surface interaction with silica   335–337
In_2O_3:Sn see indium tin oxide (ITO)
incremental gasoline, economic value   295
indium-doped tin oxide (ITO), nanotubes   259
indium tin oxide (ITO)
– and Au films   450
– EC devices   438, 444–445
– light confinement   429–430
indoor environment
– comfort   454
– importance of   435–437
infinite coordination polymer see ICP
inorganic solid supports   322
inorganic tubes, hollow   259–261
insertion/deinsertion, galvanostatic   195
insertion materials
– ideal   157
– rejected   158–160
integrated gas combined cycle coal(IGCC)-fired power plant   364–365

interfaces
- double layer capacitance 168
- electrochemical energy storage 163
- electrolyte 168–174
- functionalization 168
- macroscopic/microscopic 155
interfacial areas, heterojunctions 253
interfacial layer
- and charge injection 411–420
- XPS/UPS studies 415–420
intermediate surface diffusion (SDZ) 298
intermediate temperature solid oxide fuel cells 98
interphase formation, solid 179
ion adsorption, role of micropores 184
ion beam assisted deposition (IBAD) 206
ion solvatation shell 172
ion transport
- in EC devices 437–438
- rapid 184
ionic liquids (ILs), supported on polymeric membranes 368
IR spectroscopy, characterization of heterogeneous catalysts 321
Ir(piq)$_3$, (red dopant) structure 421
Ir(ppy)$_3$, (green dopant) structure 421
IRMOF-1
- adsorption isotherms for CO_2 389–390
- gravimetric and volumetric isotherms of CO_2 adsorption 392
iron . . . see also Fe
iron carbide see Fe_3C
iron oxide, deposition 164
iron oxide catalysts, DMAP immobilized 354
iron oxyhydroxide particles 305
irreversible capacity, origins 199–201
3-isocyanatopropyl triethoxysilane (ICPTES) 339
isomerization, naphtha reforming 288
IT-SOFC see intermediate temperature solid oxide fuel cells
ITO see indium-doped tin oxide
ITO/poly(ethylene dioxy thiophene):polystyrene sulfonate 408–410

j
jelly roll, SEM-image 257

l
lateral resolution, microcontact printing 245
lattice mismatch, self-assembled nanostructures 216

lattice nitrogen 179
layer thickness
- absorber in photovoltaics 8
- electron injection 16
- phase separation 13
- thin film superlattice 67
- see also monomolecular layer
layers
- graphene 179
- plastic like by gel preparation 160
- temperature-driven growth 163
leached gold catalysts 297
leached gold-ceria catalysts 299
LEDs see light-emitting diodes
LEP see light-emitting polymer
Li
- galvanostatic insertion/deinsertion 195
- quasi-metallic 195
- see also lithium
$LiFePO_4$, as electrode material 159
Li-ion batteries 157, 194–201
- mesoporous chromium oxides 168–169
- reversible capacity 194
Li-ion technology, development 156
Li-ions, host structures 158
^7Li nuclear magnetic resonance (^7Li-NMR) 195
Li_2ZrO_3, carbon dioxide capture 369
ligands, organic 143
light balancing 437
light confinement, indium tin oxide (ITO) 429–430
light-emitting polymer (LEP) layer 408, 412–413, 424
- current–voltage (I–V) plot 425
- thickness 426–427
light out-coupling, enhancement 429–431
light-emitting diodes, semiconductors in 34
light-emitting diodes (LEDs) 403
linear arrays, single-crystalline nanowires 249–252
linewidths, produced by stamps 245
liquid-crystal templating mechanism, cracking 294
liquid electrolytes 172
liquid phase catalysis 312
lithium . . . see also Li
lithium metal polymers (LMP), development critics 156
lithium state 195
lithography
- block copolymers 269
- nanoskiving 246–259
- soft 272

LMP *see* lithium metal polymers
loop structures   248–249
low-cost substrate, patterned paper   272
low-temperature carbons   179
low-work-function electrode (LWFE), heterojunctions   253
lowest unoccupied molecular orbital   38
lowest unoccupied molecular orbital (LUMO), phosphorescent materials   423–424
LUMO *see* lowest unoccupied molecular orbital
LWFE *see* low-work-function electrode

m
magnetic field dependence, current density   212
magnetic fields
– external   207
– YBCO film   207
magnetic flux lines, pinned and unpinned   208
magnetic nanoparticles, Asp-His bifunctional   355
material input, production costs   233
material jets, electrospinning   264
materials
– alternative   160
– nontraditional   236
MAXSORB™   373–374
MCFC *see* molten carbonate fuel cell
MCM family, of mesoporous silicas   323
MCM-41, functionalization   340
MCM41, Pd-dppf attached to   342–344
MCS *see* mesoporous calcia silicate
MEA *see* membrane-electrode assemblies
melamine, nitrogen-rich carbon precursor   191–192
membrane approaches, carbon dioxide capture   367–368
membrane architecture, in EC technology   454
membrane-electrode assemblies   25, 80, 96–98
membranes
– in fuel cells *see* fuel cells
– shadow evaporation   259
mercaptopropyl (MP) functionalities   348
mercaptopropyl trimethoxysilane (MP-TMS)   331
mesopores   184
mesoporous calcia silicate (MCS)   353
mesoporous carbons, storage capacity   171

mesoporous chromium oxides, li-ion batteries   168–169
mesoporous electrodes, porosity   168
mesoporous powders   169
mesoporous silica nanoparticles (MSN)
– CTAB   331–332
– functionalization   338–342
– MCM family   323
– organically functionalized   333
– selective surface functionalization of solid support   342–346
metal films, topographically patterned   248
metal fims, as transparent conductors   449–451
metal nanoparticles
– conversion reactions   161
– water-gas shift (WGS)   296
metal organic frameworks (MOFs)   371
metal oxide/phosphide layers, temperature-driven growth   163
metal strips   248
metal-organic frameworks, for hydrogen storage   143
metallic nanostructure production   251
metals, used for hydrotreatment   289
methanation   307–308
methanation rate   309
methane concentration, increase   362
methane synthesis rate   308
methanol
– mass and specific activities   84
– *see also* direct methanol fuel cell
methanol synthesis, nanocatalysis   298–302
microcontact printing   244–245
microdomains, Ps-B-Peo   270
micropore volume, irreversible capacity   199
micropores   184, 376–378
– electric double layer capacitors (EDLC)   182
– *see also* porous coordination polymers (PCPs)
microporosity, polymers with intrinsic   144
MIL-53
– adsorption isotherms for CO_2 390–391
– carbon dioxide adsorption at 298 K   384–385
MIL-53 structure   372
MnNDC, structure   381
α-MnO_2/MWNT composite   190
model supports   284
MOF-177   373–374
– structure   380
MOF-508b, structure   380

MOFs *see* metal organic frameworks
– hydrogen storage *see* metal-organic frameworks
molding, green nanofabrication   238–244
molding techniques   242
molecular engineering, of dye *see* dye molecular engineering
molecular modeling, carbon dioxide adsorption   389–393
molecular sieves, carbon dioxide capture   369–371
molten carbonate fuel cell   20, 79–80
monofunctional catalysts   289
monomolecular layer
– design of molecules   41
– of dye   36–37
– self-assembly   40–41
– stability   40–41
Monte Carlo methods, molecular modeling   389–390
morphology
– nanotubes   192
– of semiconductors *see* nanotubes, nanodisks
MoS$_2$ nanocluster, hydrotreatment   290
motivation, green nanofabrication   232–233
MP-TMS (mercaptopropyl trimethoxysilane)   331
MSN catalysts, bifunctionalized   351–354
multi-walled carbon nanotubes (MWNT), electrically conducting polymer (ECP)   188
multi-functionalized hetergeneous catalysts, selectivity   348–351
multi-functionalized heterogeneous catalyst system, cooperative catalysis   346–348
multi-photon emission, OLEDs and PLEDs   428–429
multicomponent nanofibers   265
multilayer structure, OLEDs and PLEDs   405–406
multiple nanowire production   254–255
MWNT *see* multi-walled carbon nanotubes

n

N . . . *see* nitrogen . . .
N$_2$ sorption analysis, characterization of heterogeneous catalysts   321
nano-architecturing, current collectors   163–168
nano-zeolite catalysts   294
nanocatalysis
– for bioenergy   308–312
– fuel production   281–318

nanoclusters, hydrotreatment   290
nanocolumns, HRTEM-images   221
nanocomposites, for hydrogen storage   148–151
nanocrystals, hierarchical assembly   268–269
nanodisks
– semiconductors *see also* nanotubes
– WO$_3$ 122–123
nanodots
– film growth   217
– STEM-image   219
nanofabrication
– energetic costs   233–234
– green   231–280
– green approaches   238–271
– structure size   231
– unconventional approaches   237
nanofiber production, electrospinning   263
nanofibers
– core/shell and hollow   265–267
– electrospinning   265
– hollow   266
nanoimprint lithography (NIL), hard mold production   238
nanolithography, dip-pen   245–246
nanomaterials
– carbon-based for energy storage   177–204
– energy perspective   205–228
– used for energy storage   158–162
nanoparticles
– areal distribution   217
– LiFePO$_4$ 159
nanopillars, soft mold production   243
nanopores
– hydrogen absorption   185
– quasi-metallic lithium   195
nanoporous carbon
– graphene layers   179
– hydrogen electrosorption   186
– turbostractic   178
nanoporous catalysts, surface-functionalized   319–355
nanoporous PCPs   376
nanorod
– Bi, band gap   62
– Bi, diffraction pattern   61
– Bi, lattice planes ordering   61
– Bi, temperature dependences   63
– *see also* nanowires
nanoscale defects
– flux pinning   207–208
– naturally occurring   212–215
– YBCO films   213

nanoskiving 253
- edge lithography 246–259
- heterojunction production 255
- production of heterojunctions 256
nanospace, inside porous structure 342–346
nanostructure
- and EC materials 439–443
- engineered "exotic" 64–66
- self-assembled 216–221
nanostructured carbons, anodes 195–196
nanostructured Cu electrodes, electrochemical performances 167
nanostructured electrodes 163
- cross section 164
nanostructured materials
- electrodes 115
- engineered "exotic" 64–66
- for hydrogen storage 139–153
- nanowire 28, 60–64
- semiconductor 41–43
- thermoelectricity 56–60
nanostructured OLEDs 403–432
nanostructured polymer heterojunctions 253–258
nanostructured semiconductor 41–43
nanostructuring, active material/electrolyte interfaces 168–174
nanotechnology 3–32
- definition 4
- for carbon dioxide capture 359–394
- key roles for energy production 6
- promising application fields 4
nanotexture 177–180
nanotube arrays, production steps 260
nanotubes
- absorption spectra 131
- carbon 10–11, 85–89
- controllable diameter 120
- electrically conducting polymer (ECP) 188
- in photoelectrochemical devices 10
- rate of hydrogen generation 119
- semiconductors 117–123
- surface sensitization 121
- TiO_2 117–122
nanowires 248
- Bi 60–62
- Bi, diameters 60
- Bi, n-type doped 61
- conjugated polymers 252
- controlled dimensions 248–259
- diameter dependences 60–64
- polymer, conjugated 252–253

- production steps 250
- Si 54, 62–64
- single-crystalline 249–252
- superlattice 55
- temperature dependences 65
naphtha reforming
- nanocatalysis 282–289
- petroleum refining 282
Ni-alumina catalyst, methanation 307
Ni metal hydrides (Ni-MH) 156
Ni oxides
- cause of optical absorption 443–444
- in foil-based EC devices 445–449
Ni–W surface, sulfur superstructure 222
NiCu precursor, fule cell technologies 104
NIL see nanoimprint lithography
NiP_2 electrodes, self-supported 163
nitrogen
- electrochemical capacitors 179
- in carbon network 191
nitrogen-enriched composites, capacitance 193
nitrogen functionalities, carbon material 180
nitrogen-rich carbon precursor, melamine 191–192
nitrous oxide concentration, increase 362
NO_x adsorption, by PCPs 394
non-covalent binding, homogeneous catalysts 324–327
non-silica support, heterogeneous catalysts 354
nonlithographic master structures 240–241
nontemplated self-assembly 267
nontraditional materials 236
nucleation, sulfur superstructure 222–223
nucleation rate, film growth 217

o

octagonal building blocks, for electrochromic (EC) oxides 439–440
OLEDs
- basic structure 404–405
- charge balance 405–411
- multilayer structure 405–406
- nanostructured 403–432
- phosphorescent materials 421–428
- PL quantum efficiency 422
- quantum confinement 405–420
- tandem 428–429
one-step carbonization of biopolymers 193
open circuit voltage, Cs_2CO_3 devices 414
open-loop structures 248–249
optical absorption, EC materials 443–445

OPV *see* organic photovoltaics
ordered bulk heterojunctions   256
organic active layer, light confinement   429–430
organic light-emitting devices *see* OLEDs
organic photovoltaics (OPV), heterojunctions   253
organic polymers, as support material   322
organically functionalized MSNs   333
oxide aerogels, vanadium   168
oxides
– li-ion batteries   168–169
– pseudocapacitive   188–190
oxidized nitrogen   180
oxycombustion
– carbon dioxide capture   364, 366
oxygen functionality   180
oxygen removal, nanocatalysis for bioenergy   310

p
P . . . *see* phospho . . .
P-PhoLEDs *see* polymer phosphorescent light-emitting diodes
palladium-1,1′-bis(diphenylphosphino)ferrocene (Pd-dppf), and MCM41   342–344
PANI *see* polyaniline
paper, patterned as a low-cost substrate   272
parallel DPN   245
particle deposition, on superconductor matrix   215
particle downsizing   160
particle replication and nonwetting templates (print), soft mold production   238
particle replication in nonwetting templates   242
particle size
– electrode material   159
– gold catalyst   297
pattern transfer, edge lithography   247
pattern width, soft lithography   273
patterned paper   272
patterning paper   271
PCN-17, structure   380
PCPs *see* porous coordination polymers
Pd catalysts, selective surface functionalization   344
Pd-dppf *see* palladium-1,1′-bis(diphenylphosphino)ferrocene
Pd-pymo-F, structure   381

PDMS molds, lifetime   244
PDMS stamp, soft lithography   272
PEDOT *see* poly-ethylenedioxythiophene
PEDOT:PSS   408–409
periodic table of the elements, electrochromism   439
perovskite-like structures, carbon dioxide capture   368–369
petroleum refining, nanocatalysis   282–318
PF *see* polyfluorene-derivative
PFO:MEH-PPV   408–411
PFPE *see* photocurable perfluoropolyether
phase changes, iron catalysts   304–305
phosphoester bond, hydrolysis   355
phosphorescent materials
– for OLEDs and PLEDs   421–428
– PL quantum efficiency   421–422
– solution-processed   422
photochromic technology   436
photocurable perfluoropolyether (PFPE), soft mold production   242
photoluminescence (PL), spectra   424–425
photovoltaic (PV) measurements, Cs_2CO_3 413–415
photovoltaic (PV) response, heterostructures   257
PHSs *see* porous hybrid solids
pinning, correlated   215
pinning effects, quantification   212
pinning enhancement   210
– nanodot   219
PL efficiency, OLEDs and PLEDs   410
PL quantum efficiency, OLEDs and PLEDs   422
planar electrode
– capacity of powder based   165
– cross section   164
plasmon resonators   250
platinum, naphtha reforming   282
platinum-alumina reforming catalyst   283
platinum-rhenium-alumina catalyst, naphtha reforming   285
PLD *see* pulsed laser deposition
PLEDs   403
– charge balance   405–411
– efficiency   410
– multilayer structure   405–406
– phosphorescent materials   421–428
– PL quantum efficiency   422
– quantum confinement   405–420
– tandem   428–429
point-like random defects, current density   211

polarization voltage 162
poly-ethylenedioxythiophene (PEDOT), electrically conducting polymer (ECP) 188
poly(dimethylsiloxane) (PDMS), soft mold production 240
polyaniline (PANI), electrically conducting polymer (ECP) 188
polyfluorene-derivative (PF), host material 423–424
polyhedral oligomeric silsesquioxanes (POSS) 330
polymer blended systems, charge balance 406–411
polymer heterojunctions, nanostructured 253–258
polymer phosphorescent light-emitting diodes (P-PhoLEDs) 422–423
polymer-based lightemitting diodes see PLEDs
polymers
– conducting 188–190
– conjugated nanowires 252–253
– soft mold production 240
– synthesis 269
polypyrrole (PPY), electrically conducting polymer (ECP) 188
polystyrene-block-poly(ethylene oxide) (Ps-B-Peo), block copolymers 269
polystyrene thermoplastic sheets, soft lithography 272
polystyrene-type resins 322
pore size
– activated carbon 171
– electric double layer capacitors (EDLC) 182–184
pores, subnanometer 172–173
porosity
– materials 169
– mesoporous electrodes 168
porous coordination polymers (PCPs)
– carbon dioxide adsorption 376–393
– CO_2 adsorption data at 195 K 377
– CO_2 adsorption data at 273 K 382
– CO_2 adsorption data at 298 K 387
– dimensionality 372
– flue gas contaminant stability 393–394
– for carbon dioxide capture 371–394
– general survey 375–376
– NO_x adsorption 394
– preliminary screening 373–375
– SO_x adsorption 394
– structural description at 195 K 377
– structural description at 273 K 382

– structural description at 298 K 387
– supernanoporous 376
– textural properties at 195 K 377
– textural properties at 273 K 382
– textural properties at 298 K 387
porous hybrid solids (PHSs) 371
porous structure, nanospace inside 342–346
POSS see polyhedral oligomeric silsesquioxanes
post-combustion, carbon dioxide capture 365–366
post-grafting silylation 328–330
powders, mesoporous 169
power consumption values, laboratory tools 235
power efficiency, PLEDs 410
power enhancement 181
PPY see polypyrrole
pre-combustion, carbon dioxide capture 364
preasphaltenes 310
preparation of graphene films, scotch-tape method 271–272
pressure swing adsorption (PSA) 375
PRINT (particle replication and nonwetting templates), production steps 243
printing 244–246
– microcontact 244–245
production tool usage 234–235
proline catalysts, heterogeneous 326
Ps-B-Peo see polystyrene-block-poly(ethylene oxide)
PSA see pressure swing adsorption
pseudo-capacitive effects 185–188, 190–194
pseudo-capacitors, carbon-based materials 185–194
pseudo-faradaic reaction, pyridinic group 191
pseudocapacitive oxides 188–190
Pt . . . see platinum . . .
pulsed laser deposition (PLD), YBCO film growth 211
PV see photovoltaic
pyridinic group, pseudo-faradaic reaction 191
pyridinic nitrogen 180
pyridonic nitrogen 180
pyrolysis temperature, electric double layer capacitors (EDLC) 182
pyrone-like structures, carbonization of biopolymers 194
pyrrolic nitrogen 180

q

QD *see* quantum dots
quantum confinement, OLEDs and PLEDs 405–420
quantum dots (QD) 268
quasi-metallic lithium, nanopores 195
quaternary nitrogen 180

r

RABIT *see* rolling assisted biaxially textured substrate
radiative forcing, change 362
random defects, current density 211
random pinning 211
rapid reaction zone (RRZ) 298
rare earth HTS 206
rate capability curves
– Bi 167
– nitrides 167
rate capability plots 166
Re . . . *see* rhenium . . .
reaction mechanisms, new, triggered by nanomaterials 161
reactivity enhancement, induced by site isolation 337–338
red dopant, Ir(piq)$_3$ 421
redox mechanism, water-gas shift (WGS) 296–297
redox process 185
refining, petroleum 282–318
reforming, naphtha 282–289
rejected insertion materials 158–160
rejected si-based electrodes 160–161
REM *see* replica molding
REN *see* renewable energies
renewable energies (REN), electrochemical energy storage 155
replica molding (REM), soft mold production 238
research targets, energy related nanoscience 312–313
reversible capacity 199
– lithium batteries 194
reversible electrosorption 185
Rh . . . *see* rhodium . . .
rhenium catalysts, naphtha reforming 285
rhodium complex
– immobilization 325, 337
rolling assisted biaxially textured substrate (RABITs) technology 206
RRZ, rapid reaction zone 298
rutile-type TiO$_2$ 159

s

S . . . *see* sulfur . . .
saccharose, HRTM images 178
SAMIM *see* solvent-assisted micromolding
saturation, with lithium 199
scanned electrospinning 264–265
schottky junctions, methanol synthesis 301
scope, green nanofabrication 236–238
scotch-tape method, preparation of graphene films 271–272
screw dislocations, defects on YBCO films 213
scum, residual layer 242
SDSP-TMS (2-(3-(trimethoxysilyl)-propyldisulfanyl)ethanesulfonic acid sodium salt) 331
SDZ, intermediate surface diffusion 298
selective aromatization 288
selective surface functionalization, of solid support 342–346
selectivity, multi-functionalized hetergeneous catalysts 348–351
self-assembled nanostructures 216–221
self-assembly, green nanofabrication 267–271
self field, YBCO film 207
self-supported electrodes 163
semiconductor fabrication, total production costs 233
semiconductor nanocrystal superlattices, stacked films 267–268
SFIL *see* step-and-flash imprint lithography (SFIL)
shadow evaporation 259–263
"ship in the bottle" heterogeneous catalyst 327
shrinky-dinks 272
Si/Al zeolites 371
Si-based electrodes, rejected 160–161
Si/C composites
– anodes 196–199
– cycleability 197
Si films 197
Si nanoparticles
– coated 160
– voltage-composition curve 161
silica–alumina zeolites, cracking 293
silica support morphology, functionalities and control of 338–342
silica supports 322–323
silica surfaces
– interaction with immobilized homogeneous catalysts 335–337

- single-site heterogeneous catalysis 324–333
silylation reaction 344–345
single-crystalline nanowires, linear arrays 249–252
single-layer MoS_2 nanocluster, hydrotreatment 290
single-site heterogeneous catalysts
- immobilization strategies 322–334
- on silica surfaces 324–333
sintering, platinum catalyst 284
site isolation, heterogeneous catalysts 337–338
site isolation effect, of AEP group 353
SIZ see stranded intermediate zone
slabs, ultramicrotome 248
slurry bubble column reactors, fischer-tropsch synthesis (FTS) 303
smart windows 436, 452–454
SMES see superconducting magnetic-energy storage
SO_x adsorption, by PCPs 394
soft lithography
- production steps using patterning paper 273
- shrinky-dinks 272
soft mold production, green nanofabrication 238
soft pattern transfer elements 240–243
solid electrolyte interphase formation 179
solid support
- selective surface functionalization 342–346
solid surfaces, single-site heterogeneous catalysis 322–334
solid-state lighting 313
- see also OLEDS
solid-state NMR spectroscopy, characterization of heterogeneous catalysts 321
solids, nanocatalysis for bioenergy 308
solvent-assisted micromolding (SAMIM)
- production steps 242
- soft mold production 238, 240
space cooling 435
specific capacitance 172–173
specific surface area (SSA) 170
spinneret 264, 266
SRES A2 emissions scenario 363
SSA see specific surface area
stacked films, semiconductor nanocrystal superlattices 267–268

steel electrode 163
step-and-flash imprint lithography (SFIL)
- hard mold production 238
- production steps 239
storage
- energy 205
- energy by carbon-based nanomaterials 177–204
stranded intermediate zone (SZ) 298
structure heights, shadow evaporation 261
structure size, nanofabrication 231
structures, open- and closed-loop 248–249
styrene, cyclopropanation 335–336
subnanometer pores 172–173
substrate, low-cost 272
sulfur coordination, catalysts 286
sulfur layers 222
sulfur superstructure, epitaxy-enabling 221–223
supercapacitors 157, 169, 180–194
- working principle 180–182
superconducting cables 205
superconducting film thickness 220
superconducting magnetic-energy storage (SMES) 205
superconductor matrix, particle deposition 215
superconductors
- energy perspective 205–228
- performance limitations 205–207
supernanoporous PCPs 376
superstructure, epitaxy-enabling 221–223
support materials, single-site heterogeneous catalysts 322–324
surface functionality
- by heteroatoms 190
- sp^2 carbons 177–180
surface functionalization, solid support 342–346
surface organometallic chemistry 329
surface structure, copper catalysts 302
surface temperature, global mean 360
surface-functionalized nanoporous catalysts 319–355
- single-site heterogeneous catalysis 320, 322–334
sustainability, three energy 229–458
synergistic catalysis, general acid and base bifunctionalized MSN catalysts 351–354
syngas, conversion 296
syngas conversion, fischer-tropsch synthesis (FTS) 302

synthesis
- bulk chemical 268
- methanol 298–302
synthesis rate, methanol synthesis 301

t
tandem structure, for OLEDs and PLEDs 428–429
temperature, global mean 360–361
template synthesis 164
- electrochemical cell 165
templates, mesoporous electrodes 168
templating mechanism, cracking 294
thermal coal liquefaction 311
thermochromic (TC) materials, transition temperature 451–453
thermochromic (TC) technology 436
thermochromic (TC) Vo$_2$-based films, in EC devices 451–452
thermoplastic sheets, soft lithography 272
thermoprogrammed desorption (TPD) 188
thin film battery, EC devices 445
thin films
- Au based 449–451
- current density 209–210
- Ps-B-Peo 270
- thermochromic (TC) Vo$_2$-based 451–452
- TiO$_2$ 442–443, 450–451
thiol functionalized MSNs 331–332
Thornton diagram 442
threading edge, defects on YBCO films 213
three energy sustainability 229–458
TiN amount, naphtha reforming 287
TiO$_2$ films
- Au based 450–451
- nanostructure 442–443
TiO$_2$ nanopillars, soft mold production 243
titanium complexes, surface immobilization 329–330
TMS see trimethylsilyl
TON see turnover number
tools, reusage of analytical instruments for fabrication 236
tools usage 234–236
topographically patterned metal films 248
topographically patterned surface, sectioning 251
TPD see thermoprogrammed desorption (TPD)
transesterification, biodiesel 319–320
transfer elements
- hard pattern 238–240
- soft pattern 240–243
transition, film and nanowire 255

transition metal complexes, surface organometallic chemistry 329
transition metal oxides
- electrically conducting polymer (ECP) 188
- support by nanotubes 189
transition metals, with oxides showing electrochromism 439
transition temperature, TC materials 451–453
transmittance, EC devices 446–448
transport, energy 205
transportation fuels, ultra-low sulfur 289
triangular nanocluster, hydrotreatment 290
3-(3′-(trimethoxysilyl)-propyldisulfanyl) propionic acid (CDSPTMS) 331
2-(3-(trimethoxysilyl)-propyldisulfanyl) ethanesulfonic acid sodium salt (SDSP-TMS) 331
trimethylsilyl (TMS), surface passivation 337–339
triphylite 158
triplet-triplet annihilation (TTA) effect 424
TTA see triplet-triplet annihilation
tubes, hollow inorganic 259–261
turnover number (TON) 288, 337
- AEP-MSNs 352–353
- DMAP-MSN catalysts 346

u
UDP see ureidopropyl
UDP-MSN, functionalization 340
UDPTMS (ureidopropyl trimethoxysilane) 339
ultra-low sulfur transportation fuels, hydrotreatment 289
ultracapacitors 169, 180
ultramicrotome 248–259
UMCM-1, structure 380
uniaxial electrospinning 265
UPS spectra
- Cs$_2$CO$_3$ 416, 419
ureidopropyl (UDP) functionalities 348
ureidopropyl trimethoxysilane (UDPTMS) 339
UV-vis spectroscopy, characterization of heterogeneous catalysts 321

v
V oxides, cause of optical absorption 444
vacuum swing adsorption (VSA) 375
vanadium oxide aerogels, electrochemical reactivity 168

vapor-deposition techniques 261
VO$_2$-based films, thermochromic 451–452
voltage-composition curve 159
– carbon-coated si nanoparticles 161
voltage output, li-ion batteries 157
volume changes, lithium storage 197
volumetric capacitance 172
volumetric capacity 171
vortex–vortex interactions, flux pinning 210
VSA *see* vacuum swing adsorption

w

W oxides
– cause of optical absorption 443–445
– in foil-based EC devices 445–449
– octahedral structure 440–441
water-gas shift (WGS)
– catalytic reaction 365
– hydrocracking 296
– nanocatalysis 296–298
wave-guide mode 429–431
wet chemical synthesis 268
WGS *see* water-gas shift
wire length, superconductors 207
work function
– Cs$_2$CO$_3$ 413–415
– effective 416

x

X-ray adsorption near-edge structure (XANES), characterization of heterogeneous catalysts 321
XPS spectra, Cs$_2$CO$_3$ 419
XPS/UPS studies, interfacial layer 415–420

y

YBCO films 206
– current density 209
– nanosize defects 213
– performance enhancement 210
– TEM images 218
YBCO growth, disturbed 214
YBCO/Y$_2$O$_3$ composite film 216
yeast 312

z

zeolite 13X 374
zeolite catalysts
– cracking 293
– hydrocracking 295
zeolites
– carbon dioxide capture 369–371
– naphtha reforming 288–289
zero-cost fabrication 271–274
Zn$_2$(CNC)$_2$dpt, structure 380
ZrO$_2$ nanodots 214